Springer Theses

Recognizing Outstanding Ph.D. Research

Aims and Scope

The series "Springer Theses" brings together a selection of the very best Ph.D. theses from around the world and across the physical sciences. Nominated and endorsed by two recognized specialists, each published volume has been selected for its scientific excellence and the high impact of its contents for the pertinent field of research. For greater accessibility to non-specialists, the published versions include an extended introduction, as well as a foreword by the student's supervisor explaining the special relevance of the work for the field. As a whole, the series will provide a valuable resource both for newcomers to the research fields described, and for other scientists seeking detailed background information on special questions. Finally, it provides an accredited documentation of the valuable contributions made by today's younger generation of scientists.

Theses are accepted into the series by invited nomination only and must fulfill all of the following criteria

- They must be written in good English.
- The topic should fall within the confines of Chemistry, Physics, Earth Sciences, Engineering and related interdisciplinary fields such as Materials, Nanoscience, Chemical Engineering, Complex Systems and Biophysics.
- The work reported in the thesis must represent a significant scientific advance.
- If the thesis includes previously published material, permission to reproduce this must be gained from the respective copyright holder.
- They must have been examined and passed during the 12 months prior to nomination.
- Each thesis should include a foreword by the supervisor outlining the significance of its content.
- The theses should have a clearly defined structure including an introduction accessible to scientists not expert in that particular field.

More information about this series at http://www.springer.com/series/8790

Markus Zinser

Search for New Heavy Charged Bosons and Measurement of High-Mass Drell-Yan Production in Proton-Proton Collisions

Doctoral Thesis accepted by
the Johannes Gutenberg University Mainz, Mainz,
Germany

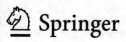

Springer

Author
Dr. Markus Zinser
Data Scientist at DB Analytics
Frankfurt am Main, Hessen, Germany

Supervisor
Prof. Dr. Stefan Tapprogge
Institut für Physik
Johannes Gutenberg University Mainz
Mainz, Germany

ISSN 2190-5053 ISSN 2190-5061 (electronic)
Springer Theses
ISBN 978-3-030-13137-1 ISBN 978-3-030-00650-1 (eBook)
https://doi.org/10.1007/978-3-030-00650-1

This Springer imprint is published by the registered company Springer Nature Switzerland AG
The registered company address is: Gewerbestrasse 11, 6330 Cham, Switzerland

Supervisor's Foreword

The understanding of the microscopic structure of matter and the related fundamental interactions has made significant progress in the last hundred years, culminating presently in the very successful Standard Model of particle physics. However this is likely not to be the last word, further insights are expected to be able to evolve this towards a more complete theory. Progress in further developing the theory describing the microcosm can directly impact our understanding of the evolution of the universe, especially at its very early stages. This thesis addresses a two-pronged approach towards further extending our knowledge and the understanding of the smallest scales. One possibility is to directly search for signatures of new physics processes, not included in the above mentioned Standard Model. An observation of new (fundamental or composite) particle states can provide valuable insight into the structure of a more complete theory. Alternatively, precision measurements of processes known within the Standard Model yield valuable information which in a complementary manner extends our knowledge. Here indirect evidence on new physics processes can be reached (typically extending to higher mass scales than accessible in a direct search) as well as the detailed understanding of processes contributing as background to searches can be improved. Using data recorded by the ATLAS experiment at the Large Hadron Collider (LHC) at CERN both avenues are pursued in this thesis. Theories extending the Standard Model often include new heavy gauge bosons, such as heavy "partners" of the well known charged W bosons (responsible e.g. for the nuclear beta decay). With the first data collected in 2015 at the highest centre-of-mass energy of 13 TeV in proton-proton collisions provided so far by the LHC, a search for resonant structures in final states containing a high transverse momentum lepton and large missing transverse energy has been performed. Albeit no indication for new physics could be found, the exclusion limits on the mass of such a heavy partner of the W boson were significantly extended. Secondly, data collected by ATLAS at a lower centre-of-mass energy of 8 TeV has been used to perform a precision measurement of differential cross-sections for the production of high mass lepton-antilepton pairs in proton-proton collisions. This process is sensitive to partonic structure of the proton, which often yields one of the dominant systematic uncertainties arising in direct searches (as well as in measurements of fundamental parameters) in proton-proton collisions. For the first time, double-differential cross-sections were studied and thanks to the very good accuracy reached, especially contraints on photon-photon induced processes were obtained (here the "photonic" content of a proton is probed). This aspect is in this thesis complemented with a phenomenological interpretation of the measurements performed. These two major and complementary results presented in this thesis are a very high quality achievement highlighting the physics potential of the ATLAS experiment at the LHC. They are presented in a concise and very comprehensible style, showing the mature scientific manner. The scientific context of the results as well as the experimental methods to obtain these are clearly described—the thesis will serve as an excellent point of reference for future Ph.D. students working on collider based particle physics.

Mainz, Germany
September 2018

Prof. Dr. Stefan Tapprogge

Abstract

The Large Hadron Collider (LHC) at CERN, at which protons collide at unprecedented center of mass energies and very high instantaneous luminosity, gives unique possibilities for precise tests of the Standard Model and to search for new physics phenomena. A precise prediction of the processes at the LHC is essential and a key role plays hereby the knowledge of the parton density functions (PDFs) of the proton.

In this thesis two analyses are presented. In the first analysis a new heavy charged gauge boson, a so-called W' boson, is searched for. Those new gauge bosons are predicted by some theories extending the Standard Model gauge group to solve some of its conceptual problems. Decays of the W' boson in final states with a lepton ($\ell^{\pm} = e^{\pm}, \mu^{\pm}$) and the corresponding (anti-)neutrino are considered. Data are used which were collected by the ATLAS experiment in the year 2015 at a center of mass energy of $\sqrt{s} = 13$ TeV. The collected data corresponds to an integrated luminosity of 3.2 fb^{-1}. The Standard Model prediction for the expected background is estimated with Monte Carlo simulations and methods based on data. The resulting spectrum of the transverse mass is tested, using statistical methods, for differences between data and the Standard Model predictions. No significant deviation from the Standard Model predictions is found and masses of a Sequential Standard Model W' boson below 4.07 TeV are excluded with a confidence level of 95%.

In the second analysis a measurement of the double-differential cross section of the process $pp \rightarrow Z/\gamma^* + X \rightarrow \ell^+\ell^- + X$, including also a $\gamma\gamma$ induced contribution, at a center of mass energy of $\sqrt{s} = 8$ TeV is presented. The measurement is performed in an invariant mass region of 116 to 1500 GeV as a function of invariant mass and absolute rapidity of the $\ell^+\ell^-$-pair as well as a function of invariant mass and pseudorapidity separation of the $\ell^+\ell^-$-pair. The analyzed data were recorded by the ATLAS experiment in the year 2012 and correspond to an integrated luminosity of 20.3 fb^{-1}. It is expected that the measured cross sections are sensitive to the PDFs at very high values of the Bjorken-x scaling variable and to the photon structure of the proton. In this thesis the measurement of the cross sections for the decay into an e^+e^--pair is discussed in detail. The measured cross sections are combined with a measurement for the decay into a $\mu^+\mu^-$-pair to reduce the systematic and statistical uncertainties. The combined cross sections are compared to theory predictions and studies of the sensitivity of the measurement are carried out. It is shown that, with help of this measurement, the uncertainty of the photon PDF can be strongly reduced.

Contents

Chapter 1
Preamble

The concept of elementary particles forming all matter is very old and goes back to the atomic hypothesis first formulated by Democritus around 400 BC. It took around 2300 years until technological progress allowed for the discovery of the first elementary particle. It was J.J. Thompson who discovered in 1897 the electron by showing that cathode rays were actually made of negatively charged particles [1]. The following discovery of the atomic nucleus in 1911 and later the proton by Rutherford [2] and in 1932 of the neutron by Chadwick [3] seemed to complete the picture and to allow the explanation of matter formed in atoms by fundamental particles.

In the 1960s in several experiments numerous new particles were discovered. The desire to group these particles in a systematic way lead to the introduction of the quark model [4, 5], which described many of these new particles as bound states of quarks. The largest success of that model in this time was the prediction of a new bound state, the Ω^-, which was discovered in 1964 [6]. At the end of the 1960s results from electron-nucleon scattering showed that also the proton has an internal structure [7, 8]. This observation was a further strong argument for the quark model.

In the following years, technical progress allowed the construction of particle accelerators with higher and higher energies and lead to the discovery of more and more new particles. In 1979, the gluon, the mediator of the strong force, was discovered at the electron-positron collider PETRA (Positron–Elektron Tandem Ring Anlage) at DESY (Deutsches Elektronen-Synchrotron) [9–12]. Four years later, the mediators of the weak force, the Z and $W^{\pm}$ bosons were discovered at CERNs (Conseil Européen pour la Recherche Nucleaire) Super-Proton-Antiproton Synchrotron ($Sp\bar{p}S$) collider [13–16].

These successes in the search for the understanding of the structure of matter were a consequence of an interplay between experimental observations and predictions made by theoreticians, which lead to the development of the Standard Model of elementary particle physics. It can describe the structure of matter with fundamental building blocks and explains the elementary processes for three of the four funda-

M. Zinser, *Search for New Heavy Charged Bosons and Measurement of High-Mass Drell-Yan Production in Proton-Proton Collisions*, Springer Theses,
https://doi.org/10.1007/978-3-030-00650-1_1

mental forces. The Standard Model is a very powerful theory and its predictions are verified up to highest precision. Even though the Standard Model is very successful, there are observations, e.g., dark matter or the matter-antimatter asymmetry, which it cannot explain. Thus, extensions of the Standard Model are needed. Many of these predict the appearance of new particles with masses in the TeV range. Those particles can be searched for by performing indirect or direct searches. The history of particle physics has proven that measurements at particle colliders operating at the high-energy frontier is a very promising way of searching for these particles.

The Large Hadron Collider (LHC), a proton-proton accelerator at CERN in Geneva, is a powerful machine which allows to search for new physics phenomena and to test the predictions of the Standard Model at the highest yet reached energy scales. For these tests and measurements, precise predictions of the processes at the LHC are needed. To obtain a high level of accuracy for the predictions, a very good understanding of the structure of the proton is essential. In this context the knowledge of the parton distribution functions (PDFs) of the proton plays a key role.

In this thesis two analyses are presented. First a search for a high-mass resonance decaying into a final state with a charged lepton ($\ell^{\pm} = e^{\pm}, \mu^{\pm}$) and the corresponding (anti-)neutrino is performed. Those hypothetic particles are called $\wp$ bosons and are assumed to behave similar to the known Standard Model W boson. The search is performed with the first data taken by the ATLAS experiment at proton-proton collisions at a center of mass energy of 13 TeV in the year 2015 and thus probing a new not yet reached energy scale.

In a second analysis a measurement of the Drell–Yan process $pp \rightarrow Z/\gamma^{*} + X \rightarrow \ell^{+}\ell^{-} + X$ ($\ell = e, \mu$), also including a $\gamma\gamma$ induced contribution, is performed with data taken by the ATLAS experiment at a center of mass energy of the proton-proton collisions of $\sqrt{s} = 8$ TeV. A double-differential cross section is measured at high invariant masses of the lepton pair ($m_{\ell\ell} > 116$ GeV). The measurement is performed as a function of absolute rapidity of the lepton pair or absolute pseudorapidity separation of the leptons and invariant mass. Such a measurement can help to improve the parton distribution functions of the proton at high momentum fractions x. In particular sensitivity to the PDFs of the antiquarks and photons in the proton is expected. The former are not well constrained at high values of x while the latter is in general largely unconstrained.

This thesis is structured as follows. Part I addresses in Chaps. 2 and 3 the theoretical foundations and predictions needed for this work. In part II the LHC and the ATLAS experiment are described in Chaps. 4 and 5. The reconstruction and identification of particles with the ATLAS experiment is subsequently described in Chap. 6.

In part III the search for new physics in the final state of an electron or muon and missing transverse momentum is presented. Chapter 7 briefly motivates the search followed by a discussion of the analysis strategy in Chap. 8. The analysis, including the event and object selection, the background determination, the estimation of systematic uncertainties, and the comparison of the expected background to the data, is presented in Chap. 9. The data are subsequently analyzed using statistical methods in Chap. 10. Part III ends with Chap. 11, where a summary of the obtained results and an outlook on expected results in the future is given.

Part IV presents the measurement of the high-mass Drell–Yan cross section. Its structure is similar to part III. First the measurement is briefly motivated in Chap. 12 and the analysis strategy is discussed in Chap. 13. The actual analysis of the electron-positron channel, including the event and object selection, the background determination, and ending with a comparison of data and expected signal and background, is presented in Chap. 14. The methodology for the cross section measurement, systematic uncertainties, and the obtained results for the high-mass Drell–Yan cross section in the electron-positron channel are discussed in Chap. 15. Afterwards, in Chap. 16, the measurement of the muon channel is briefly discussed. It has not been carried out in the context of this thesis but is used as an input for the calculation of a combined cross section in Chap. 17. The calculated combined cross section is in the same chapter also compared to theoretical predictions and interpreted in terms of sensitivity to PDFs. The part ends with Chap. 18, where a summary of the obtained results and an outlook on possible measurements in the future is given.

A general summary of all results is finally given in Chap. 19.

References

1. Thomson JJ (1897) Cathode rays. Philos Mag 44:293–316. https://doi.org/10.1080/14786449708621070
2. Rutherford E (1911) The scattering of alpha and beta particles by matter and the structure of the atom. Philos Mag 21:669–688. https://doi.org/10.1080/14786440508637080
3. Chadwick J (1932) Possible existence of a neutron. Nature 129:312. https://doi.org/10.1038/129312a0
4. Gell-Mann M (1964) A schematic model of baryons and mesons. Phys Lett 8:214–215. https://doi.org/10.1016/S0031-9163(64)92001-3
5. Zweig G (1964) An SU(3) model for strong interaction symmetry and its breaking. Version 2. In: Lichtenberg D, Rosen SP (eds) Developments in the quark theory of hadrons. Vol. 1. 1964–1978, pp 22–101. http://inspirehep.net/record/4674/files/cern-th-412.pdf
6. Barnes VE (1964) Observation of a Hyperon with Strangeness -3. Phys Rev Lett 12:204–206. https://doi.org/10.1103/PhysRevLett.12.204
7. Breidenbach M (1969) Observed Behavior of Highly Inelastic electron-Proton Scattering. Phys Rev Lett 23:935–939. https://doi.org/10.1103/PhysRevLett.23.935
8. Bloom ED (1969) High-energy inelastic e p scattering at 6-degrees and 10-degrees. Phys Rev Lett 23:930–934. https://doi.org/10.1103/PhysRevLett.23.930
9. Brandelik R (1979) Evidence for planar events in e+ e- Annihilation at high-energies. Phys Lett B 86:243–249. https://doi.org/10.1016/0370-2693(79)90830-X
10. Barber DP (1979) Discovery of three jet events and a test of quantum chromodynamics at PETRA energies. Phys Rev Lett 43:830. https://doi.org/10.1103/PhysRevLett.43.830
11. Berger C (1979) Evidence for gluon bremsstrahlung in e+ e- Annihilations at high-energies. Phys Lett B 86:418–425. https://doi.org/10.1016/0370-2693(79)90869-4
12. Bartel W (1980) Observation of planar three jet events in e+ e- Annihilation and evidence for gluon bremsstrahlung. Phys Lett B 91:142–147. https://doi.org/10.1016/0370-2693(80)90680-2
13. Arnison G (1983) Experimental observation of isolated large transverse energy electrons with associated missing energy at $\sqrt{(s)} = 540$ GeV. Phys Lett B 122:103–116. https://doi.org/10.1016/0370-2693(83)91177-2

14. Banner M (1983) Observation of single isolated electrons of high transverse momentum in events with missing transverse energy at the CERN anti-p p collider. Phys Lett B 122:476–485. https://doi.org/10.1016/0370-2693(83)91605-2
15. Arnison G (1983) Experimental observation of lepton pairs of invariant mass around 95-GeV/c^2 at the CERN SPS collider. Phys Lett B 126:398–410. https://doi.org/10.1016/0370-2693(83)90188-0
16. Bagnaia P (1983) Evidence for Z0 $\rightarrow$ e + e− at the CERN anti-p p collider. Phys Lett B 129:130–140. https://doi.org/10.1016/0370-2693(83)90744-X

Part I
Introduction to Theory

Chapter 2
Theory Foundations

In the first part of this chapter, a brief introduction into the Standard Model of particle physics and its interactions is given. This is followed by a discussion of the formalism which is needed to describe proton–proton (pp) collisions. Also the extraction of the needed ingredients to predict the outcome of these collisions is described, followed by a discussion of the Drell–Yan and photon induced process. Finally, the limitations and problems of the Standard Model are discussed and some theories which aim to solve these limitations are presented. The chapter ends with a discussion of models predicting new physics in the final state of a charged lepton and neutrino. The discussion follows to large parts the discussion in [1].

Throughout this thesis, the convention $\hbar = c = 1$ is used, therefore masses and momenta are quoted in units of energy, electron volts (eV).

2.1 The Standard Model of Particle Physics

2.1.1 Overview of the Fundamental Particles and Interactions

The Standard Model (SM) of particle physics [2] is one of the most successful models in physics. It describes the dynamics and interactions of all currently known elementary particles and three of the four fundamental interactions very precisely. So far, the Standard Model survived every experimental test.

In our current understanding, matter consists of point-like particles with half integer spin, called fermions. Gauge bosons with integer spin mediate the fundamental forces between these fermions.

M. Zinser, *Search for New Heavy Charged Bosons and Measurement of High-Mass Drell-Yan Production in Proton-Proton Collisions*, Springer Theses,
https://doi.org/10.1007/978-3-030-00650-1_2

The three fundamental forces described by the Standard Model are the electromagnetic, the weak, and the strong interaction. The incorporation of the gravitational force is yet an unresolved challenge. However, its strength is, at the energy scales probed so far, negligible at the subatomic scale and its incorporation therefore not necessary to precisely describe the fundamental interactions. The massless photon (γ) is the mediator of the electromagnetic interaction. It couples to particles which carry electric charge but does not carry an electric charge itself. As the photon is massless, the electromagnetic force has an infinite range. The weak interaction is mediated by three different gauge bosons, the electric positively and negatively charged $W^{\pm}$-bosons, and the neutral Z-boson. The weak bosons couple to particles with a third component of the weak isospin T_3. The Z boson couples in addition also to particles that carry an electric charge. The $W^{\pm}$ bosons are the mediators of the charged current and for example responsible for the β-decay of atomic nuclei, while the Z boson is the mediator of the neutral current. The weak interaction is very short-ranged, as the three gauge bosons are very heavy ($m_W \approx 80.4$ GeV, $m_Z \approx 91.2$ GeV). The strong interaction is mediated by eight different massless and electrically neutral gluons (g). They couple to particles carrying the so-called color charge which occurs in three different types: red (r), green (g) and blue (b). Gluons carry color charge themselves and as a result couple to each other. This leads, despite the fact that they are massless, to a short range of the strong interaction. Table 2.1 lists again all gauge bosons of the Standard Model.

All fermions interact with the weak force.[1] They can be divided into three generations of leptons and quarks.

The leptons do not undergo strong interactions, since they do not carry color charge. The electron (e), muon (μ), and tau (τ) carry the electric charge $Q/e = -1$ (e is the elementary charge). They interact therefore both, electromagnetically and weakly, whereas neutrinos (ν) do not carry an electric charge and thus interact only weakly. Neutrinos are initially treated as massless particles in the Standard Model although the discovery of neutrino oscillations has proven that they have a non-vanishing mass [3]. The mass of the neutrinos has not yet been directly measured. An upper limit of <2 eV (95% confidence level) on the anti-electron neutrino mass has been set by measuring the endpoint of the electron energy spectrum for the tritium

Table 2.1 Overview of the forces described by the Standard Model and their gauge bosons

Interaction	Boson	Mass [GeV]	Corresponding charge
Electromagnetic	Photon (γ)	0	Electric charge
weak	$W^{\pm}$	≈ 80.4	Weak isospin
	Z	≈ 90.2	Weak isospin/electric charge
Strong	Gluon (g)	0	Color charge (r, g, b)

[1] There is a difference whether a fermion has left-handed or right-handed chirality. This is discussed in more detail in Sect. 2.1.5.

β-decay [4]. Indirect limits from astrophysical observations indicate that the sum of all three neutrino masses must be less than 0.3 eV [5].

Quarks carry a charge of $Q/e = +2/3$ or $Q/e = -1/3$ and interact via all three forces, since they also carry a color charge. They can be separated into six different flavors.

The mass of all fermions rises in the same order as their generation and varies from (excluding the neutrinos) $\approx$0.5 MeV to $\approx$170 GeV. Their masses therefore span many orders of magnitude. For the quarks, two different definitions of the masses exist. The current quark mass which is the mass of the quark itself and the constituent mass which is the mass of the quark plus the gluon field surrounding it. For the heavy quarks (c, b, t) these are almost the same, whereas there are large differences for the light quarks (u, d, s). The current masses of the light quarks cannot be directly measured and thus have large uncertainties of up to 30%.

Every fermion exists as a particle as well as an antiparticle. Both have the same mass and differ in the sign of the additive quantum numbers. For example, the electron carries an electric charge of $Q/e = -1$ whereas its antiparticle, the positron, carries a charge $Q/e = +1$. The fermions of the 2nd and 3rd generation can decay via the weak force into fermions of the lower generations. Quarks have never been observed as free particles but occur only in bound states. Those composite particles are referred to as hadrons and can be classified into two groups: Mesons consist of a quark and an anti-quark and baryons consist of three quarks or three anti-quarks. All hadrons are colorless, i.e., they are color singlet states, which are realized by combining either a color and an anticolor for the mesons or all three colors or anticolors for the baryons. Recently also combined states with four and five (anti-)quarks, so-called tetra- and pentaquarks, have been observed [6–8]. Table 2.2 shows a listing of all leptons and quarks with their charges.

2.1.2 Mathematical Structure of the Standard Model

The mathematical structure of the Standard Model is given by a gauge quantum field theory [10]. All fundamental particles are described as excitations of quantum fields which are defined at all points in space time. Fermions are described by fermion fields $\psi(x)$, also known as (Dirac-) spinors, and gauge bosons are described by vector fields $A_\mu(x)$. The dynamics of the fundamental fields are determined by the Lagrangian density $\mathcal{L}$ (short Lagrangian). The Dirac equation for a free fermionic field describing a fermion with a mass m is given by

$$(i\gamma^\mu \partial_\mu - m)\psi(x) = 0, \tag{2.1}$$

and the Lagrangian for this field is given by

$$\mathcal{L} = \bar{\psi}(x)(i\gamma^\mu \partial_\mu - m)\psi(x), \tag{2.2}$$

Table 2.2 Fermions of the Standard Model, divided into leptons and quarks. Given are the name, the symbol, some quantum numbers and their masses [9]. The masses of the particles are rounded and given without any uncertainties. For the light quarks (u, d) the current mass is given. Antiparticles are not listed explicitly. The third component of the weak isospin is given for the left-handed fermions. All right-handed fermions have $T_3 = 0$

Leptons						
Generation	Name	Symbol	Color	T_3	Q/e	Mass
1.	Electron	e^-	No	$-1/2$	-1	0.511 MeV
	Electron neutrino	ν_e	No	$+1/2$	0	< 2 eV
2.	Muon	μ^-	No	$-1/2$	-1	105.6 MeV
	Muon neutrino	ν_μ	No	$+1/2$	0	< 0.19 MeV
3.	Tau	τ^-	No	$-1/2$	-1	1776.8 MeV
	Tau neutrino	ν_τ	No	$+1/2$	0	< 18.2 MeV
Quarks						
Generation	Name	Symbol	Color	T_3	Q/e	Mass
1.	Up	u	Yes	$+1/2$	2/3	2.3 MeV
	Down	d	Yes	$-1/2$	$-1/3$	4.8 MeV
2.	Charm	c	Yes	$+1/2$	2/3	1.3 GeV
	Strange	s	Yes	$-1/2$	$-1/3$	95 MeV
3.	Top	t	Yes	$+1/2$	2/3	173.2 GeV
	Bottom	b	Yes	$-1/2$	$-1/3$	4.2 GeV

where γ^μ are the gamma matrices and $\bar{\psi} = \psi^\dagger \gamma^0$.

The Lagrangian $\mathcal{L}$ is gauge invariant under global transformations of the group $U(1)$. Thus $\mathcal{L}$ is invariant under the following transformation:

$$\psi(x) \rightarrow \psi'(x) = e^{i\alpha}\psi(x), \tag{2.3}$$

where α is a global phase with the same value at every point in space time. The symmetry is called local, if α has different values for different points x in space time. Performing the local phase transformation

$$\psi(x) \rightarrow \psi'(x) = e^{i\alpha(x)}\psi(x), \tag{2.4}$$

changes the Lagrangian by

$$\delta\mathcal{L} = -\bar{\psi}(x)\gamma^\mu \partial_\mu \alpha(x)\psi(x). \tag{2.5}$$

It is thus not gauge invariant under local transformations of the group $U(1)$. The gauge invariance can be restored by replacing

$$\partial_\mu \to \mathcal{D}_\mu = \partial_\mu + ieA_\mu(x), \tag{2.6}$$

where $\mathcal{D}_\mu$ is called the covariant derivative and e is, according to Noether's theorem [11], a conserved charge of the particle described by the field $\psi(x)$. It can be identified with the usual electric charge. $A_\mu(x)$ is an introduced gauge field which transforms under the phase transformation as

$$A_\mu(x) \to A'_\mu(x) = A_\mu(x) - \frac{1}{e}\partial_\mu \alpha(x). \tag{2.7}$$

The requirement of the local gauge invariance for a free fermionic field leads to the introduction of the bosonic gauge field $A_\mu(x)$ which can be identified as photon field, the mediator of the electromagnetic force. Inserting Eq. 2.6 in 2.2 leads to the following Lagrangian:

$$\mathcal{L} = \bar{\psi}(x)(i\gamma^\mu \partial_\mu - m)\psi(x) - e\bar{\psi}(x)\gamma^\mu A_\mu(x)\psi(x). \tag{2.8}$$

The second term hereby corresponds to an interaction of the fermionic field with the photon field. As photons are observed as free particles, a kinetic term for the photon field has to be added. The full Lagrangian of Quantum Electrodynamics (QED), the theory describing the electromagnetic interactions reads then as:

$$\mathcal{L}_{QED} = \bar{\psi}(x)(i\gamma^\mu \partial_\mu - m)\psi(x) - e\bar{\psi}(x)\gamma^\mu A_\mu(x)\psi(x) - \frac{1}{4}F_{\mu\nu}F^{\mu\nu}, \tag{2.9}$$

where $F_{\mu\nu}$ is the electromagnetic field tensor:

$$F_{\mu\nu} = \partial_\mu A_\nu(x) - \partial_\nu A_\mu(x). \tag{2.10}$$

The complete Standard Model Lagrangian can be made invariant under a local symmetry transformation of the group $SU(2)_L \times U(1)_Y \times SU(3)_C$ by introducing additional bosonic gauge fields which can then be identified as mediators of the fundamental forces. The number of bosonic gauge fields needed to be introduced is equal to the number of generators of the symmetry group.

2.1.3 Feynman Formalism

Feynman diagrams [12] are pictorial representations of the mathematical expressions of the amplitudes of fundamental processes. The Feynman formalism will be discussed on the example of Bhabha scattering, the scattering of an electron-positron pair $e^+e^- \to e^+e^-$ which is a simple process of QED. Figure 2.1 shows the fundamental Feynman vertex of QED. Solid lines represent charged leptons, whereas the curved line represents a photon.

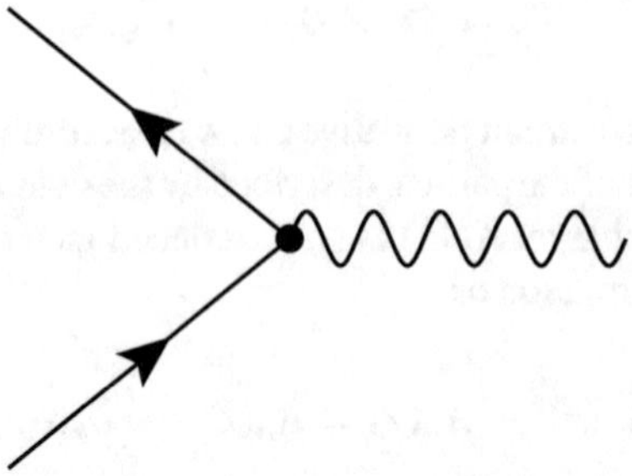

Fig. 2.1 Feynman diagram representing the fundamental interaction of Quantum Electrodynamics. The solid lines represent charged leptons, whereas the curved line represents a photon

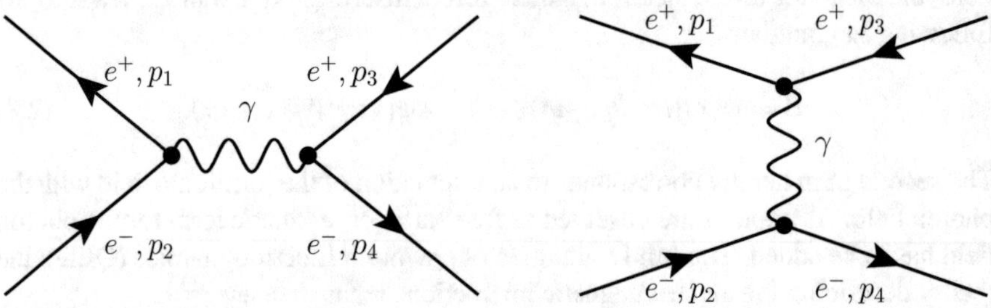

Fig. 2.2 Leading order Feynman diagrams for the Bhabha scattering process

Feynman diagrams[2] for the Bhabha scattering process, as shown in Fig. 2.2, can be constructed from these fundamental vertices. From the Lagrangian $\mathcal{L}_{QED}$, the Feynman rules can be determined to calculate the matrix element amplitude $\mathcal{M}$ contributing from such a diagram.

All possible Feynman diagrams contributing to a process have to be considered for the calculation of the exact amplitude. Each vertex in a diagram is contributing with e. The diagrams shown in Fig. 2.2 represent the leading order (LO) amplitudes and are contributing with e^2. The first diagram is a so-called *s-channel* diagram, where the electron and positron annihilate into a photon which decays again into an electron-positron pair. The second diagram is a so-called *t-channel* diagram, where the electron and positron scatter via the exchange of a photon. Both diagrams lead to contributions which differ in their kinematic behavior. At an electron-positron collider, the former diagram would lead to e^+e^--pairs which have a larger angle to the beam axis while the latter would lead to e^+e^--pairs closer to the beam axis.

Figure 2.3 shows two higher order contributions at next-to-leading order (NLO) and next-to-next-to-leading order (NNLO). The first diagram is a virtual higher order correction at NNLO (also called vacuum polarization) since it has the same final state as the leading order diagram. The second diagram is a real higher order correction at NLO since an additional photon is emitted which also occurs in the final state.

[2]In this thesis the time axis is always along the abscissa.

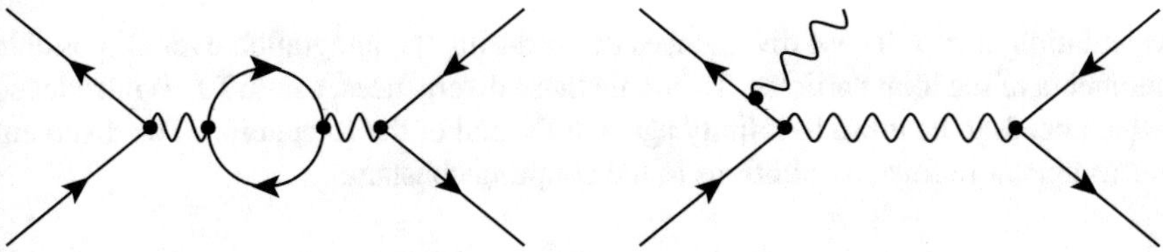

Fig. 2.3 Higher order Feynman diagrams

A matrix element amplitude $\mathcal{M}$ is defined for a single point in phase space. To obtain a cross section σ of a process, the amplitude has to be integrated over the available phase space. Fermi's golden rule [13] can be used which connects the amplitudes $\mathcal{M}$ and the available phase space to calculate the cross section. For the Bhabha process it can be expressed as:

$$\sigma = \frac{S}{4\sqrt{(p_1 \cdot p_2)^2 - (m_1 m_2)^2}} \int |\mathcal{M}|^2 (2\pi)^4 \delta^4(p_1 + p_2 - p_3 - p_4) \times \prod_{j=3}^{4} 2\pi\delta(p_j^2 - m_j^2)\Theta(p_j^0)\frac{d^4 p_j}{(2\pi)^4}. \tag{2.11}$$

Here, p_1 and p_2 denote the four momenta of the incoming particles whereas p_3 and p_4 denote the four momenta of the outgoing particles. The factor $\delta^4(p_1 + p_2 - p_3 - p_4)$ ensures four-momentum conservation and the particles are forced to be on their mass shell by the factor $\delta(p_j^2 - m_j^2)$. The Θ-function ensures that the outgoing particles have positive energies. S is a factor correcting for double counting in case there are identical particles in the final state. When calculating $|\mathcal{M}|^2$, also additional terms occur which account for the interference of both diagrams.

If the contribution $\sigma^{(n)}$ from all diagrams of the same order n is calculated, the complete cross section of a process can be written down as an expansion in powers of α_{em}:

$$\sigma = \sum_{n=1}^{A} \sigma^{(n)} \alpha_{em}^n, \tag{2.12}$$

A is the highest order to which the coefficients $\sigma^{(n)}$ are known and $\alpha_{em} = e^2/4\pi$ the electromagnetic fine structure constant. For an exact calculation all possible Feynman diagrams would have to be taken into account. Calculations at lower orders are only a good approximation for the total cross section if α_{em} is small and therefore the amplitudes from higher orders are suppressed accordingly. This is in case of QED a valid assumption, where $\alpha_{em} \approx 1/137$. However, in the case of the strong interaction, the coupling constant α_s can become very large and a calculation using Eq. 2.12 is no longer appropriate.

When calculating a cross section there are virtual loop diagrams of higher order in α_{em} (for example the left diagram in Fig. 2.3) which lead to divergences in the calculation of $\sigma^{(n)}$. These divergences occur during the integration over all possible momenta of the loop particles. To handle these divergences, a cutoff Λ is introduced which needs to be taken to infinity again at the end of the calculation. The divergent terms appear in form of additions to the coupling constant:

$$\alpha_{em,physical} = \alpha_{em} + \delta\alpha_{em}. \tag{2.13}$$

The additions $\delta\alpha_{em}$ are infinite in the limit $\Lambda \to \infty$. The actual bare coupling constant α_{em} must therefore contain compensating infinities to obtain the physical value which is measured in experiment. The coupling constant is so-called 'renormalized'. An illustrative explanation is that the infinite charge is screened by charges coming from vacuum polarization in such a way that the measured charge is finite.

Renormalization leads, due to finite correction terms independent of Λ, to a dependency of the coupling constant[3] on the scale of momentum transfer Q^2 and an unphysical renormalization scale μ_R^2. The cross section σ now also depends on μ_R due to the dependency of α_{em}. Since μ_R is an unphysical quantity, the physical result σ must be independent of the choice of μ_R, which leads to the equation:

$$\mu_R \frac{d}{d\mu_R}\sigma(\mu_R) = 0. \tag{2.14}$$

This equation holds exactly if $\sigma(\mu_R)$ is calculated up to all orders. If this is applied on a finite order approximation, the numerical result will depend on the choice unphysical scale μ_R. The dependency on the choice of μ_R gets lower when higher orders are calculated and can be interpreted as a theoretical uncertainty on the knowledge of σ.

2.1.4 The Strong Interaction

Quantum Chromodynamics (QCD) [14] is the theory describing the strong interactions. It is a gauge field theory that describes the strong interactions of colored quarks and gluons. The corresponding symmetry group is the $SU(3)_C$, which has $N_C^2 - 1 = 8$ generators[4] which can be represented by the Gell-Mann matrices λ_i, $i = 1, 2, 3, \ldots, 8$. The requirement of $\mathcal{L}_{QCD}$ to be local gauge invariant under transformations of the group $SU(3)_C$ leads to following Lagrangian of the QCD

$$\mathcal{L}_{QCD} = \sum_q \bar{\psi}_{q,a}(i\gamma^\mu(D_\mu)_{ab} - m_q\delta_{ab})\psi_{q,b} - \frac{1}{4}G^A_{\mu\nu}G^{\mu\nu}_A, \tag{2.15}$$

[3] There are additional loop diagrams which lead to the running of the masses of the leptons.

[4] $N_C = 3$ is the number of color charges.

where repeated indices are summed over. $\psi_{q,a}$ are the quark-fields of flavor q and mass m_q, with a color-index a or b that runs over all three colors. The covariant derivative is given by

$$(\mathcal{D}_\mu)_{ab} = \partial_\mu \delta_{ab} + i g_s \frac{\lambda^C_{ab}}{2} \mathcal{A}^C_\mu. \tag{2.16}$$

The gauge fields $\mathcal{A}^C_\mu$ correspond to the gluons fields, with C running over all eight kinds of gluons. The quantity g_s is the QCD coupling constant which can be redefined to an effective "fine-structure constant" for QCD by $\alpha_s = g_s^2/4\pi$. It is usual in literature to call also α_s the strong coupling constant. Finally, the gluon field strength tensor is given by

$$G^A_{\mu\nu} = \partial_\mu \mathcal{A}^C_\nu - \partial_\nu \mathcal{A}^\rho_\mu - g_s f_{ABC} \mathcal{A}^B_\mu \mathcal{A}^C_\nu \qquad [\lambda^A, \lambda^B] = i f_{ABC} \lambda^C, \tag{2.17}$$

where f_{ABC} are the structure constants of the $SU(3)_C$ group. The last term in the gluon field tensor occurs due to the non-Abelian structure of the $SU(3)_C$ group. It corresponds to the self coupling between the gluons.

At LO the dependency of α_s on Q^2 can be written as

$$\alpha_s(Q^2) = \frac{\alpha_s(\mu^2)}{1 + \alpha_s(\mu^2)\beta_0 \ln \frac{Q^2}{\mu^2}} \qquad \beta_0 = \frac{33 - 2N_f}{12\pi}, \tag{2.18}$$

where μ^2 is a reference scale where α_s is known. The factor β_0 is the leading order coefficient of the perturbative expansion of the β-function [14] which predicts the running of α_s, and N_f the number of quark flavors contributing at this scale Q^2. The QCD coupling constant decreases for high values of Q^2 (small distances) which leads to quasi free quarks. This behavior is called "asymptotic freedom". At small values of Q^2 (large distances) the coupling constant increases. If α_s is in the order of unity observables cannot any longer be calculated as an expansion in powers of α_s. The value for α_s at the scale of the mass of the Z boson is $\alpha_s(m_Z^2) = 0.1181 \pm 0.0013$ [9]. QCD at a scale where α_s is small enough to calculate observables perturbatively according to Eq. 2.12 is called perturbative QCD. The scale where α_s gets greater than unity and perturbative expansions start to diverge is called $\Lambda_{QCD} \approx 220$ MeV.[5] Figure 2.4 shows a summary of measurements of α_s as a function of the energy scale Q.

The properties of QCD can be further understood by introducing an effective potential between two quarks in a meson. It is empirically found that the potential has a Coulomb behavior $\propto 1/r$ at short distances and a linear rising potential $\propto r$ at larger distances. Hence, if two quarks are tried to be separated it is from a certain distance on energetically favorable to produce a new quark-antiquark-($q\bar{q}$)-pair out of the vacuum. These can then build hadrons, new colorless bound states. The process of building these colorless states is called hadronization. This feature of QCD is called "confinement", meaning that there are no free quarks and gluons. If a high energetic

[5]For using all flavors up to the b-quark.

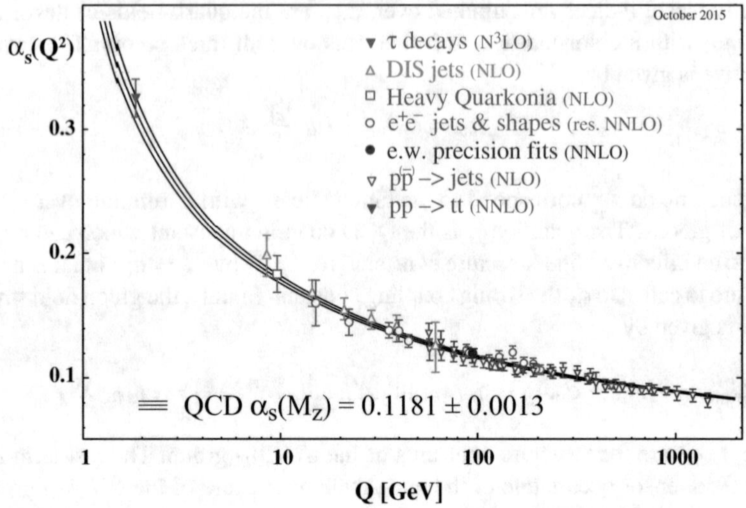

Fig. 2.4 Summary of measurements of α_s as a function of the energy scale Q. Figure taken from Ref. [9]

quark or gluon is produced, it can loose energy by radiating additional gluons, up to an energy scale where confinement occurs and hadrons are formed. This leads to a collimated shower of hadrons which is also called jet.

2.1.5 The Electroweak Interaction and Spontaneous Symmetry Breaking

Historically the electromagnetic and weak interactions were treated as two separate theories. An unification of these two theories, the electroweak theory, was developed by Glashow, Salam and Weinberg [15–17].

Based on the observation that the charged current only couples to left-handed particles, the quantum number of the weak isospin T can be introduced. The weak forces are now constructed in such a way that the corresponding gauge bosons couple to the third component of the weak isospin T_3. By exploiting the isospin formalism [18], left-handed fermions can be grouped into doublets with $T = 1/2$ and thus $T_3 = \pm 1/2$. All right-handed fermions form a singlet with $T = 0$, $T_3 = 0$ and as a result they do not undergo charged current interactions. To describe the electromagnetic interaction, which couples to both left-handed and right-handed particles, the weak hypercharge Y_w is introduced. Analogous to the Gell-Mann–Nishijima formula [19], the electric charge Q and the third component of the weak isospin T_3 can be related to the weak hypercharge Y_w by:

$$Q = T_3 + \frac{Y_\mathrm{W}}{2}. \tag{2.19}$$

The corresponding symmetry group related to the weak isospin is the group $SU(2)_L$ which has three generators $T_i = \sigma_i/2$, given by the Pauli matrices σ_i. L in this context stands for left-handed. The three bosonic vector fields corresponding to these generators are W_μ^1, W_μ^2 and W_μ^3 and only couple to left-handed particles. The symmetry group associated to the weak hypercharge is the group $U(1)_Y$ which has one generator and thus one gauge field B_μ. These two groups build up the symmetry group of the electroweak theory $SU(2)_L \times U(1)_Y$. The requirement of local gauge invariance under this symmetry group leads to the following Lagrangian:

$$\mathcal{L}_{EW} = \sum_j \bar{\psi}_j^L i\gamma^\mu \mathcal{D}_\mu \psi_j^L + \sum_{j,\sigma} \bar{\psi}_{j\sigma}^R i\gamma^\mu \mathcal{D}_\mu \psi_{j\sigma}^R - \frac{1}{4} W_{\mu\nu}^i W_i^{\mu\nu} - \frac{1}{4} B_{\mu\nu} B^{\mu\nu}, \tag{2.20}$$

where j is the generation index, ψ^L are the doublets of the left-handed fermion fields and ψ^R are the right-handed fermion fields with the component for the flavor σ. $\mathcal{D}_\mu$ is the covariant derivative:

$$\mathcal{D}_\mu = \partial_\mu - ig\frac{\text{œ}}{2}\mathbf{W}\text{-} + ig'\frac{Y_\mathrm{W}}{2} B_\mu. \tag{2.21}$$

There are two coupling constants, g for $SU(2)_L$ and g' for $U(1)_Y$. The corresponding field strength tensors are

$$\begin{aligned} W_{\mu\nu}^i &= \partial_\mu W_\nu^i - \partial_\nu W_\mu^i + g\epsilon_{ijk} W_\mu^j W_\nu^k, \\ B_{\mu\nu} &= \partial_\mu B_\nu - \partial_\nu B_\mu. \end{aligned} \tag{2.22}$$

The symmetry group $SU(2)$ is a non-Abelian group which leads to a self coupling of the W fields. This is shown by the third term of the corresponding field strength tensor, which couples these components. The physical mass eigenstates $W_\mu^\pm$ can be obtained via a linear combination of W_μ^1 and W_μ^2:

$$W_\mu^\pm = \frac{1}{\sqrt{2}}(W_\mu^1 \mp iW_\mu^2) \tag{2.23}$$

and the Z boson Z_μ and photon field A_μ via a rotation of the fields W_μ^3 and B_μ about the weak mixing angle θ_W

$$\begin{pmatrix} A_\mu \\ Z_\mu \end{pmatrix} = \begin{pmatrix} \cos\theta_W & \sin\theta_W \\ -\sin\theta_W & \cos\theta_W \end{pmatrix} \begin{pmatrix} B_\mu^0 \\ W_\mu^3 \end{pmatrix}. \tag{2.24}$$

In 1983, the $W^\pm$ and Z boson have been discovered at the Sp$\bar{\text{p}}$S collider at CERN [20–23]. Mass terms would violate the electroweak gauge symmetry, hence the $W^\pm$ and Z boson also have to be massless, as they are a linear combination of the massless fields W_μ^1, W_μ^2, W_μ^3 and B_μ^0. This is in contrast to the experimental observation. The electroweak gauge symmetry therefore has to be broken.

By using the mechanism of spontaneous symmetry breaking, the W and Z bosons can acquire mass while the photon remains massless. This is done by introducing a single complex scalar doublet field

$$\Phi(x) = \begin{pmatrix} \phi^+(x) \\ \phi^0(x) \end{pmatrix}, \tag{2.25}$$

called Higgs field [24], with its Lagrangian

$$\mathcal{L}_H = (\mathcal{D}_\mu \Phi)^\dagger (\mathcal{D}^\mu \Phi) - V(\Phi), \tag{2.26}$$

where the potential $V(\phi)$ is given by

$$V(\Phi) = -\mu^2 \Phi^\dagger \Phi + \frac{\lambda}{4}(\Phi^\dagger \Phi)^2. \tag{2.27}$$

The potential is invariant under the local gauge transformations of $SU(2)_L \times U(1)_Y$. It is constructed in such a way that $V(\Phi)$ has for $\mu^2 > 0$ and $\lambda > 0$ a degenerate ground state $\Phi^\dagger \Phi = -\frac{4\mu^2}{\lambda} = v^2$ with a non-vanishing vacuum expectation value v. The ground state $\langle \Phi \rangle = \frac{1}{\sqrt{2}} \binom{0}{v}$ can now be chosen in such a way that the $SU(2)_L \times U(1)_Y$-symmetry is broken to $U(1)_{EM}$. If Φ is expanded around the vacuum expectation value [25], it is found to have the following form:

$$\Phi(x) \approx \frac{1}{\sqrt{2}} \begin{pmatrix} 0 \\ v + H(x) \end{pmatrix}. \tag{2.28}$$

The field $H(x)$ describes a physical neutral scalar, called Higgs boson with the mass $m_H = \mu\sqrt{2}$. In July 2012 a new boson consistent with the Higgs boson was observed by ATLAS [26] and CMS [27] at the LHC. Its mass has until now been measured to be 125.09 ± 0.24 GeV[28].

The three additional degrees of freedom of Φ are absorbed, leading to mass terms for three out of four physical gauge bosons:

$$m_W = \frac{1}{2} v g \qquad m_Z = \frac{1}{2} v \sqrt{g^2 + g'^2}. \tag{2.29}$$

The photon remains massless. As $\mathcal{L}_H$ is invariant under the local gauge transformations of the electroweak symmetry group, adding $\mathcal{L}_H$ does therefore not break the gauge symmetry of the electroweak Lagrangian. At the same time the W and Z boson have obtained a mass. The ratio of the masses of the massive bosons can at leading order (LO) be expressed as

$$\cos\theta_W \approx \frac{m_W}{m_Z}, \tag{2.30}$$

and the relation of the coupling constants can be expressed as

$$g \sin \theta_W = g' cos\theta_W = e. \tag{2.31}$$

These relation can be tested within the Standard Model. Also the masses of the fermions, which were also required to be massless, can be explained by a Yukawa coupling to the scalar Higgs field. In the unitarity gauge the Lagrangian has the simple form:

$$\mathcal{L}_{Yukawa} = -\sum_f m_f \bar{\psi}_f \psi_f - \sum_f \frac{m_f}{v} \bar{\psi}_f \psi_f H. \tag{2.32}$$

Hence, the fermions couple to the Higgs field with a coupling constant equal to their mass. This relation and other properties of the Higgs mechanism have still to be measured precisely. So far all measurements are in agreement with the predictions made by the electroweak theory and the Higgs mechanism.

2.2 The Phenomenology of Proton–Proton Collisions

Protons are baryons and therefore composite objects whose complicated dynamics cannot be calculated in the framework of QCD. This compositeness complicates the description of a proton–proton collision with respect to a collision at a lepton collider, where point-like particles collide. The quantum number of the baryons are given, according to the quark-model, by the three valence quarks. For the proton these are two u-quarks and one d-quark. The valence quarks are bound by the exchange of gluons. During this exchange, several processes can occur. For instance a gluon can split into a $q\bar{q}$-pair. These dynamically changing quarks are called sea quarks, since they form a "sea" of $q\bar{q}$-pairs. Also the valence quarks or a gluon itself can radiate a gluon. All objects in the proton, gluons, valence- and sea-quarks are named partons. These processes often happen below the scale of Λ_{QCD} and can hence not be described with perturbative QCD. A phenomenological model has therefore been developed to describe hadron-hadron collisions.

During inelastic hadron-hadron collisions, the objects do not interact as a whole, but only the partons inside of the hadrons. As a consequence, not the whole center of mass energy of the colliding protons is available in a proton–proton collision. Figure 2.5 shows a schematic view of such a scattering process. Two protons A and B collide and the partons a and b, which carry a momentum fraction x_a and x_b of the proton, scatter in the hard scattering process with a cross section $\hat{\sigma}$. The probability to find a parton with a given x inside the proton is parametrized by the parton distribution functions (PDF) $f_{a/A}(x_a)/f_{b/B}(x_b)$.

Quarks and gluons produced in a collision are colored objects and cannot exist as free particles. They will radiate further gluons or split into $q\bar{q}$-pairs leading to a cascade of partons, called *parton shower*. These parton showers stop once a scale

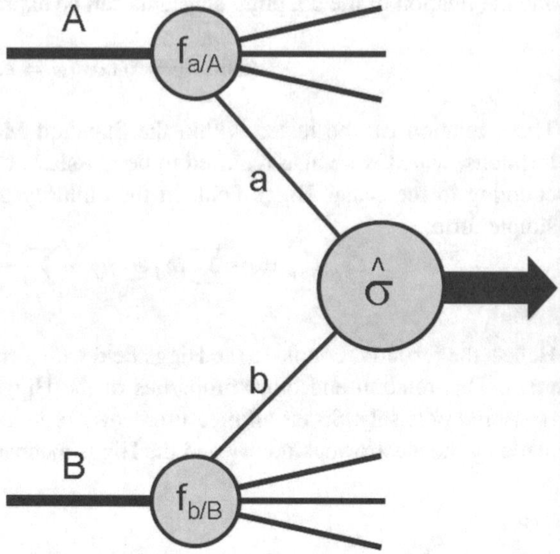

Fig. 2.5 Schematic view of a hard scattering process with a cross section $\hat{\sigma}$. The incoming protons are labeled with A and B, the scattered partons with the momentum fraction $x_{a,b}$ of the proton are labeled as a and b. The probability to find these partons at a given momentum fraction x is parametrized by the parton distribution functions $f_{a/A}(x_a)/f_{b/B}(x_b)$. Figure taken from Ref. [29]

is reached at which hadrons form. This process is called *hadronization*. Also the protons have taken out a parton and are thus left in a colored state leading to the color production of additional partons. These partons are together with initial- and final-state radiation and multiple interactions of partons inside the protons referred to as *underlying event*. A proton–proton collision is therefore a complicated process which does not only require the understanding of the hard scattering process but also all further processes which are occurring. These processes are described in more detail in Sect. 3.2. In the following only the calculation of the hard scattering process will be discussed.

2.2.1 The Structure of Protons

As discussed previously, the hadron-hadron cross section for an inelastic hard scattering process cannot be calculated directly with perturbative QCD, since physics processes of all scales in Q^2 are involved. It was first pointed out by S.D. Drell and T.-M. Yan [30] that the parton model which was developed by Feynman [31] to describe lepton-hadron scattering, can be extended to also describe hadron-hadron scattering. The main idea is to separate the perturbatively calculable short distance interactions and the non perturbative long distance interactions. The part calculable with perturbative QCD is given by the subprocess cross section $\hat{\sigma}$, which can be calculated using the Feynman rules discussed in Sect. 2.1.3. The non-perturbative part has to be described by functions which cannot be calculated but have to be extracted from measurements. These functions parametrize the probability to find a parton of

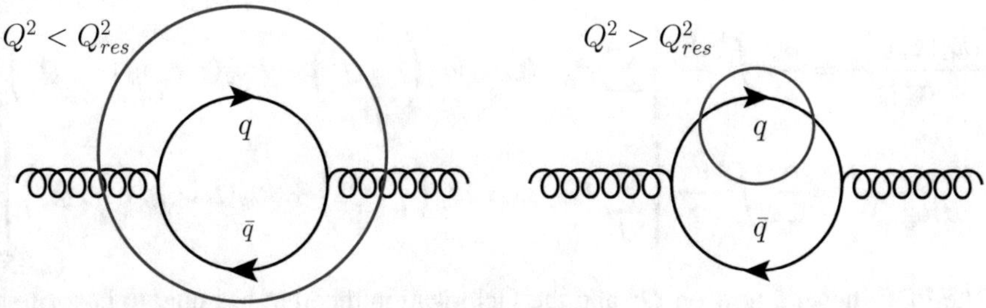

Fig. 2.6 Diagram of a gluon that splits into a quark-antiquark pair, which annihilates back to a gluon. The blue circle indicates the resolution due to the Q^2 of the process. In the left case the quark-antiquark pair cannot be resolved whereas in the right case it can be resolved

a certain flavor at a certain momentum fraction x of the hadron. The factorization theorem can then be used to calculate the proton–proton cross section σ_{AB} for a specific hard process $\hat{\sigma}_{ab \to X}$:

$$\sigma_{AB} = \sum_{a,b} \int dx_a dx_b \; f_{a/A}(x_a) f_{b/B}(x_b) \, \hat{\sigma}_{ab \to X}(x_a, x_b). \tag{2.33}$$

The PDFs $f_{a/A}(x_a)/f_{b/B}(x_b)$ have, besides the dependency on x, a dependency on the Q^2 value at which a certain process takes places. An illustrative explanation can be seen in Fig. 2.6. A higher momentum transfer results in a higher spatial resolution. If the Q^2 of the process, which corresponds to a certain resolution, is below a certain Q^2_{res}, additional substructures cannot be resolved. If the momentum transfer is above this scale, additional processes can be resolved. This fact leads to a dependency of the PDFs on the momentum transfer Q^2, since in the latter case the probability to find additional partons is higher.

The probability for a parton i to emit a parton f or to undergo a splitting that yields a parton f is described by the corresponding Altarelli–Parisi [32] splitting functions $P_{if}(z)$, where $1 - z$ is the fraction of momentum carried by the emitted parton. These splitting functions can be expressed as perturbative expansions:

$$P_{if}(z, \alpha_s) = P_{if}^{(0)}(z) + \frac{\alpha_s}{2\pi} P_{if}^{(1)}(z) + \cdots . \tag{2.34}$$

They are at the moment calculated up to NLO and NNLO [33]. The splitting functions have four different forms: P_{qq}– a quark radiates a gluon, P_{qg}– a quark radiates a quark and becomes a gluon, P_{gq}– a gluon radiates a quark and becomes a quark, P_{gg}– a gluon radiates a gluon. The dependency of the parton distributions q_i and g on Q^2 can be determined using the splitting functions with the DGLAP equations[6]:

[6]Dokshitzer–Gribov–Lipatov–Altarelli–Parisi equations.

$$\frac{\partial q_i(x,Q^2)}{\partial \log Q^2} = \frac{\alpha_s}{2\pi}\int_x^1 \frac{dz}{z}\left\{\sum_j P_{q_iq_j}(z,\alpha_s)q_j\left(\frac{x}{z},Q^2\right) + P_{q_ig}(z,\alpha_s)g\left(\frac{x}{z},Q^2\right)\right\}$$

$$\frac{\partial g(x,Q^2)}{\partial \log Q^2} = \frac{\alpha_s}{2\pi}\int_x^1 \frac{dz}{z}\left\{\sum_j P_{gq_j}(z,\alpha_s)q_j\left(\frac{x}{z},Q^2\right) + P_{gg}(z,\alpha_s)g\left(\frac{x}{z},Q^2\right)\right\}. \tag{2.35}$$

The PDFs depend now on Q^2 and the factorization theorem has now to be written as:

$$\sigma_{AB} = \sum_{a,b}\int dx_a dx_b\, f_{a/A}(x_a,\mu_F^2) f_{b/B}(x_b,\mu_F^2) \times [\hat{\sigma}_0 + \alpha_s(\mu_R^2)\hat{\sigma}_1 + \cdots]_{ab\to X}. \tag{2.36}$$

Here μ_F is the factorization scale, which can be thought of as the scale that separates the long- and short-distance physics. The partonic cross section $\hat{\sigma}$ is now also expressed as a perturbative expansion in α_s. Formally the cross section calculated in all orders of perturbation theory is independent from the choice of the parameters μ_R and μ_F. However, in the absence of a complete set of higher order corrects, it is necessary to make a specific choice. Different choices will lead to different numerical results which is a reflection of the theoretical uncertainty. The partonic cross section and the splitting functions have to have the same order in α_s, to be consistent.

2.2.2 Determination of Parton Distribution Functions

The full x dependency of the PDFs can currently not be predicted. Thus this dependency has to be extracted somewhere else, usually from global QCD fits to several measurements. Most important for the determination are the results from deep inelastic scattering (DIS) where a proton is probed by a lepton. The most precise measurement of protons was done by the H1 [34] and ZEUS [35] experiments at the HERA accelerator. These measurements are predominantly at low x and cannot distinguish between quarks and antiquarks. There are also DIS measurements done at fixed-target experiments, e.g. [36], which are at higher x. Jet data from collider experiments, e.g. [37, 38], cover a broad range on x and Q^2 and are especially important for the high x gluon distribution.

To extract the x-dependence from these measurements, first a scale Q_0 has to be chosen at which a generic functional form of the parametrization for the quark and gluon distributions is used

$$F(x, Q_0^2) = Ax^B(1-x)^C P(x; D, \ldots). \tag{2.37}$$

The parameters B and C are physically motivated. Parameter B is associated to the behavior at small-x while C is associated to the large-x valence counting rules. However, they are not sufficient enough to describe either quark or gluon distributions. Thus the term $P(x; D, \ldots)$ is a suitable smooth function which adds more flexibility, depending on the number of parameters. The parametrization scale Q_0 is often chosen to be in the range 1–2 GeV. This is above the region where α_s is large and not in the region where the extracted PDF is used. The functional form with a set of start parameters is evolved in Q^2 and convoluted with the partonic cross section to predict a cross section which can be compared to the actual measurements. For the measured and calculated cross sections a χ^2 is calculated. The starting parameters are now deduced by minimizing the χ^2. Once these parameters are determined, the PDFs can, starting from the parametrization scale, be evolved to any Q^2 using the DGLAP equations.

The extracted PDFs have uncertainties corresponding to the experimental uncertainties of the measurements used for the global fit. These uncertainties can be propagated to uncertainties on the deduced parametrization parameters. However, the propagation of these uncertainties to the PDFs cannot be done straight forward, since some of the parametrization parameters are highly correlated. To calculate sets of uncertainties which are uncorrelated and can be directly propagated, often the Hessian method is used [39]. In this method the $n \times n$ covariance matrix (n is the number of parameters) is build for the up and down variation of the parameters by either 68 or 90% confidence level. This matrix can then be rotated into an orthogonal eigenvector basis. The result are $2n$ eigenvector sets (one set for the up and one set for the down variation) which allow the uncorrelated propagation of the fit uncertainties. These eigenvector sets for up (X_i^+) and down (X_i^-) variation can then be combined to an asymmetric uncertainty ΔX_{max}^+ and ΔX_{max}^- on the PDF or an observable using the PDF with following formula:

$$\begin{aligned} \Delta X_{max}^+ &= \sqrt{\sum_{i=1}^{2n} [\max(X_i^+ - X_0, X_i^- - X_0, 0)]^2} \\ \Delta X_{max}^- &= \sqrt{\sum_{i=1}^{2n} [\max(X_0 - X_i^+, X_0 - X_i^-, 0)]^2}, \end{aligned} \tag{2.38}$$

where X_i^+ and X_i^- are the respective up and down variations of source i and X_0 denotes the central value. ΔX_{max}^+ adds in quadrature the PDF uncertainty contributions that lead to an increase of the observable X, and ΔX_{max}^- the PDF uncertainty contributions that lead to a decrease.

Additional uncertainties arise from the chosen parametrization at Q_0 and the value of α_s used in the evolution. There are different approaches for the treatment of these uncertainties.

The extraction of these PDFs is usually done by different groups of theorists and experimentalists specialized to this topic. The extracted PDFs are then made public in a certain order of α_s which is given by the order of the splitting functions used for the DGLAP evolution. Figure 2.7 shows the NNLO PDF with its corresponding

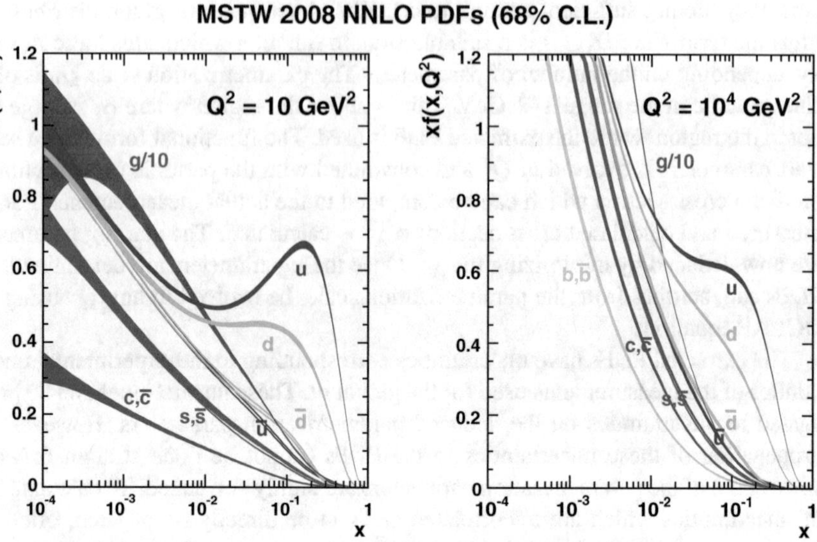

Fig. 2.7 MSTW2008NNLO PDF set as a function of Bjorken-x for quarks and gluons (divided by a factor 10) shown at a scale of $Q^2 = 10\ \text{GeV}^2$ on the left and $Q^2 = 10^4\ \text{GeV}^2$ on the right. The uncertainty of the PDFs is indicated by an uncertainty band. Figure taken from Ref. [40]

uncertainties extracted by the MSTW group [40]. The distributions of quarks and gluons at $Q^2 = 10\ \text{GeV}^2$ and $Q^2 = 10^4\ \text{GeV}^2$ are shown. The distributions show an increase with decreasing x due to the increasing contributions from the sea. At higher x around $\approx 1/3$ the u and d distributions have a peak which corresponds to valence part of the proton. At higher Q^2 these peaks are getting less significant and the sea part is contributing more to higher values of x.

2.2.3 The Drell–Yan Process

The Drell–Yan process [30] is the production of a lepton pair $\ell^+\ell^-$ at a hadron collider by quark-antiquark annihilation. In the basic Drell–Yan process, the $q\bar{q}$-pair annihilates to a virtual photon $q\bar{q} \to \gamma^* \to \ell^+\ell^-$. From now on this process is discussed for the case of a decay into an electron-positron pair. The cross section for this process at leading order can easily be obtained from the fundamental QED $e^+e^- \to \mu^+\mu^-$ cross section, with the addition of appropriate color and charge factors [29]:

$$\hat{\sigma}(q\bar{q} \to \gamma^* \to e^+e^-) = \frac{4\pi\alpha^2}{3\hat{s}} \frac{1}{N_C} Q_q^2, \tag{2.39}$$

where Q_q is the charge of the quarks, $\hat{s}$ the squared center of mass energy of the incoming partons and $1/N_C = 1/3$ is a color factor, taking into account that only three color combinations are possible since the intermediate state has to be colorless. The partonic center of mass energy is equal to the virtuality of the photon and the invariant mass of the electron-positron pair:

$$\sqrt{\hat{s}} = m_{\gamma^*} = m_{e^+e^-} = \sqrt{(p_{e^+} + p_{e^-})^2}, \tag{2.40}$$

where p_{e^+} and p_{e^-} are the momentum four vectors of the positron and electron, respectively. Hence, looking at the invariant mass of the lepton pair the cross section has a strongly falling behavior $\hat{\sigma} \propto 1/m^2_{e^+e^-}$. If $m_{e^+e^-} \approx m_Z$, the process can also take place via the exchange of a Z boson $q\bar{q} \to Z \to e^+e^-$, leading to a Breit–Wigner resonance in the spectrum of the invariant mass near m_Z. These two possible processes, the exchange via a virtual photon and the exchange via a Z boson interfere.

The four vectors of the incoming partons can be written as (assuming $m_{parton} = 0$)

$$p_a^\mu = \frac{\sqrt{s}}{2}(x_a, 0, 0, x_a), \quad p_b^\mu = \frac{\sqrt{s}}{2}(x_b, 0, 0, -x_b), \tag{2.41}$$

where s is the squared center of mass energy of the hadrons which is related to the partonic quantity by $\hat{s} = x_a x_b s$. Using the four vectors, the rapidity $y_{e^+e^-} = \frac{1}{2}\log(\frac{E+p_z}{E-p_z})$ of the e^+e^--pair can be expressed as

$$y_{e^+e^-} = \frac{1}{2}\log\left(\frac{x_a}{x_b}\right), \tag{2.42}$$

and hence

$$x_a = \frac{m_{e^+e^-}}{\sqrt{s}}e^{+y_{e^+e^-}}, \quad x_b = \frac{m_{e^+e^-}}{\sqrt{s}}e^{-y_{e^+e^-}}. \tag{2.43}$$

Thus different invariant masses $m_{e^+e^-}$ and different rapidities $y_{e^+e^-}$ probe different values of the parton x.

Figure 2.8 shows the relationship between the variables x and Q^2 and the kinematic variables corresponding to a final state of mass M and produced with rapidity y. Also shown are the regions of phase space each experiment can reach.

The DIS experiments have access to lower values of Q^2, HERA probes lower values of x and fixed target experiments higher values of x. The kinematic plane for the LHC is shown for a center of mass energy of $\sqrt{s} = 7$ TeV. A broad range in both variables, x and Q^2, is covered by the LHC. The measurement of the Drell–Yan process starting at invariant masses above the Z-resonance ($m_{e^+e^-} > 116$ GeV) probes values of $x > 10^{-2}$ when going up to higher rapidities even reaching approximately values of $x \approx 10^{-3}$ to $x \approx 1$. Since for Drell–Yan production an antiquark is needed, a cross section measurement is especially sensitive to the $\bar{u}$- and $\bar{d}$-distributions at higher x.

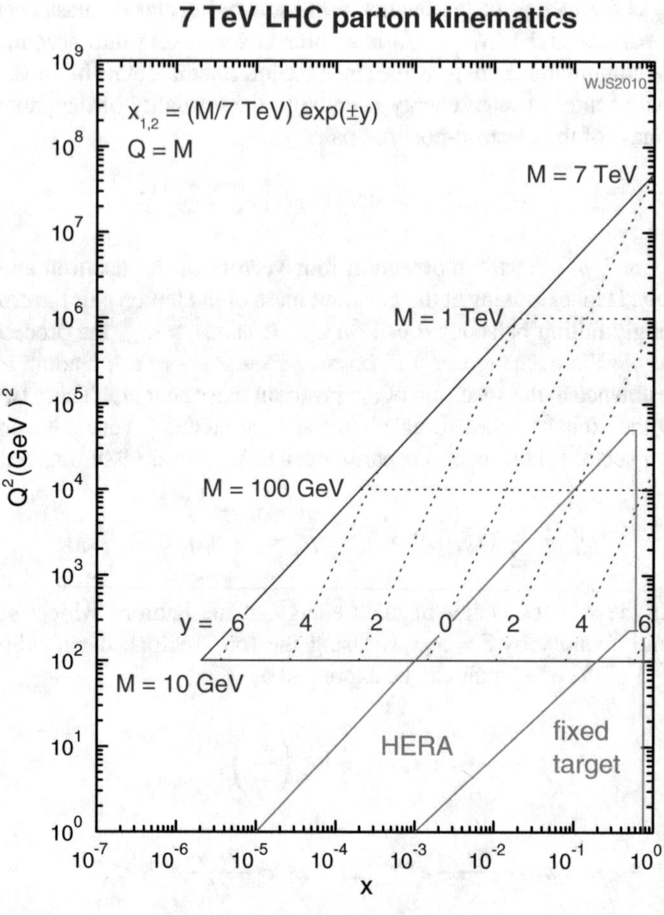

Fig. 2.8 Graphical representation of the relationship between parton (x, Q^2) variables and the kinematic variables corresponding to a final state of mass M produced with rapidity y at the LHC collider with $\sqrt{s} = 7$ TeV. Figure taken from Ref. [41]

For the Drell–Yan process usually the factorization scale and renormalization scale are set to the mass of the process $\mu_R = \mu_F = m_{e^+e^-}$. This convention is also used in this analysis for all theoretical calculations.

2.2.4 The Photon Induced Process

So far only the production of lepton pairs $\ell^+\ell^-$ via the Drell–Yan process has been discussed. The quarks in the proton carry electric charges themselves and hence cannot only radiate gluons but also photons. This means that besides the partonic

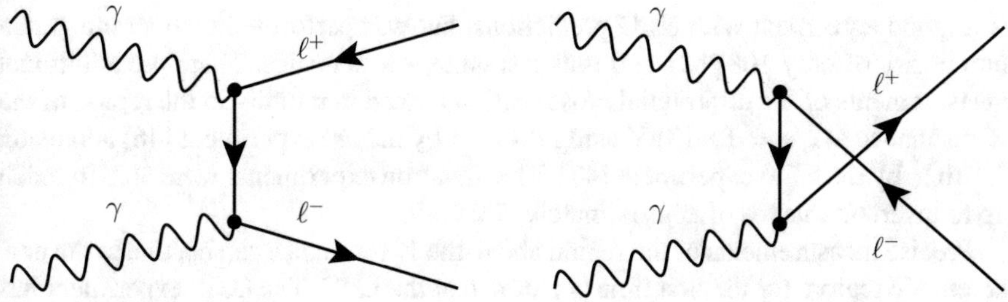

Fig. 2.9 Leading order Feynman diagrams for the photon induced production of dilepton pairs

structure of the proton, there is also a photonic structure. These photons can, via the so-called photon induced process, also produce lepton pairs. The leading order Feynman diagrams for this process are shown in Fig. 2.9. The left diagram corresponds to a t-channel diagram and the right diagram to a u-channel diagram. The photon induced process has at leading order no s-channel diagram and therefore different kinematic properties than the Drell–Yan process. Lepton pairs are produced via scattering of the two photons and hence the photon induced process has a higher contribution at small angles with respect to the direction of the incoming photons.

The photon part of the proton can be accounted for by introducing a photon PDF and calculating the evolution in Q^2 with modified DGLAP equations. Many of the standard PDFs used at the LHC do not include this photon part. Only three PDFs are currently published which also take into account this part: the MRST2004qed PDF set [42], the NNPDF2.3qed PDF set [43], and the CT14qed PDF set [44].

The $\gamma\gamma$-initiated contribution becomes a significant part of the dilepton production at high invariant masses. The knowledge of this process is therefore an important input for analyses searching for a heavy resonance decaying into lepton pairs.

The proton itself is also charged and can radiate photons which can produce lepton pairs via elastic scattering of the protons. However, the photon induced process in this thesis refers to the inelastic scattering of the protons which is dominant at high invariant masses of the dilepton pair.

2.2.5 *Recent Results*

The Drell–Yan process was measured at several hadron-hadron colliders, but the region above the Z-resonance was only measured by the experiments at the Tevatron collider[7] and the LHC. The CDF experiment at the Tevatron has measured the double-differential cross section binned in invariant mass and rapidity for the region 66 GeV $< m_{e^+e^-} <$ 116 GeV and $m_{e^+e^-} >$ 116 GeV at $\sqrt{s} = 1.8$ TeV [45]. The measurement

[7]Tevatron is a proton-antiproton collider at Fermilab and was operated at $\sqrt{s} = 1.8$ TeV and $\sqrt{s} = 1.96$ TeV.

is in good agreement with NLO predictions, but was performed with an integrated luminosity of only $108\,\mathrm{pb}^{-1}$ and thus has quite low statistics. There are additional measurements of the differential cross section binned in rapidity in the region of the Z-resonance at $\sqrt{s} = 1.96$ TeV using $0.4\,\mathrm{fb}^{-1}$ by the D0 experiment [46] and using $2.1\,\mathrm{fb}^{-1}$ by the CDF experiment [47]. The Tevatron experiments were able to reach up to invariant masses of approximately 500 GeV.

Precise measurements of the region above the Z-resonance can be, due to the new kinematic region, for the first time performed at the LHC. The CMS experiment has measured the invariant mass spectrum of the Drell–Yan process at $\sqrt{s} = 7$ TeV up to 600 GeV using $36\,\mathrm{pb}^{-1}$ of data [48]. Additionally there are two measurements of the differential cross section at $\sqrt{s} = 7$ TeV and $\sqrt{s} = 8$ TeV in two mass windows from 120 to 200 GeV and 200 to 1500 GeV, binned in rapidity and using $4.5\,\mathrm{fb}^{-1}$ and $19.7\,\mathrm{fb}^{-1}$ of data [49, 50]. A measurement of the differential cross section at $\sqrt{s} = 7$ TeV binned in invariant mass up to 1.5 TeV using $4.9\,\mathrm{fb}^{-1}$ of data has been performed by the ATLAS experiment [51]. The latter three analyses for the first time pointed out the importance of the $\gamma\gamma$-initiated processes at high invariant masses. The CMS measurement treated this contribution as a background to obtain the pure Drell–Yan contribution, while the ATLAS measurement treated this contribution as part of the signal. This allowed to use the measurement as an input for the determination of the photon PDF by the NNPDF collaboration [43].

2.3 Physics Beyond the Standard Model

2.3.1 Limitations of the Standard Model

The Standard Model of particle physics is one of the best tested theories in physics. Based on the concepts of the Standard Model many predictions were made, for example the existence of the top-quark or the Higgs bosons, and later on confirmed by experimental observations. Also the parameters of the Standard Model are partially tested very precisely. The anomalous magnetic moment of the electron shows for example agreement between theory and the experimentally measured value to more than 10 significant figures [52, 53]. Although not yet proven to be wrong, the Standard Model has several conceptual problems or is not able to describe observed phenomena. Some of these phenomena/problems are in the following discussed.

Neutrino mass The observation of neutrino-oscillations has proven that neutrinos have to have a mass. The Standard Model does not include a right-handed spinor for neutrinos. As long as neutrinos do not have mass, there is also no reason to include it. If neutrinos are Dirac particles,[8] then a mass term requires the inclusion of the right-handed spinor into the Standard Model. However, it seems unsatisfactory to have a state included in the Standard Model which does not undergo any of the interactions.

[8]Meaning that they have spin 1/2 and both, a neutrino and an anti-neutrino exists.

In addition, there is the question why the mass of the different neutrinos does not lie in the range of the masses of the other leptons, but is several orders of magnitude smaller.

Dark matter and dark energy The measurement of the rotation velocity of the luminous matter in galaxies as a function of the distance from the galactic center showed that the velocity is roughly constant from a certain distance on [54]. This observation is in contrast to the expectation that the velocity should decrease with the distance due to the lower gravitational attraction. A simple explanation for this behavior is the existence of a non-luminous dark matter halo in the galaxy. The existence of this dark matter is by now supported by many other astrophysical observations like measurements of the cosmic microwave background [55] or the observation of gravitational lensing [56]. Due to its observed properties, a dark matter candidate cannot couple via the strong or electromagnetic interaction. The only candidate within the Standard Model would therefore be the neutrino, since it only acts through the weak interaction. However, measurements have shown that neutrinos can only account for a small part of the dark matter and hence a suitable dark matter candidate is still missing. Even less understood is the existence of dark energy. Observations show that the expansion of the universe is accelerating. For the acceleration some kind of energy is needed, called dark energy. Measurements of the cosmic microwave background [57] show, that our universe contains to about 27% dark matter, 68% dark energy, and only 5% ordinary matter. The Standard Model of particle physics explains therefore only about 5% of the energy content of the universe and can clearly not be the final answer.

Grand unification The unification of the electric and magnetic force to the electromagnetic force and the further unification with the weak force to the electroweak force have motivated the idea that all observed forces are different manifestations of the same force. A further unification is also motivated by the experience that unified theories have a lot of predictive power. A grand unification theory is a theory that proposes a single gauge symmetry as source for all Standard Model interactions. In 1974 Glashow and Georgi proposed a model based on the group $SU(5)$ [58] which is the simplest group containing the Standard Model. However, these type of theories often predict the decay of the proton which has not yet been observed. A complete theory should not only unify the three forces of the Standard Model but also include gravity.

Matter-antimatter asymmetry In the big bang, particles and antiparticles should have been created in almost the same amount. Considering only the amount of CP violation predicted by the Standard Model, they should have annihilated almost completely again. However, a much larger amount of matter is observed in our universe which requires additional CP violation which cannot be accounted for by the Standard Model.

The hierarchy problem The previous arguments have shown that there are good reasons to believe that the Standard Model is only an effective theory. Effective theories are usually valid up to some energy scale Λ at that new physics occurs which needs to be described by a more complete theory. In the Standard Model, the Higgs mass receives corrections with $\mathcal{O}(\Lambda^2)$. If the Standard Model is valid up to a very high energy scale, then the corrections are much larger than the actual mass of the Higgs. This means that the bare mass parameter of the Higgs in the Standard Model must be fine tuned in a way that almost completely cancels the quantum corrections. This precise tuning seems to be unnatural and the corresponding philosophical problem is called hierarchy or naturalness problem.

2.3.2 Theories Beyond the Standard Model

Theories which extend the Standard Model do solve some of the problems described above and they are usually referred to as theories beyond the Standard Model (BSM). Some of the most prevalent theories are in the following briefly introduced.

Supersymmetry In Supersymmetry (SUSY) [59] a symmetry is introduced which relates the two basic classes of particles: bosons and fermions. Each particle is associated with a superpartner, all fermions with a superpartner of integer spin and each boson with a superpartner of half integer spin. Some particles can have more than one superpartner, the Higgs boson has for example several superpartners which can also be charged. If supersymmetry would be an exact symmetry, these pairs superpartners should have the same mass and internal quantum numbers (besides spin). However, this is not what is observed by experiments, for example no particle with spin 0 and the mass of the electron has been observed. Supersymmetry must therefore be a broken symmetry and their superpartners differ in mass. The symmetry breaking leads to the introduction of a lot of new parameters in the theory which is an often criticized feature of the theory.

Supersymmetry is able to solve a lot of the problems mentioned above. In supersymmetry each particle has a quantum number related to it, called R-parity. All Standard Model particles have R-parity of $+1$ while all supersymmetric particles have R-parity of -1. In R-parity conserving theories, the lightest supersymmetric particle (LSP) cannot further decay into Standard Model particles and is therefore a natural candidate for dark matter. In the Standard Model, the weak, strong, and electromagnetic coupling constant fail to unify at high energies. Supersymmetry can lead to the unification of the coupling constants at the scale of about 10^{16} GeV and therefore to a unification of all three forces. At the same time, the supersymmetric particles contribute to the loop corrections of the Higgs mass with a negative sign. They cancel therefore the corrections and lead naturally to the Higgs mass which is observed in experiment. The latter is also one of the largest weak spots of the theory. In order to avoid further fine tuning of the bare Higgs mass, supersymmetry needs to be realized at the scale of electroweak symmetry breaking ($\mathcal{O}(100\,\text{GeV})$). The higher the scale of supersymmetry is, the more fine tuning needs to be applied. Since no supersymmetric particles have been found so far by the LHC experiments, the scale must be above the scale of $\mathcal{O}(1\,\text{TeV})$. This leads to the fact that even with supersymmetry, fine tuning is most likely needed to some extend.

Extra dimensions Another class of extensions to the Standard Model are theories extending the usual $(3 + 1)$-dimensional space time. In these theories, a $(3 + \delta + 1)$-dimensional space time is assumed with δ additional spacial dimensions. The usual $(3 + 1)$-dimensional space time is in these theories referred to as *brane* on which the Standard Model lives. Another formulation of the hierarchy problem is the question why the scale of gravity is so much higher ($\mathcal{O}(10^{19}\,\text{GeV})$) than the scale of electroweak symmetry breaking. These classes of theories solve the hierarchy problem by explaining that gravity is so weak compared to all the other forces since only the graviton[9] can propagate to these extra spacial dimensions. Two prominent theories are a model proposed by Arkani-Hamed, Dimopoulos, and Dvali (ADD model) [60] and the Arkani-Hamedmodel [61]. Since no extra dimensions are observed, these need to be compactified. This compactification is achieved in these models by different mechanisms.

Left-right symmetry In the electroweak theory parity violation is constructed by forming $SU(2)$ doublets from the left-handed fermion fields whereas the right-handed fields are $SU(2)$ singlets. In left-right symmetric models [62, 63], also the right-handed fields form doublets of a $SU(2)$ symmetry, leading to the following symmetry:

$$SU(2)_R \times SU(2)_L \times U(1). \tag{2.44}$$

In the Standard Model the group $SU(2)_L \times U_Y(1)$ contains the $W^\pm$ and Z bosons which are obtaining mass via spontaneous symmetry breaking. Similarly, the group $SU(2)_R$ leads to three additional bosons, namely $W_R'^\pm$ and Z'. Since parity violation of the weak interaction is an experimental fact this symmetry must be broken:

$$SU(2)_R \times SU(2)_L \times U(1) \rightarrow SU(2)_L \times U(1)_Y. \tag{2.45}$$

Via spontaneous symmetry breaking also the bosons of the $SU(2)_R$ group can obtain a mass. The requirement of parity violation at the energy scales probed so far leads to the requirement that these bosons must be heavy. In this case, the interaction via the $W_R'^\pm$ bosons is suppressed at low energies, and parity violation is restored.

An interesting feature of this model is the so-called seesaw-mechanism [64]. The left-right symmetric model, breaking the left-right symmetry gives a large mass to the right-handed neutrino. The right-handed neutrino mass is at the same time related to the left-handed neutrino mass, making it very light. This mechanism therefore explains the very small left-handed neutrino masses observed in experiment.

String theory In string theory [65] point-like particles are replaced by one-dimensional objects called strings. String theory is not, like the other mentioned models, based on a quantum field theory. Instead it is a new mathematical framework which is related to quantum field theory. String theory naturally includes gravity

[9] The graviton is the hypothetical mediator of the gravitational force.

and often also supersymmetry. String theory can therefore solve a variety of problems but it is on the other hand difficult to make predictions. A lot of assumptions need to be made to be able to make predictions which can be tested by experiments.

2.3.3 New Physics with a Charged Lepton and a Neutrino in the Final State

In the following some models are introduced which lead to a signature in the final state containing a charged lepton ($\ell^{\pm} = e^{\pm}, \mu^{\pm}$) and missing transverse momentum caused by a particle leaving the detector unseen (see Sect. 6.5).

2.3.3.1 Dark Matter Models

In many models dark matter is assumed to interact via the weak force. They contain particles called weakly interacting massive particle (WIMP). If these WIMPs are existing, they can be produced in pairs at the LHC via a yet unknown interaction. These WIMP pairs would leave the detector unseen. To detect an event it is necessary to have some signature measured by the detector to trigger the event. Hence, searches are performed in which a gluon is radiated via initial state radiation which triggers the event. The signature for these kind of events would then contain a jet from the radiated gluon plus missing transverse momentum from the WIMP pair. It is also possible that a W boson is radiated via initial state radiation decaying subsequently into a lepton and neutrino. This would lead to a final state containing a charged lepton and missing transverse momentum. Usually, the Standard Model interaction with the dark matter is expressed with an effective field theory as a four-point contact interaction [66–69]. Limits have been set for various operator types [70] for the effective field theory by searches performed at a center of mass energy of 8 TeV by both, the ATLAS [71] and CMS [72] experiment. The limits obtained especially constrain the dark matter - Standard Model interaction cross section at low dark matter masses, where direct detection experiments are dominated by background. The searches in the final state containing a lepton lead in general to weaker limits compared to searches containing a jet. The reason is the low cross section for radiating a W boson compared to radiating a gluon. An exception are models in which constructive interference enhances the cross section for W radiation.

With rising center of mass energy at the LHC, the interaction between Standard Model and dark matter might be probed and the effective field theory approach is questionable. For the searches performed at $\sqrt{s} = 13$ TeV, this approach has therefore been replaced by simplified models [73] in which the interaction is mediated via a neutral Z' boson with its mass as a free parameter. It has also recently been pointed out that the models in which constructive interference enhanced the cross section are not valid as they are violating the electroweak gauge symmetry [74]. They are therefore not any longer studied at the LHC.

2.3.3.2 Supersymmetry Models

Supersymmetry predicts the existence of charginos ($\tilde{\chi}^{\pm}$), which are electrically charged fermions. They are linear combinations of a wino (superpartner of the W boson) and the electrically charged higgsino (superpartners of the Higgs boson). At the LHC charginos can be produced with neutralinos ($\tilde{\chi}^0$, linear combination of the electrically neutral superpartners of the electroweak bosons and Higgs boson). By marking choices for some of the parameters in Supersymmetry, different simplified models can be derived. In some models [75] charginos can further decay to a charged lepton and neutrino while the neutralinos decay completely invisible to a Sneutrino (superpartner of the neutrino) and neutrino. This leads to a topology containing a charged lepton and missing transverse momentum. These models have so far not been probed by analysis searching in the final state of a lepton and missing transverse momentum.

2.3.3.3 W^* Model

In references [76, 77] a model is discussed which extends the $SU(2)_L$ group by a $SU(3)_L$ group in order to solve the hierarchy problem. These models predict the existence of Z^* and W^* bosons which have a magnetic type coupling to fermions. The W^* boson decays like the Standard Model W boson via $W^* \rightarrow \ell\nu_\ell$. It is predicted to be in the TeV range (in order to solve the hierarchy problem). A search for these bosons has been performed by the ATLAS at the center of mass energy of 8 TeV and masses of the W^* boson above 3.21 TeV are excluded with 95% confidence level [71].

2.3.3.4 Left-Right-Symmetric Model

As discussed previously, in the left-right symmetric model, the $W_R'^{\pm}$ bosons obtain a mass via spontaneous symmetry breaking of the group $SU(2)_R \times SU(2)_L \times U(1)$. The mass of the $W_R'^{\pm}$ bosons has to be high compared to the scale at which parity violation has been probed yet. The purely left-handed and purely right-handed couplings lead to identical cross sections as long as the handedness of the couplings to the quarks and leptons are equal. Hence, the $W_R'^{\pm}$ can be thought of as a heavier version of the Standard Model W bosons, if it decays to a charged lepton and a light neutrino. If it decays to a heavy right-handed neutrino, the neutrino will further decay to visible particles leading to a different final state.

2.3.3.5 Sequential Standard Model

In the Sequential Standard Model (SSM) [78], a W' boson is introduced which has the same quantum numbers and couplings to fermions as the Standard Model W boson and no couplings to the W and Z bosons. The W' boson is hence a copy of the Standard Model boson, only with a higher mass and width. The latter is approximately 3% of the pole mass. There is no particular theoretical motivation for the SSM and hence it cannot be expected that this model is in any kind realized by nature. It is rather a reference model for gauge bosons arising from new, broken gauge symmetries. It is hence often used as a benchmark for presenting exclusion limits and comparing between experiments. Limits on the mass of a W' boson have been set by the ATLAS and CMS collaboration using the data recorded at $\sqrt{s} = 8$ TeV. The ATLAS experiment excluded with 95% confidence level masses below 3.24 TeV [71] and the CMS experiment below 3.28 TeV [72].

Since the SSM W' boson can have same the final state as the W boson, interference effects need to be considered between these two processes [79]. The CMS collaboration has studied the impact of the interference on the exclusion limits [72] and found a large impact on the exclusion limit of up to 700 GeV when accounting for these effects. However, typically these effects are not included as the interference depends strongly on the coupling which depends on the considered model. To be as model independent as possible, interference effects are excluded throughout this thesis.

References

1. Zinser M (2013) Double differential cross section for Drell-Yan production of high-mass e^+e^--pairs in pp collisions at $\sqrt{s} = 8$ TeV with the ATLAS experiment. MA thesis, Mainz U., 2013-08-07. http://inspirehep.net/record/1296478/files/553896852$_$CERN-THESIS-2013-258.pdf
2. Halzen F, Martin A (1984) Quarks and leptons: an introductory course in modern particle physics. Wiley, New York. isbn: 978-0471887416
3. Fukuda Y et al (1998) Evidence for oscillation of atmospheric neutrinos. Phys Rev Lett 81: 1562–1567. https://doi.org/10.1103/PhysRevLett.81.1562, arXiv:hep-ex/9807003 [hep-ex]
4. Bonn J (2002) Limits on neutrino masses from tritium β decay. Nucl Phys Proc Suppl 110:395–397. https://doi.org/10.1016/S0920-5632(02)01520-7
5. Goobar A et al (2006) A new bound on the neutrino mass from the sdss baryon acoustic peak. JCAP 0606: 019. https://doi.org/10.1088/1475-7516/2006/06/019, arXiv:astro-ph/0602155 [astro-ph]
6. Aaij R et al (2014) Observation of the resonant character of the $Z(4430)^-$ state. Phys Rev Lett 112.22: 222002. https://doi.org/10.1103/PhysRevLett.112.222002, arXiv:1404.1903 [hep-ex]
7. Ablikim M et al (2013) Observation of a charged Charmoniumlike structure in $e^+e^- \rightarrow \pi^+\pi^- J/\psi$ at $\sqrt{s} = 4.26$ GeV. Phys Rev Lett 110: 252001. https://doi.org/10.1103/PhysRevLett.110.252001, arXiv: 1303.5949 [hep-ex]
8. Aaij R et al (2015) Observation of J/ψ resonances consistent with pentaquark states in $\Lambda_b^0 \rightarrow J/\psi K^- p$ decays. Phys Rev Lett. 115: 072001. https://doi.org/10.1103/PhysRevLett.115.072001, arXiv:1507.03414 [hep-ex]

9. Beringer J (2012) Review of particle physics (RPP). Phys Rev D86:010001. https://doi.org/10.1103/PhysRevD.86.010001
10. Peskin ME, Schroeder DV (1995) An introduction to quantum field theory. Addison-Wesley, Boston. isbn: 978-0-201-50397-5
11. Noether E (1918) Invariante Variationsprobleme. ger. In: Nachrichten von der Gesellschaft der Wissenschaften zu Göttingen, Mathematisch-Physikalische Klasse. pp. 235–257. http://eudml.org/doc/59024
12. Feynman RP (1949) The theory of positrons. Phys Rev 76. https://doi.org/10.1103/PhysRev.76.749
13. Orear J, Fermi E (1950) Nuclear physics: a course given by enrico fermi at the university of Chicago. University of Chicago Press, Chicago
14. Brock R et al (1994) Handbook of perturbative QCD; Version 1.1. Rev Mod Phys
15. Glashow S (1961) Partial symmetries of weak interactions. Nucl Phys 22:579–588. https://doi.org/10.1016/0029-5582(61)90469-2
16. Salam A (1968) Weak and electromagnetic interactions. Conf Proc C680519:367–377
17. Weinberg S (1967) A Model of Leptons. Phys Rev Lett 19:1264–1266. https://doi.org/10.1103/PhysRevLett.19.1264
18. Heisenberg W (1932) Über den Bau der Atomkerne. I. Zeitschrift für Physik 77: 1–2. https://doi.org/10.1007/BF01342433
19. Griffiths D (2008) Introduction to elementary particles. Wiley, New York. isbn: 978-3-527-40601-2
20. Arnison G (1983) Experimental observation of isolated large transverse energy electrons with associated missing energy at sqrt(s) = 540 GeV. Phys Lett B 122:103–116. https://doi.org/10.1016/0370-2693(83)91177-2
21. Banner M (1983) Observation of single isolated electrons of high transverse momentum in events with missing transverse energy at the CERN anti-p p collider. Phys Lett B 122:476–485. https://doi.org/10.1016/0370-2693(83)91605-2
22. Arnison G (1983) Experimental observation of lepton pairs of invariant mass around 95-GeV/c^2 at the CERN SPS collider. Phys Lett B 126:398–410. https://doi.org/10.1016/0370-2693(83)90188-0
23. Bagnaia P (1983) Evidence for Z0→ e+ e- at the CERN anti-p p collider. Phys Lett B 129:130–140. https://doi.org/10.1016/0370-2693(83)90744-X
24. Higgs PW (1964) Broken symmetries, massless particles and gauge fields. Phys Lett 12:132–133. https://doi.org/10.1016/0031-9163(64)91136-9
25. Hollik W (2006) Electroweak theory. J Phys Conf Ser 53:7–43. https://doi.org/10.1088/1742-6596/53/1/002
26. Aad G et al (2012) Observation of a new particle in the search for the Standard Model Higgs boson with the ATLAS detector at the LHC. Phys Lett B716: 1–29. https://doi.org/10.1016/j.physletb.2012.08.020. arXiv:1207.7214 [hep-ex]
27. Chatrchyan S et al (2012) Observation of a new boson at a mass of 125 GeV with the CMS experiment at the LHC. Phys Lett B716: 30–61. https://doi.org/10.1016/j.physletb.2012.08.021. arXiv:1207.7235 [hep-ex]
28. Aad G et al (2015) Combined measurement of the higgs boson mass in *pp* collisions at $\sqrt{s} =$ 7 and 8 TeV with the ATLAS and CMS experiments. Phys Rev Lett 114: 191803. https://doi.org/10.1103/PhysRevLett.114.191803, arXiv:1503.07589 [hep-ex]
29. Campbell JM, Huston J, Stirling W (2007) Hard interactions of quarks and gluons: a primer for LHC physics. Rep Prog Phys 70: 89. https://doi.org/10.1088/0034-4885/70/1/R02, arXiv:hep-ph/0611148 [hep-ph]
30. Drell S, Yan T-M (1971) Partons and their applications at high-energies. Ann Phys 66: 578. https://doi.org/10.1016/0003-4916(71)90071-6
31. Feynman RP (1969) The behavior of hadron collisions at extreme energies. Conf Proc C690905:237–258
32. Altarelli G, Parisi G (1977) Asymptotic freedom in parton language. Nucl Phys B126:298. https://doi.org/10.1016/0550-3213(77)90384-4

33. Vogt A, Moch S, Vermaseren J (2004) The Three-loop splitting functions in QCD: the singlet case. Nucl Phys B691: 129–181. https://doi.org/10.1016/j.nuclphysb.2004.04.024, arXiv:hep-ph/0404111 [hep-ph]
34. Adloff C et al (2000) Measurement of neutral and charged current cross-sections in positron proton collisions at large momentum transfer'. Eur Phys J C13: 609–639. https://doi.org/10.1007/s100520050721. arXiv:hep-ex/9908059 [hep-ex]
35. Chekanov S et al (2001) Measurement of the neutral current cross-section and F(2) structure function for deep inelastic e + p scattering at HERA. Eur Phys J C21: 443–471. https://doi.org/10.1007/s100520100749, arXiv:hep-ex/0105090 [hep-ex]
36. Towell R et al (2001) Improved measurement of the anti-d/anti-u asymmetry in the nucleon sea. Phys Rev D64: 052002. https://doi.org/10.1103/PhysRevD.64.052002, arXiv:hep-ex/0103030 [hep-ex]
37. Aad G et al (2011) Measurement of inclusive jet and dijet cross sections in proton-proton collisions at 7 TeV centre-of-mass energy with the ATLAS detector. Eur Phys J C71: 1512. https://doi.org/10.1140/epjc/s10052-010-1512-2, arXiv:1009.5908 [hep-ex]
38. Aad G et al (2015) Measurement of four-jet differential cross sections in $\sqrt{s} = 8$ TeV proton–proton collisions using the ATLAS detector. JHEP 12: 105. https://doi.org/10.1007/JHEP12(2015)105, arXiv:1509.07335 [hep-ex]
39. Pumplin J et al (2001) Uncertainties of predictions from parton distribution functions. 2. The Hessian method. Phys Rev D65: 014013. https://doi.org/10.1103/PhysRevD.65.014013, arXiv:hep-ph/0101032 [hep-ph]
40. Martin et al AD (2009) Parton distributions for the LHC. Eur Phys J C 63: 189. https://doi.org/10.1140/epjc/s10052-009-1072-5. arXiv:0901.0002 [hep-ph]
41. Stirling W Private communication
42. Martin AD et al (2005) Parton distributions incorporating QED contributions. Eur Phys J C 39: 155. https://doi.org/10.1140/epjc/s2004-02088-7, arXiv:hep-ph/0411040 [hep-ph]
43. Ball RD et al (2013) Parton distributions with QED corrections. Nucl Phys B877: 290–320. https://doi.org/10.1016/j.nuclphysb.2013.10.010, arXiv:1308.0598 [hep-ph]
44. Schmidt C et al (2015) CT14QED PDFs from isolated photon production in deep inelastic scattering. arXiv: 1509.02905 [hep-ph]
45. Bodek A (2001) Measurement of d sigma/d y for high mass drell-yan e+ e- pairs at CDF. Int J Mod Phys A16S1A: 262–264. arXiv: hep-ex/0009067 [hep-ex]
46. Abazov V et al (2007) Measurement of the shape of the boson rapidity distribution for $p\bar{p} \rightarrow Z/gamma* \rightarrow e^+e^- + X$ events produced at $\sqrt{s}$ of 1.96-TeV. Phys Rev D76: 012003. https://doi.org/10.1103/PhysRevD.76.012003, arXiv:hep-ex/0702025 [HEP-EX]
47. Aaltonen TA et al (2010) Measurement of $d\sigma/dy$ of Drell-Yan e^+e^- pairs in the Z Mass Region from $p\bar{p}$ Collisions at $\sqrt{s} =$ 1:96 TeV. Phys Lett B692: 232–239. https://doi.org/10.1016/j.physletb.2010.06.043. arXiv:0908.3914 [hep-ex]
48. Chatrchyan S et al (2011) Measurement of the Drell-Yan cross section in pp collisions at $\sqrt{s} =$ 7 TeV. JHEP 1110: 007. https://doi.org/10.1007/JHEP10(2011)007, arXiv:1108.0566 [hep-ex]
49. Chatrchyan S et al (2013) Measurement of the differential and double-differential Drell-Yan cross sections in proton-proton collisions at $\sqrt{s} = 7$ TeV. JHEP 12: 030. https://doi.org/10.1007/JHEP12(2013)030, arXiv:1310.7291 [hep-ex]
50. Khachatryan V et al (2015) Measurements of differential and double-differential Drell-Yan cross sections in proton–proton collisions at 8 TeV. Eur Phys J C75.4: 147. https://doi.org/10.1140/epjc/s10052-015-3364-2, arXiv:1412.1115 [hep-ex]
51. Aad G et al (2013) Measurement of the high-mass Drell-Yan differential cross-section in pp collisions at sqrt(s)= 7 TeV with the ATLAS detector. Phys Lett B725: 223–242. https://doi.org/10.1016/j.physletb.2013.07.049, arXiv:1305.4192 [hep-ex]
52. Hanneke D, Hoogerheide SF, Gabrielse G (2011) Cavity control of a single-electron quantum cyclotron: measuring the electron magnetic moment. Phys Rev A83: 052122. https://doi.org/10.1103/PhysRevA.83.052122, arXiv: 1009.4831 [physics.atom-ph]
53. Aoyama T et al (2012) Tenth-order QED contribution to the electron g-2 and an improved value of the fine structure constant. Phys Rev Lett 109: 111807. https://doi.org/10.1103/PhysRevLett.109.111807. arXiv:1205.5368 [hep-ph]

54. Rubin VC, Ford WK Jr (1970) Rotation of the andromeda nebula from a spectroscopic survey of emission regions. Astrophys J 159:379–403. https://doi.org/10.1086/150317
55. Hinshaw G et al (2009) Five-year Wilkinson microwave anisotropy probe (WMAP) observations: data processing, sky maps, and basic results. Astrophys J Suppl 180: 225–245. https://doi.org/10.1088/0067-0049/180/2/225, arXiv:0803.0732 [astro-ph]
56. Clowe D et al (2006) A direct empirical proof of the existence of dark matter. Astrophys J 648: L109–L113. https://doi.org/10.1086/508162, arXiv:astro-ph/0608407 [astro-ph]
57. Ade PAR et al (2014) 'Planck 2013 results. I. overview of products and scientific results. Astron Astrophys 571: A1. https://doi.org/10.1051/0004-6361/201321529, arXiv:1303.5062 [astro-ph.CO]
58. Georgi H, Glashow SL (1974) Unity of all elementary particle forces. Phys Rev Lett 32:438–441. https://doi.org/10.1103/PhysRevLett.32.438
59. Martin SP (1997) A Supersymmetry primer. [Adv Ser Direct High Energy Phys 18: 1(1998)]. https://doi.org/10.1142/9789812839657_0001,10.1142/9789814307505_0001, arXiv:hep-ph/9709356 [hep-ph]
60. Arkani-Hamed N, Dimopoulos S, Dvali GR (1998) The hierarchy problem and new dimensions at a millimeter. Phys Lett B429: 263–272. https://doi.org/10.1016/S0370-2693(98)00466-3, arXiv: hep-ph/9803315 [hep-ph]
61. Randall L, Sundrum R (1999) A large mass hierarchy from a small extra dimension. Phys Rev Lett 83: 3370–3373. https://doi.org/10.1103/PhysRevLett.83.3370, arXiv:hep-ph/9905221 [hep-ph]
62. Senjanovic G, Mohapatra RN (1975) Exact left-right symmetry and spontaneous violation of parity. Phys Rev D 12:1502. https://doi.org/10.1103/PhysRevD.12.1502
63. Maiezza A et al (2010) Left-right symmetry at LHC. Phys Rev D82: 055022. https://doi.org/10.1103/PhysRevD.82.055022, arXiv:1005.5160 [hep-ph]
64. Mohapatra RN, Senjanovic G (1980) Neutrino mass and spontaneous parity violation. Phys Rev Lett 44:912. https://doi.org/10.1103/PhysRevLett.44.912
65. Polchinski J (1994) What is string theory? NATO Advanced Study Institute: Les Houches Summer School, Session 62: fluctuating geometries in statistical mechanics and field theory Les Houches, France, August 2–September 9 1994. arXiv: hep-th/9411028 [hep-th]
66. Beltran M et al (2010) Maverick dark matter at colliders. JHEP 09: 037. https://doi.org/10.1007/JHEP09(2010)037, arXiv:1002.4137 [hep-ph]
67. Rajaraman A et al (2011) LHC bounds on interactions of dark matter. Phys Rev D84: 095013. https://doi.org/10.1103/PhysRevD.84.095013, arXiv:1108.1196 [hep-ph]
68. Fox PJ et al (2012) Missing energy signatures of dark matter at the LHC. Phys Rev D85: 056011. https://doi.org/10.1103/PhysRevD.85.056011, arXiv:1109.4398 [hep-ph]
69. Cotta RC et al (2013) Bounds on dark matter interactions with electroweak Gauge Bosons. Phys Rev D88: 116009. https://doi.org/10.1103/PhysRevD.88.116009, arXiv:1210.0525 [hep-ph]
70. Goodman J et al (2010) Constraints on dark matter from colliders. Phys Rev D82: 116010. https://doi.org/10.1103/PhysRevD.82.116010, arXiv:1008.1783 [hep-ph]
71. Aad G et al (2014) Search for new particles in events with one lepton and missing transverse momentum in *pp* collisions at $\sqrt{s} = 8$ TeV with the ATLAS detector. JHEP 09: 037. https://doi.org/10.1007/JHEP09(2014)037, arXiv:1407.7494 [hep-ex]
72. Khachatryan V et al (2015) Search for physics beyond the standard model in final states with a lepton and missing transverse energy in proton-proton collisions at sqrt(s) = 8 TeV. Phys Rev D91.9: 092005. https://doi.org/10.1103/PhysRevD.91.092005, arXiv:1408.2745 [hep-ex]
73. Abercrombie D et al (2015) In: Boveia A et al (ed) Dark matter benchmark models for early LHC Run-2 searches: report of the ATLAS/CMS dark matter forum. arXiv: 1507.00966 [hep-ex]
74. Haisch U, Re E (2015) Simplified dark matter top-quark interactions at the LHC. JHEP 06: 078. https://doi.org/10.1007/JHEP06(2015)078. arXiv:1503.00691 [hep-ph]
75. Arina C et al (2015) Constraints on sneutrino dark matter from LHC Run 1. JHEP 05: 142. https://doi.org/10.1007/JHEP05(2015)142, arXiv:1503.02960 [hep-ph]

76. M. V. Chizhov and G. Dvali. ‘Origin and Phenomenology of Weak-Doublet Spin-1 Bosons’. In: *Phys. Lett*. B703 (2011), pp. 593–598. doi: https://doi.org/10.1016/j.physletb.2011.08.056.arXiv: 0908.0924 [hep-ph]
77. Chizhov MV, Dvali G (2011) Origin and phenomenology of weak-doublet spin-1 bosons. Phys Lett B703: 593–598. https://doi.org/10.1016/j.physletb.2011.08.056, arXiv:0908.0924 [hep-ph]
78. Altarelli G, Mele B, Ruiz-Altaba M (1989) Searching for new heavy vector bosons in $p\bar{p}$ colliders. Z Phys C45: 109 [Erratum: Z Phys C47,676 (1990)]. https://doi.org/10.1007/BF01552335, https://doi.org/10.1007/BF01556677
79. Accomando E et al (2012) Interference effects in heavy W’-boson searches at the LHC. Phys Rev D85: 115017. https://doi.org/10.1103/PhysRevD.85.115017, arXiv:1110.0713 [hep-ph]

Chapter 3
Theoretical Predictions and Simulation

In the following chapter first the theoretical tools used in this thesis are discussed and afterwards the principle of simulating a physics event for a proton–proton collision is introduced.

3.1 Theoretical Predictions

In the following tools for the calculation of W/Z predictions are briefly discussed.
FEWZ FEWZ [1–3] can calculate the double-differential production of dilepton pairs via the neutral current (production of a Z) and the charged current (production of a W) Drell–Yan process. It is able to make predictions at NNLO in QCD and to include all spin correlations and finite-width effects. In case of the neutral current Drell–Yan process it can also calculate electroweak corrections up to NLO. Acceptance cuts can be defined to calculate the cross sections in a certain region of phase space. FEWZ is the main tool for the theory predictions used in the analysis presented in part IV.
VRAP VRAP [4] can, like FEWZ, calculate the fully double-differential cross section for the neutral current and charged current Drell–Yan process as a function of invariant mass and rapidity. It is able to make predictions at NNLO in QCD but does not include further electroweak corrections. VRAP is used in part III to calculate higher order QCD corrections for the W and Z processes.
SANC The SANC program implements calculations for the complete NLO QCD and electroweak corrections for the charged current and neutral current Drell–Yan process. Hence, it provides in contrast to the FEWZ program also electroweak corrections for the W process. It is used in part III to calculate electroweak higher order corrections for both processes.

M. Zinser, *Search for New Heavy Charged Bosons and Measurement of High-Mass Drell-Yan Production in Proton-Proton Collisions*, Springer Theses,
https://doi.org/10.1007/978-3-030-00650-1_3

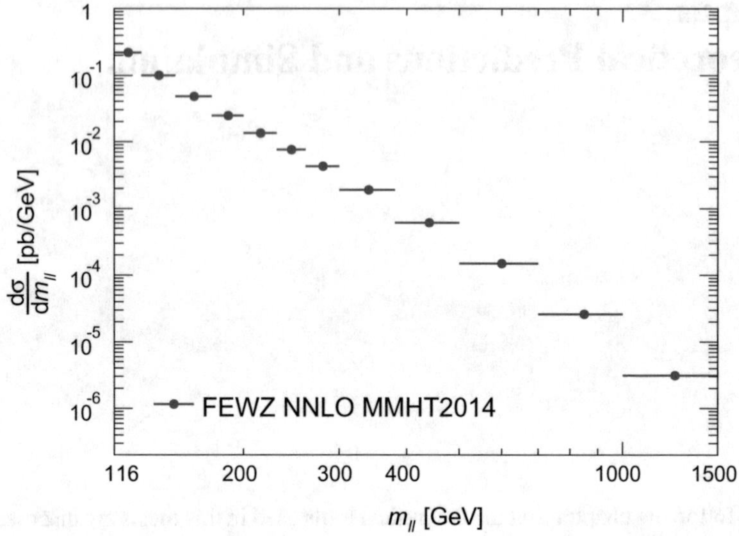

Fig. 3.1 Differential cross section prediction, binned in invariant mass. The prediction was calculated with FEWZ using the MMHT2014NNLO PDF [9]. Data taken from Ref. [10]

Further programs which are able to calculate the differential charged current and neutral current Drell–Yan cross sections are DYNNLO [5] (at NNLO), HORACE [6] (at NLO), and WZGRAD [7, 8] (at NLO). Figure 3.1 shows the differential neutral current Drell–Yan cross section as a function of invariant mass as calculated with FEWZ using the MMHT2014 PDF set [9] in the range 116 GeV $< m_{\ell\ell} <$ 1500 GeV. The cross section is steeply falling over about five orders of magnitude towards higher invariant masses.

3.2 Physics Simulation

Physical processes are simulated to compare the observed data to predictions from theory. The simulation is done on an event-by-event basis and can be separated into two steps. First, the physics simulation of all involved particles is performed and thereafter the detector response to the particles is simulated. The simulated data sets can then be reconstructed like actual recorded data. In this chapter the first step is discussed (the latter steps are discussed separately in Sect. 5.9).

The generation of the physics process can be further divided into five main steps:

1. Hard process
2. Parton shower
3. Underlying event
4. Hadronization
5. Unstable particle decays

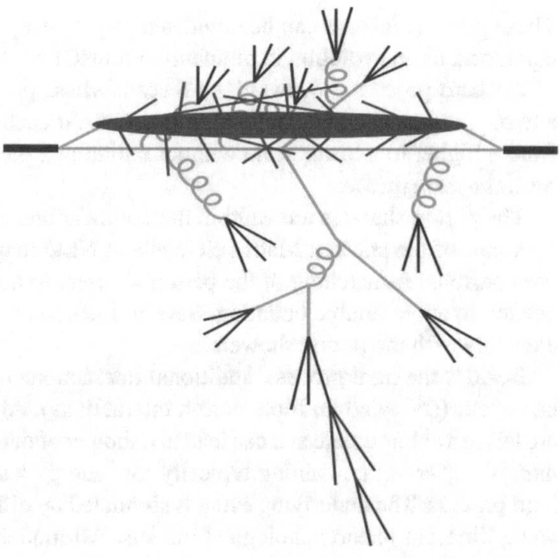

Fig. 3.2 Diagram showing the structure of a proton–proton collision, where the different colors indicate the different stages involved in the event generation. Figure taken from Ref. [11]

Figure 3.2 illustrates the different steps of the simulation, where the color corresponds to these steps listed above.

At the beginning the matrix element of the *hard process* is calculated. This involves the calculation of the probability distribution of the hard scatter process from perturbation theory. The calculated probability distribution is then convoluted with the PDFs of the incoming partons. Here choices have to be made for the factorization and renormalization scales and for the PDF. With the resulting probability distribution, four vectors of the outcoming particles can be generated using a random generator. Due to the random generation process, programs doing this, are called Monte Carlo generators. The set of events generated by such a Monte Carlo generator are often referred to as Monte Carlo (MC) samples. The calculations of the hard process are performed at a certain order in perturbation theory. Currently Monte Carlo generators exist which perform the calculation of the matrix elements and the subsequent convolution with the PDFs at LO or NLO. Additional phase space restrictions can be imposed on the generation of the four vectors of the particles. This can become useful to ensure sufficient statistics in specific regions of phase space which are most relevant for the analysis the MC sample is used for. If not already done by the Monte Carlo generator itself, additional real photon emission (final state radiation) of the outcoming particles can be simulated by external programs. In ATLAS often PHOTOS is used [12].

The initial incoming and outgoing partons involved in the hard process are colored particles and thus can radiate further gluons or gluons can split into a $q\bar{q}$-pair. In case of an incoming parton this process is called initial state radiation (ISR) and in case of an outgoing parton final state radiation (FSR). These newly produced partons can then split or radiate themselves further gluons which leads to an extended shower.

These *parton showers* can be simulated step-by-step, with the help of the DGLAP equations, as an evolution in momentum transfer starting from the momentum scale of the hard process downwards to a scale where perturbation theory breaks down and the partons become confined in hadrons. At each step the probability to evolve from a higher to a lower scale without radiating a gluon or splitting is given by the *Sudakov form factor*.

The parton showers are valid in the collinear and soft limit[1] which describes the bulk part of the shower. Matrix elements at NLO may also include the radiation of hard partons. A matching of the parton showers to the matrix elements is therefore needed to avoid double counting. Several methods exist which combine the matrix elements with the parton showers.

Besides the hard process, additional interactions of other partons in the protons can occur (so-called multiple parton interactions). Also the remnants of the proton are left in a colored state and can lead to a shower of partons. This leads to an so-called *underlying event*, containing typically low energy hadrons, which contaminate the hard process. The underlying event is simulated by different Monte Carlo generators using different phenomenological models. All models have free parameters which need to be tuned on data.

At the scale where perturbation theory breaks down, *hadronization* models simulate the transition of colored particles into hadrons, which are in the end measured in the detector. The two main models in use are the *string model* [13, 14] and the *cluster model* [15]. The former transforms the partonic systems directly into hadrons while the latter constructs an intermediate stage of cluster objects. These phenomenological models include parameters which have to be tuned on data.

In the end, many of the produced hadrons are not stable and thus also decays have to be simulated.

3.2.1 Event Generators

There are several different Monte Carlo generators available which can handle all or a part of the event generation steps. The generators used in this thesis are in the following briefly introduced.

PYTHIA is a general-purpose event generator. It has been developed over the last 30 years and was used extensively for e^+e^-, ep, $p\bar{p}$, and pp physics at LEP, HERA, the Tevatron, and the LHC. It is one of the most used generators for physics studies at the LHC. PYTHIA 6 [16] is based on Fortran 77, while its successor PYTHIA 8 [17] is a completely rewritten C++ version. Both generators are heavily used inside ATLAS. While PYTHIA 8 provides a more modern version it is a bit less tested and evolved than PYTHIA 6.

[1]The largest part of the radiation happens under small angles (collinear) and for low energies of the radiated partons (soft). Here divergences occur which are handled by the parton shower models.

Unlike other event generators, PYTHIA does not have automated code generation for processes. Instead over 200 hard-coded Standard Model and beyond Standard Model processes are implemented. PYTHIA can perform the whole event generation process from the calculation of the hard process at LO up to the parton shower, hadronization, underlying event modeling, and particle decays. PYTHIA is designed to allow for external input and can be interfaced to other event generators. This is often used to calculate the hard process at higher orders with other programs and subsequently let PYTHIA do the modeling of the parton shower, hadronization, underlying event, and particle decays.

POWHEG [18] is unlike PYTHIA not a general-purpose event generator, i.e. it is not able to simulate the whole chain from calculating the hard process up to the particle decays. However, it is able to calculate the hard scatter process with NLO accuracy. It can be interfaced to all modern event generators via the Les Houches Event interface [19]. In ATLAS the POWHEG generator is often interfaced with PYTHIA to perform the modeling of the parton showers, hadronization, modeling of the underlying event, and particle decays.

SHERPA is a general-purpose event generator, capable of simulating the physics of lepton-lepton, lepton-hadron, and hadron-hadron collisions as well as photon induced processes. In recent Sherpa versions [20] it was made possible to calculate the matrix elements at NLO. In Sherpa not a physics process is specified but the final state which should be generated. Sherpa then automatically calculates all needed matrix elements and performs the event generation. It is also possible to generate final states with additional jets.

MC@NLO [21, 22] is an event generator which, as the name indicates, can calculate the hard process at NLO. It uses its own algorithm for the parton showering and also includes spin correlations for most processes. For the modeling of the underlying event, MC@NLO is typically interfaced to HERWIG++.

HERWIG/HERWIG++ [23, 24] is another general-purpose event generator. The HERWIG++ program (written in C++) evolved from the HERWIG program (written in Fortran 77). The generator automatically generates the hard process up to LO and simulates decays with full spin correlations. It also provides sophisticated hadronic decay models, particularly for bottom hadrons and τ leptons.

References

1. Li Y, Petriello F (2012) Combining QCD and electroweak corrections to dilepton production in the framework of the FEWZ simulation code. Phys Rev D86: 094034. https://doi.org/10.1103/PhysRevD.86.094034, arXiv: 1208.5967 [hep-ph]
2. Melnikov K, Petriello F (2006) Electroweak gauge boson production at hadron colliders through $\mathcal{O}(\alpha_s^2)$. Phys Rev D 74: 114017. https://doi.org/10.1103/PhysRevD.74.114017, arXiv: hep--ph/0609070 [hep-ph]

3. Gavin R et al (2011) FEWZ 2.0: A code for hadronic Z production at next-to-next-to-leading order. Comput Phys Commun 182: 2388. https://doi.org/10.1016/j.cpc.2011.06.008, arXiv: 1011.3540 [hep-ph]
4. Anastasiou C, Dixon L, Melnikov K, Petriello F (2004) High precision QCD at hadron colliders: Electroweak gauge boson rapidity distributions at NNLO. Phys Rev D69: 094008. https://doi.org/10.1103/PhysRevD.69.094008, arXiv: hep-ph/0312266 [hep-ph]
5. Catani S et al (2009) Vector boson production at hadron colliders: a fully exclusive QCD calculation at NNLO'. Phys Rev Lett 103: 082001. https://doi.org/10.1103/PhysRevLett.103.082001, arXiv: 0903.2120 [hep-ph]
6. Carloni Calame CM et al (2007) Precision electroweak calculation of the production of a high transverse-momentum lepton pair at hadron colliders. JHEP 10: 109. https://doi.org/10.1088/1126-6708/2007/10/109, arXiv: 0710.1722 [hep-ph]
7. Baur U, Keller S, Wackeroth D (1999) Electroweak radiative corrections to W boson production in hadronic collisions. Phys Rev D59: 013002. https://doi.org/10.1103/PhysRevD.59.013002, arXiv: hep-ph/9807417 [hep-ph]
8. Baur U et al (2002) Electroweak radiative corrections to neutral current Drell-Yan processes at hadron colliders. Phys Rev D65: 033007. https://doi.org/10.1103/PhysRevD.65.033007, arXiv: hep-ph/0108274 [hep-ph]
9. Harland-Lang LA et al (2015) Parton distributions in the LHC era: MMHT 2014 PDFs. Eur Phys J C75.5: 204. https://doi.org/10.1140/epjc/s10052-015-3397-6, arXiv: 1412.3989 [hep-ph]
10. ATLAS Collaboration. http://atlas.web.cern.ch/Atlas/GROUPS/PHYSICS/PAPERS/STDM-2014-06/
11. Seymour MH, Marx M (2013) Monte Carlo event generators. arXiv: 1304.6677 [hep-ph]
12. Golonka P, Was Z (2006) PHOTOS Monte Carlo: a precision tool for QED corrections in Z and W decays. Eur Phys J C45: 97. https://doi.org/10.1140/epjc/s2005-02396-4, arXiv: hep-ph/0506026
13. Andersson B (1983) Parton fragmentation and string dynamics. Phys Rep 97:31–145. https://doi.org/10.1016/0370-1573(83)90080-7
14. Andersson B (1997) The Lund model. Camb Monogr Part Phys Nucl Phys Cosmol 7:1–471
15. Fox GC, Wolfram S (1980) A model for parton showers in QCD. Nucl Phys B 168:285–295. https://doi.org/10.1016/0550-3213(80)90111-X
16. Sjostrand T, Mrenna S, Skands PZ (2006) PYTHIA 6.4 physics and manual. JHEP 05: 026. https://doi.org/10.1088/1126-6708/2006/05/026, arXiv: hep-ph/0603175 [hep-ph]
17. Sjostrand T, Mrenna S, Skands PZ (2008) A brief introduction to PYTHIA 8.1. Comput Phys Commun 178: 852. https://doi.org/10.1016/j.cpc.2008.01.036, arXiv: 0710.3820 [hep-ph]
18. Alioli S, Nason P, Oleari C, Re E (2010) A general framework for implementing NLO calculations in shower Monte Carlo programs: the POWHEG BOX. JHEP 06: 043. https://doi.org/10.1007/JHEP06(2010)043, arXiv: 1002.2581 [hep-ph]
19. Alwall J et al (2007) A Standard format for Les Houches event files. Comput Phys Commun 176: 300–304. https://doi.org/10.1016/j.cpc.2006.11.010, arXiv: hep-ph/0609017 [hep-ph]
20. Gleisberg T et al (2009) Event generation with SHERPA 1.1. JHEP 02: 007. https://doi.org/10.1088/1126-6708/2009/02/007, arXiv: 0811.4622 [hep-ph]
21. Frixione S, Webber BR (2002) Matching NLO QCD computations and parton shower simulations. JHEP 06: 029. https://doi.org/10.1088/1126-6708/2002/06/029, arXiv: hep-ph/0204244 [hep-ph]
22. Frixione S, Nason P, Webber BR (2003) Matching NLO QCD and parton showers in heavy avor production. JHEP 0308: 007. https://doi.org/10.1088/1126-6708/2003/08/007, arXiv: hep-ph/0305252 [hep-ph]
23. Corcella G et al (2001) HERWIG 6: an event generator for hadron emission reactions with interfering gluons (including supersymmetric processes). JHEP 01: 010. https://doi.org/10.1088/1126-6708/2001/01/010, arXiv: hep-ph/0011363 [hep-ph]
24. Bahr M et al (2008) Herwig++ physics and manual. Eur Phys J C 58: 639. https://doi.org/10.1140/epjc/s10052-008-0798-9, arXiv: 0803.0883 [hep-ph]

Part II
The Experimental Setup

Chapter 4
The Large Hadron Collider

The Large Hadron Collider (LHC) [1] is a particle accelerator located at the European Laboratory for Particle Physics CERN near Geneva in Switzerland. It was designed to reach very high center of mass energies and luminosities for the discovery of new physics beyond the Standard Model and for the precise measurement of the Standard Model parameters in yet inaccessible regions of phase space.

The LHC can be operated with two types of beams, proton beams and heavy ion beams.[1] The main physics program is based on proton-proton collisions, where energies up to 7 TeV per proton beam and luminosities up to $10^{34}\,\mathrm{cm}^{-2}\mathrm{s}^{-1}$ are foreseen. It is currently the particle collider with the highest reach in center of mass energy.

A short overview about the CERN accelerator complex will be given in Sect. 4.1. The key accelerator parameters during the 2012 and 2015 operation are afterwards discussed in Sect. 4.2 and the experiments at the LHC are introduced in Sect. 4.3.

4.1 Accelerator Complex

For being filled into the LHC, the protons have to be first accelerated by a chain of pre-accelerators. Figure 4.1 shows the CERN accelerator complex with the LHC and all its pre-accelerators. The maximum energy per beam, the circumference of the accelerator and the year of its initial-startup are shown. Accelerators which do not serve as an input to the LHC are not shown. Protons are produced by ionizing hydrogen and afterwards transferred to the accelerator chain. The accelerator chain starts with the Linac2, a linear accelerator in which the protons are accelerated in bunches on a length of 33 m to an energy of 50 MeV. Subsequently, the proton bunches are running through a chain of circular accelerators: the Booster, the Proton

[1] Typically lead ions are used.

M. Zinser, *Search for New Heavy Charged Bosons and Measurement of High-Mass Drell-Yan Production in Proton-Proton Collisions*, Springer Theses,
https://doi.org/10.1007/978-3-030-00650-1_4

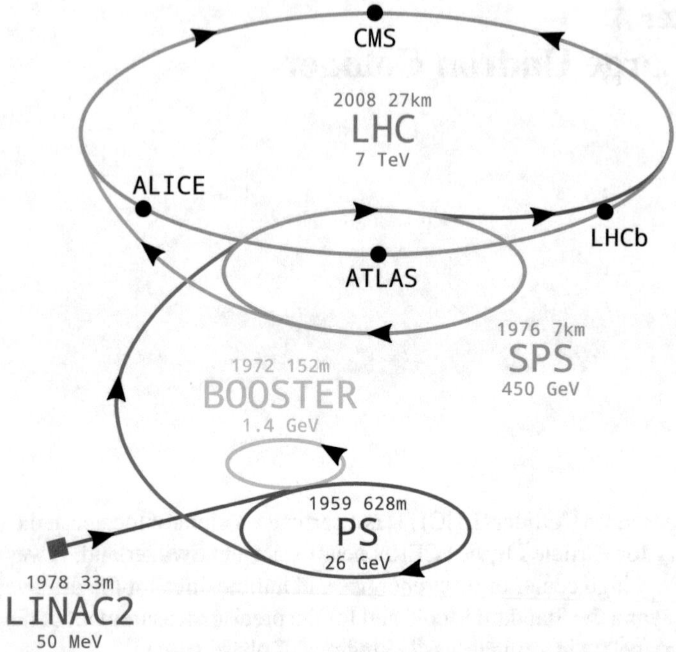

Fig. 4.1 The CERN accelerator complex. For a description of the proton acceleration chain see text. Figure taken from Ref. [2]

Synchrotron (PS) and the Super Proton Synchrotron (SPS), after which they are filled (both in a clockwise and counterclockwise direction) with an energy of 450 GeV per proton into the LHC. Here, the protons are further accelerated to energies up to 6.5 TeV.[2]

The LHC is installed in a 27 km long tunnel, which is up to 175 m beneath the surface, that was originally build for the LEP collider.[3] Each proton bunch is accelerated by eight superconducting radiofrequency cavities. The proton bunches are kept on the circular trajectory by 1232 superconducting dipole magnets. The LHC dipoles use niobium-titanium (NbTi) cables and are operated at a temperature of 1.9 K. The magnetic field of the dipole magnets reaches up to 8.3 T. It is pointing in opposite directions in the two beam pipes, since the protons in the two beam pipes are orbiting in opposite directions. A total of 392 quadrupole magnets are used to focus the proton beam. The two proton beams are circulating in two beam pipes and are brought to collision at four interaction points.

[2]Energies up to 7 TeV are foreseen but have not been reached until August 2016.

[3]**L**arge **E**lectron **P**ositron collider.

4.2 LHC Performance

The key parameter for a collider is, besides its beam energy, the luminosity it is able to deliver. The instantaneous luminosity directly relates the event rate of a physics process to its cross section: $R = \mathcal{L} \cdot \sigma$. The instantaneous luminosity can be calculated from accelerator parameters using the following formula:

$$\mathcal{L} = \frac{N_p^2 n_b f_r \gamma}{4\pi \epsilon_n \beta^*} F, \tag{4.1}$$

where N_p is the number of protons per bunch, and n_b the number of bunches per proton beam. The relativistic γ-factor and the revolution frequency f_r enter the numerator. The revolution frequency for the LHC is about 11.2 kHz. Additionally important for the instantaneous luminosity are the normalized transverse emittance ϵ_n and the value of the beta function at the interaction point β^*. The latter two parameters describe the brightness of the beam. The factor F accounts for a geometrical correction due to the crossing angle under which the beams are brought to collision. It was similar for the 2012 and 2015 operation and in the range of about 0.7–0.8. Table 4.1 lists the parameters which were obtained during the 2012 and 2015 operation of the LHC. Given are also the design parameters. The LHC was in 2012 operated with a spacing of the proton bunches of 50 ns. This is twice the spacing for which the LHC was designed for. The LHC can in principle hold up to 2808 proton bunches. Due to the larger bunch spacing only up to 1374 bunches were filled during the 2012 operation. After the 2012 operation, a two year long shutdown took place, in which the magnets of the LHC were upgraded for collisions at higher center of mass energies. During 2015, the LHC was for the first time regularly operated with the nominal bunch spacing of 25 ns and it was therefore possible to fill up to 2244 bunches into the machine. The peak luminosities at a start of a fill[4] reached up to $7.7 \times 10^{33}\,\mathrm{cm}^{-2}\mathrm{s}^{-1}$ during 2012 and about $5.0 \times 10^{33}\,\mathrm{cm}^{-2}\mathrm{s}^{-1}$ during 2015. This is about 80% of the design goal of the LHC. The higher peak luminosity was reached in 2012, despite the lower number of bunches. The reason is the easier operation mode which resulted in a brighter beam when compared to 2015. In 2012 an emittance of $\epsilon_n = 2.5$ μm and $\beta^* = 60$ cm was achieved while these parameters were $\epsilon_n = 3.5$ μm and $\beta^* = 80$ cm in 2015. The LHC will further improve its performance during the data taking and in June 2016 the design luminosity was reached for the first time. In the year 2012, an integrated luminosity of $\mathcal{L}_{int} = \int \mathcal{L} dt = 22.8\ \mathrm{fb}^{-1}$ has been delivered by the LHC in a data taking period from April to December. In 2015, a lot of time went into development and commissioning of the LHC for the first 25 ns collisions and the first operation at $\sqrt{s} = 13$ TeV. The delivered integrated luminosity in the period from May to November was therefore with 4.2 fb^{-1} smaller than what

---

[4]The instantaneous luminosity is decreasing with time as the beam width is getting larger with time and the number of protons is decreasing due to the inelastic collisions.

Table 4.1 LHC parameters during the 2012 and 2015 operation [3, 4]. Also given are the design values

Year	E_{Beam} [TeV]	N_p	n_b	ϵ_n [μm]	β^* [cm]	Bunch spacing [ns]	Peak luminosity [$cm^{-2}s^{-1}$]
2012	4	1.7×10^{11}	1374	2.5	60	50	7.7×10^{33}
2015	6.5	1.15×10^{11}	2244	3.5	80	25/50	5.0×10^{33}
Design	7	1.15×10^{11}	2808	3.75	55	25	1.0×10^{34}

was achieved in 2012. During a bunch crossing usually multiple inelastic proton-proton collisions occur. The additional inelastic collisions are also called "pile-up". In 2012 (2015) a mean of about 21 (14) collisions occurred. The number is in 2012 substantially larger due to the higher instantaneous luminosity.

4.3 Experiments at the LHC

Four main experiments are localized in caverns around the LHC ring. Two of the experiments, the ATLAS[5] experiment [5] and the CMS,[6] experiment [6] are build as general purpose experiments to cover a wide range of the physics program available at the LHC. The LHCb experiment [7] focuses on physics involving bottom quarks. The proton beams at the LHCb interaction point are less focused as the experiment was designed for lower luminosities of $2 \times 10^{32}\,cm^{-2}s^{-1}$. The ALICE[7] experiment [8] was primarily designed to study heavy-ion collisions.

In addition, there are three smaller experiments allocated at the LHC. The TOTEM[8] experiment [9] is close to the CMS experiment and aims to measure protons from elastic collisions which escape the CMS experiment. It is also used to monitor the LHC luminosity. The LHCf[9] experiment [10] is installed 140 m away from the ATLAS experiment. Its main purpose is to study neutral pions to test Monte Carlo models for proton showering, as they are used for the simulation of cosmic rays in the earth atmosphere. The last detector is the MoEDAL[10] experiment [11], which is an extension of the LHCb experiment, and used to search for example for magnetic monopoles.

[5]**A** **T**oroidal **L**HC **A**pparatu**S**.

[6]**C**ompact **M**uon **S**olenoid.

[7]**A** **L**arge **I**on **C**olliding **E**xperiment.

[8]**Tot**al **E**lastic and Diffractive Cross Section **M**easurement.

[9]**LHC** **f**orward.

[10]**Mo**nopole and **E**xotics **D**etector **a**t the **L**HC.

References

1. Evans L, Bryant P (2008) LHC Machine. JINST 3:S08001. https://doi.org/10.1088/1748-0221/3/08/S08001.
2. Endner O Private communication
3. Lamont M (2013) Status of the LHC. J Phys Conf Ser 455:012001. https://doi.org/10.1088/1742-6596/455/1/012001
4. Papotti Giulia. https://indico.cern.ch/event/448109/contributions/1942059/attachments/1216261/1793941/2015overview_paper.pdf
5. Aad G et al (2008) The ATLAS Experiment at the CERN Large Hadron Collider. JINST 3:S08003. https://doi.org/10.1088/1748-0221/3/08/S08003.
6. Chatrchyan S et al (2008) The CMS experiment at the CERN LHC. JINST 3:S08004. https://doi.org/10.1088/1748-0221/3/08/S08004.
7. Alves AA Jr et al (2008) The LHCb Detector at the LHC. JINST 3:S08005. https://doi.org/10.1088/1748-0221/3/08/S08005.
8. Aamodt K et al (2008) The ALICE experiment at the CERN LHC. JINST 3:S08002. https://doi.org/10.1088/1748-0221/3/08/S08002.
9. Anelli G et al (2008) The TOTEM experiment at the CERN Large Hadron Collider. JINST 3:S08007. https://doi.org/10.1088/1748-0221/3/08/S08007.
10. Adriani O et al (2008) The LHCf detector at the CERN Large Hadron Collider. JINST 3:S08006. https://doi.org/10.1088/1748-0221/3/08/S08006.
11. Pinfold J et al (2009) Technical design report of the MoEDAL experiment. Technical report CERN-LHCC-2009-006. MoEDAL-TDR-001. https://cds.cern.ch/record/1181486

References

Chapter 5
The ATLAS Experiment

The ATLAS experiment [1] is one of the four main experiments at the LHC. It is a general purpose detector, built at one of the four interaction points. ATLAS was constructed to measure precisely electrons, positrons, photons, muons, and jets in large kinematic regions, to allow tests of the Standard Model and searches for new particles. It consists of several layers of different detector systems, which surround the beam axis. An overview of the ATLAS experiment is shown in Fig. 5.1.

The coordinate system used in ATLAS and some commonly used kinematic variables are described in Sect. 5.1. A brief overview about the ATLAS experiment and its detector systems is given in Sect. 5.2. Here also the coordinate system used in ATLAS is introduced. The tracking systems, the calorimeters and the muon spectrometers are afterwards discussed in more detail in the Sects. 5.3 to 5.5. The multi-level trigger system and data acquisition system is addressed in Sect. 5.6. The data acquisition and processing, and the luminosity estimation are described in Sects. 5.7 and 5.8.

5.1 Coordinate System of ATLAS

The coordinate system used by ATLAS is a right handed Cartesian coordinate system with its origin at the interaction point, where the protons collide. The positive x-axis points towards the center of the LHC ring and the y-axis upwards to the surface. Thus the z-axis points counter-clockwise along the beam axis. The azimuthal angle ϕ is defined around the beam axis in the x-y plane. The range of ϕ is going from $-\pi$ to π with $\phi = 0$ pointing towards the direction of the x-axis. Hence, the range 0 to π describes the upper half plane of the detector whereas $-\pi$ to 0 describes the lower half plane. Instead of a polar angle θ, which is measured from the positive z-axis, it is convenient to use the pseudorapidity η. It can be calculated from θ using $\eta = -\ln(\tan(\theta/2))$. All detector dimensions are given in terms of η.

M. Zinser, *Search for New Heavy Charged Bosons and Measurement of High-Mass Drell-Yan Production in Proton-Proton Collisions*, Springer Theses,
https://doi.org/10.1007/978-3-030-00650-1_5

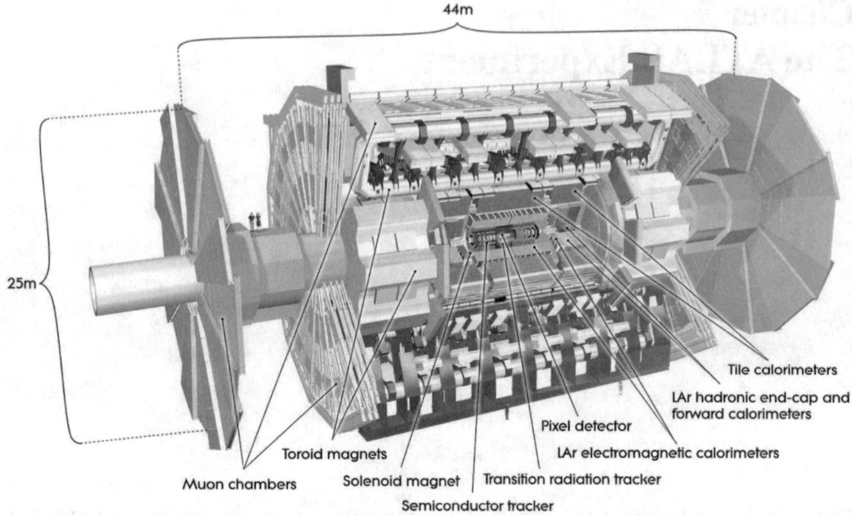

Fig. 5.1 Cut-away view of the ATLAS experiment. The dimensions of the detector are 25 m in height and 44 m in length. The overall weight of the detector is approximately 7000 tonnes. Figure taken from Ref. [1]

5.1.1 Common Kinematic Variables

The rapidity of a massive particle is defined as

$$y = \frac{1}{2} \ln \left(\frac{E + p_z}{E - p_z} \right), \tag{5.1}$$

where E is the energy of the particle and p_z its longitudinal momentum. The rapidity is a measure for the boost of a particle along the beam axis. Differences $\Delta\theta$ are, in contrast to differences of the rapidity Δy, not Lorentz invariant under boosts along the beam axis. The pseudorapidity is for massless particles equal to the rapidity, which is in good approximation valid for many particles at the LHC energies. Hence, also pseudorapidity differences $\Delta\eta$ are in good approximation invariant under boots along the beam axis.

The transverse momentum p_T, transverse energy E_T, and the missing transverse energy E_T^{miss} are commonly used and measured in the x-y plane. Transverse momentum and energy are defined by

$$p_T = \sqrt{p_x^2 + p_y^2} \tag{5.2}$$

and

$$E_T = \sqrt{p_T^2 + m^2} \tag{5.3}$$

For mass-less particles p_T and E_T are the same. The incoming partons within a proton have to first approximation only a momentum parallel to the beam axis. Momentum conservation requires therefore the vectorial sum of all momenta in the transverse plane to be zero. This can be exploited to indirectly measure particles, like neutrinos, which leave the detector unseen. Hence, the missing transverse momentum is given by the negative vector sum of all reconstructed transverse momenta

$$\vec{p}_{\mathrm{T}}^{\mathrm{miss}} = -\sum_i \vec{p}_{\mathrm{T,i}}^{\mathrm{rec}}. \tag{5.4}$$

The missing transverse energy is then defined as $E_{\mathrm{T}}^{\mathrm{miss}} = |\vec{p}_{\mathrm{T}}^{\mathrm{miss}}|$.

In different aspects, the distance ΔR in the η,ϕ-plane is used and defined as

$$\Delta R = \sqrt{\Delta\eta^2 + \Delta\phi^2}. \tag{5.5}$$

5.2 Overview of ATLAS

The inner detector is the tracking system of ATLAS (a more detailed description can be found in Sect. 5.3) and the closest detector to the beam axis. It has a coverage up to $|\eta| = 2.5$ and consists of three subsystems, first the pixel detector, followed by the Semi Conductor Tracker (SCT) and Transition Radiation Tracker (TRT). A solenoidal magnetic field of 2 T makes it possible to measure the transverse momentum of charged particles. The inner detector is additionally designed to measure vertices and identify electrons.

Following are the electromagnetic and hadronic calorimeters which are used to measure the energy of particles. As electromagnetic calorimeter a liquid argon sampling-calorimeter is used up to $|\eta| < 3.2$. A scintillator tile calorimeter is used as hadronic calorimeter covering the range $|\eta| < 1.7$. The hadronic endcap calorimeters cover $1.5 < |\eta| < 3.2$ and use, like the electromagnetic counterparts, liquid argon technology. Finally, there is the liquid argon forward calorimeter, covering the range $3.1 < |\eta| < 4.9$, which is used for measuring both, electromagnetic and hadronic objects. A more detailed description of the calorimeter system can be found in Sect. 5.4.

The calorimeter is surrounded by the muon spectrometer which consists of a toroid system, separated into a long barrel and two inserted endcap magnets, and tracking chambers. The toroid system has an air-core and generates a strong magnetic field with strong bending power in a large volume within a light and open structure. There are three layers of tracking chambers. These components of the muon spectrometer have a coverage up to $|\eta| = 2.7$ and define the overall dimension of the ATLAS experiment. The muon system also includes trigger chambers, covering a range up to $|\eta| = 2.4$. A more detailed description of the muon system can be found in Sect. 5.5.

A multi-level trigger system (a more detailed description is given in Sect. 5.6) is used to reduce the rate of pp collisions ($\approx$40 MHz) to a rate which can be processed and stored ($\mathcal{O}(100)$ Hz). To reduce this rate, the trigger system has to select events which are of special interest. The first trigger stage, the Level-1 (L1) trigger, is a hardware based system and uses a subset of the total detector information to make the decision whether to continue processing an event or not. This reduces the rate already down to approximately 100 kHz. The subsequent software based trigger stages reduce the rate further to the needed $\mathcal{O}(100)$ Hz.

The main difference between the ATLAS experiment and the CMS experiment is the magnet system. While CMS has a single solenoidal magnet with a field strength of 3.8 T, ATLAS has a solenoidal magnet with a strength of 2 T plus a toroidal magnet system. The CMS design choice leads to a higher momentum resolution for tracks due to the higher magnetic field in the tracking system but also imposes strict requirements on the design of other detector parts due to an iron return yoke for the magnetic field. The lower field strength of the ATLAS solenoid is compensated by the additional toroidal magnet outside of the calorimeters. Another difference between ATLAS and CMS is the design of the electromagnetic calorimeter. ATLAS uses a liquid-argon sampling calorimeter while CMS uses a homogeneous calorimeter constructed from crystals of lead tungstate. Both design principles lead to comparable momentum and energy resolutions.

During a two year shutdown between end of 2012 and beginning of 2015, the ATLAS trigger and data acquisition system was upgraded and an additional layer was inserted into the pixel detector.

5.3 Tracking System

The inner detector [2] (ID) is the ATLAS tracking system and is shown in Fig. 5.2. It consists of three subsystems which are mounted around the beam axis. The superconducting solenoid [3], which produces the magnetic field of 2 T, needed for the momentum measurement of charged particles, has a length of 5.3 m and a diameter of 2.5 m. With the solenoidal magnetic field and the inner detector components a momentum resolution of $\sigma_{p_T}/p_T \approx 0.05\%\ p_T\,[\text{GeV}] \oplus 1\%$ can be achieved. The subsystems of the inner detector are in the following described in more detail. Its prescription follows largely Chap. 4 of Ref. [1].

5.3.1 Pixel Detector

The pixel detector [4] is one of the two precision tracking detectors, with a coverage of $|\eta| < 2.5$. It is the innermost layer of the inner detector and has in the 2012 configuration a distance to the beam axis of $R = 50.5$ mm. In the central region, three layers of silicon pixel modules are cylindrical mounted around the beam axis, while in the endcap regions three discs each are mounted perpendicular to the beam axis. Its purpose is the measurement of particle tracks with a very high resolution, to reconstruct the interaction point (primary vertex) and secondary vertices from

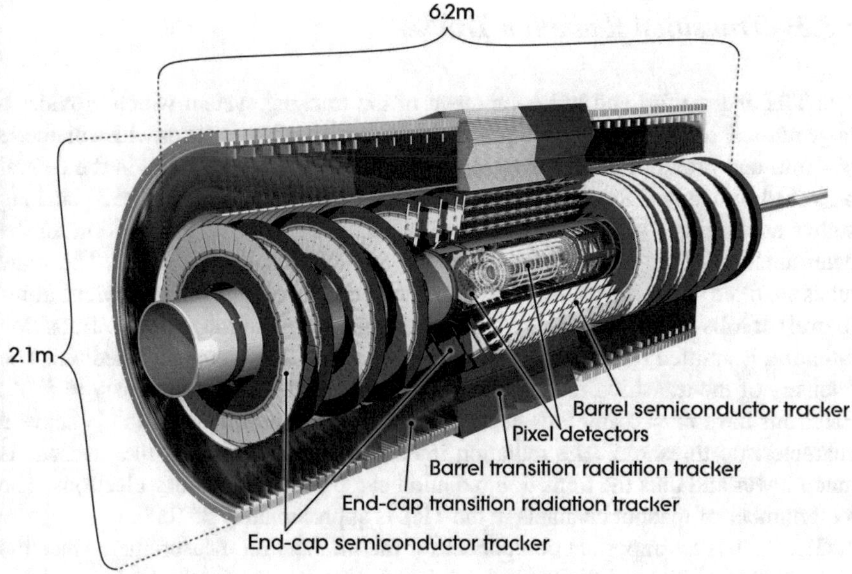

Fig. 5.2 Cut-away view of the ATLAS Inner detector. Figure taken from Ref. [1]

the decay of long-lived particles. The innermost layer of the pixel detector is called b-layer because of its importance to reconstruct the secondary vertices of decaying B-hadrons. The pixel modules have dimensions of $50\times400\,\mu\text{m}^2$. The position resolution is $10\,\mu\text{m}$ in the R-ϕ plane and $115\,\mu\text{m}$ in $z(R)$ for the central (endcap) region. Due to this fine granularity, around 80.4 million readout channels are needed.

During the two year long LHC shutdown after the 2012 data taking, a fourth pixel layer, the so-called insertable b-layer (IBL) [5] was inserted between the beam pipe and the pixel detector. This fourth layer reduces the distance to the beam pipe down to $R = 25.7$ mm and mainly improves the capability of reconstructing secondary vertices.

5.3.2 *Semi-conductor Tracker*

The semi conductor tracker is mounted in a distance of 299 to 514 mm from the beam axis, and is the second layer of the inner detector. It is a silicon microstrip detector covering the region $|\eta| < 2.5$. In the central region eight strip layers are used which are joined to four layers of small-angle (40 mrad) stereo strips to allow the measurement of both coordinates. In the endcap region nine discs on each side are installed, using two radial layers of strips each. The SCT is designed such that each particle within its coverage traverses through all four double layers. The spatial resolution of the SCT is $17\,\mu\text{m}$ in the R-ϕ plane and $580\,\mu\text{m}$ in the $z(R)$ for the central (endcap) region. The SCT has approximately 6.3 million readout channels.

5.3.3 Transition Radiation Tracker

The TRT is the third and last component of the tracking system which provides a large number of hits (typically 36 per track). It consists of straw tubes with a diameter of 4 mm and provides coverage up to $|\eta| = 2.0$. The straw tubes are in the central region 144 cm long and parallel to the beam axis. In the endcap region the 37 cm long straws are arranged radially in wheels. The TRT provides R-ϕ information for the determination of the transverse momentum with an accuracy of 130 μm. The straw tubes are filled with a Xe-based gas mixture and interleaved with polypropylene fibres (barrel) or foils (endcaps), which serve as the transition radiation material. Transition radiation is emitted by this material, if charged particles traverse this medium. The intensity of the transition radiation is proportional to the Lorentz factor $\gamma = E/m$. Electrons have $m \approx 0$ and thus at high energies the transition radiation is above a characteristic threshold. The radiation intensity for heavy objects like hadrons is much lower and thus the transition radiation can be used to identify electrons. The total number of readout channels in the TRT is approximately 351000.

The TRT is an important component for the momentum measurement since the high number of hits and the larger track length compensate for the lower precision per point compared to the silicon detectors.

5.4 The Calorimeter System

The energy of particles (except muons and neutrinos) is measured in ATLAS with sampling calorimeters, in which layers of passive and active material alternate. When incident particles like electrons, positrons, hadrons or photons traverse the calorimeter, they interact with the material in the calorimeter. In the dense passive layers, these incident particles lead to particle showers. The deposited energy of these showers, also called clusters, can be measured in the active layers and allows conclusion about the energy of the incident particle.

There is a difference between electromagnetic and hadronic showers and thus there are separate calorimeters, one for electrons, positrons, and photons and one for hadrons. In electromagnetic calorimeters, electrons and positrons are radiating photons via Bremsstrahlung which then further produce electrons and positrons via pair production, leading to a cascade of particles which is stopped by ionization, while photons are first producing electrons and positrons via pair production.

Ionization electrons are produced by passage of charged particles. They drift to electrodes and produce electrical currents proportional to the deposited energy. The initial energy E_0 of the incident electron, positron or photon decreases exponentially with $E(x) = E_0 e^{-x/X_0}$ until it is completely stopped. The parameter X_0 is called the radiation length which is material dependent. The hadronic showering process is dominated by a succession of inelastic hadronic interactions via the strong force. A characteristic quantity for the length of a hadronic shower is the absorption length λ. Hadronic showers are typically longer and broader than electromagnetic ones and thus the hadronic calorimeter is placed after the electromagnetic one.

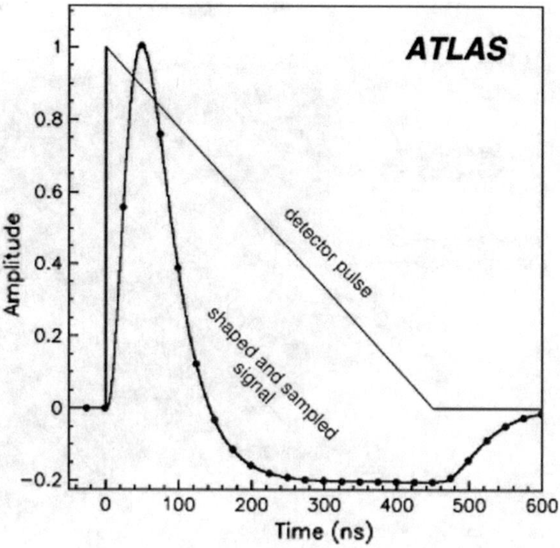

Fig. 5.3 Signal of an electromagnetic calorimeter barrel cell. The triangle shaped signal is the signal before shaping, the other line represents the signal after shaping. The dots show the sampling points in a 25 ns spacing. Figure taken from Ref. [7]

The detector signal is a triangular pulse with a very fast rise ($\approx$1 ns) and a long tail (several hundreds ns) during the time the ionization electrons are drifting towards the electrodes. Since the relevant information is the pulse height and position (time) of the peak, ideally only one sample at the exact moment that the signal reaches its maximum should be necessary. However, in practice, the time variation would give larger imprecisions in the energy measurements. To solve this issue, a pulse shaping is performed. The amplitude and the peak time can then be extracted from multiple sampling points. Figure 5.3 shows the signal shape before and after shaping. Due to the long drift time, multiple pulses from subsequent beam crossings may be overlaid. This phenomenon is called "out-of-time" pileup. Simultaneous proton-proton collisions at the same bunch crossing lead furthermore to so-called "in-time" pile-up. The so-called undershoot, the tail of the signal that falls below the zero line, helps to minimize the out-of-time pile-up impact of the energy measurement. Further information can be found in Ref. [6].

Figure 5.4 shows a cut-away view of the calorimeter system of ATLAS. The electromagnetic and hadronic calorimeters are in the following described in more detail. Its prescription follows largely on this chapter of Ref. [1].

5.4.1 Electromagnetic Calorimeter

For the electromagnetic calorimeter of ATLAS [6], lead is used in the region $|\eta| < 3.2$ as an absorber medium and liquid argon as an active medium. The electrodes to measure the energy deposited in the liquid argon and the lead absorbers are build in

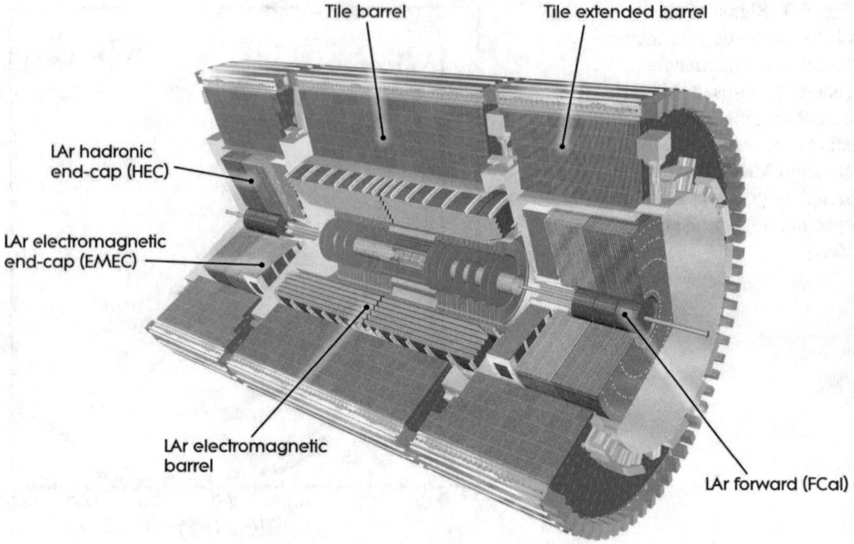

Fig. 5.4 Cut-away view of the ATLAS calorimeter system. Figure taken from Ref. [1]

an accordion geometry, in order to provide complete and uniform coverage in ϕ. The thickness of the absorber plates varies with η in such a way that the energy resolution is optimal [8]. The electromagnetic calorimeter consists of four different regions. First, there is the central part up to $|\eta| = 1.475$, called barrel calorimeter, which has a thickness of at least $22X_0$. The barrel calorimeter consists of 16 modules, each covering an angle of $\Delta\phi = 22.5°$. In the region $1.375 < |\eta| < 3.2$ there is the endcap calorimeter which is again separated into the "outer wheel" $1.375 < |\eta| < 2.5$ and the "inner wheel" $2.5 < |\eta| < 3.2$. The forward calorimeter, which is also used for the measurement of hadrons, is in the region $3.1 < |\eta| < 4.9$.

The part of the calorimeter which is intended for precision measurements ($|\eta| < 2.5$) is separated into three layers. Figure 5.5 shows the three layers and the accordion geometry of a single module of the electromagnetic barrel calorimeter. Upstream of the first layer, there is in the range $|\eta| < 1.8$ the so-called presampler which is a 11 mm thick layer of liquid argon. It has the purpose to estimate the energy lost in front of the calorimeter. The first layer has a granularity of 0.0031×0.0982 in $\eta \times \phi$. The cells are also called "strips", due to the fine segmentation in η. They allow to distinguish close by particles that enter the calorimeter, e.g., two photons from a π_0 decay. The second layer has a more coarse granularity of 0.025×0.0245 in $\eta \times \phi$. It has a thickness of $16X_0$ and is thus intended to measure the bulk part of the energy. The third layer has again a much coarser granularity and the purpose to correct for the overlap of the energy deposition in the following hadronic calorimeter. Including the presampler cells, a barrel module features 3424 and a module in the endcap roughly 4000 readout cells.

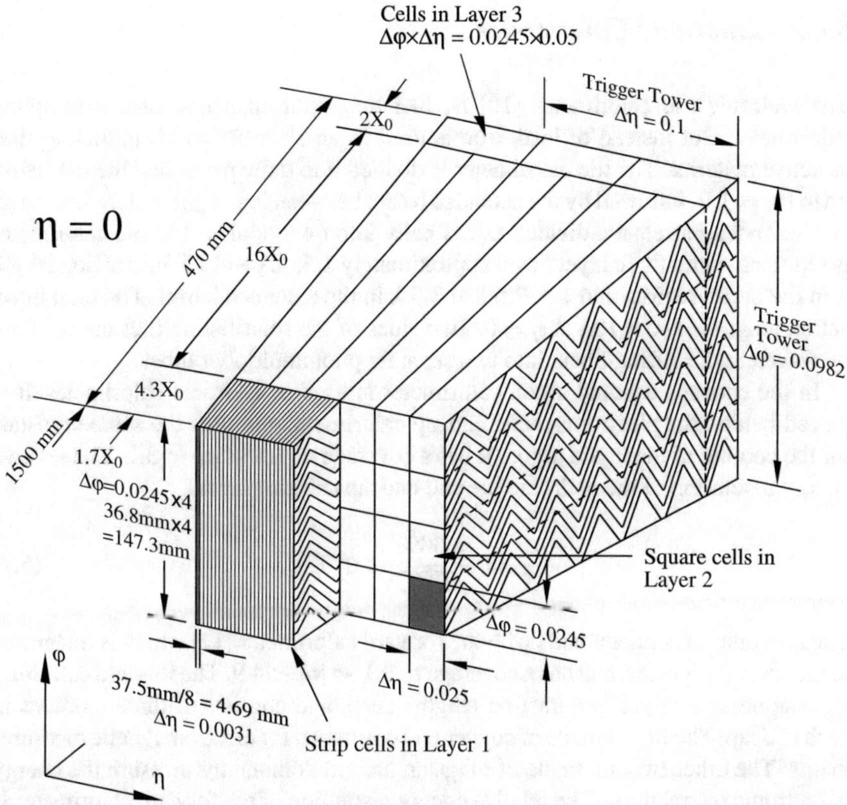

Fig. 5.5 Sketch of a barrel module where the different layers and the accordion geometry is visible. Also shown is the granularity in η and ϕ of the cells for each of the tree layers. Figure taken from Ref. [1]

The relative energy resolution in the electromagnetic calorimeter can the parameterized in the following way [9]:

$$\frac{\sigma_E}{E} = \frac{a}{\sqrt{E}} \oplus \frac{b}{E} \oplus c, \tag{5.6}$$

where a, b and c are η-dependent parameters; a is the sampling term, b is the noise term, and c is the constant term. The sampling term contributes mostly at low energy and has at low $|\eta|$ a design value of about $10\%/\sqrt{E[\text{GeV}]}$. At large $|\eta|$ it is expected to worsen as the amount of material in front of the calorimeter increases. The noise term is about $350 \times \cosh\eta$ MeV for a typical cluster in the barrel calorimeter and for a mean number of interactions per bunch crossing $\langle\mu\rangle = 20$. At high $|\eta|$ it is dominated by the pile-up noise. At higher energies the sampling term and the noise term become less important and the relative energy resolution tends asymptotically to the constant term, c, which has a design value of 0.7%. The constant term is originating from the calibration of the calorimeter.

5.4.2 Hadronic Calorimeter

The hadronic tile calorimeter [10] is, like the electromagnetic one, a sampling calorimeter. But instead of lead, iron is used as an absorber and scintillating tiles as active material. The tile calorimeter is divided into three parts, first the tile barrel up to $|\eta| = 1.0$, followed by the extended barrel between $0.8 < |\eta| < 1.7$. The barrel and extended barrels are divided azimuthally into 64 modules. The tile calorimeter is segmented into three layers with approximately 1.5, 4.1 and 1.8 interaction length λ in the barrel region, and 1.5, 2.6 and 3.3 λ in the extended barrel. The total interaction length is about 10λ at $\eta = 0$. Two sides of the scintillating tiles are read out by wavelength shifting fibers into two separate photomultiplier tubes.

In the endcaps a liquid argon calorimeter is used as hadronic calorimeter. It is placed behind the electromagnetic endcap calorimeter and uses the same cryostats for the cooling of the liquid argon. It has a coverage of $1.5 < |\eta| < 3.2$. The relative energy resolution of the hadronic tile and endcap calorimeter is

$$\frac{\sigma_E}{E} = \frac{50\%}{\sqrt{E[\text{GeV}]}} \oplus 3\%. \tag{5.7}$$

The hadronic calorimeter ends with the forward calorimeter [11] which is integrated in the endcap cryostats and has a coverage of $3.1 < |\eta| < 4.9$. The forward calorimeter is approximately 10 interaction lengths deep, and consists of three modules in each endcap. The first is made of copper and optimized for electromagnetic measurements. The other two are made of tungsten and predominantly measure the energy of hadronic interactions. The relative energy resolution of the forward calorimeter is

$$\frac{\sigma_E}{E} = \frac{100\%}{\sqrt{E[\text{GeV}]}} \oplus 10\%. \tag{5.8}$$

5.5 Muon System

Muons are the only particles which can traverse the calorimeters unstopped.[1] While traversing the calorimeters they only deposit a small amount of energy (typically about 3 GeV). For the measurement and identification of the muons a system of trigger and high-precision tracking chambers [12] is used which is placed outside of the calorimeters. The measurement of the muon momentum is based on the magnetic deflection of the muon tracks in the large superconducting air-core toroid magnets [13]. In the range $|\eta| < 1.4$, magnetic bending is provided by the large barrel toroid [14]. It provides a bending power of 1.5 to 5.5 Tm. In the region $1.6 < |\eta| < 2.7$, the tracks are bent by two smaller endcap magnets [15] which are inserted into both ends of the barrel toroid. Here the bending power is approximately 1.0 to 7.5 Tm. Each

[1] Besides neutrinos which leave the detector unseen.

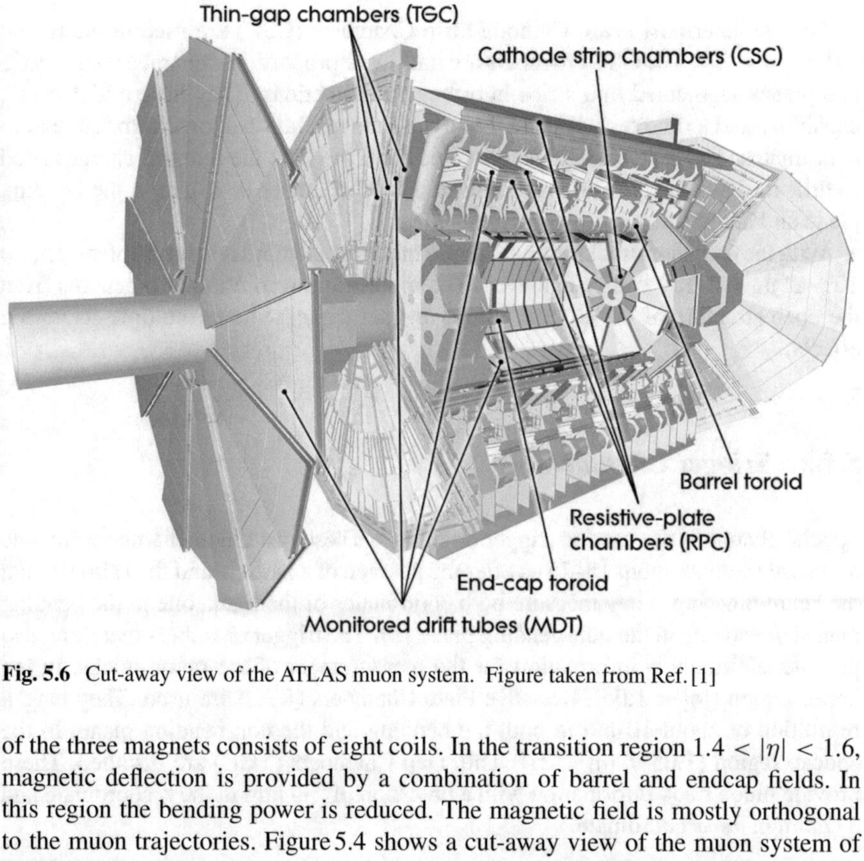

Fig. 5.6 Cut-away view of the ATLAS muon system. Figure taken from Ref. [1]

of the three magnets consists of eight coils. In the transition region $1.4 < |\eta| < 1.6$, magnetic deflection is provided by a combination of barrel and endcap fields. In this region the bending power is reduced. The magnetic field is mostly orthogonal to the muon trajectories. Figure 5.4 shows a cut-away view of the muon system of ATLAS. The different subsystems for triggering and precision measurement are in the following described in more detail. Its prescription follows largely Chap. 6 of Ref. [1] (Fig. 5.6).

5.5.1 Precision-Tracking Chambers

The purpose of the precision-tracking chambers is to precisely determine the muon track in the bending plane (η). The precise measurement of the η position which allows for the precise determination of the muon momentum is performed in the barrel region by the Monitored Drift Tube (MDT) chambers. They are located in three layers between and on the eight coils of the superconducting barrel toroid magnet. The second and the third layer have a coverage of $|\eta| < 2.7$ while the first layer has a coverage of $|\eta| < 2.0$. In the center of the detector ($|\eta| \approx 0$), a gap has been left to allow for services to the solenoid magnet, the calorimeters and the inner detector. The MDT chambers consist of three to eight layers of drift tubes which achieve an average resolution of 80 μm per tube and about 35 μm per chamber.

For the innermost layer, Cathode-Strip Chambers (CSC) are used in the region $2.0 < |\eta| < 2.7$. The CSC chambers are multiwire proportional chambers with cathode planes segmented into strips in orthogonal directions. They have a higher rate capability and a time resolution of 7 ns. The time resolution allows to measure both coordinates. The ϕ coordinate is measured from the time the induced charges need to drift to the cathode. The resolution of a CSC chamber is 40 µm in the bending plane and about 5 mm in the transverse plane.

With the precision-tracking chambers, a muon momentum resolution of $\sigma_{p_T}/p_T = 10\%$ at $p_T = 1$ TeV can be achieved. The momentum resolution also benefits from the open structure of the air-core toroid magnet which reduces multiple-scattering effects.

5.5.2 *Trigger Chambers*

Special chambers are used to trigger on muons. These fast muon chambers are able to provide signals about 15–25 ns after the passage of a particle and thus allow to tag the beam-crossing. They measure both coordinates of the track, one in the bending plane (η) and one in the non-bending plane (ϕ). The trigger chambers therefore also provide additional ϕ information for the measurement of the muon tracks. In the barrel region ($|\eta| < 1.05$), Resistive Plate Chambers (RPC) are used. They have a resolution of about 10 mm in both the bending and the non-bending plane. In the endcap region ($1.05 < |\eta| < 2.4$) Thin Gap Chambers (TGC) are installed. These provide muon track information with a precision of 2–7 mm in the η coordinate and 3–7 mm in the ϕ coordinate.

5.6 The Trigger System

The ATLAS trigger system [16] is divided into multiple levels. It has the important task to reduce the event rate from 40 MHz to about $\mathcal{O}(100)$ Hz at which events (that will have an average size of the order of 1 MB) can be written to mass storage. The first trigger stage, the Level-1 (L1) trigger, is implemented using custom-made electronics. The decision of the L1 trigger is seeded to the High Level Trigger (HLT), which is based on software algorithms running on a processor farm, and in which a decision is formed whether an event is written to the mass storage.

5.6.1 *Level-1 Trigger*

The task of the ATLAS L1 trigger is to perform a fast rate reduction, by selecting only events with interesting signatures. The L1 trigger has a time window of about

2.5 μs to make a decision whether an event is further processed or not. While the L1 trigger performs this decision, all detector information is stored in temporary pipeline memories. Information from both, the fast muon trigger chambers and the calorimeters are used in the first trigger stage. The L1 trigger can be divided into three parts: the L1 calorimeter trigger (L1Calo), the L1 muon trigger (L1Muon) and the decision part in the central trigger processor (CTP).

The calorimeter trigger relies heavily on firmware-programmable FPGAs. On-detector electronics provide separately the sum of analog signals of hadronic and electromagnetic calorimeter cells in a window of approximately $\Delta\eta \times \Delta\phi = 0.1 \times 0.1$. Such an energy sum is called trigger tower. Its dimension is shown for an electromagnetic trigger tower in Fig. 5.5. The analog calorimeter signals are first digitized by the preprocessor (PPr) in fast 10-bit ADCs. The digitized signals are afterwards converted into transverse energy values E_{T} using look-up tables. The preprocessor also identifies the bunch-crossing of the energy deposition. The energy depositions are then transmitted to both the jet/energy processor (JEP) and cluster processor (CP). The CP subsystem identifies electron, photon, and τ-lepton candidates, by searching for local energy maxima above a certain programmable threshold. These regions of interest (RoI) have a size of 2×2 trigger towers and are defined by a sliding-window technique (see for example Sect. 6.2.1). Also isolation criteria can be required for the candidates. The JEP subsystem uses the same technique and defines jet candidates with sizes of $\Delta\eta \times \Delta\phi = 0.4 \times 0.4$, $\Delta\eta \times \Delta\phi = 0.6 \times 0.6$, and $\Delta\eta \times \Delta\phi = 0.8 \times 0.8$. The jet/energy processor also evaluates the total scalar transverse energy and the missing transverse energy of each event, based on all cells over the acceptance of $|\eta| < 4.9$.

The muon trigger consists of three different parts. The RPC and TGC trigger first identify muon candidates which fall into six p_{T} windows from 5 GeV to 35 GeV by using a simple tracking algorithm. The basic principle of this algorithm is to require a coincidence in position of the hits in the different trigger stations. The coincidence is required to be in a window around the extrapolated track to straight to the interaction point. The width of the window is related to the p_{T} threshold to be applied. A large width corresponds to a low p_{T} muon with a large curvature of the track while a large width corresponds to a high p_{T} muon with a straight track. The information of the RPC and TGC are finally combined in the muon to CTP interface (MuCTPI) and sent to the CTP.

The central trigger processor is finally performing the Level-1 event decision based on multiplicities of high-p_{T} objects sent from the calorimeter trigger, the MuCTPI, and using the information on the global energy sums.

The first trigger reduces the event rate from about 40 MHz to about 75 kHz. Regions of interest defined by the L1 are sent to the High Level Trigger.

During the shutdown after the 2012 data taking, an additional system was installed. The L1 topological processor (L1Topo) was placed between the L1Calo and L1Muon systems and the CTP. It is able to combine information from the L1Calo and L1Muon system and to compute complex quantities like invariant masses and angular variables. Based on these quantities additional information for the trigger decision is sent to the CTP.

5.6.2 High Level Trigger

The ATLAS HLT [17] consists of the Level-2 (L2) trigger and the event filter (EF). Both are pure software based triggers which are running on processor farms. The L2 trigger uses the full granularity and precision of all detector systems, but only in the regions of interest that were defined by the L1 calorimeter and muon trigger. For the L2 trigger also tracks are reconstructed using track reconstruction algorithms (see Sect. 6.1). The L2 trigger has about 10 ms for its decision and reduces the event rate to about 3 kHz. A further reduction to the required rate of $\mathcal{O}(100)$ Hz is done by the EF which is seeded by the decisions of the L2. The event filter reconstructs the complete event using all available information and already applies several calibrations, corrections and identification criteria to the physics objects (see Chap. 6). The events are sorted into different streams which correspond to the physics objects triggering the event. For events that pass also the last trigger level, the information of all sub-detectors is recorded.

The presented approach has the disadvantage that both, the L2 and the EF were running on different computing farms. If the L2 requests partial information for an event and a positive decision is taken, this information has to be re-requested as a part of the full event information. The Level-2 and EF have therefore been merged during the shutdown of the LHC into a single HLT. This system is still processing the event in multiple stages by first requesting only partial information and then, if a positive decision is taken, evaluating the whole event information. The biggest advantage of the new system is that this process is now happening at a single machine and information is not duplicated.

5.7 Data Acquisition and Processing

5.7.1 Data Acquisition

During the process of making a trigger decision all information of the detector has to be stored. Therefore all signals of the detector components are digitized and buffered in such a way that they are available in case of a positive trigger decision. Each detector component has an on-detector buffer pipeline, which allows to buffer the data during the L1 trigger decision. Once an event is accepted by the L1 trigger, the data from the pipelines are transferred off the detector via 1574 readout links. There the signals are digitized and transferred to the data acquisition (DAQ) system. The first stage of the DAQ system, the readout system, stores the data temporarily in local buffers. The stored data in the RoI's is then subsequently solicited by the L2/HLT trigger system (2012 configuration/2015 configuration). Those events selected by the L2/HLT trigger are then transferred to the event-building system, where the whole event is reconstructed and subsequently sent to the event filter for the final decision. In the 2015 configuration of the trigger system this was also done by the HLT. The information of accepted event is then stored in the so-called RAW data format on magnetic tapes in the CERN computer center.

The ATLAS data taking is steered by the RunControl system (RC) [18]. The system communicates with all the different detector components presented above. Once all parts of the ATLAS detector are ready and the LHC declares stable beams, a data taking run can be started. To each run a unique run number is assigned. A data taking run usually corresponds to a single fill of the LHC and therefore typically represents a period of hours up to a day. These runs are further divided into luminosity blocks which correspond to a data taking time of approximately a minute. In these luminosity blocks the instantaneous luminosity is approximately constant. The luminosity blocks can later be flagged according to whether a problem with one of the subsystems occurred. The luminosity blocks in which all detector parts of the ATLAS experiment that are important for physics analyses are running are listed in the so-called Good Runs List. The runs in which the LHC delivered stable conditions are grouped in periods. The periods are labeled alphabetically and typically have a length of days to weeks.

5.7.2 Data Processing

The further processing and reprocessing of the data happens in the LHC Computing Grid [19, 20]. The Grid is a network of many computer clusters organized in several levels, so-called Tiers. The Tier-0 is the CERN computer center which applies reconstructions algorithms (see Chap. 6) and calibrations to the data. The whole information on detector level is transformed into information on object level into a data format called Event Summary Data (ESD). These ESD are distributed to the Tier-1 centers, which are located around the world and provide storage space for the data as well as additional processing power, e.g., for recalibration of the data. Additionally a copy of the raw data is distributed among the Tier-1 centers. From the ESD, the Analysis Object Data (AOD) are derived, which only contain information about specific physics objects which are needed for the analysis, like electrons, muons, jets or photons. From the AOD level on, the analysis model differ for the two analyses presented in this thesis.

In the analysis using data from the year 2012, a further extraction of the AODs to the Derived Physics Data (DPD) is done. The DPDs are transferred to the Tier-2 centers, which provide processing power for physics analysis and Monte Carlo production. For the analysis needed data can be copied to local Tier-3 centers. Such a Tier-3 is the local *mainzgrid* which is part of the computing cluster *mogon* [21]. Data in the D3PD format, a special type of DPD, is used for the analysis. D3PDs store the information into ROOT Ntuples. ROOT Ntuples are a commonly used data format in high energy physics. The program ROOT [22] is a statistical analysis framework which is also used in this analysis. It provides the possibility of analyzing data and has various possibilities to visualize data in histograms. All shown histograms in this thesis were produced using ROOT.

The above procedure includes the production of various data formats. If a problem at AOD level is found, all subsequent production steps need to be repeated. This procedure was found to be not optimal and during the shutdown of the LHC a new analysis model was installed. It was found that most physicists prefer to use ROOT for the final analysis step. The AODs were therefore modified in a way to be readable by ROOT. The new format was called xAOD. To further reduce the physical size of the data, each analysis defines an analysis specific preselection that is applied to the xAODs. Besides the preselection of events, also the information for an event is reduced to the information vital for the analysis. The resulting preselected data format is called derived xAOD (DAOD).[2]

5.8 Luminosity Measurement

For a pp collider the luminosity can be determined by

$$\mathcal{L} = \frac{R_{inel}}{\sigma_{inel}}, \tag{5.9}$$

where R_{inel} is the rate of inelastic collisions and σ_{inel} is the pp inelastic cross section. For a storage ring operating at a revolution frequency f_r and with n_b bunch pairs colliding per revolution, the luminosity can be rewritten as

$$\mathcal{L} = \frac{\mu n_b f_r}{\sigma_{inel}}, \tag{5.10}$$

where μ is the number of average inelastic interactions per bunch crossing. ATLAS monitors the delivered luminosity by measuring μ with several detectors and several different algorithms. These algorithms are for instance based on counting inelastic events or the number of hits in the detector. When using different detectors and algorithms, the measured μ_{meas} has to be corrected with the efficiency and acceptance of the detector and algorithm, to obtain $\mu = \mu_{meas}/\epsilon$. In the same way is $\sigma = \sigma_{meas}/\epsilon$. The calibration of the luminosity scale for a particular detector and algorithm is equivalent to determining the cross section σ_{meas}.

Equation (4.1) can be rewritten in the following form

$$\mathcal{L} = \frac{n_b f_r N_1 N_2}{2\pi \Sigma_x \Sigma_y}, \tag{5.11}$$

where N_1 and N_2 is the number of protons in beam one or two and Σ_x and Σ_y characterizes the horizontal and vertical convolved beam width. Combining Eqs. (5.10) and (5.11) leads to:

[2]The format DAOD_EXOT9 was used for the W' search.

$$\sigma_{meas} = \mu_{meas} \frac{2\pi \Sigma_x \Sigma_y}{n_1 n_2} \tag{5.12}$$

The luminosity detectors are calibrated to the inelastic cross section using beam-separation scans, also known as van der Meer (vdM) scans [23]. In a vdM scan, the beams are separated in steps of known distances. Measuring μ_{meas} during a vdM scan as a function of the beam separation in x or y leads to a Gaussian distribution width a width equal to Σ_x or Σ_y. The parameters can therefore be extracted by a fit. The product $N_1 N_2$ is determined by beam current measurements and provided by the LHC group. A more detailed description of the algorithms and sub-detectors used for luminosity determination can be found in [24].

The systematic uncertainty for the determination, which is obtained by comparing the results from the different sub-detectors and methods is for the 2012 data set 1.9% [25] and for the 2015 data set 5% [26]. The largest uncertainties are coming from the vdM calibration and from an uncertainty arising from the extrapolating of the conditions during the vdM scan to the nominal high-luminosity conditions.

5.9 Detector Simulation

In Sect. 3.2, the simulation of a physics event was discussed. This simulation was independent from the detector. The simulation of the detector and the response of the detector to the physics event has to be simulated separately to be able to compare the simulation with data. The program GEANT4 [27] is used for the detector simulation.

The generators produce events in a standard HepMC format [28]. These files contain the truth information of an event, i.e., a history of the simulated interaction from incoming to outgoing particles. The generator only simulates prompt decays (e.g. W, Z bosons or τ-leptons) and stores all particles as outgoing particles that are expected to propagate through the detector. In a first step these files are read in and GEANT4 simulates the path of the generated particles through the detector. Therefore a detailed model of the detector, including all details about geometry and materials used as well as details about the magnetic fields, is implemented. The interaction of the particles with the matter of the detector is entirely simulated. Additional produced particles, like photons from Bremsstrahlung and particles in an electromagnetic or hadronic shower, are also propagated through the detector. The result is a precise record of the amount of energy deposited in which part of the detector at which time. This information is written to so-called hit files.

In a second step, these hit files are read in and the response of the detector components to the deposited energy and the electronics of the readout system is simulated. Therefore also effects like calibrations or dead readout channels are simulated, to simulate conditions as they are present for data taking. To save computing time, the simulation of the detector is only performed for the single hard-scattering process which was simulated by the event generator. To account for pile-up, various types of simulated events are read in, and hits from each are overlaid. The simulated pile-up profile can be set during the digitization step and thus also changed without performing the whole simulation step again. The information is stored in the same way as for data taking, additionally truth information about the particles produced during the simulation is added.

On the digitized detector information, the L1 decision is emulated and the HLT and the reconstruction is run. No events are discarded but the trigger decision is emulated. The reconstruction is identical for the simulation and the data, with exception that truth information is only available in simulation.

Whereas the physics simulation is in comparison quite fast, the simulation of the detector takes a significantly longer time. For instance, the simulation of an event $pp \rightarrow W^{\pm} + X \rightarrow e^{\pm}\nu_e + X$ takes about 19 min [29]. The by far longest time is hereby needed for the simulation of the hits. Of all detector parts, the simulation of the electromagnetic and hadronic showers takes the largest fraction of this time.

5.9.1 Correction of the Pile-Up Profile

A good quantity for the in-time pile-up is the number of primary vertices[3] n_{PV}, as it is a direct measurement for the number of inelastic collisions and as the inner detector is fast enough to not be affected by out-of-time pile-up. A quantity which is also sensitive to the in-time pile-up is the number of interactions $\langle\mu\rangle$ averaged over one bunch train[4] and a luminosity block. These quantities are strongly dependent on the settings of the LHC, like the number of protons in a bunch and the spacing between different proton bunches. Since the physics simulation takes partially place before or during the time of data taking, the parameters for the pile-up distribution of the final data are not known. Thus approximate distributions are simulated that are meant to be matched to the actual data. To adjust the simulation, every event is reweighted using a reweighting tool[5] provided by ATLAS [30]. After the data taking, for most of the MC samples the digitization step has been repeated and a pile-up profile which is very similar to the profile in data has been simulated. For a better description, still a reweighting is performed. Figure 5.7 shows the mean interactions per bunch crossing for the 2012 data and as simulated in MC. The left side shows the distribution as simulated for pre-data-taking production and the left side the distribution for the post-data-taking production. The pre-data-taking distribution for the simulation of the 2015 data look similar.

All MC samples used in this thesis have been reweighted to the distribution in data. It has been found difficult to describe both, the $\langle\mu\rangle$ and the n_{PV} distributions in data equally well with MC, thus the reweighting was adjusted to better fit the n_{PV} distribution.

---

[3] Number of vertices with more than two tracks.

[4] A bunch train consists of 72 proton bunches.

[5] *PileupReweighting-00-03-18* has been used for the analysis described in part III and *PileupReweighting-00-02-12* for the analysis described in part IV.

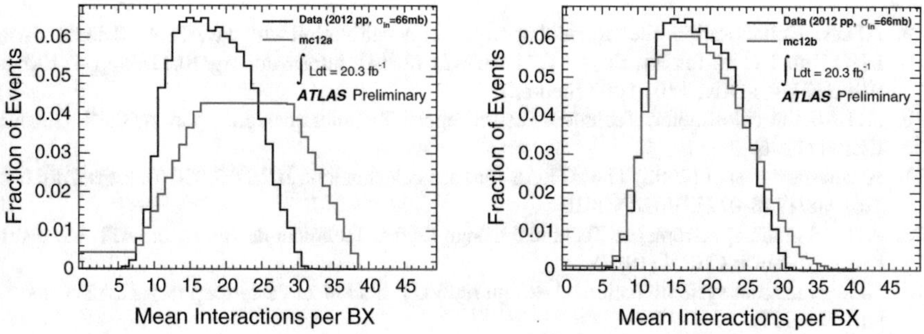

Fig. 5.7 Distribution for the mean interactions per bunch crossing. The distribution for the data taken in 2012 is shown in black and the distribution used in Monte Carlo is shown in red. The left side shows the Monte Carlo distribution for the pre-data-taking production and the right side for the post-data-taking production. Figure taken from Ref. [31]

5.9.2 Correction of the Vertex Distribution

The z-position of the vertices follows a Gaussian distribution with the mean at the interaction point and a width of about 50 mm. Differences have been found between the simulated distribution and the distribution in data. In some analyses a reweighting is therefore performed to match the simulated distribution to the distribution in data. Both analyses presented in this thesis are not expected to be sensitive to this effect, therefore no reweighting has been performed.

References

1. Aad G et al (2008) The ATLAS experiment at the CERN large Hadron Collider. JINST 3:S08003. https://doi.org/10.1088/1748-0221/3/08/S08003
2. Haywood S et al (1997) ATLAS inner detector: Technical design report, 2. Technical design report ATLAS. Geneva: CERN
3. ATLAS central solenoid: Technical design report. Technical design report ATLAS. Electronic version not available. Geneva: CERN (1997)
4. Wermes N, Hallewel G (1998) ATLAS pixel detector: Technical design report. Technical design report ATLAS. Geneva: CERN
5. Capeans M et al (2010) ATLAS insertable B-layer Technical design report. Technical report CERN-LHCC-2010-013. ATLAS-TDR-19. Geneva: CERN. https://cds.cern.ch/record/1291633
6. ATLAS liquid-argon calorimeter: Technical design report. Technical design report ATLAS. Geneva: CERN (1996)
7. Nikiforou N (2013) Performance of the ATLAS Liquid Argon Calorimeter after three years of LHC operation and plans for a future upgrade. In: Proceedings, 3rd international conference on advancements in nuclear instrumentation measurement methods and their applications (ANIMMA 2013). https://doi.org/10.1109/ANIMMA.2013.6728060, arXiv: 1306.6756 [physics.ins-det]. https://inspirehep.net/record/1240499/files/arXiv:1306.6756.pdf
8. ATLAS calorimeter performance: Technical design report. Technical design report ATLAS. Geneva: CERN (1996)

9. ATLAS Collaboration. Electron and photon energy calibration with the ATLAS detector using LHC Run 1 data. In: Eur Phys J C74.10:3071 (2014). https://doi.org/10.1140/epjc/s10052-014-3071-4, arXiv: 1407.5063 [hep-ex]
10. ATLAS tile calorimeter: Technical design report. Technical design report ATLAS. Geneva: CERN (1996)
11. Artamonov A et al (2008) The ATLAS forward calorimeters. JINST 3:02010. https://doi.org/10.1088/1748-0221/3/02/P02010
12. ATLAS muon spectrometer: Technical design report. Technical design report ATLAS. distribution. Geneva: CERN (1997)
13. ATLAS magnet system: Technical design report, 1. Technical design report ATLAS. Geneva: CERN (1997)
14. Badiou JP et al (1997) ATLAS barrel toroid: Technical design report. Technical design report ATLAS. Electronic version not available. Geneva: CERN
15. ATLAS end-cap toroids: Technical design report. Technical design report ATLAS. Electronic version not available. Geneva: CERN (1997)
16. Casadei D (2012) Performance of the ATLAS trigger system. J Phys Conf Ser 396:012011. https://doi.org/10.1088/1742-6596/396/1/012011
17. Jenni P et al (2003) ATLAS high-level trigger, data-acquisition and controls: Technical design report. Technical design report ATLAS. Geneva: CERN
18. Lehmann Miotto G (2010) Configuration and control of the ATLAS trigger and data acquisition. Nucl Instrum Methods A623:549–551. https://doi.org/10.1016/j.nima.2010.03.066
19. Eck C et al (2005) LHC computing Grid: Technical design report. Version 1.06 (20 June 2005). Technical design report LCG. Geneva: CERN. https://cds.cern.ch/record/840543
20. Jones R (2003) ATLAS computing and the GRID. Nucl Instrum Methods A502:372–375. https://doi.org/10.1016/S0168-9002(03)00446-7
21. High Performance Computing Group, University of Mainz. https://hpc.uni-mainz.de
22. Brun R, Rademakers F (1997) ROOT: an object oriented data analysis framework. Nucl Instrum Methods A389:81–86. https://doi.org/10.1016/S0168-9002(97)00048-X
23. Balagura V (2011) Notes on van der Meer Scan for Absolute Luminosity Measurement. Nucl Instrum Methods A654:634–638. https://doi.org/10.1016/j.nima.2011.06.007, arXiv: 1103.1129 [physics.ins-det]
24. Aad G et al (2013) Improved luminosity determination in pp collisions at $\sqrt{s} = 7$ TeV using the ATLAS detector at the LHC. Eur Phys J C73.8:2518. https://doi.org/10.1140/epjc/s10052-013-2518-3, arXiv: 1302.4393 [hep-ex]
25. Luminosity Group T (2012) Preliminary Luminosity Determination in pp Collisions at $\sqrt{s} = 8$ TeV using the ATLAS detector in 2012. Technical report ATL-COM-LUM-2012-013. Geneva: CERN
26. ATLAS Physics Modeling Group. https://twiki.cern.ch/twiki/bin/view/Atlas/LuminosityForPhysics (Internal documentation)
27. Agostinelli S (2003) GEANT4: a simulation toolkit. Nucl Instrum Methods A506:250–303. https://doi.org/10.1016/S0168-9002(03)01368-8
28. Dobbs M, Hansen JB (2001) The HepMC C++ Monte Carlo event record for high energy physics. Comput Phys Commun 134:41–46. https://doi.org/10.1016/S0010-4655(00)00189-2
29. Aad G et al (2010) The ATLAS Simulation Infrastructure. In: Eur Phys J C70:823–874. https://doi.org/10.1140/epjc/s10052-010-1429-9, arXiv: 1005.4568 [physics.ins-det]
30. Physics Analysis Tools. Pileup Reweighting. https://twiki.cern.ch/twiki/bin/viewauth/~AtlasProtected/ExtendedPileupReweighting
31. ATLAS Data Preparation Group. https://twiki.cern.ch/twiki/bin/view/AtlasPublic/DataPrepGenPublicResults

Chapter 6
Particle Reconstruction and Identification in ATLAS

In the following chapter, the reconstruction and identification of tracks (Sect. 6.1), electrons, muons, jets (Sects. 6.2 to 6.4) and of the missing transverse momentum (Sect. 6.5) is discussed.

6.1 Track Reconstruction

Aim of the track reconstruction is to reconstruct the path of a charged particle through the inner detector. In case of a muon, also the path through the muon spectrometer is relevant. However, this part is discussed in Sect. 6.3 and in the following only the reconstruction of tracks in the inner detector is discussed. In this section the concept of the track reconstruction is summarized. Tracks are reconstructed in the inner detector using a sequence of algorithms. More details on the algorithms can be found in Ref. [1].

In a first step, hits in the pixel detector and the SCT are transformed into three dimensional space points. For the SCT a hit from either side of the module is required to obtain both coordinates. The hits in the TRT are transformed into drift circles using the timing information. A track seed is formed from a combination of space points in the three pixel layers and the first SCT layer. These track candidates are then extended up to the fourth layer of the SCT by using a Kalman-filter [2] which takes into account material corrections. The track candidates within the acceptance of the TRT are then fitted and extended by the TRT hits. This first step of reconstruction aims mainly towards the reconstruction of tracks from prompt particles.[1]

[1]Prompt particles are defined as particles with a mean lifetime of greater than 3×10^{-11} s directly produced in a *pp* interaction or from the subsequent decays or interactions of particles with a lifetime shorter than 3×10^{-11} s.

M. Zinser, *Search for New Heavy Charged Bosons and Measurement of High-Mass Drell-Yan Production in Proton-Proton Collisions*, Springer Theses,
https://doi.org/10.1007/978-3-030-00650-1_6

Subsequently an algorithm starts from hits in the TRT that have not been associated to any track yet. It extends the track inwards by adding hits from the SCT and pixel detector. This second step mainly reconstructs tracks from converted photons and long-lived particles, which do not necessarily have a hit in the inner most layers of the detector.

After all tracks are fitted, vertex finder algorithms are used to assign the tracks to their vertices. A reconstructed vertex is required to have at least two tracks associated to it. The primary vertex is defined as the vertex with the highest $\sum p_{\mathrm{T}}^2$ of all tracks. The position of the proton beam is measured by monitoring the primary vertex position over a certain time and taking the beam position from the mean of the Gaussian distribution. The vertices are afterwards re-reconstructed with the position of the proton beam as an additional measurement. Resolutions achieved for vertices are about 23 μm in transverse direction and 40 μm along the beam axis [3]. After the vertex reconstruction,additional algorithms search for secondary vertices and photon conversions.

A more detailed description of the track reconstruction and of the performance during the 2012 data taking is given in Ref. [4].

Important quality criteria for tracks are the transverse and longitudinal impact parameters d_0 and z_0. The transverse impact parameter d_0 is the distance in the transverse plane of the track from the position of the proton beam or the primary vertex. In this thesis requirements are placed on the d_0 significance, which is defined as d_0 divided by its uncertainty. Non-prompt particles are expected to have a larger d_0 and a requirement on this parameter is hence a possibility to distinguish prompt particles from non-prompt particles. The longitudinal impact parameter z_0 is the distance of the track from the vertex along the beam axis. It is usually defined with respect to the primary vertex.

6.2 Electrons

Since electrons and positrons only differ in the curvature of their tracks, but have the same signature, positrons are in this thesis from now on also denoted as electrons. The electron reconstruction, identification, trigger, calibration, and corrections which have to be applied to the simulation are in the following discussed. The section concentrates on electrons in the region $|\eta| < 2.47$ which are meant for precision measurements. Electrons reconstructed in the forward calorimeters are not discussed.

6.2.1 Reconstruction

Reconstruction of an electron candidate starts always from an energy deposition (cluster) in the electromagnetic calorimeter. To search for such a cluster, a *sliding-window* algorithm [5] is used. The electromagnetic calorimeter is first divided into an

η-ϕ-matrix with $N_\eta = 200$ and $N_\phi = 256$. Thereby matrix elements with a concrete size of the calorimeter cells in the second layer ($\Delta\eta \times \Delta\phi = 0.025 \times 0.0245$) are formed. In a first step a window of the size 3×5, in units of 0.025×0.0245 in $\eta \times \phi$ space, runs over the matrix and searches for an energy deposition with a transverse energy above 2.5 GeV. In a second step, tracks are searched which match the identified clusters. The distance between the track impact point and the cluster center is required to satisfy $|\Delta\eta| < 0.05$. To account for radiation losses due to Bremsstrahlung an asymmetric $\Delta\phi$ cut is chosen. Track impact point and cluster center have to have a $\Delta\phi < 0.1$ on the side where the extrapolated track bends, and $\Delta\phi < 0.05$ on the other side. The cluster is discarded as an electron candidate if no track is matched to it. If there is more than one track matching to the cluster, the ones with hits in the pixel detector and SCT are preferred and the one with the smallest ΔR to the cluster is chosen. After matching the track, the electron cluster is rebuilt using a 3×7 (5×5) window in the barrel (endcap). The larger window in ϕ in the barrel region is chosen to account for radiation losses due to Bremsstrahlung. The cluster energy is then determined by summing the estimated energy deposited in the material before the electromagnetic calorimeter, the measured energy deposited in the cluster, the estimated energy deposited outside the cluster, and the estimated energy deposited beyond the electromagnetic calorimeter. A more detailed description of the electron reconstruction is given in Refs. [6, 7].

6.2.2 *Identification*

A large part of the reconstructed electron candidates are not real electrons. The falsely reconstructed electrons, which are dominantly jets, have to be rejected. It is at the same time necessary to make sure that a sufficient amount of real electrons is kept. ATLAS provides two different electron identification methods, the first is based on cuts of track and shower shape variables [8], and the second is based on a likelihood approach [9]. Three different levels of identification *loose*, *medium* and *tight* are defined for both approaches. These three identification levels are optimized in such a way that a signal efficiency of 90% for *loose*, 80% for *medium*, and 70% for *tight* is achieved, whereas the background rejection is getting higher from *loose* to *tight*. The levels are designed in a way that the *medium* identification level fully contains the *loose* level as well as the *tight* level contains *medium*. The three identification levels for both approaches are in the following briefly introduced and explained. Figure 6.1 shows a schematic view of the electron reconstruction and identification principle.

6.2.2.1 Cut-Based Identification

The *loose* identification level imposes restrictions on the ratio between the transverse energy in the electromagnetic and hadronic calorimeter to reject jets which would cause a high energy deposition in the hadronic calorimeter. If the energy deposited

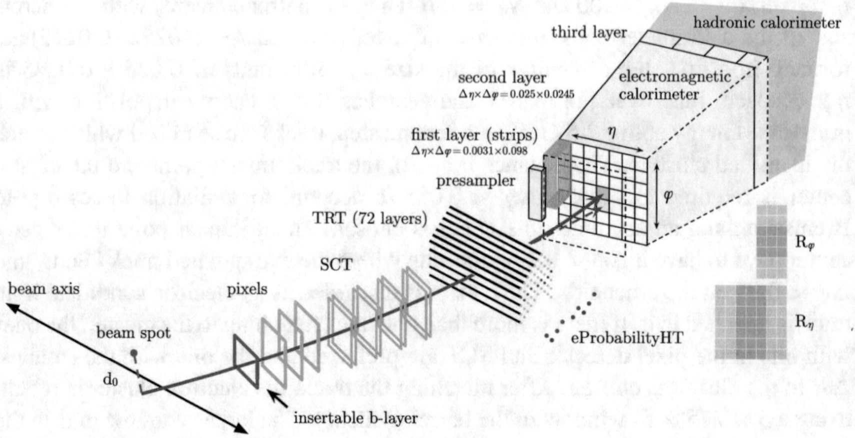

Fig. 6.1 Schematic view of the electron reconstruction and identification. Figure taken from Ref. [9]

in the first layer of the electromagnetic calorimeter is more than 0.5% of the total deposited energy, further cuts on the first layer deposition are imposed. The total shower width in the first layer w_{stot} is defined as

$$w_{\text{stot}} = \sqrt{\frac{\sum_i E_i (i - i_{\max})^2}{\sum_i E_i}}, \tag{6.1}$$

where i is the index of the strip in the first layer and $i_{\max}$ the index of the strip with the highest energy. Typically w_{stot} is defined summing over 20 strips in η. Jets have broader showers than electrons and thus can be rejected by restricting the shower width towards lower values. A jet can contain π^0 mesons which decay dominantly into two photons, leading to two nearby energy depositions in the electromagnetic calorimeter. To reject photons from such a decay $\pi^0 \rightarrow \gamma\gamma$, a second maximum in the energy deposition of the first layer can be searched. The quantity E_{ratio} is the difference between highest and second highest energy deposition in one of the strips, divided by its sum. If the difference between these energies is below a certain value, then the candidate is assumed to originate from a π^0 decay and is rejected. In the second layer of the electromagnetic calorimeter, restrictions on the ratio R_η between the energies deposited in a window of 3×7 to the window 7×7 are imposed. By restricting the ratio to higher values, it is ensured that not a broad symmetric shower is selected, like typical for hadronic showers, but a shower broad in ϕ like it is expected due to radiated Bremsstrahlung.[2] A similar quantity, sensitive to the same issue, is the lateral shower width in the second layer $w_{\eta,2}$. It is defined by

[2]The shower is expected to be broader in ϕ due to the radiated photons from Bremsstrahlung, which are measured nearby in ϕ to the electron cluster.

$$w_{\eta,2} = \sqrt{\frac{\sum_i E_i \eta_i^2}{\sum_i E_i} - \left(\frac{\sum_i E_i \eta_i}{\sum_i E_i}\right)^2}, \tag{6.2}$$

where i is the index of the cell in the second layer. The sum is calculated within a window of 3 × 5 cells. To ensure the matching between the chosen track and the cluster, it can be required that the distance $\Delta\eta_1$ of the impact point of the track and the η of the cluster in the first layer is below a certain value. The track is also required to have a sufficient amount of hits in the pixel (n_{Pixel}) and SCT (n_{Si}) detector.

The *medium* identification level imposes the same cuts as the *loose* identification level but uses partially tighter restrictions. Additionally to ensure that the shower barycenter is in the second layer, the ratio between the energy in the third layer to the complete cluster energy (f_3) is restricted. This cut is only imposed to clusters with a p_{T} lower than 80 GeV, since for growing p_{T} the barycenter moves towards the hadronic calorimeter. Electrons should cause transition radiation in the TRT above a certain threshold. For electrons within the acceptance of the TRT it is required that a sufficient amount of the TRT hits are such high-threshold hits (F_{HT}). To reject tracks which are coming from secondary vertices or photon conversions, it is required that the associated track has a hit in the first layer of the pixel detector (n_{Blayer}). Finally, to ensure that the track originates from a primary vertex, a cut is imposed on the transverse impact parameter d_0.

The *tight* identification level imposes the same cuts as the *medium* identification level with again partially tighter restrictions. To ensure that the track and the cluster belong to the same physics object, a cut is made on the ratio E/p of the measured energy and the measured momentum. To tighten the matching between track and cluster, an additional $\Delta\phi$ cut is imposed and to further constrain the track quality a minimum number of hits in the TRT (n_{TRT}) is required. Electron candidates which are flagged by a specific algorithm as objects which are coming from a photon conversion are also rejected.

6.2.2.2 Likelihood-Based Identification

The likelihood-based identification is a multivariate analysis technique that simultaneously evaluates several properties of the electron candidates when making a selection decision. A probability density function is created using n variables with discriminating power (signal vs. background), where $\vec{x}$ is the set of variables for a given electron. The same set of variables is used as in the cut-based approach described above. An exception are the variables E/p, w_{stot}, $|\Delta\phi|$, and the number of hits of the track. Each electron is assigned with a discriminant value $d_{\mathcal{L}}$:

$$d_{\mathcal{L}} = \frac{\mathcal{L}_S}{\mathcal{L}_S + \mathcal{L}_B}, \qquad \mathcal{L}_S(\vec{x}) = \prod_{i=1}^{n} P_{S,i}(x_i), \tag{6.3}$$

where $P_{S,i}(x_i)$ is the value of the signal probability density function of the ith variable evaluated at x_i. $\mathcal{L}_B$ is defined in the same way as $\mathcal{L}_S$, where $P_{B,i}(x_i)$ refers to the background probability function. The likelihoods are optimized on a Monte Carlo sample with $Z \rightarrow ee$ (signal) events and simulated events containing two jets (background). Differences of the width or position in the distributions of the variables $\vec{x}$ are observed between data and simulation. Hence, during the optimization, these distributions are corrected to match the distributions in data.

Cuts, which obtain the target electron efficiencies, are placed on the likelihood discriminant and the number of track hits. This ensures that the three different identification levels are a subset of each other. During the 2015 data taking, an efficiency loss at high p_{T} was observed for the *tight* identification level. This problem was not observed for the *medium* identification level. As a solution, the cut on the likelihood discriminant was reduced above $p_{\mathrm{T}} = 125$ GeV to the cut value of the *medium* level. To provide additional discrimination in this region, additional cuts, similar to those used in the cut-based approach, are placed on E/p and w_{stot}. The likelihood-based approach allows for better background rejection for a given signal efficiency than the cut-based identification that would use selection criteria sequentially on each variable. It needs on the other hand a better understanding of the differences between data and simulation.

6.2.2.3 Isolation

Isolation requirements can be imposed, since single electrons should produce a shower located in a rather small region, whereas jets produce broader showers. The sum of the energy in a region around the cluster center larger than a certain radius ΔR, can be used to discriminate between isolated electron candidates, e.g., from W or Z decays and non-isolated electron candidates from jets. Similar isolation requirements can be imposed on the tracks surrounding an electron candidate. Such isolation requirements are not imposed by the three identification levels and can be applied additionally to electron candidates.

6.2.3 *Electron Trigger*

Triggers are used to select events containing electrons. All electron triggers are seeded by a Level-1 trigger requiring one or multiple electromagnetic calorimeter trigger objects above a certain energy threshold. To further enhance the purity of the data and to reduce the rate of the triggers, identification requirements are placed at the HLT level. These identification criteria are similar to the offline requirements but always slightly looser to avoid efficiency losses. Triggers exist for both, the cut-based approach and the likelihood-based approach. In case of the cut-based identification, also triggers meant for triggering photons can be used, as the identification level imposed on these does not include track information and places the same requirements on the electromagnetic clusters as the electron triggers.

6.2.4 Electron Energy Correction

The energy of an electron candidate is built from the energy of a cluster of cells in the electromagnetic calorimeter. It is calibrated to the original electron energy using multivariate techniques which are based on simulation. The calibration procedure is described in detail in Ref. [7].

To further calibrate the reconstructed energy of the electrons in data, η-dependent corrections are applied to recalibrate the energy. The corrections are small and below one percent with an accuracy on the order of 0.1%. They were obtained by selecting a sample of Z- and J/ψ- candidates. The corrections are then derived by comparing these resonances in data and Monte Carlo simulation. For the recalibration, corrections[3] were used, obtained by the electron performance group of ATLAS [7].

In the Monte Carlo simulation a too optimistic energy resolution of the electromagnetic calorimeter is assumed. For this reason, the simulated energy gets smeared by a correction following a Gaussian distribution. The width of the Gaussian distribution is determined by selecting a sample of $Z \rightarrow ee$ and $J/\Psi \rightarrow ee$ candidates and comparing the reconstructed width of the invariant mass distribution in data and simulation. The corrections to the energy resolution are determined by the ATLAS electron performance group [7]. The same software tools as for the electron energy calibration are used. The corrections are on the order of one per mille, with slightly higher corrections around the transition region between the detector barrel and the detector endcaps. The accuracy of these corrections is on the order of 0.01%.

6.2.5 Efficiency Corrections

The probability to select a real electron in the analysis is the product of the efficiencies of four main steps, namely the application of the trigger algorithms, the reconstruction of the electron object and the specific electron identification and isolation criteria. For these four steps the efficiency in data and in simulation show small differences. To correct for these differences, efficiency corrections are derived which are defined as $w_{SF} = \epsilon_{data}/\epsilon_{MC}$, where ϵ is the efficiency of a certain identification step.

The efficiency in data ϵ_{data} is measured in a sample of Z candidates which is obtained using a so called "tag and probe method". In this method an electron candidate with a very strict identification is selected and called tag. Then a second electron candidate, called probe, is selected which builds with the tag a pair with an invariant mass in a window around the Z-peak. With this probe the efficiency is studied. This method provides a clean sample of probe electrons, since the region of the Z-peak is dominated by real electrons. The efficiency in simulation is simply measured, by using the same tag and probe method on a Monte Carlo simulating $pp \rightarrow Z/\gamma^* + X \rightarrow e^+e^- + X$. All efficiency correction weights are derived by

[3] *ElectronPhotonFourMomentumCorrection-00-01-46* has been used for the analysis described in part III and *ElectronPhotonFourMomentumCorrection-00-00-34* for the analysis described in part IV.

the ATLAS electron performance group [8, 9]. A tool[4] is provided which contains the efficiency corrections. The derived efficiency corrections weights are binned in electron p_{T} and η. They typically deviate from one on the order of one percent and are applied as weight on a single object basis.

6.3 Muons

The muon reconstruction, identification, trigger, and corrections which have to be applied to the simulation are in the following discussed. This section concentrates on muons used in the analysis of the 2015 data set. The reconstruction of muons in the 2012 data set is very similar but slightly different in some places. It is described in more detail in Ref. [10].

6.3.1 Reconstruction

The muon reconstruction starts with the track reconstruction in the muon spectrometer. A Hough transform [11] is used to search for hits aligned on a trajectory in the MDT and trigger chambers. In the CSC a slightly different algorithm is used. Muon track candidates are then built by fitting the hits found in the muon system. The fit starts from the middle layers of the spectrometer and then extends to the inner and outer layers.

Different muon reconstruction algorithms then combine the information from the muon spectrometer with the information in the inner detector and calorimeters. At the beginning, track reconstruction in the inner detector and the muon spectrometer is performed independently. The main algorithm performs a global refit of the tracks in the inner detector and muon spectrometer. During the fit, muon spectrometer hits may be added or removed to allow for a better track quality. Most muons are reconstructed by starting from the muon spectrometer track and extending it into the inner detector. In a complementary approach the fitting procedure starts in the inner detector. Muons reconstructed by this algorithm are referred to as *combined muons*.

Additional reconstruction algorithms exist which combine single tracks in the inner detector with the energy deposition in the calorimeter[5] (*calorimeter-tagged muons*), a muon track in the inner detector is extrapolated to a single hit in the muon spectrometer (*segment-tagged muons*), or tracks in the muon spectrometer to not match any track in the inner detector (*extrapolated muons*). The muon quality of these reconstruction algorithms is in general lower. They are described in more detail in Ref. [12].

[4] *ElectronEfficiencyCorrection-00-01-42* has been used for the analysis described in part III and *ElectronEfficiencyCorrection-00-00-50* for the analysis described in part IV.

[5] Muons deposit an energy of about 3 GeV when traversing the calorimeter. The deposited energy is independent of the muon momentum.

6.3.2 Identification

A part of the reconstructed muon candidates originate from pion and kaon decays. These muons are considered as background and are suppressed by applying quality requirements on the muon candidate. The requirements are designed to have at the same time a high efficiency for prompt muons and to guarantee a robust momentum measurement.

Muon candidate which originate from in-flight decays of charged hadrons often show the presence of a distinctive "kink" in the reconstructed track. As a consequence they have a lower fit quality and the momentum measured in the inner detector and the muon spectrometer may not be compatible. Hence, the normalized χ^2 of the track fit is one of the variables which can be used for the discrimination of prompt muons and background muons. The q/p significance, defined as the absolute value of the difference between the ratio of the charge and momentum of the muon track measured in the inner detector and in the muon spectrometer divided by the sum in quadrature of the corresponding uncertainties, is another quantity which can be used. A similar quantity is ρ', the absolute value of the difference between the transverse momentum measurements in the inner detector and muon spectrometer divided by the p_{T} of the combined track. To ensure a good track quality, specific requirements on the number of hits in the inner detector and muon spectrometer are used. A muon candidate is required to have at least one hit in the pixel detector, five hits in the SCT, and that 10% of the TRT hits originally assigned to the track are included in the combined fit. The latter requirement is only applied to muon candidates inside the acceptance of the TRT.

Four muon identification selections (*loose*, *medium*, *tight*, *high-p_{T}*) are defined by the ATLAS muon performance group. The *medium* identification is the default muon definition used in ATLAS. The *high-p_{T}* identification level aims to maximize the momentum resolution for muons with $p_{\mathrm{T}} > 100$ GeV. It was specifically designed for the analysis presented in part III and for the search of a heavy dimuon resonance (Z'). These two identification levels are in the following described in more detail, the *loose* and *tight* identification levels are described in more detail in Ref. [12].

The *medium* identification level includes only *combined muons* and *extrapolated muons*. The former are required to have at least three hits in at least two MDT layers, except for the muons within $|\eta| < 0.1$, where one layer is required. The latter are only used in the region $2.5 < |\eta| < 2.7$ which is not relevant for this thesis. In addition a requirement of q/p significance smaller than 7 is imposed.

The *high-p_{T}* identification level only includes *combined muons*. All requirements imposed by the *medium* selection are also applied here. On top of these criteria each muon is required to have at least hits in three precisions layers of the muon spectrometer. For a precise momentum measurement, the inner tracking detector and the muon spectrometer chambers have to be precisely aligned. All muon candidates whose tracks in the muon spectrometer fall into poorly aligned chambers of the muon spectrometer according to their tracks' $\eta - \phi$ coordinates are also rejected. The alignment is usually ensured by taking data without the toroidal magnet and exploiting the straight track pointing from the inner tracking detector to the muon spectrometer.

For some newly installed chambers the alignment was not yet fully understood and the regions are therefore vetoed. The vetoed regions are the barrel/endcap overlap region $1.01 < |\eta| < 1.1$ and the regions $1.05 < |\eta| < 1.3$ for $0.21 < |\phi| < 0.57$, $1.00 < |\phi| < 1.33$, $1.78 < |\phi| < 2.14$ and $2.57 < |\phi| < 2.93$.

6.3.2.1 Isolation

Isolation requirements can be imposed, since non-prompt muons are often accompanied by other charged particles contained in a jet. The sum of the track p_{T} in a region around the muon track larger than a certain radius ΔR, can be used to discriminate between prompt muons, e.g., from W or Z decays and non-prompt muons in jets, e.g., from hadron decays.

6.3.3 *Muon Trigger*

Triggers are used to select events containing muons. All muon triggers are seeded by a Level-1 trigger requiring one or multiple muon candidates, reconstructed from the trigger chamber hits as described in 5.6. In the HLT, the information from the MDT chambers is added and a track from the muon spectrometer hits is formed using a simple parameterized function. The track in the muon spectrometer is then combined with the closest track in the inner detector. If the muon candidate still passes a certain p_{T} requirement, a track reconstruction similar to the offline reconstruction is performed. The selection of the muon candidates is in the end based on the multiplicity, transverse momentum, and track isolation requirements.

6.3.4 *Muon Momentum Scale and Resolution Corrections*

Similarly to the electrons, corrections are derived to correct for differences of the muon momentum scale and resolution observed between data and simulation. In contrast to the electron calibration, no corrections are applied to data. The corrections are derived by comparing the position and width of the Z- and J/ψ-resonance in data and Monte Carlo simulation. As the corrections are only applied to simulation, the position of the Z- and J/ψ-resonance in data does not necessarily match the PDG value which was used in the simulation. The corrections are binned in muon η and are typically in the per mille range for the momentum scale (with an accuracy on the order of 0.1%) and in the low percent range for the resolution.

6.3.5 *Efficiency Corrections*

Similar to the efficiency corrections for electrons, also efficiency corrections for the muon reconstruction, isolation, and trigger efficiency are derived. They are calculated using a tag and probe method selecting a sample of Z- and J/ψ-candidates. The corrections are binned in η and ϕ, no strong p_T dependency was observed. A detailed description of the efficiencies derived for the muons in the simulation of the 2015 data set can be found in Ref. [12]. The corrections are on the order of a few percent.

6.4 Jets

The jet reconstruction, identification, trigger, and corrections which have to be applied are discussed in the following. The section follows largely the discussion in Ref. [13].

6.4.1 Reconstruction

Jets are collimated bundles of hadrons emerging from the fragmentation of high energetic partons. Hence, they are depositing their energy in the calorimeters. The main part of the energy is usually measured in the hadronic part of the calorimeter. Clusters of energy deposits in the calorimeter are built from topological connected calorimeter cells that contain a significant signal above noise [5, 14]. These clusters are called "topo-clusters". The ATLAS hadronic calorimeter is a non-compensating calorimeter, i.e. the energy measured for hadronic showers does not correspond to the true energy. The measured clusters are therefore reconstructed at the electromagnetic scale. Techniques exist (local cluster weighting) which correct for this, but are not used in this thesis and thus not further discussed. The energy clusters are subsequently used as an input for jet finding algorithms.

The main jet finding algorithm in ATLAS is the anti-k_t algorithm [15]. It fulfills the requirements for a jet algorithm to be collinear and infrared safe, i.e., its result does not change significant if small angle or low energy gluon emission appeared. The basic idea of this algorithm is to introduce distances d_{ij} between objects i and j and d_{iB} between object i and the beam (B). If the smallest distance is d_{ij}, the two objects are recombined, if the smallest distance is d_{iB}, i is considered as a jet and removed from the list of considered objects. The calculation of the distances is repeated until no objects are left. The definition of the distances is:

$$d_{ij} = \min(k_{t,i}^{-2}, k_{t,j}^{-2})\frac{\Delta_{ij}^2}{R^2}, \qquad d_{iB} = k_{t,i}^{-2}, \tag{6.4}$$

where $\Delta_{ij}^2 = (y_i - y_j)^2 + (\phi_i - \phi_j)^2$ and $k_{t,i}$, y_i, and ϕ_i are the transverse momentum, rapidity, and azimuth angle of particle i, respectively. R is the radius parameter determining the size of the jets. In this thesis only jets with a radius parameter $R = 0.4$ are used.

6.4.2 Jet Energy Calibration

The jets defined by the anti-k_t algorithm use hadronic clusters calibrated to the electromagnetic energy scale. Hence, they have to be calibrated to obtain the four momentum of the jet. The jet energy scale is therefore calibrated in several steps. First, a correction is applied to account for the energy offset caused by pile-up interactions. In a next step an origin correction is applied to the jet direction, to make the jet pointing back to the primary vertex instead of the nominal interaction point. Then, a Monte Carlo based correction is applied to the jet energy. Finally, a residual correction, derived from in situ, is applied to jets reconstructed in data. The methodology of deriving these corrections is described in Ref. [13]. The corrections are provided by the ATLAS jet performance group [16] and implemented in a software tool.[6]

6.4.3 Identification

Jets can originate from events in which one of the protons collides with the residual gas within the beam pipe or with material outside of the ATLAS detector, for example the collimators. These jets are considered as background, since they are not originating from a *pp* collision. Another source of background are cosmic-ray muons and calorimeter noise. Quality criteria are defined by the ATLAS jet performance group to reject these background jets. The criteria are based on the quality of the energy reconstruction, the jet energy deposits in the direction of the shower development, and reconstructed tracks matched to the jet. The identification criteria are discussed in more detail in Ref. [17]. Jets from background events usually make up a very small fraction.

To differentiate jets from pile-up vertices and jets from the hard-scatter vertex, a multivariate technique is used. The jet-vertex-tagger [18] uses a combination of track-based variables to quantity the likelihood that a jet originates from a hard-scatter process. The values of the tagger range from 0 (most likely pile-up jet) to 1 (most likely hard-scatter jet).

6.4.4 Jet Trigger

At the Level 1 trigger, jets are built from trigger towers using a sliding-window algorithm. In case of a positive trigger decision, the jets are refined in the HLT using reconstruction algorithms similar to those described above.

---

[6] *JetCalibTools-00-04-61* has been used for the analysis presented in part III.

6.4.5 *Correction of the Simulation*

A potential difference can emerge from the jet energy resolution in data and simulation. However, in both, the simulation describing the 2012 data and simulation describing the 2015 data, these differences were found to be negligible [16]. No further corrections are therefore applied but systematic uncertainties might come along with the jet energy resolution in the simulation.

6.5 Missing Transverse Momentum

Neutrinos leave the detector unseen and can only be reconstructed indirectly by reconstructing the missing transverse momentum $E_{\mathrm{T}}^{\mathrm{miss}}$, the momentum imbalance of the event. The measurement of $E_{\mathrm{T}}^{\mathrm{miss}}$ is mainly relevant for the thesis presented in part III. The procedure described in the following section is therefore following the procedure used for the 2015 data set. The information of this section is based on the Refs. [19, 20].

The $E_{\mathrm{T}}^{\mathrm{miss}}$ reconstruction process uses reconstructed, calibrated objects to estimate the transverse momentum imbalance in an event. An object based reconstruction provides a better $E_{\mathrm{T}}^{\mathrm{miss}}$ resolution than a simple sum of calorimeter cells or tracks, as object specific calibrations can be applied. The components of $E_{\mathrm{T}}^{\mathrm{miss}}$ of an event is defined in the following way

$$E_{\mathrm{x(y)}}^{\mathrm{miss}} = E_{\mathrm{x(y)}}^{\mathrm{miss},e} + E_{\mathrm{x(y)}}^{\mathrm{miss},\gamma} + E_{\mathrm{x(y)}}^{\mathrm{miss},\tau} + E_{\mathrm{x(y)}}^{\mathrm{miss},jets} + E_{\mathrm{x(y)}}^{\mathrm{miss},\mu} + E_{\mathrm{x(y)}}^{\mathrm{miss},soft} \tag{6.5}$$

The terms for the charged leptons, photons, and jets correspond to the calibrated momenta[7] for the respective objects. Analysis specific selections are applied to these objects. The calorimeter energy depositions are associated in a specific order to avoid overlap between objects. An energy deposition already used will not be considered for any other object. The order is as follows: electrons, photons, hadronically decaying τ's, jets and finally muons. Tracks which are not considered to any of the objects are combined in the track soft term $E_{\mathrm{x(y)}}^{\mathrm{miss},soft}$. A different possibility would be to sum all calorimeter cells which have not been associated to any of the objects. However, the track based soft term was found to be more robust against pile-up and is as a consequence used as default in the analysis presented in part III. The analysis presented in part IV uses in some places an $E_{\mathrm{T}}^{\mathrm{miss}}$ variable for cuts and further checks of the analysis performance. The $E_{\mathrm{T}}^{\mathrm{miss}}$ performance is here not crucial, hence a calorimeter based soft term is used and no further calibration was applied to the objects. From the components the magnitude $E_{\mathrm{T}}^{\mathrm{miss}}$ and azimuthal angle ϕ^{miss} can be calculated as

[7]In case of the hadronically decaying τ-leptons only the hadronic jet is calibrated and no correction is applied for the momentum carried away by the neutrino.

$$E_{\mathrm{T}}^{\mathrm{miss}} = \sqrt{(E_{\mathrm{x}}^{\mathrm{miss}})^2 + (E_{\mathrm{y}}^{\mathrm{miss}})^2}, \qquad \phi^{\mathrm{miss}} = \arctan(E_{\mathrm{y}}^{\mathrm{miss}}/E_{\mathrm{x}}^{\mathrm{miss}}). \tag{6.6}$$

The τ, photon and track soft term selection is in the following briefly discussed. The selection of the other objects is discussed in more detail in Sect. 9.2.

6.5.1 Photon Selection

The photon identification exploits the evolution of the electromagnetic showers. Different cut-based and likelihood-based identification levels exist similar to those described for electrons. Only photons passing a *tight* identification level are used for the $E_{\mathrm{T}}^{\mathrm{miss}}$ calculation. Furthermore, the photons have to have $p_{\mathrm{T}} > 25$ GeV and $|\eta| < 2.37$. Photons falling within the transition region of the barrel and endcap electromagnetic calorimeters $1.37 < |\eta| < 1.52$ are also discarded. A more detailed description of the photon identification can be found in Ref. [21].

6.5.2 τ Selection

Hadronically decaying τ's may be differentiated from jets based on their low track multiplicity and narrow shower shape. These and other discriminating characteristics are combined in a Boosted Decision Tree. The τ candidates entering the $E_{\mathrm{T}}^{\mathrm{miss}}$ selection are passing a *medium* identification level. Furthermore, they have to have $p_{\mathrm{T}} > 20$ GeV and $|\eta| < 2.5$. Tau candidates falling within the transition region of the barrel and endcap electromagnetic calorimeters $1.37 < |\eta| < 1.52$ are also discarded. A more detailed description of the τ identification can be found in Ref. [22].

6.5.3 Track Soft Term Selection

For the calculation of the track soft term, only tracks are considered which have $p_{\mathrm{T}} > 0.5$ GeV and which lie within the acceptance of the inner detector ($|\eta| < 2.5$). Furthermore, all tracks are required to have at least 7 hits in the pixel detector and SCT. Only tracks originating from the primary vertex (largest $\sum p_{\mathrm{T}}^2$) are considered and required to have a transverse impact parameter d_0 of less than 1.5 mm. It is also required that the d_0 significance is less than 3.

Tracks which satisfy these selection criteria and are not associated to any of the reconstructed objects passing the selection criteria are used. The tracks are excluded if they are within $\Delta R = 0.05$ of an electron or photon cluster, or within $\Delta R = 0.2$ of a hadronically-decaying τ. Tracks associated to a muon are replaced by the combined track fit of the muon spectrometer and inner detector. Since the tracks are matched to the primary vertex, the track soft term is relatively insensitive to pile-up. It does however, not include contributions from soft neutral particles and from regions outside of the inner detector acceptance ($|\eta| > 2.5$).

References

1. Cornelissen T et al (2007) Concepts, design and implementation of the ATLAS new tracking (NEWT). Technical report ATL-SOFT-PUB-2007-007. ATL-COM-SOFT-2007-002. Geneva: CERN. https://cds.cern.ch/record/1020106
2. Fruhwirth R (1987) Application of Kalman filtering to track and vertex fitting. Nucl Instrum Methods A262:444–450. https://doi.org/10.1016/0168-9002(87)90887-4
3. Pagan Griso S et al (2012) Vertex reconstruction plots: collision performance plots for approval. Technical report ATL-COM-PHYS-2012-474. Geneva: CERN. https://cds.cern.ch/record/1445579
4. Performance of the ATLAS inner detector track and vertex reconstruction in the high pile-up LHC environment. Technical report ATLAS-CONF-2012-042. Geneva: CERN (2012). https://cds.cern.ch/record/1435196
5. Lampl W et al (2008) Calorimeter clustering algorithms: description and performance. Technical report ATL-LARG-PUB-2008-002. ATL-COM-LARG-2008-003. Geneva: CERN. https://cds.cern.ch/record/1099735
6. Improved electron reconstruction in ATLAS using the Gaussian sum filter-based model for bremsstrahlung. Technical report ATLAS-CONF-2012-047. Geneva: CERN (2012). https://cds.cern.ch/record/1449796
7. ATLAS Collaboration. Electron and photon energy calibration with the ATLAS detector using LHC Run 1 data. Eur Phys J C74.10:3071 (2014). https://doi.org/10.1140/epjc/s10052-014-3071-4, arXiv: 1407.5063 [hep-ex]
8. ATLAS Collaboration. Electron efficiency measurements with the ATLAS detector using the 2012 LHC proton-proton collision data. In: ATLAS-CONF-2014-032 (2014). https://cds.cern.ch/record/1706245
9. Electron efficiency measurements with the ATLAS detector using the 2015 LHC proton-proton collision data. Technical report ATLAS-CONF-2016-024. Geneva: CERN (2016). https://cds.cern.ch/record/2157687
10. ATLAS Collaboration. Measurement of the muon reconstruction performance of the ATLAS detector using 2011 and 2012 LHC proton-proton collision data. Eur Phys J C74.5:3130 (2014). https://doi.org/10.1140/epjc/s10052-014-3130-x, arXiv: 1407.3935 [hep-ex]
11. Illingworth J, Kittler J (1988) A survey of the hough transform. In: Computer vision, graphics, and image processing 44.1:87–116. ISSN: 0734-189X. https://doi.org/10.1016/S0734-189X(88)80033-1, http://www.sciencedirect.com/science/article/pii/S0734189X88800331
12. Aad G et al (2016) Muon reconstruction performance of the ATLAS detector in proton-proton collision data at $\sqrt{s}$=13 TeV. Eur Phys J C76.5:292. https://doi.org/10.1140/epjc/s10052-016-4120-y, arXiv: 1603.05598 [hep-ex]
13. Aad G et al (2015) Jet energy measurement and its systematic uncertainty in proton-proton collisions at $\sqrt{s} = 7$ TeV with the ATLAS detector. Eur Phys J C75:17. https://doi.org/10.1140/epjc/s10052-014-3190-y, arXiv: 1406.0076 [hep-ex]
14. Aad G et al (2016) Topological cell clustering in the ATLAS calorimeters and its performance in LHC Run 1. arXiv: 1603.02934 [hep-ex]
15. Cacciari M, Salam GP, Soyez G (2008) The Anti-k(t) jet clustering algorithm. JHEP 04:063. https://doi.org/10.1088/1126-6708/2008/04/063, arXiv: 0802.1189 [hep-ph]
16. Collaboration A (2015) Jet calibration and systematic uncertainties for jets reconstructed in the ATLAS detector at $\sqrt{s} = 13$ TeV. Technical report ATL-PHYS-PUB-2015-015. Geneva: CERN. https://cds.cern.ch/record/2037613

17. Selection of jets produced in 13 TeV proton-proton collisions with the ATLAS detector. Technical report ATLAS-CONF-2015-029. Geneva: CERN (2015). http://cds.cern.ch/record/2037702
18. Tagging and suppression of pileup jets with the ATLAS detector. Technical report ATLAS-CONF-2014-018. Geneva: CERN (2014). http://cds.cern.ch/record/1700870
19. ATLAS Collaboration. Performance of missing transverse momentum reconstruction for the ATLAS detector in the first proton-proton collisions at at $\sqrt{s} = 13$ TeV. In: ATL-PHYS-PUB-2015-027 (2015). https://cds.cern.ch/record/2037904
20. ATLAS Collaboration. Expected performance of missing transverse momentum reconstruction for the ATLAS detector at $\sqrt{s} = 13$ TeV. In: ATL-PHYS-PUB-2015-023 (2015). https://cds.cern.ch/record/2037700
21. Aaboud M et al (2016) Measurement of the photon identification efficiencies with the ATLAS detector using LHC Run-1 data. arXiv: 1606.01813 [hep-ex]
22. Aad G et al Identification and energy calibration of hadronically decaying tau leptons with the ATLAS experiment in pp collisions at $\sqrt{s} = 8$ TeV. Eur Phys J C75.7:303 (2015). https://doi.org/10.1140/epjc/s10052-015-3500-z, arXiv: 1412.7086 [hep-ex]

Part III
Search for New Physics in Final States with One Lepton Plus Missing Transverse Momentum at $\sqrt{s} = 13$ TeV

Chapter 7
Motivation

The conceptual problems of the Standard Model and open questions like the nature of dark matter motivate the search for new physics. New physics models which try to solve these problems or find an explanation for the open questions are often an extension of the Standard Model. These new physics models almost always predict new particles that can be searched for. The hierarchy problem motivates the appearance of new physics at the TeV scale. While there are very well-elaborated theories, like Supersymmetry, which often make rather explicit predictions about the particles that are expected, there are at the same time a lot of theories which are still in the conceptual phase. Many theories beyond the Standard Model extend the Standard Model gauge groups. These new gauge groups lead to new gauge bosons that can behave similar as the W and Z bosons in the Standard Model. Some theories which predict a new heavy charged boson have been introduced in Sect. 2.3.3. It should also be considered that it is possible that the true extension of the Standard Model has not yet been thought of.

The new heavy charged bosons might decay into a charged lepton ($\ell^{\pm} = e^{\pm}, \mu^{\pm}$) and the corresponding (anti-)neutrino. This is a very clear final state which can be probed in a more general way. Hence, the presented analysis aims to search for deviations of the data from the expected Standard Model background without making too many model specific assumptions. As a consequence, a generic heavy charged gauge boson, the Sequential Standard Model W', is used as a reference model. In this model, the W' boson has the same couplings as the Standard Model W but no couplings to W and Z bosons.

Figure 7.1 shows the ratio of the exclusion limit on the cross section of a W' boson and the predicted SSM cross section as a function of the W' pole mass resulting from previous analyses performed by the CDF [1] collaboration at the Tevatron

M. Zinser, *Search for New Heavy Charged Bosons and Measurement of High-Mass Drell-Yan Production in Proton-Proton Collisions*, Springer Theses,
https://doi.org/10.1007/978-3-030-00650-1_7

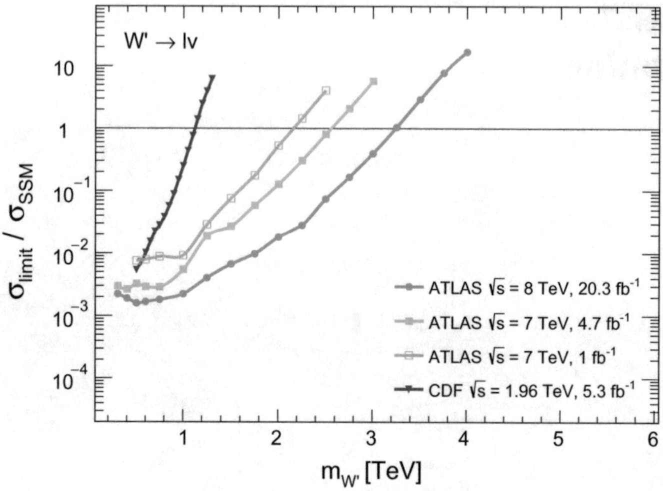

Fig. 7.1 Normalized cross section limits ($\sigma_{\text{limit}}/\sigma_{\text{SSM}}$) for W' bosons as a function of mass for this measurement and from previous CDF [1] and ATLAS [3–5] searches. The cross section calculations assume the W' has the same couplings as the SM W boson. The region above each curve is excluded at 95% CL

accelerator[1] and the ATLAS collaboration with data taken at a center of mass energy of 7 and 8 TeV [3–5]. So far SSM W' masses below 3.24 TeV have been excluded by the ATLAS collaboration and below 3.28 TeV by the CMS collaboration [6] with 95% confidence level (CL).

The increase in center of mass energy from 1.96 TeV at Tevatron to 7 TeV at the LHC significantly increased the mass reach of these analyses even when using less integrated luminosity. The presented analysis will be performed with the first data delivered by the LHC at $\sqrt{s} = 13$ TeV. It is therefore probing a new energy regime and a significant boost of sensitivity at the highest masses is expected. It is at the same time also possible that new physics appears at lower masses with much lower cross sections than predicted by the SSM. The analysis presented in this thesis will therefore also try to improve the sensitivity in this range.

The ATLAS W' group analyzing the 2015 data was lead by myself where I performed the complete analysis of the electron channel and the statistical interpretation of both channels. The analysis of the muon channel has afterwards been added for this thesis. The results first have been made public in a conference note [7] (December 2015) and a publication [8] (June 2016) which has been submitted to Phys. Lett. B.

The analysis is structured as follows. In Chap. 8 the analysis strategy is introduced. Chapter 9 is the main part of this work and describes in Sect. 9.1 the Monte Carlo samples followed by the description of the electron and muon channel event selection in Sect. 9.2. Section 9.3 discusses the estimation of the multijet background which

[1] Also an analysis from the D0 collaboration exists [2] but it provides not the needed information to be included in Fig. 7.1.

is arising from jets being misidentified as electron or muon and the extrapolation of several backgrounds towards high transverse mass is described in Sect. 9.4. The estimation of the systematic uncertainties is discussed in Sect. 9.5 and the chapter closes with a comparison of the selected data with the expected background in Sect. 9.6. Chapter 10 starts with a discussion of the statistical framework in Sect. 10.1 and the obtained transverse mass spectra in Sect. 10.2. In Sect. 10.3 a quantification of potential observed excesses is performed by a likelihood ratio test and an exclusion limit on the cross section of the SSM W' is determined using a Bayesian approach and compared to previous results in Sect. 10.4. The discussion of the analysis ends with a conclusion and an outlook in Chap. 11.

References

1. T. A.e. a. CDF Collaboration (2011) Search for a new heavy gauge boson W0 with electron + missing ET event signature in $p\bar{p}$ collisions at ps = 1:96 TeV. Phys Rev D83: 031102. https://doi.org/10.1103/PhysRevD.83.031102, arXiv:1012.5145
2. Abazov VM et al (2008) Search for W0 bosons decaying to an electron and a neutrino with the D0 detector. Phys Rev Lett 100: 031804. https://doi.org/10.1103/PhysRevLett.100.031804, arXiv: 0710.2966
3. Collaboration A (2011) Search for a heavy gauge boson decaying to a charged lepton and a neutrino in 1 fb^1 of pp collisions at $\sqrt{s} = 7$ TeV using the ATLAS detector. Phys Lett B705: 2846. https://doi.org/10.1016/j.physletb.2011.09.093, arXiv: 1108.1316
4. Collaboration A (2011) Search for high-mass states with one lepton plus missing transverse momentum in proton-proton collisions at $\sqrt{s} = 7$ TeV with the ATLAS detector. Phys Lett B701: 5069. https://doi.org/10.1016/j.physletb.2011.05.043, arXiv: 1103.1391
5. Aad G et al (2014) Search for new particles in events with one lepton and missing transverse momentum in pp collisions at $\sqrt{=}8TeV$ with the ATLAS detector. JHEP 09: 037. https://doi.org/10.1007/JHEP09(2014)037, arXiv: 1407.7494
6. Khachatryan V et al (2015) Search for physics beyond the standard model in final states with a lepton and missing transverse energy in proton-proton collisions at sqrt(s) = 8 TeV. Phys Rev D 91(9): 092005. https://doi.org/10.1103/PhysRevD.91.092005, arXiv: 1408.2745
7. Search for new resonances in events with one lepton and missing transverse momentum in pp collisions at $\sqrt{s} = 13$ TeV with the ATLAS detector. Technical report, ATLAS-CONF-2015-063. Geneva: CERN (2015). https://cds.cern.ch/record/2114829
8. Aaboud M et al (2016) Search for new resonances in events with one lepton and missing transverse momentum in pp collisions at $\sqrt{s} = 13$ TeV with the ATLAS detector. arXiv: 1606.03977

Chapter 8
Analysis Strategy

The invariant mass distribution is the most direct way to search for a new particle, as a new particle will lead to a resonance at its pole mass. However, the invariant mass $m_{\ell\nu}$ of the $\ell\nu$-pair cannot be reconstructed, as the information about the momentum of the neutrino can only be reconstructed indirectly and only in the transverse plane, by reconstructing the missing transverse momentum $E_{\mathrm{T}}^{\mathrm{miss}}$. A way to indirectly access this information is to reconstruct the transverse mass m_{T} of the event

$$m_{\mathrm{T}} = \sqrt{2 p_{\mathrm{T}} E_{\mathrm{T}}^{\mathrm{miss}} (1 - \cos \Delta\phi_{\ell\nu})},$$

where $\Delta\phi_{\ell\nu}$ is the azimuthal angle between the lepton and $E_{\mathrm{T}}^{\mathrm{miss}}$ in the transverse plane and p_{T} the transverse momentum of the lepton. The transverse mass was the kinematic observable that played a key role in the discovery of the W boson at the Sp$\bar{\mathrm{p}}$S collider [1, 2] and later in its precise measurement at the Tevatron collider [3–5]. Figure 8.1 shows the $m_{\ell\nu}$ and m_{T} distribution for four simulated SSM W' bosons with different pole masses. The histograms for the different pole masses are normalized to the same area to allow for shape comparisons. The Breit–Wigner resonance, which is visible in the $m_{\ell\nu}$ spectrum, is leading to a clear signature at the pole mass of the W'. This signature is significantly diluted in the m_{T} spectrum. A peak with a sharp edge at $m_{\mathrm{T}} = m_{W'}$ can be observed instead of a resonant distribution around the pole mass of the W'. This sharp edge can be used to indirectly measure the pole mass. A transverse mass close to the edge is only reconstructed if the W' is produced at rest. In this case the lepton and neutrino decay back-to-back and therefore $\cos \Delta\phi_{\ell\nu} \approx -1$. Furthermore, lepton and neutrino need to have $\eta \approx 0$, so that the p_{T} and $E_{\mathrm{T}}^{\mathrm{miss}}$ are maximized and $m_{\mathrm{T}} \approx m_{W'}$. Every other configuration will lead to a lower transverse mass. A significant part of the signal is therefore contributing to a tail at lower values of m_{T}. The signal shape in the m_{T} distribution can be described by a Jacobian peak

M. Zinser, *Search for New Heavy Charged Bosons and Measurement of High-Mass Drell-Yan Production in Proton-Proton Collisions*, Springer Theses,
https://doi.org/10.1007/978-3-030-00650-1_8

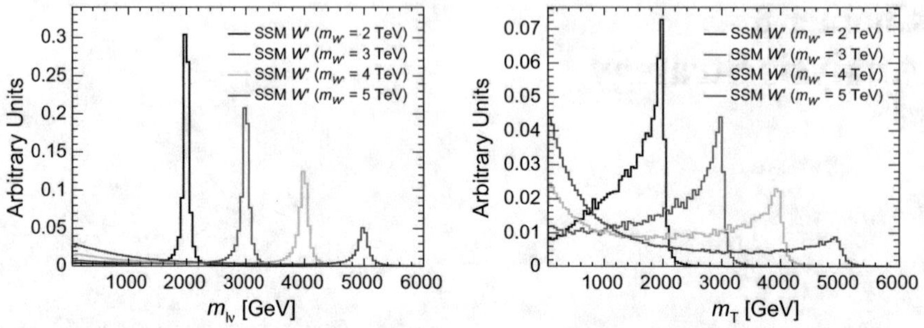

Fig. 8.1 Generated invariant mass $m_{\ell\nu}$ (left) and generated transverse mass m_{T} (right) distribution a W' with four different pole masses. The histograms for the different pole masses are normalized to its area to allow for shape comparison

$$\frac{dN}{dm_{\mathrm{T}}} \propto \frac{m_{\mathrm{T}}}{m_{W'}^2} \frac{2 - \left(\frac{m_{\mathrm{T}}}{m_{W'}}\right)^2}{\sqrt{1 - \left(\frac{m_{\mathrm{T}}}{m_{W'}}\right)^2}}.$$

Another important effect can be observed when comparing the signal shapes for different pole masses. While 90% of the contribution of a W' with a mass of 2 TeV is still contained in a window of $m_{\ell\nu} \pm 500$ GeV around its pole mass, this number reduces to 26% for a W' with a mass of 5 TeV. The rest of the signal is contributing to a low mass tail which is getting more pronounced for higher pole masses of the W'. The low mass tail is even larger in the m_{T} distribution. Here only 50% (9%) of the events of a W' with a mass of 2 TeV (5 TeV) are contributing to the same window. The reason for this enhancement for higher pole masses is the low parton-parton-luminosity for very high invariant masses. A quark-antiquark pair with very high Bjorken-x is needed to produce a W' at very high invariant mass $m_{\ell\nu}$. The probability to find a quark-antiquark pair at very high Bjorken-x is very low while it is enhanced at lower values of Bjorken-x.

The general strategy of this analysis is to select events with high $E_{\mathrm{T}}^{\mathrm{miss}}$ and a single electron or muon with high p_{T}. Figure 8.2 shows an event display of an event with a very high transverse mass of 1.95 TeV. The event contains an electron with very high transverse momentum of $p_{\mathrm{T}} = 1.01$ TeV. The green towers depict the energy deposits in the EM calorimeter. For this event the reconstructed $E_{\mathrm{T}}^{\mathrm{miss}}$ is 0.94 TeV and therefore well balanced with the p_{T} of the electron. The decay products of a heavy particle produced at rest typically have very similar p_{T}. The direction of the $E_{\mathrm{T}}^{\mathrm{miss}}$ is indicated by the red dashed line. Electron and $E_{\mathrm{T}}^{\mathrm{miss}}$ are back-to-back in the transverse plane, leading to the very high transverse mass of the event.

The m_{T} distribution of the data will be compared to the expected backgrounds from SM processes. The leading SM background for this analysis is off-shell[1] $W \to \ell\nu$

[1] Off-shell means in this sense the production far away from the mass of the W.

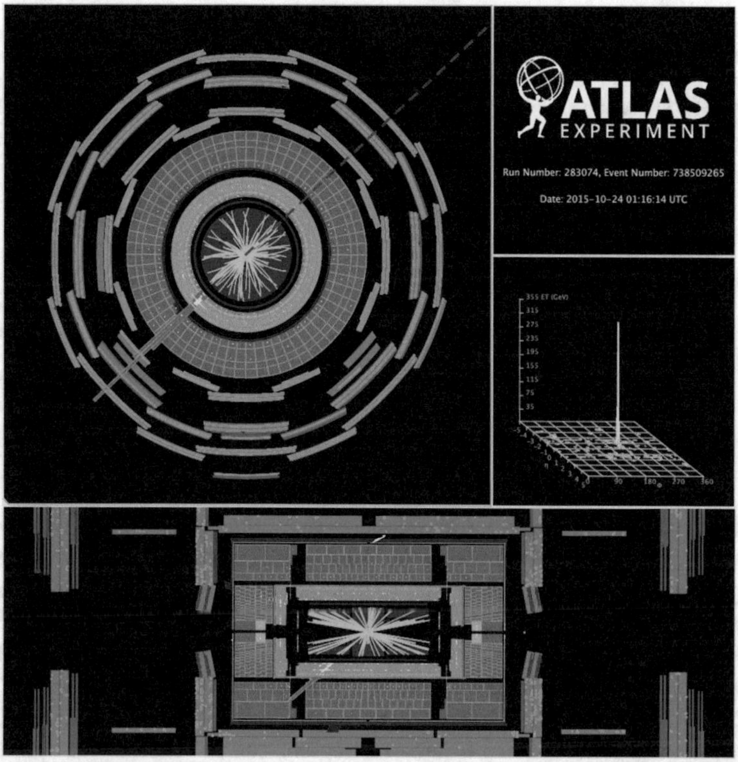

Fig. 8.2 Event display for an event recorded in 2015 with high m_T in the electron channel. The event contains an electron with $p_T = 1.01\,\text{TeV}$, $\eta = -0.9$, $\phi = -2.4$. The event has a missing transverse momentum $E_T^{\text{miss}} = 0.94\,\text{TeV}$, and a transverse mass of $m_T = 1.95\,\text{TeV}$. The red dashed line represents the E_T^{miss} direction. The green towers depict the energy deposits in the EM calorimeter of the electron. Tracks with $p_T > 0.5\,\text{GeV}$ reconstructed in the inner detector are shown in blue. Figure taken from Ref. [6]

production which leads to an identical final state as a W' boson. Figure 8.3 shows $W' \to e\nu$ signals for four different pole masses on top of the leading $W \to e\nu$ background. The Jacobian peak of the W boson is visible at around $m_T \approx 80\,\text{GeV}$. The background is steeply falling over several orders of magnitude towards higher m_T. A W' boson would become apparent as an resonant excess in the data above the SM background at very high transverse mass. The interference between the W and W' boson are very model dependent and are therefore not simulated, as they would contradict the idea of performing a model independent generic search. When searching for such an excess, the shape of the signal can be exploited. In the presence of an excess, statistical methods are used to estimate its significance, whereas in absence of an excess a limit on the mass of a W' boson can be calculated.

The subleading background is coming from final states containing a top- and antitop-quark. The top- and antitop-quarks will decay dominantly via $t \to Wb$, there-

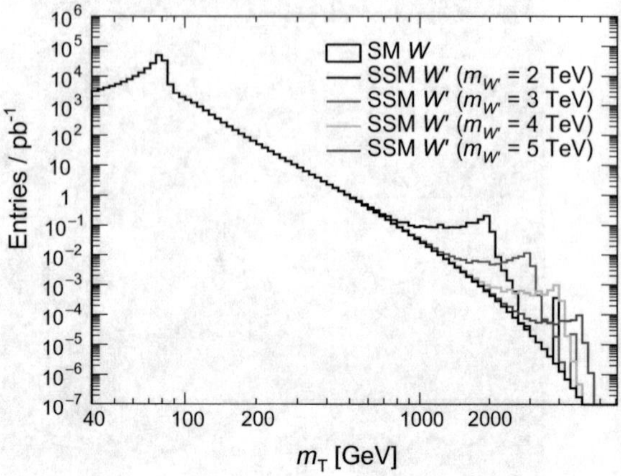

Fig. 8.3 Generated transverse mass m_{T} spectrum of SSM W' signals for four different pole masses on top of the leading Standard Model W background

fore leading to a W boson plus an additional b- or $\bar{b}$-quark. An additional background arises from Z/γ^* production which can lead to two electrons or muons with high transverse momentum while $E_{\mathrm{T}}^{\mathrm{miss}}$ can be mis-reconstructed when one of the leptons is outside of the detector acceptance or mismeasured. The decay $Z/\gamma^* \rightarrow \tau\tau$ can lead to real $E_{\mathrm{T}}^{\mathrm{miss}}$ and electrons or muons if at least one of the τ-leptons decays leptonically $\tau \rightarrow \ell\nu_\tau\nu_\ell$, where $\ell = e, \mu$. As W and Z/γ^* are contributing to the background, also all diboson processes WW, WZ and ZZ constitute relevant backgrounds. Backgrounds leading to multiple leptons can be reduced by requiring the presence of exactly one lepton in the event. In addition, a background arises from objects being misidentified as electrons or muons. This background is very different in size and origin for the two analysis channels. In the electron channel this background mainly originates from jets being misidentified as electrons and $E_{\mathrm{T}}^{\mathrm{miss}}$ coming from an imbalance in the event as the jet energy was not fully taken into account or mismeasured. In the muon channel the background originates from real muons which are coming from secondary decays of b-mesons. Most of these secondary muons can be rejected by requiring the muon to originate very closely from the production vertex. MC simulation reliably predicts all backgrounds containing real leptons and will be used to estimate these. The background containing misidentified leptons needs to be extracted from data as MC simulation in general fails to describe the probability that a lepton is misidentified.

Initially a blinded analysis was performed by rejecting all events in data with $m_{\mathrm{T}} > 500$ GeV. The approval to include the events in the blinded region was given by the ATLAS collaboration after presenting the analysis strategy and good understanding of the data in the region $m_{\mathrm{T}} < 500$ GeV. In this thesis the analysis is therefore presented without any blinding requirements.

References

1. Arnison G (1983) Experimental observation of isolated large transverse energy electrons with associated missing energy at sqrt(s) = 540 GeV. Phys Lett B 122:103–116. https://doi.org/10.1016/0370-2693(83)91177-2 Feb
2. Banner M (1983) Observation of Single Isolated Electrons of High Transverse momentum in events with missing transverse energy at the CERN anti-p p collider. Phys Lett B 122:476–485. https://doi.org/10.1016/0370-2693(83)91605-2
3. Aaltonen T (2012) Precise measurement of the W-boson mass with the CDF II detector. Phys Rev Lett 108:151803. https://doi.org/10.1103/PhysRevLett.108.151803, arXiv: 1203.0275
4. Abazov VM (2012) Measurement of the W boson mass with the D0 detector. Phys Rev Lett 108:151804. https://doi.org/10.1103/PhysRevLett.108.151804, arXiv: 1203.0293
5. Aaltonen TA et al (2013) Combination of CDF and D0 W-boson mass measurements. Phys Rev D 88(5):052018. https://doi.org/10.1103/PhysRevD.88.052018, arXiv: 1307.7627
6. Search for new resonances in events with one lepton and missing transverse momentum in pp collisions at $\sqrt{s} = 13$ TeV with the ATLAS detector. Technical report, ATLAS-CONF-2015-063. Geneva: CERN (2015). https://cds.cern.ch/record/2114829

Chapter 9
Analysis

The following chapter describes the analysis of the electron and muon channel. First all MC samples used for the analysis are discussed in Sect. 9.1. The data used and the event and lepton selection criteria are discussed in Sect. 9.2. The determination and validation of the multijet background is discussed in Sect. 9.3 and the extrapolation of the MC and multijet background towards high transverse mass is presented in Sect. 9.4. The systematic uncertainties on the background estimation are discussed in Sect. 9.5. Finally, the selected data are compared to the Standard Model background expectation in Sect. 9.6.

9.1 Monte Carlo Simulations

The following section contains a description of the Monte Carlo samples used in this analysis. The first part describes the signal simulations and a second part the simulations for background processes. All MC samples used in this analysis have been centrally provided by the ATLAS collaboration [1].

9.1.1 Simulated Signal Processes

The search will be performed over a large range in m_{T}. Therefore simulated signal samples for a large range of pole masses are needed to test the signal hypothesis. Producing signal samples for every pole mass tested becomes very computational intensive. Instead a single "flat" W' sample has been produced at leading order (LO) using PYTHIA 8.183 [2] and the NNPDF2.3 LO PDF set [3]. For this signal sample the Breit–Wigner term has been removed from the event generation. This leads to the production of a flat falling spectrum, similar to the off-shell tail of the W process. In addition, the square of the matrix element has been divided by a function of $m_{\ell\nu}$ to

M. Zinser, *Search for New Heavy Charged Bosons and Measurement of High-Mass Drell-Yan Production in Proton-Proton Collisions*, Springer Theses,
https://doi.org/10.1007/978-3-030-00650-1_9

avoid a fast drop in cross section as a function of $m_{\ell\nu}$. The final-state photon radiation (QED FSR) and the modeling of the parton showering and hadronization is handled by PYTHIA. The W' has the same couplings as the SM W boson. Interference effects between the W and W' are not included and the decay into a WZ pair is not allowed. The interference effects and the couplings to the Standard Model bosons are very model dependent and would contradict the idea of performing a model independent generic search. The decay $W' \to tb$ is allowed if kinematically possible. The signal samples are produced separately for the process $W' \to e\nu$ and $W' \to \mu\nu$ requiring the invariant mass $m_{\ell\nu}$ to be larger than 25 GeV.

The resulting samples which are approximately flat in $\log(m_{\ell\nu})$ can be reweighted to any pole mass $m_{W'}$ using the methodology described in Appendix A. Figure 9.1 shows on the left side the invariant mass $m_{\ell\nu}$ and on the right side the transverse mass m_{T} of the flat signal sample before and after reweighting to pole masses of $m_{W'} = 2, 3, 4$ and 5 TeV. In addition, signal samples with a fixed mass have been produced using the same MC setup to validate the reweighting procedure. Very good agreement between the reweighted samples and the samples with fixed mass can be observed over the whole $m_{\ell\nu}$ and m_{T} range. Detailed information about the samples can be found in the appendix in Table B.1.

The calculation of the matrix element of the hard scattering process is done in PYTHIA at LO in QCD. Theory correction factors are provided by the ATLAS collaboration [4] which correct for differences between the LO calculation and calculations including higher orders in QCD. These correction factors are obtained by a polynomial fit to the ratio of the W cross section as a function of $m_{\ell\nu}$ at LO and NNLO calculated using VRAP [5]. For the LO cross section the same PDF set as for the MC generation with PYTHIA is used (NNPDF2.3LO), while for the NNLO calculation the CT14NNLO PDF set [6] is used. The renormalization (μ_R) and factorization scales (μ_F) are set equal to the value of $m_{\ell\nu}$ at which the cross section is calculated. It is assumed that the higher order corrections derived for the W process are also valid

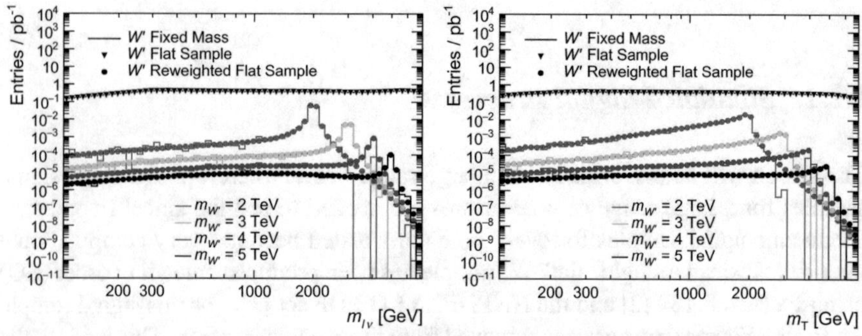

Fig. 9.1 Truth invariant mass and truth transverse mass spectrum of the W' signal samples generated with a fixed mass and the flat samples reweighted to the same mass. The black triangles show the flat sample before reweighting. In red, green, blue and black four different pole masses are shown for the validation samples (solid line) and the reweighted flat sample (dots)

for the W' process as they are very similar. The corrections are given as a function of $m_{\ell\nu}$ separately for positive and negative charged W' bosons. The resulting correction factors are 1.34/1.37, 1.42/1.35 and 1.23/1.10 for positive/negative charged W' at 0.5, 2 and 4 TeV, respectively. The effect from missing QCD diagrams is very similar for positively and negatively charged W' bosons. The differences between W^+ and W^- are resulting from using different PDFs for the LO and NNLO calculation.

9.1.2 *Simulated Background Processes*

9.1.2.1 *W* Production

The main background in this analysis is arising from the SM W production $pp \rightarrow W + X \rightarrow \ell\nu + X$. The background is originating from off-shell produced W bosons and steeply falling towards higher m_{T}. The background is simulated using POWHEG- BOX v2 [7] with the CT10 PDF set [8] for the matrix element of the hard scattering process. PYTHIA 8.186 [2] is used for the modeling of the parton shower, hadronization and particle decays and QED FSR is simulated using Photos [9]. The cross section for off-shell W production is strongly falling as a function of the invariant mass. Very large statistics would be needed to sufficiently populate the distribution at high invariant masses and therefore high m_{T}. Thus, the background is produced in 19 slices in invariant mass $m_{\ell\nu}$ to save computing time. The slices are starting from 120 GeV $< m_{\ell\nu} <$ 180 GeV and reach up to $m_{\ell\nu} >$ 5000 GeV. Samples for all three lepton flavors are generated at NLO, separated into W^+ and W^-. The number of generated events for each sample reaches from at low mass 500,000 down to 50,000 at high mass, corresponding to an integrated luminosity of 15.6 fb^{-1} for W^+ (22.5 fb^{-1} for W^-) to $3.25 \cdot 10^8$ fb^{-1} for W^+ ($8.11 \cdot 10^8$ fb^{-1} for W^-). Additionally, inclusive W^+ and W^- samples have been generated over the whole mass range using the same MC setup. For the inclusive sample 30 million events are generated for W^+ (40 million for W^-). The cross section times branching ratio σBr for each lepton generation is 11.3 nb for $W^+ \rightarrow \ell^+\nu$ and 8.3 nb for $W^- \rightarrow \ell^-\bar{\nu}$. Events generated with an invariant mass of $m_{\ell\nu} >$ 120 GeV are rejected to avoid overlap between the inclusive samples and the mass-binned samples. Detailed information about the MC samples can be found in the appendix in Tables B.2, B.3 and B.4.

Figure 9.2 shows on the left side the resulting invariant mass and on the right side the resulting transverse mass spectrum m_{T} for the $W^+ \rightarrow e^+\nu_e$ background. The colored lines show the individual mass bins and the black line shows the resulting sum of all samples scaled up by a factor of two for easier visibility. The samples provide sufficient statistics up to very high m_{T} of several TeV.

Theory correction factors to reweight the underlying cross section generated by POWHEG from NLO to NNLO in QCD are derived using VRAP. The corrections are derived in the same way as for the signal samples (see Sect. 9.1.1). The resulting QCD correction factors are 1.03/1.04, 1.02/1.01 and 1.09/0.87 for W^+/W^- at 0.5, 2 and 4 TeV, respectively. In addition, correction factors are derived to correct

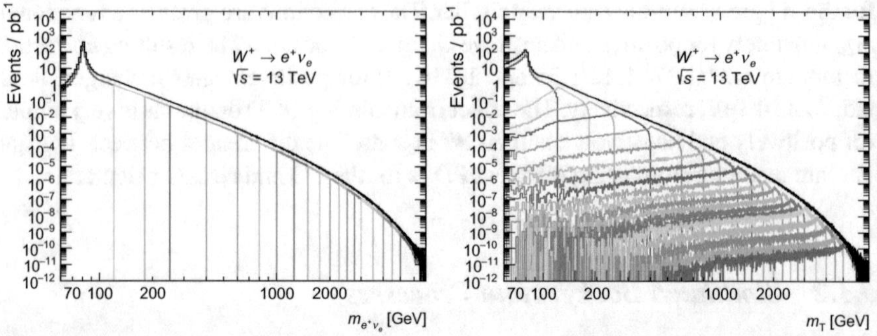

Fig. 9.2 Generated invariant mass $m_{e^+\nu_e}$ and transverse mass spectrum m_T of the inclusive and invariant mass-binned $W^+ \to e^+\nu_e$ MC samples. The colored lines show the different mass slices and the black line the sum of all samples, scaled up by a factor of two for easier visibility

for electroweak (EW) effects beyond QED FSR. The correction is calculated separately for both charges using MCSANC [10, 11] and includes other higher order EW effects, namely initial state radiation (ISR), ISR/FSR interference, and Sudakov logarithm single-loop corrections [12]. The resulting EW correction factors are 0.95/0.95, 0.86/0.86 and 0.81/0.80 for W^+/W^- production at 0.5, 2 and 4 TeV, respectively. The corrections have been provided by the ATLAS collaboration [4].

9.1.2.2 Z/γ^* Production

Neutral current Drell–Yan production ($pp \to Z/\gamma^* + X \to \ell\ell + X$) is an additional source of high-p_T leptons. If one of the leptons is outside of the detector acceptance, large values of $E_\mathrm{T}^\mathrm{miss}$ can occur due to the mis-balance in the event. Real $E_\mathrm{T}^\mathrm{miss}$, caused by neutrinos, can occur from the decay $Z/\gamma^* \to \tau\tau$ where the τ-leptons decay further. The Monte Carlo samples are generated at NLO for all three lepton flavors using the same setup as for the W process. An inclusive sample with 20 million events, corresponding to an integrated luminosity of $L_\mathrm{int} = 10.5\ \mathrm{fb}^{-1}$, is extended to high $m_{\ell\ell}$ with samples binned in $m_{\ell\ell}$. The bin ranges are the same as for the W background. The cross section times branching ratio of the inclusive sample for one lepton generation is $\sigma Br = 1.9\,\mathrm{nb}$ and therefore about a factor ten smaller than σBr for the W background.

Theory correction factors, to correct for EW and NNLO QCD effects, are derived in the same way as for the W background [4]. The resulting QCD correction factors are 1.04, 1.02 and 0.94 for masses at 0.5, 2 and 4 TeV, respectively. The resulting EW correction factors are 0.99, 0.92 and 0.88 for masses at 0.5, 2 and 4 TeV, respectively. Detailed information about the samples can be found in the appendix in Tables B.5, B.6 and B.7.

9.1.2.3 Top-Quark Production

Produced top- and antitop-quarks dominantly decay into b- or $\bar{b}$-quarks under emission of a W boson. These W bosons can then further decay into leptons. Three different processes of top- and antitop-quark production are considered: top-antitop pair production, single (anti-)top production, and single (anti-)top production in association with a W. These backgrounds are from now on referred to as top-quark background. All processes including top/antitop-quarks are simulated at NLO in QCD using POWHEG v2 with the CT10 PDF set for the matrix element of the hard scattering process. PYTHIA 6.428 [13] is used for the modeling of the parton shower, hadronization and particle decays and QED FSR is simulated using Photos. The produced samples are all filtered at the generation stage for events in which at least one of the W bosons decays into a charged lepton.

The dominant process is the production of top-antitop pairs $pp \to t\bar{t} + X \to W^+bW^-\bar{b} + X$. The MC sample is normalized to a cross section of $\sigma_{t\bar{t}} = 832^{+20}_{-29}(scale) \pm 35(PDF + \alpha_S)$ pb as calculated with the Top++2.0 [14] assuming a top-quark mass $m_t = 172.5$ GeV. The cross section is calculated at NNLO in QCD, including soft-gluon resummation to next-to-next-to-leading-log order (see Ref. [14] and references therein). The first uncertainty comes from the independent variation of the factorization and renormalization scales, μ_F and μ_R, while the second one is associated to variations in the PDF and α_S, following the PDF4LHC [15] prescription.[1] Varying the top-quark mass by ± 1 GeV leads to an additional systematic uncertainty of ± 23 pb, which is also added in quadrature.

Single top production in association with a W can lead up to two leptons and $E_{\mathrm{T}}^{\mathrm{miss}}$ from the W decay. The MC sample is normalized to a NLO cross section, including soft-gluon resummation to next-to-next-to-leading-log, of $\sigma_{tW} = 71.7 \pm 3.8$ pb [16].

Single top production can lead up to one lepton and $E_{\mathrm{T}}^{\mathrm{miss}}$ from the W decay. The MC samples are normalized to a NLO cross section, including soft-gluon resummation to next-to-next-to-leading-log, of $\sigma_t = 136 \pm 4$ pb and $\sigma_{\bar{t}} = 81.0 \pm 2.4$ pb [17]. All higher order cross sections have been provided by the ATLAS collaboration [18, 19]. Detailed information about the samples can be found in the appendix in Table B.8.

9.1.2.4 Diboson Production

The smallest background component is arising from the production of WW-, WZ- or ZZ-boson pairs. These background processes are simulated at NLO in QCD using SHERPA 2.1.1 [20] with the CT10 PDF set. SHERPA also models the parton shower, hadronization, particle decays and QED FSR. While for other MC generators the physics process is specified, for Sherpa only the final state is specified and all pro-

[1] The PDF4LHC prescription for calculating the PDF uncertainties is to take the envelope of the uncertainties from the MSTW2008 68% CL NNLO, CT10 NNLO and NNPDF2.3 5f FFN PDF sets.

cesses contributing to this final state are simulated. MC samples for the following final states have been generated: $\ell\ell\ell\ell$, $\ell\ell\ell\nu$, $\ell\ell\nu\nu$, $\ell\nu\nu\nu$, $\ell\nu qq$, $\ell\ell qq$. The $\ell\nu qq$ final state has with 49.8 pb the largest cross section. Detailed information about the samples can be found in the appendix in Table B.8.

9.2 Data and Selection Criteria

The following section contains a description of the data set which is used in the analysis and all selection criteria applied at the event level and to the electrons and muons found in the event. Finally, the signal efficiency for the presented selection and the resolution of the relevant kinematic quantities is discussed.

9.2.1 Data

The data used in this analysis was delivered by the LHC at $\sqrt{s} = 13$ TeV and recorded by the ATLAS experiment. The data taking period was from June to November 2015 and the recorded data set corresponds to a total integrated luminosity of 3.9 fb^{-1}. From June to July data was taken from collisions with a 50 ns spacing between the proton bunches in the LHC. From August onwards the spacing was reduced to 25 ns, which is the LHC design value. Only data taken with a bunch spacing of 25 ns are used for this analysis, as analyzing data for both conditions would also need MC samples for both conditions and as the integrated luminosity for the 50 ns data set is small. The left plot in Fig. 9.3 shows the sum of the integrated luminosity delivered by the LHC, recorded by ATLAS and ready for physics analyses for the data taking period

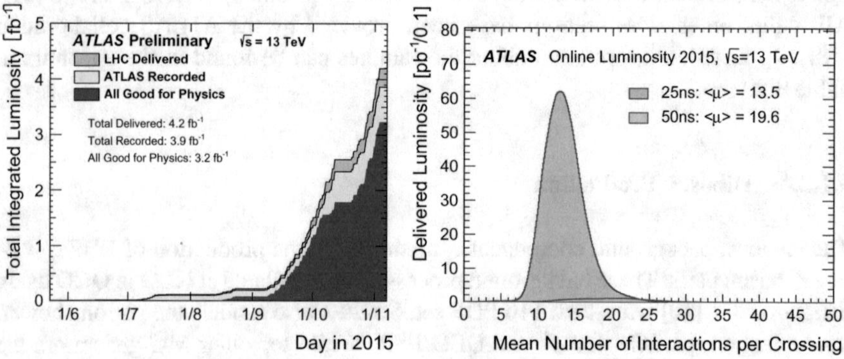

Fig. 9.3 The left side shows the sum of integrated luminosity for data taking in 2015 by day. In green the sum of the integrated luminosity delivered by the LHC is shown. The sum of the integrated luminosity record by ATLAS is shown in yellow. The plot of the right side shows the number of interactions per bunch crossing separately for the 50 and 25 ns bunch spacing conditions. Figures are taken from Ref. [21]

in the year 2015. The data taken with 25 ns bunch spacing corresponds to a total integrated luminosity of 3.8 fb^{-1} and a mean number of simultaneous collisions of $\langle\mu\rangle = 13.5$. The right plot in Fig. 14.3 shows the number of simultaneous collisions per proton bunch crossing separately for the 25 and 50 ns bunch spacing conditions.

9.2.2 Event Cleaning

The data has been preselected, in order to reduce the amount of data to analyze and the required amount of disk space. Only events which contain at least one reconstructed electron or muon with $p_T > 50$ GeV are used. The electron has to fulfill also at least one of the three likelihood identification levels for electrons (see Sect. 6.2.2).

The 2015 data set is divided into the periods A to J (see Sect. 5.7.1 for more details on the structure of the data). Periods D-J correspond to the data with 25 ns bunch spacing and were used for this analysis. All events used for this analysis have to be in a luminosity block which is part of the Good Runs List.[2] In addition, events are discarded in which a noise burst was observed in the electromagnetic or hadronic calorimeter. Such a noise burst could fake energy depositions and would make an accurate energy measurement impossible. The information of the SCT might be corrupted if an ionizing particle with high energy traverses electronic modules. Events with such corrupted tracking information are also rejected. It might be sometimes necessary to restart the trigger system during data taking. During such a restart events might not have the complete detector information and are therefore also rejected. Finally, events which have no reconstructed vertex with at least two tracks associated to it are also rejected. Events with incomplete detector information, corrupted tracking information, noise bursts in the calorimeter or no reconstructed vertex make up less than a per mille of the total events. Table 9.1 shows the number of events remaining after each cleaning cut.

The integrated luminosity after requiring the events to pass all quality requirements is 3.2 fb^{-1}. Hence, this is the number quoted as the integrated luminosity for the data set. The sum of the integrated luminosity ready for physics analysis is shown in blue in Fig. 9.3. The largest reduction of data is coming from the requirement of a run to be in the Good Runs List. Overall it corresponds to a reduction of 0.7 fb^{-1}. A large part of the reduction is coming from two runs in which the inner most layer of the pixel detector was turned off and runs in which the toroidal magnetic field was turned off to record data for alignment studies of the muon spectrometer.

---

[2]The Good Runs List used in this analysis is: data15_13TeV.periodAllYear_DetStatus-v73-pro19-08_DQDefects-00-01-02_PHYS_StandardGRL_All_Good_25ns.xml.

Table 9.1 Number of events which remain after each selection cut. Preselected data was used, where one electron or muon candidate with $p_T > 50$ GeV was required. The electron candidates have to fulfill in addition any of the three likelihood identification criteria

Selection cut	Number of events
Event passes Good Runs List	35,213,358
Veto on noise burst in the electromagnetic calorimeter	35,204,177
Veto on noise burst in the hadronic calorimeter	35,204,177
Veto on incomplete tracking information	35,203,517
Veto on events during trigger restart	35,203,517
Event has at least one vertex with more than two tracks	35,203,276

9.2.3 Electron Selection

Events in the electron selection are required to pass at least one of three triggers which require a single electron.[3] The first trigger requires the electron to fulfill the *medium* likelihood identification criteria and to have $p_T > 24$ GeV. It is seeded by a Level-1 trigger which requires the energy in the hadronic calorimeter behind the electromagnetic cluster to be below a certain threshold at the first trigger stage.[4] This criteria is turned off for energies above 50 GeV. The second trigger requires the electron to fulfill the *medium* likelihood identification criteria and to have $p_T > 60$ GeV. The third trigger requires the electron to fulfill the *loose* likelihood identification criteria and to have $p_T > 120$ GeV. The efficiency of the three triggers with respect to the final signal selection is about 95% at $p_T = 50$ GeV and rises up to 99% for $p_T > 500$ GeV (see Fig. C.1 in the appendix for details). Additional information about the electron identification can be found in Sect. 6.2.2. The E_T requirement of the first trigger stage is 20 GeV for all three triggers.

All electrons are considered which are reconstructed by a reconstruction algorithm which first looks for an energy deposition in the electromagnetic calorimeter and then searches for a track matching to this energy deposition. A more detailed description of the electron reconstruction is given in Sect. 6.2.1. The electron candidates have to be detected in the central detector region of $|\eta| < 2.47$, in order to have tracking information available. The tracking detectors have a coverage up to $|\eta| = 2.5$, the region of $|\eta| < 2.47$ is chosen to ensure that the electromagnetic shower caused by the electron is contained in the region $|\eta| < 2.5$. In addition, electron candidates which are in the transition region $1.37 < |\eta| < 1.52$ between the barrel and endcap electromagnetic calorimeter are rejected, as these candidates have a worse energy resolution. The η information for this cut is chosen to be the η information from the electromagnetic shower in the second layer of the electromagnetic calorimeter,

[3] e24_lhmedium_L1EM20VH or e60_lhmedium or e120_lhloose.

[4] The isolation requires the energy in the hadronic calorimeter in 2×2 cells behind the energy deposition in the electromagnetic calorimeter to be less than 20 GeV – 22 GeV (depending on the region in η).

as energy resolution is the motivation for these restrictions. Electron candidates are rejected that are measured in a detector region which was known to not work properly at that time. This excludes electron candidates in regions where for instance some electronic component was broken or problems with the high-voltage supply occurred. Less than 0.1% of the reconstructed electron candidates are affected by this quality requirement. The p_T threshold for the electron candidates is chosen to be 5 GeV above the threshold at which no isolation is applied at the first trigger stage, i.e., $p_T > 55$ GeV. The cut is chosen to ensure that no threshold effects affect the trigger efficiency. The p_T of the electron candidate is determined by taking the energy measurement from the calorimeter and usually taking the η position from the track measurement. The track of the electron candidate has to have a d_0 significance below 5. The d_0 significance is the distance of the track from the position of the proton beam in the transverse plane divided by its uncertainty. A restriction ensures that the electron candidates are originating from the collision vertex and are not coming from secondary particle decays. More information about the track reconstruction is provided in Sect. 6.1. The efficiency for electrons with $p_T > 55$ GeV from a W decay to fulfill the d_0 significance cut is above 99.8%. All electron candidates have to pass the tight likelihood identification criteria in order to reduce background from other processes faking the electron signature. The efficiency of the identification criteria is about 93, 96 and 92% for electrons with a p_T of 55 GeV, 300 GeV and 2 TeV, respectively. Jets usually have a wider energy deposition in the calorimeter and share their momentum with several tracks. The energy deposition in the calorimeter and the track of the electron candidate are both required to be isolated, in order to further reduce background originating from jets. The sum of the calorimeter transverse energy deposits in the isolation cone of a size $\Delta R = 0.2$ (excluding the electron energy deposition itself) divided by the electron p_T is used as a discrimination criterion. For the track-based isolation, the scalar sum of the p_T of all tracks (excluding the electron track itself) inside a cone with a size of $\Delta R = 10\,\text{GeV}/p_T$ and a maximum value of $\Delta R = 0.2$ around the electron track, divided by the electron p_T has to be below a given cut value. Both, the cut values for calorimeter and track-based isolation criteria are tuned for an overall efficiency of 98% independent of the p_T of the electron [22]. Table 9.2 shows the number of events with at least one electron remaining after each selection cut. The largest event reduction is coming from the requirement of the trigger, the minimum p_T cut and the requirement of the identification criteria.

9.2.4 Muon Selection

Events in the muon selection are required to pass a trigger which requires a muon with $p_T > 50$ GeV. The muon has to have hits in all three stations of the muon trigger. This leads to an efficiency of about 70% in the central barrel region ($|\eta| < 1.05$) and about 80% in the endcap region ($1.05 < |\eta| < 2.4$). The lower efficiency when compared to the electron triggers is coming from a limited coverage and efficiency of the muon trigger chambers. The largest gaps of the muon trigger system are caused by the large

Table 9.2 Number of events with at least one electron remaining after each selection cut

Selection cut	Number of events
After event selection	35,203,276
Event passes trigger requirements	7,920,490
At least one object is reconstructed as an electron candidate by a specific algorithm	7,920,451
At least one electron with $\|\eta\| < 2.47$, which is not in the transition region $1.37 < \|\eta\| < 1.52$	7,803,716
At least one electron fulfilling the object quality check	7,797,886
At least one electron with $p_{\mathrm{T}} > 55$ GeV	5,076,009
At least one electron with d_0 sig. < 5	4,958,497
At least one electron fulfilling the *tight* likelihood identification	3,459,409
At least one electron is fulfilling the isolation requirements	3,159,429

coils of the toroidal magnet and by the feet structure on which the ATLAS detector is placed.

All muon candidates have to be reconstructed by the standard ATLAS muon reconstruction which is described in more detail in Sect. 6.3.1. An explicit cut of $|\eta| < 2.5$ (coverage of the inner detector) is applied to all muon candidates in the analysis while an implicit cut of $|\eta| < 2.4$ is applied by requiring a single muon trigger which has a coverage up to $|\eta| = 2.4$. Special importance in this analysis is given to the quality of the muon candidate. At very high p_{T}, muon candidates will have a very straight track and it becomes difficult to measure precisely the momentum. A good momentum resolution is needed to reconstruct the $E_{\mathrm{T}}^{\mathrm{miss}}$ and m_{T} and therefore see a clear signal. An additional problem are badly reconstructed muon candidates which can fake muons with a very high p_{T} and lead to a mis-balance in the event, causing $E_{\mathrm{T}}^{\mathrm{miss}}$ in the opposite direction of the muon candidate. This would lead to an event with very high m_{T} which is basically indistinguishable from a signal event. Hence, it is of special importance to ensure that the muon is very well measured. All muons are required to fulfill the *high-p_{T}* identification level [23]. It is designed to maximize the momentum resolution of the muon and is described in more detail in Sect. 6.3.2. The p_{T} resolution of muons in the barrel fulfilling the selection is 13, 21 and 24% for muons with a p_{T} of 500 GeV, 1.5 TeV and 3 TeV respectively. This is about a factor of two better compared to the standard muon selection. The requirement of three precision hits in the muon spectrometer instead of two reduces the selection efficiency by about 20% across all p_{T}, but this loss is justified given the substantial improvement in resolution. In order to reduce muons from secondary particle decays, a requirement of d_0 significance < 3 is placed. Cosmic muons can traverse the detector and lead to background which is not covered by the MC simulations. To reject these cosmic muons a cut is placed on the longitudinal distance Δz_0 of the track of the inner detector with respect to the vertex with the highest $\sum p_{\mathrm{T}}^2$. The longitudinal distance Δz_0 is multiplied with $\sin(\theta)$ to avoid rejecting muons with an expected larger error in the

Table 9.3 Number of events with at least one muon remaining after each selection cut

Selection cut	Number of events
After event selection	35,203,276
Event passes trigger requirement	6,691,568
At least one object is reconstructed as a muon candidate by a specific algorithm	6,691,538
At least one muon with $p_{\mathrm{T}} > 55\,\mathrm{GeV}$	4,490,407
At least one muon with $\lvert\eta\rvert < 2.5$ fulfilling the high-p_{T} selection	3,756,572
At least one muon with d_0 sig. < 3	2,998,862
At least one muon with $\lvert\Delta z_0\rvert \sin\theta < 10\,\mathrm{mm}$	2,995,400
At least one muon is fulfilling the isolation requirements	2,131,237

more forward region. A cut of $\lvert\Delta z_0\rvert \sin(\theta) < 10$ mm has been found to sufficiently reject cosmic muons. A track isolation is required in order to reduce background from muons coming from heavy flavor decays in a jet. The sum over the track p_{T}'s in an isolation cone around the muon (excluding the muon itself) divided by the muon p_{T} is required to be below a p_{T} dependent cut, tuned for 99% efficiency. The size ΔR of the isolation cone is defined as 10 GeV divided by the muon p_{T} and has a maximum size of $\Delta R = 0.3$. Table 9.3 shows the number of events with at least one muon remaining after each selection cut. The largest event reduction is coming from the requirement of the trigger, the minimum p_{T} cut and the requirement of the identification and isolation criteria.

9.2.5 *Common Selection*

All events are required to have exactly one muon or electron fulfilling the selection mentioned above. Furthermore, events are vetoed if they contain any additional electron or muon passing a loosened version of the above selections. The p_{T} cut for these additional electrons or muons is reduced to 20 GeV. The electrons are only required to fulfill the *medium* likelihood selection and the muons to pass the *medium* muon selection. The veto on events with additional electrons or muons is placed to reject background arising from top, diboson or Z/γ^* events which can contain multiple electrons or muons. The veto also ensures orthogonality between the electron and muon selection.

The calculation of the missing transverse momentum is based on the selected electrons, muons, photons, τ-leptons and jets which are found in the event. Electrons or muons fulfilling the signal selection are used. Photons with $p_{\mathrm{T}} > 25$ GeV, which fulfill a photon identification[5] requirement and $\lvert\eta\rvert < 2.37$ are selected. Taus with

[5] *Tight*.

Table 9.4 Number of events remaining after the common selection cuts

Selection cut	Number of events	
	$e\nu$ selection	$\mu\nu$ selection
After lepton selection	3159429	2131237
Event passes additional lepton veto	2818769	1789317
Event passes $E_{\mathrm{T}}^{\mathrm{miss}} > 55\,\mathrm{GeV}$	399536	343666
Event passes $m_{\mathrm{T}} > 110\,\mathrm{GeV}$	177592	176801

$p_{\mathrm{T}} > 20\,\mathrm{GeV}$, which fulfill a τ identification requirement[6] and $|\eta| < 2.5$ are selected. Photon and τ candidates which are in the transition region $1.37 < |\eta| < 1.52$ between barrel and endcap are also excluded. The selection of photons and taus has not been further optimized since the typical contribution to the $E_{\mathrm{T}}^{\mathrm{miss}}$ value is very small. Jets used in the $E_{\mathrm{T}}^{\mathrm{miss}}$ calculation are reconstructed using the anti-k_{t} algorithm [24] with a distance parameter of 0.4 and $p_{\mathrm{T}} > 20\,\mathrm{GeV}$. The jet-vertex-tagger technique is used to separate jets from the hard scatter process from pile-up jets in the central region of the detector ($|\eta| < 2.4$). Details about this technique can be found in Sect. 6.4. The value of the tagger has to be below 0.64 for jets with $p_{\mathrm{T}} < 50\,\mathrm{GeV}$. Jets with higher p_{T} are unlikely to originate from pile-up processes. Reconstructed tracks not belonging to any of these physics objects are also added to the value of $E_{\mathrm{T}}^{\mathrm{miss}}$. The contribution from these tracks is called soft term. A cut of $E_{\mathrm{T}}^{\mathrm{miss}} > 55\,\mathrm{GeV}$ is placed on all events to reduce background from processes which do not contain a neutrino. Finally, the transverse mass m_{T} of the event is calculated and required to be above $110\,\mathrm{GeV}$. Table 9.4 shows the number of events remaining after the common selection cuts for both selections.

9.2.6 *Selection Efficiency*

The total rejection of all backgrounds is above 95%, mainly due to the high kinematic cuts of $p_{\mathrm{T}} > 55\,\mathrm{GeV}$, $E_{\mathrm{T}}^{\mathrm{miss}} > 55\,\mathrm{GeV}$ and $m_{\mathrm{T}} > 110\,\mathrm{GeV}$. Detailed tables for the efficiency of each selection step are given for the backgrounds and W' signals with different pole masses in Appendix C. The trigger, identification, and isolation efficiency for both channels as a function of η, ϕ, and p_{T} is shown in the same appendix in Fig. C.1. Figure 9.4 shows the acceptance for the kinematic cuts ($p_{\mathrm{T}} > 55\,\mathrm{GeV}$, $E_{\mathrm{T}}^{\mathrm{miss}} > 55\,\mathrm{GeV}$, $m_{\mathrm{T}} > 110\,\mathrm{GeV}$) and the acceptance times efficiency for the electron and muon channel selection versus the pole mass of a W'. The acceptance, which is shown in black, is defined as all candidates which are generated within the chosen kinematic cuts divided by all generated candidates. The acceptance for a W' with $m_{W'} = 150\,\mathrm{GeV}$ is about 45% due to the stringent kinematic cuts and rises up to about 98% for masses around $m_{W'} = 1.5\,\mathrm{TeV}$. From $2.5\,\mathrm{TeV}$ onwards the acceptance starts to slowly decrease to 85% at $m_{W'} = 6\,\mathrm{TeV}$. The reason for the

[6] *Medium.*

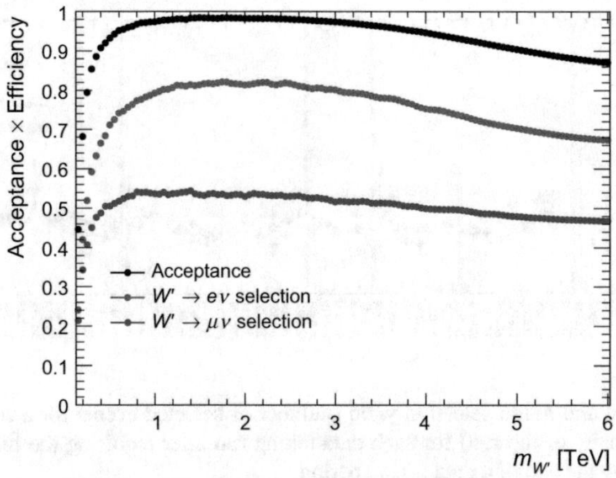

Fig. 9.4 Total signal acceptance (black) and acceptance times efficiency (red, blue) versus SSM W' pole mass for the SSM W' model. The acceptance is defined as the number of generated events within the kinematic cuts ($p_T > 55$ GeV, $E_T^{\text{miss}} > 55$ GeV, $m_T > 110$ GeV) divided by the number of all generated events. The acceptance times efficiency is defined as the number of all selected candidates in the range 110 GeV $< m_T <$ 7000 GeV divided by the number of all generated candidates

drop in acceptance is the contribution of the signal at low transverse mass which becomes more and more pronounced for higher $m_{W'}$ (as shown in Fig. 8.1). The acceptance times efficiency is defined as the number of all selected candidates in the range 110 GeV $< m_T <$ 7000 GeV divided by the number of all generated candidates. The kinematic cuts for both channels are the same, hence the acceptance for both channels is the same and all differences are resulting from differences in the selection efficiency. The efficiency of the muon selection is for most pole masses about 20–30% lower than the efficiency of the electron selection. The main reason for the large differences is the lower trigger efficiency of the muon trigger and the lower efficiency of the muon high-p_T selection. Figure 9.5 shows the yield (number of selected events for a run divided by the integrated luminosity of the run) for the electron and muon selection for each data taking run. The yield should be constant for all data taking runs. A significant difference would indicate a problem during data taking. No significant difference is found for any of the runs.

9.2.7 E_T^{miss}, p_T and m_T Resolution

Figure 9.6 shows the p_T (top left), E_T^{miss} (top right) and m_T (bottom) resolution as a function of their corresponding generated quantity. The resolution was determined by taking the difference between the measured quantity and the generated Born level truth quantity divided by the truth quantity. The Born level truth definition does not consider losses due to the QED final state radiation and Bremsstrahlung. The resulting distribution is therefore non-Gaussian as it has tails from large radiation.

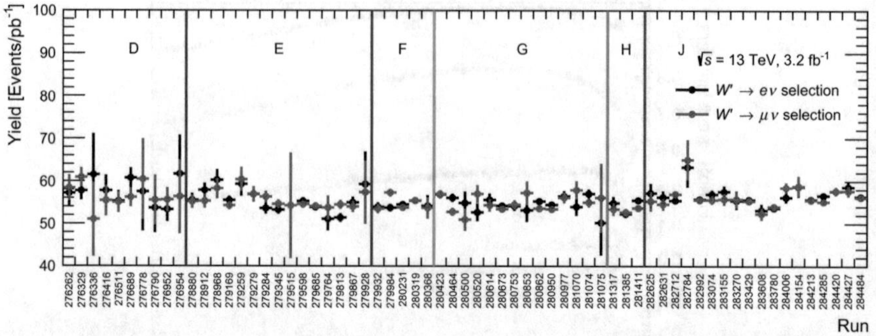

Fig. 9.5 Electron and muon selection yield (number of selected events for a run divided by the integrated luminosity of the run) for each data taking run after requiring the final selection. The vertical lines show the end of a data taking period

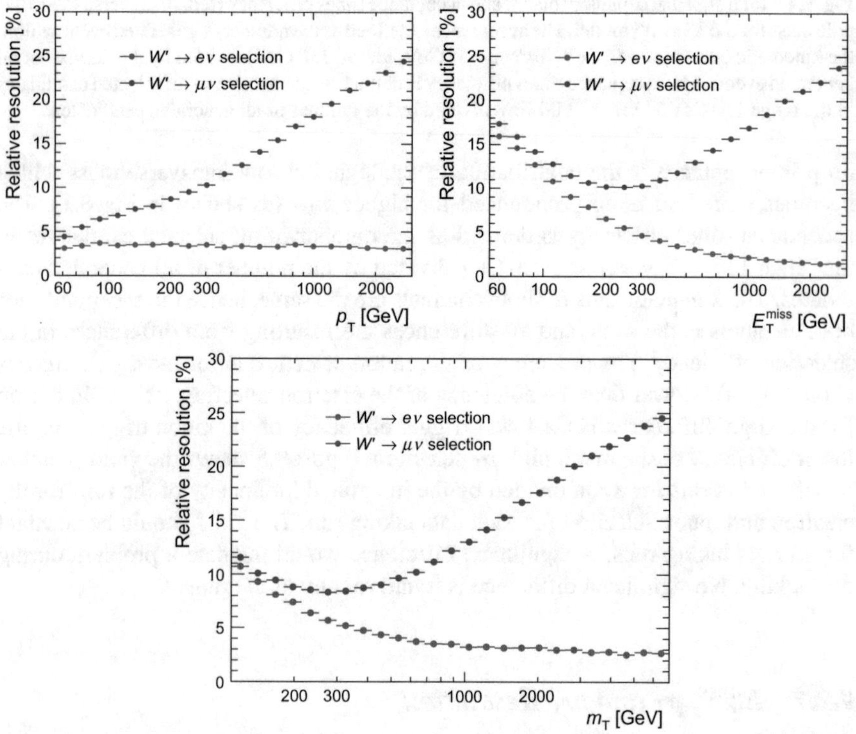

Fig. 9.6 The relative p_T (top left), E_T^{miss} (top right) and m_T resolutions are shown as a function of their generated quantities. Both the muon resolution (blue) and the electron resolution (red) are shown. The resolutions were calculated using the flat W' signal MC sample

The RMS of the distribution is taken as relative resolution. Both, the resolution for the muon channel and the electron channel are shown. They were calculated using the flat W' signal MC sample.

The electron p_{T} resolution is about 3% at 55 GeV and stays constant towards higher p_{T}. The p_{T} resolution is dominated by the relative energy resolution which is getting better at higher energies (see Sect. 5.4.1). The observed resolution is constant since it is defined with respect to the Born level truth p_{T}. Final state radiation and Bremsstrahlung at large angles is not included in the measurement of the electron energy. The probability of Bremsstrahlung is getting larger towards higher p_{T} and therefore the improving energy resolution is canceled by the higher probability of Bremsstrahlung, leading to the observed constant resolution. The muon p_{T} resolution is about 4% at 55 GeV and then rises up to about 24% at 3 TeV. The muon p_{T} measurement fully relies on the track measurement. The higher the muon p_{T} is, the straighter is the muon track. The p_{T} measurement therefore becomes more and more difficult. The effect of defining the resolution with respect to the Born level is small as the probability for a muon to radiate a photon is much smaller than for an electron. The resolution for $E_{\mathrm{T}}^{\mathrm{miss}} = 55$ GeV is about 16% in the electron channel and 17% in the muon channel. For this quantity the resolution of the soft terms and jets limits the resolution. The small difference is coming from the better p_{T} resolution in the electron channel. The resolution of the soft terms and jets becomes less important, the higher the $E_{\mathrm{T}}^{\mathrm{miss}}$ is. The resolution therefore improves. The higher the $E_{\mathrm{T}}^{\mathrm{miss}}$, the more important becomes the measurement of the lepton p_{T}. At very high $E_{\mathrm{T}}^{\mathrm{miss}}$ values the behavior becomes therefore basically the same as for the lepton p_{T}. Both the p_{T} and $E_{\mathrm{T}}^{\mathrm{miss}}$ resolution affect the m_{T} resolution. Both channels have a very similar resolution of about 11–12% at 110 GeV, getting better to about 7% at around 200 GeV in the electron channel and 9% in the muon channel. The p_{T} resolution then takes over and dominates also the m_{T} resolution. At $m_{\mathrm{T}} = 4$ TeV, the electron channel has a very good resolution of about 3%, whereas the muon channel has a much worse resolution of 22%.

9.3 Determination of the Multijet Background

This section describes how the multijet background is estimated. For most background processes the MC samples which were introduced in Sect. 9.1 are used. An additional background arises from misidentified leptons. This background is not well described in MC and has to be measured in data. The method which is used to measure this background is introduced in Sect. 9.3.1 and its implementation for the electron and muon channel is discussed in Sect. 9.3.2 for the former and in Sect. 9.3.3 for the latter.

9.3.1 Matrix Method

This background arising from misidentified leptons is quite different for the two channels. In the electron channel the background consists mainly of jets misidentified as electrons, whereas in the muon channel non-prompt muons from heavy-flavor decays make up a large part. Since the background originates in both channels from QCD final states it will be further denoted as multijet background. Even though the multijet backgrounds in both channels are very different, the matrix method can be used for both of them. The idea of this method is to loosen some of the identification criteria for electrons or muons and measure the efficiency for these objects to pass the signal selection (also denoted as "tight" selection). The efficiency gives a handle on the contribution from misidentified leptons in the signal selection. It is defined for real leptons (called real efficiency r) and fake leptons (called fake efficiency f) separately:

$$r = \frac{N^{\text{real}}_{\text{tight}}}{N^{\text{real}}_{\text{loose}}}, \quad f = \frac{N^{\text{fake}}_{\text{tight}}}{N^{\text{fake}}_{\text{loose}}}. \tag{9.1}$$

$N^{\text{real}}_{\text{loose}}/N^{\text{fake}}_{\text{loose}}$ are the number of real/fake leptons passing the loosened selection and $N^{\text{real}}_{\text{tight}}/N^{\text{fake}}_{\text{tight}}$ are the number of real/fake leptons passing the signal selection. The real efficiency r is usually well described in MC contrary to the fake efficiency f which is typically measured from data in a fake enriched control region.

The true background in a given bin of a distribution can be separated into events from real leptons N_R and events from fake leptons N_F. The lepton identification gives a priori no handle to estimate these truth quantities. The number of events in a given bin passing the loosened selection can be split into the number of events N_L failing the signal selection and the number of events N_T passing the signal selection. The real and fake efficiencies provide a connection between these truth quantities and measurable quantities:

$$\begin{pmatrix} N_T \\ N_L \end{pmatrix} = \begin{pmatrix} r & f \\ 1-r & 1-f \end{pmatrix} \begin{pmatrix} N_R \\ N_F \end{pmatrix} \tag{9.2}$$

The relevant part for the measurement of the multijet background is given in the first line of the matrix equation:

$$N_T = rN_R + fN_F\,.$$

The events passing the signal selection N_T are composed of a part originating from real leptons rN_R and a part originating from fake leptons fN_F. An equation for the truth quantities can be obtained by inverting the matrix

$$\begin{pmatrix} N_R \\ N_F \end{pmatrix} = \frac{1}{r(1-f)-f(1-r)} \begin{pmatrix} 1-f & -f \\ r-1 & r \end{pmatrix} \begin{pmatrix} N_T \\ N_L \end{pmatrix}$$

An equation for the number of fake leptons which pass the signal selection follows then by insertion:

$$fN_F = \frac{f}{r-f}\left(r(N_L + N_T) - N_T\right). \tag{9.3}$$

This equation only contains the measurable quantities N_T and N_L and the measurable efficiencies r and f, hence can be used to compute the multijet background. The fake efficiencies and real efficiencies depend in general on kinematic properties like p_{T} or η of the lepton. They can therefore be binned in these variables to take these dependencies into account. The background will in this case be calculated on an event by event basis. The estimation of the multijet method becomes more and more stable the larger the gap in discrimination between the loosened selection and the signal selection is. It has to be always ensured that the signal selection is a subset of the loosened selection. This requirement leads to the need of slightly modifying the $E_{\mathrm{T}}^{\mathrm{miss}}$ definition for the computation of the multijet background. As only leptons passing the signal selection are added to the $E_{\mathrm{T}}^{\mathrm{miss}}$ calculation, the resulting $E_{\mathrm{T}}^{\mathrm{miss}}$ value of an event passing the loose selection differs whether a lepton is passing the signal selection or not. Therefore, the signal selection might not be a subset of the loose selection as it will end up in a different bin of, e.g., m_{T} or $E_{\mathrm{T}}^{\mathrm{miss}}$. Hence, all leptons passing the loosened selection are added to the computation of the $E_{\mathrm{T}}^{\mathrm{miss}}$ value for the multijet background.

9.3.2 *Multijet Background in the Electron Channel*

A typical contribution to the multijet background in the electron channel is coming from misidentified light-flavor jets. These jets typically contain a lot of charged and neutral pions. The neutral pions decay mainly via $\pi^0 \to \gamma\gamma$. At the energies of the LHC these decay products are usually highly boosted and can lead to a narrow electromagnetic energy deposition. The track of the charged pions can be associated with the energy deposition of the neutral photons and therefore result in a jet being misidentified as an electron. A powerful criterion to reject this kind of background is the association of the track and the energy deposition and isolation criteria. Further backgrounds can arise from photons converting into an e^+e^--pair in one of the inner most layers of the tracking detectors and electrons from secondary particle decays. The majority of the multijet background events will have low values of $E_{\mathrm{T}}^{\mathrm{miss}}$. Hence, the region at low $E_{\mathrm{T}}^{\mathrm{miss}}$ values can be used as a control region as the multijet background is strongly enhanced. Large values of $E_{\mathrm{T}}^{\mathrm{miss}}$ can appear if the energy of the jet is mismeasured.

The loosened selection in the electron channel is restricted by the trigger requirements. Two different regions are defined since triggers with two different identification levels are used. The selection is loosened to the *medium* likelihood identification level in the region 55 GeV $< p_T <$ 125 GeV and to the *loose* likelihood identification level for $p_T \geq$ 125 GeV. No isolation criteria is applied in the loosened selection.

9.3.2.1 Measurement of the Real Efficiency

The real efficiency $r = N^{\text{real}}_{\text{tight}}/N^{\text{real}}_{\text{loose}}$ is measured from MC, since it is usually well modeled in the simulation. The same efficiency corrections as discussed in Sect. 6.2.5 are applied to account for small differences of the identification and isolation efficiencies between data and MC. Hence, the real efficiency measured from MC is effectively matched to the real efficiency which would be measured in data. The W background MC provides a large sample of real electrons which can be used to measure r. The electron is required to be reconstructed within a cone of $\Delta R < 0.2$ around the generated electrons to avoid dilution from misidentified jets. The real efficiency binned in η and p_T is shown in Fig. 9.7. The efficiency rises from 94% at $p_T = 55$ GeV to about 97% for $p_T > 100$ GeV before it slightly drops again for $p_T > 300$ GeV. A variation from 92.5 to 96% for the region $p_T < 125$ GeV and from 93 to 98% for the region $p_T \geq 125$ GeV is observed for different detector regions in η. The statistics of the MC sample does not allow to extract the real efficiencies in fine bins of η and p_T. Therefore, the real efficiencies are binned in p_T for three different detector regions: $0.0 < |\eta| < 1.37$ (central calorimeter), $1.52 < |\eta| < 2.01$ (coverage of the TRT), $2.01 < |\eta| < 2.47$. The real efficiencies as they are applied to the data are shown in Fig. 9.8.

9.3.2.2 Measurement of the Fake Efficiency

The fake efficiency $f = N^{\text{fake}}_{\text{tight}}/N^{\text{fake}}_{\text{loose}}$ cannot be reliably calculated with simulation and therefore needs to be measured in data. The measurement can be performed

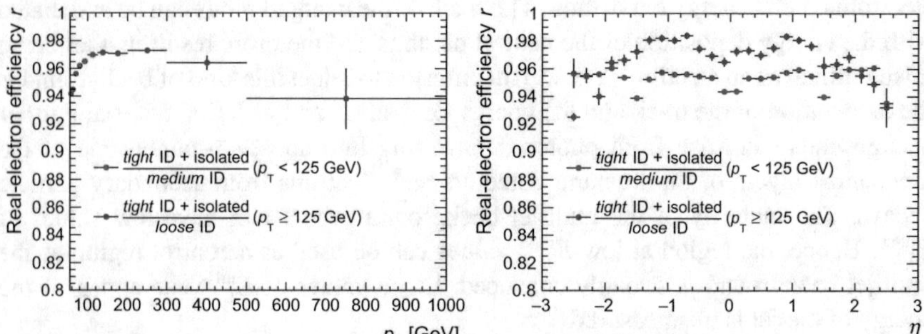

Fig. 9.7 Real-electron efficiency $r = N^{\text{real}}_{\text{tight}}/N^{\text{real}}_{\text{loose}}$ with its statistical uncertainties as a function of p_T (left) and η (right) determined from the W MC samples. The real-electron efficiency is shown separately for the region $p_T < 125$ GeV (red) and $p_T \geq 125$ GeV (blue)

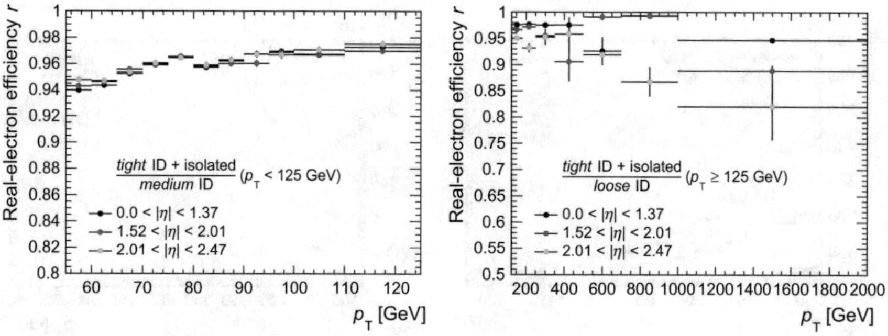

Fig. 9.8 Real-electron efficiency $r = N^{\mathrm{real}}_{\mathrm{tight}}/N^{\mathrm{real}}_{\mathrm{loose}}$ with its statistical uncertainties as a function of p_{T} for three different detector regions in $|\eta|$. The real-electron efficiency was determined from the W MC samples and is shown for the region $p_{\mathrm{T}} < 125$ GeV on the left side and for the region $p_{\mathrm{T}} \geq 125$ GeV on the right side

in a fake enriched control region which is in addition chosen to be orthogonal to the signal selection. The multijet background is distributed mainly at low values of $E_{\mathrm{T}}^{\mathrm{miss}}$. Hence, the $E_{\mathrm{T}}^{\mathrm{miss}}$ cut of the analysis is inverted by requiring $E_{\mathrm{T}}^{\mathrm{miss}} < 55$ GeV to define this fake enriched control region. The m_{T} cut is removed to further increase the background contribution. The remaining dilution from real electrons in this control region is mainly coming from the W and Z/γ^* processes. The dilution from Z/γ^* can be reduced by applying a veto on all events in which two electron candidates with $p_{\mathrm{T}} > 20$ GeV fulfill the *medium* likelihood identification criteria or fulfill the *loose* likelihood identification and have an invariant mass of 66 GeV $< m_{ee} <$ 116 GeV. The remaining dilution from real electrons can be corrected using MC. The dominant background arises from a di-jet topology. To further increase this contribution it is required that at least one jet with $p_{\mathrm{T}} > 40$ GeV, not overlapping with the electron candidate ($\Delta R_{e,jet} > 0.2$), is found in the event. All additional cuts are the same as in the signal selection.

The dilution, binned in p_{T}, η and $E_{\mathrm{T}}^{\mathrm{miss}}$ is shown for the denominator $N^{\mathrm{fake}}_{\mathrm{loose}}$ in Fig. 9.9 and for the numerator $N^{\mathrm{fake}}_{\mathrm{tight}}$ in Fig. 9.10. For the *loose* identification selection a cut of $p_{\mathrm{T}} \geq 125$ GeV is applied whereas for the *medium* identification selection a cut of $p_{\mathrm{T}} > 55$ GeV is applied.

The dilution from real electrons in the four regions is very different. The *loose* identification selection shows as expected the least dilution from around 6% in the endcap region to around 30% in the central region (also varying with p_{T} and $E_{\mathrm{T}}^{\mathrm{miss}}$). This behavior is expected since the dominant part of the background is di-jet production which is to a large extend a t-channel process and is therefore more pronounced in the forward direction. The dilution is higher in the *medium* likelihood identification selection, varying between around 30% in the endcap region and around 65% in the central region. The dilution is, as expected, the highest in the signal selection, as it has the highest background rejection. The contribution from real electrons is ranging

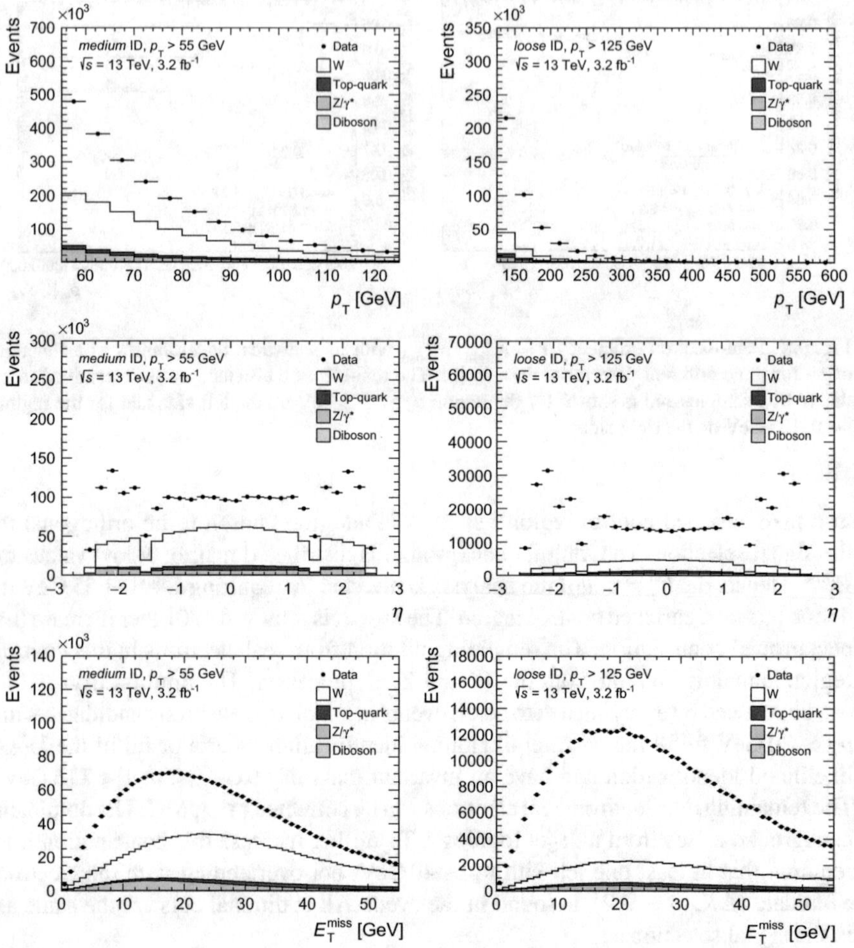

Fig. 9.9 Distribution of data with its statistical uncertainties, binned in η, p_T and E_T^{miss}, in the fake enriched control region for the two denominator categories. The left row shows the distributions for the region $p_T > 55$ GeV and the right row for the region $p_T > 125$ GeV. The sources of real electron dilution are added on top of each other

from around 45% up to 80% for the region $p_T > 55$ GeV. It is slightly decreased from around 35% up to 65% in the region $p_T \geq 125$ GeV.

The fake efficiencies, corrected for the real electron contamination, are shown in Fig. 9.11 as a function of η, p_T, E_T^{miss} and $|\Delta\phi_{e,E_T^{miss}}|$. The fake efficiency does depend on all these observables. In p_T it varies from about 50% at $p_T = 55$ GeV to about 63% at $p_T = 125$ GeV. From $p_T = 125$ GeV on it drops to about 20% and stays constant up to very high p_T. The drop at $p_T = 125$ GeV is caused by the looser selection in the denominator. The fake efficiencies also depend on η as different detector regions have different discrimination power. It also varies strongly in $|\Delta\phi_{e,E_T^{miss}}|$, the angle

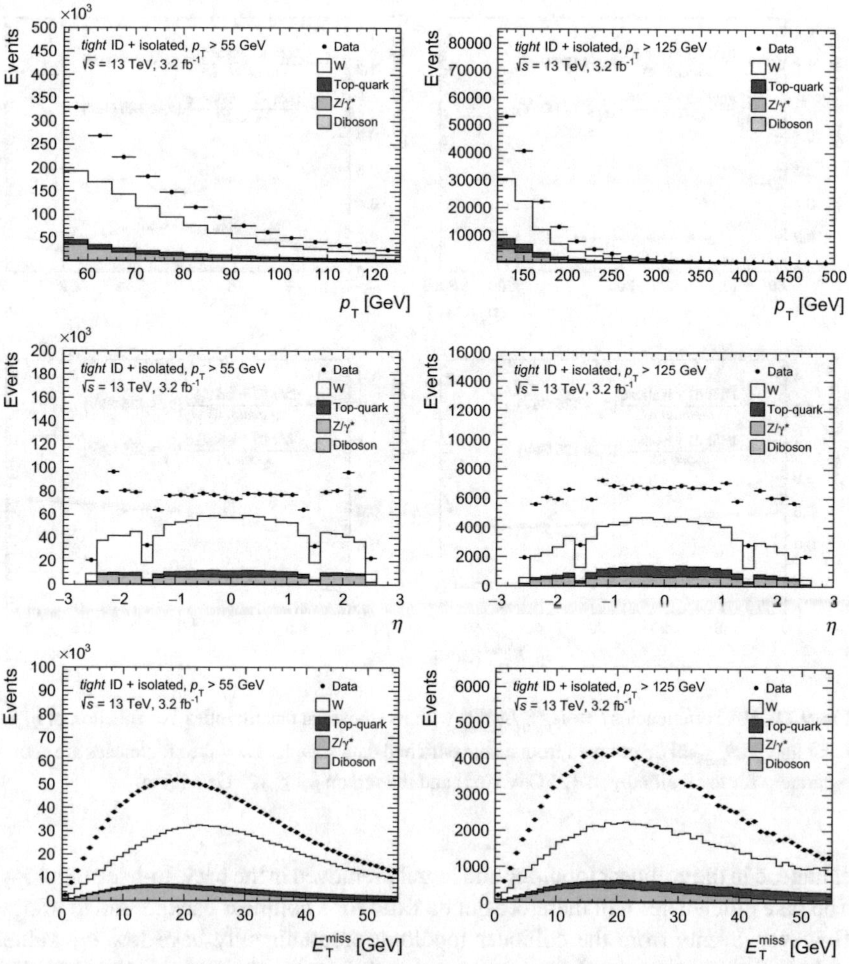

Fig. 9.10 Distribution of data with its statistical uncertainties, binned in η, p_T and E_T^{miss}, in the fake enriched control region for the two numerator categories. The left row shows the distributions for the region $p_T > 55$ GeV and the right row for the region $p_T > 125$ GeV. The sources of real electron dilution are added on top of each other

between the electron and E_T^{miss} in the transverse plane. The collinear topology (low $|\Delta\phi_{e,E_T^{miss}}|$ values) has a much smaller fake efficiency than the back-to-back topology ($|\Delta\phi_{e,E_T^{miss}}| \approx \pi$). The fake efficiencies can be binned in these variables to account for the dependencies. Problematic is only the dependency in E_T^{miss}. The fake efficiencies are measured in the region $E_T^{miss} < 55$ GeV and applied in the region $E_T^{miss} > 55$ GeV. In order for this transition from the control region into the signal region to be valid, a constant fake efficiency is needed. Figure 9.12 shows the E_T^{miss} dependency of the fake efficiencies in a back-to-back (left) and collinear topology (right). The dependency is

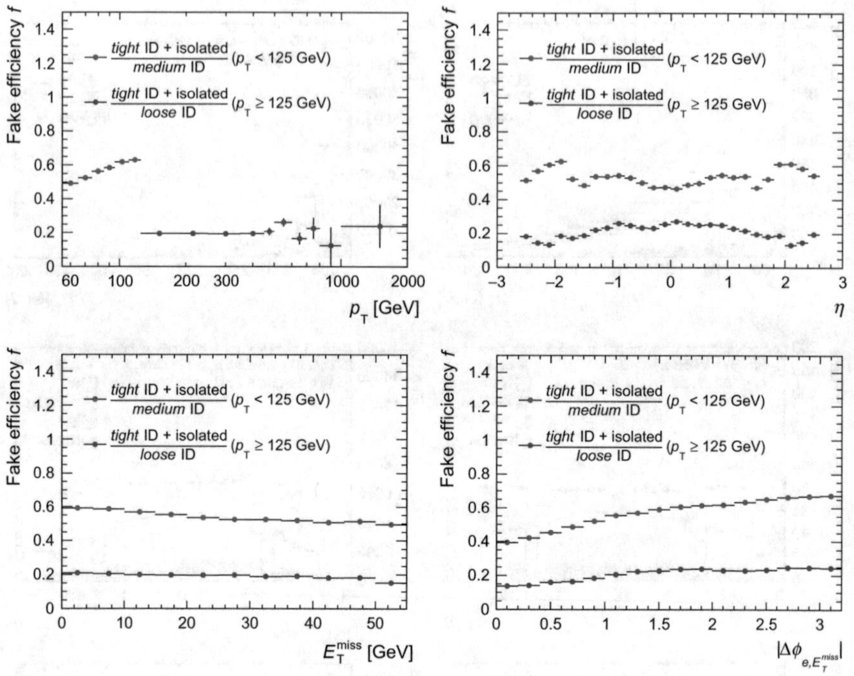

Fig. 9.11 Fake efficiencies $f = N_{\text{tight}}^{\text{fake}}/N_{\text{loose}}^{\text{fake}}$ with its statistical uncertainties as a function of η, p_{T}, $E_{\text{T}}^{\text{miss}}$ and $|\Delta\phi_{e,E_{\text{T}}^{\text{miss}}}|$ determined from a fake enriched data sample. The fake efficiencies are shown separately for the region $p_{\text{T}} < 125$ GeV (red) and the region $p_{\text{T}} \geq 125$ GeV (blue)

enhanced in the collinear topology and largely removed in the back-to-back topology. The fake efficiencies will therefore not be valid for a collinear background topology. However, events from the collinear topology predominantly have low m_{T} values and are therefore removed from the signal region by requiring $m_{\text{T}} > 110$ GeV. The remaining $E_{\text{T}}^{\text{miss}}$ dependency for the back-to-back topology will be addressed with a systematic uncertainty.

The statistics of the data is not good enough to calculate the fake efficiencies in all three variables simultaneously. Instead two fake efficiencies are calculated, the first one binned in $|\Delta\phi_{e,E_{\text{T}}^{\text{miss}}}|$ and p_{T} and the second one binned in $|\Delta\phi_{e,E_{\text{T}}^{\text{miss}}}|$ and $|\eta|$. The former fake efficiency is shown in Fig. 9.13 and the latter in Fig. 9.14. The average of those two fake efficiencies is calculated and applied to data. The fake efficiencies are in general larger for higher values of $|\Delta\phi_{e,E_{\text{T}}^{\text{miss}}}|$ and in the region $p_{\text{T}} < 125$ GeV slightly increasing towards higher $|\eta|$ values. They range from about 35% to about 75% binned in $|\eta|$. A slight increase of f with increasing p_{T} is observed. The p_{T} binned fake efficiencies range from about 35% to about 80%. In the region $p_{\text{T}} \geq 125$ GeV, the behavior changes and smaller values are measured for large $|\eta|$. The fake efficiencies are here in general smaller, ranging from about 10% to about 35%. No strong p_{T} dependence is observed besides for $p_{\text{T}} > 500$ GeV, where these

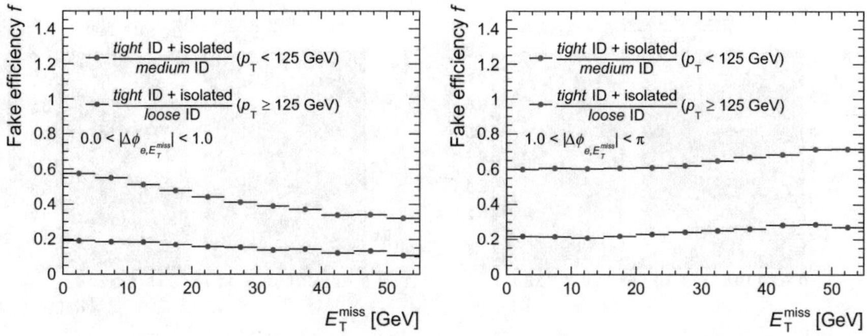

Fig. 9.12 Fake efficiency $f = N^{\text{fake}}_{\text{tight}}/N^{\text{fake}}_{\text{loose}}$ with its statistical uncertainties as a function of $E_{\text{T}}^{\text{miss}}$ in two different regions of $\Delta\phi_{e,E_{\text{T}}^{\text{miss}}}$. The fake efficiencies are shown separately for the region $p_{\text{T}} < 125\,\text{GeV}$ (red) and the region $p_{\text{T}} > 125\,\text{GeV}$ (blue)

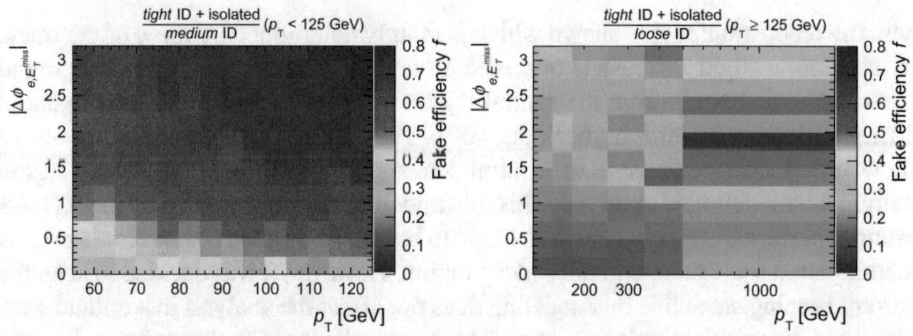

Fig. 9.13 Fake efficiency as a function of p_{T} in different $|\Delta\phi_{e,E_{\text{T}}^{\text{miss}}}|$ bins, determined from a fake enriched data sample. The fake efficiency for the region $p_{\text{T}} < 125\,\text{GeV}$ is shown on the left hand side and for $p_{\text{T}} \geq 125\,\text{GeV}$ on the right hand side

are assumed to come from the limited statistical precision. The fake efficiency binned in p_{T} ranges from 13 to 30% for $p_{\text{T}} < 500\,\text{GeV}$ and from about 5% to about 40% for $p_{\text{T}} > 500\,\text{GeV}$.

9.3.2.3 Validation of the Multijet Background

The validity of the estimated multijet background can be studied in a control region in which this background is enhanced. The $E_{\text{T}}^{\text{miss}}$ and m_{T} cuts are released in order to significantly increase the background contribution from multijet events. Figure 9.15 shows the η, ϕ, p_{T}, $E_{\text{T}}^{\text{miss}}$, $p_{\text{T}}/E_{\text{T}}^{\text{miss}}$ and m_{T} distributions in this control region. A more or less flat distribution is observed in data as a function of η, with dips around $|\eta| \approx 1.4$, which corresponds to the transition region between central and endcap electromagnetic calorimeter. A small number of electrons are still observed in this region, as the cut is placed on the η position of the electromagnetic shower and

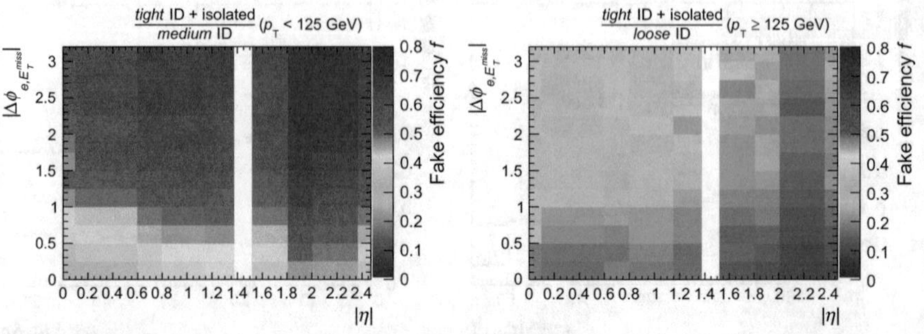

Fig. 9.14 Fake efficiency as a function of $|\eta|$ in different $|\Delta\phi_{e,E_{\mathrm{T}}^{\mathrm{miss}}}|$ bins, determined from a fake enriched data sample. The fake efficiency for the region $p_{\mathrm{T}} < 125$ GeV is shown on the left hand side and for $p_{\mathrm{T}} \geq 125$ GeV on the right hand side

here the reconstructed η is shown which is mainly determined by the η of the track. In addition, a slight increase is observed around $|\eta| \approx 2$. The multijet background contributes to about 25% in the central region of the detector. It increases to about 50% contribution towards higher values of η. The data and background expectations show very good agreement in the central detector region while in the endcap region a slight mismodeling is observed. This mismodeling has been studied by the ATLAS electron performance group and was found to be caused by a too coarse binning of the identification scale factors in the endcap region. However, it was decided to keep the current binning since this mismodeling does not affect the analysis in a critical way. The ϕ spectrum is flat for data and for all backgrounds, showing that no large detector effects cause a local increase of the multijet background due to a higher misidentification rate. However, a slight mismodeling is observed in the region $-2.0 < \phi < 0.0$ which is caused by systematic effects from calorimeter shape distortions on the track matching between track and energy deposition. Since the identification efficiency corrections are averaged over ϕ, it does not correct the ϕ dependence. The p_{T} spectrum is steeply falling towards higher values of p_{T}. The ratio between data and background expectation is slightly increasing in the region 55 GeV $< p_{\mathrm{T}} <$ 125 GeV. This trend is caused by the strong p_{T}-dependence of the fake efficiency in that region. The dependency is not fully taken into account as the fake efficiency binned in p_{T} is averaged with the η dependent fake efficiency. For $p_{\mathrm{T}} \geq 125$ GeV an excellent agreement is observed. The data distribution peaks at $E_{\mathrm{T}}^{\mathrm{miss}} \approx 25$ GeV, at the same place where the W background is peaking. The p_{T} values of the neutrino from the W decay are usually peaking at $\approx$40 GeV, half the mass of the W-boson. The position of the peak is shifted towards lower values of $p_{\mathrm{T}} \approx 25$ GeV by imposing a cut of $p_{\mathrm{T}} > 55$ GeV on the electron. The multijet background peaks at event smaller $E_{\mathrm{T}}^{\mathrm{miss}}$ values around 20 GeV. Here the contribution from the multijet background is about 60%, while it is much smaller for higher values of $E_{\mathrm{T}}^{\mathrm{miss}}$. The multijet distribution

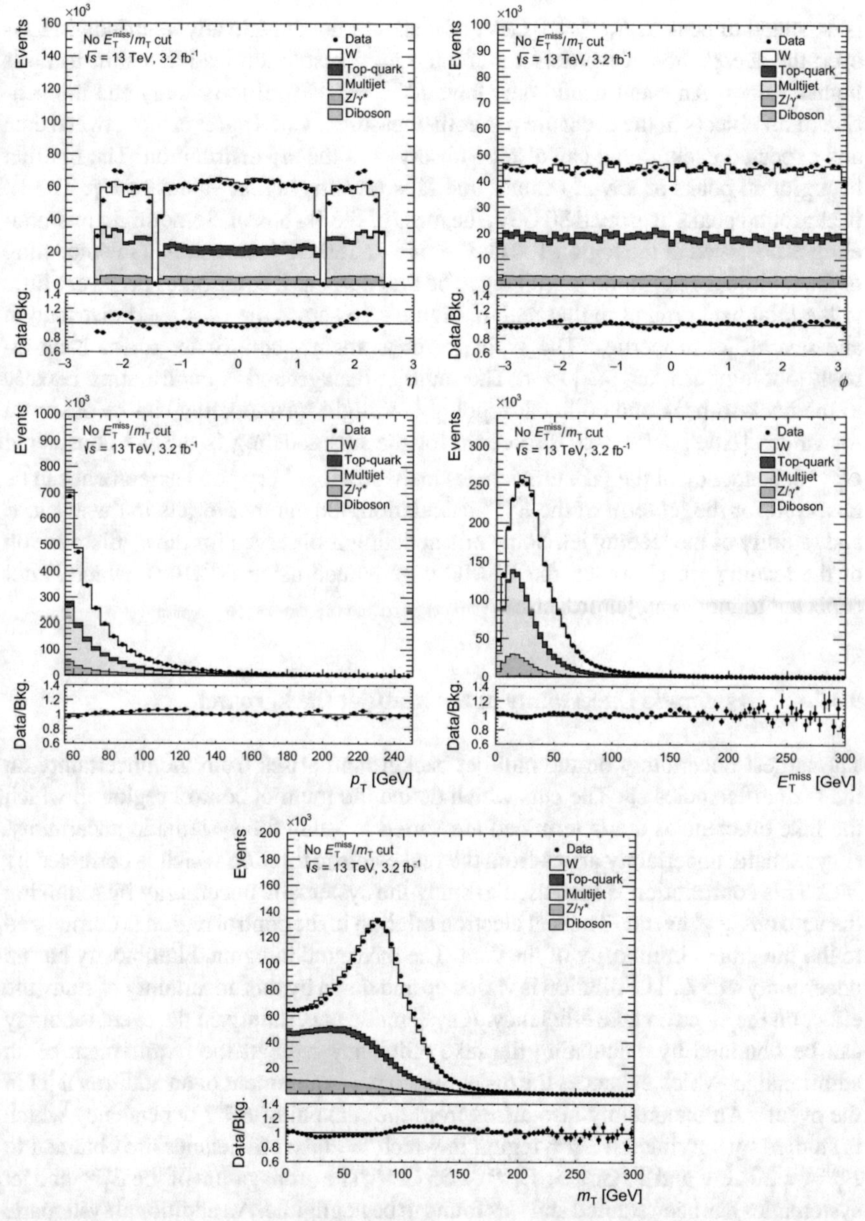

Fig. 9.15 The η, ϕ, p_T, E_T^{miss}, and m_T distributions for events satisfying all selection criteria for the electron multijet control region (E_T^{miss} and m_T cut released). The distributions are compared to the stacked sum of all expected backgrounds

is expected to peak at $E_{\mathrm{T}}^{\mathrm{miss}} = 0\,\mathrm{GeV}$, as it should not contain any neutrinos.[7] However, the acceptance of the detector and the $E_{\mathrm{T}}^{\mathrm{miss}}$ resolution lead to a shift towards higher values. An event would only have $E_{\mathrm{T}}^{\mathrm{miss}} = 0\,\mathrm{GeV}$ if the energy and momentum of all objects in the event are perfectly measured. Good agreement between data and expected background can also be observed in the m_{T} distribution. The multijet background peaks at low m_{T} values and falls towards higher values of m_{T}. The W background peaks at around 80 GeV, the mass of the W-boson. Some slight mismodeling is observed in the region $100\,\mathrm{GeV} < m_{\mathrm{T}} < 180\,\mathrm{GeV}$. However, a mismodeling of the multijet background is unlikely to be the cause, as it contributes only very little to the total background in that region. Figure 9.16 shows the $|\Delta\phi_{e,E_{\mathrm{T}}^{\mathrm{miss}}}|$ distribution and several jet properties. The W background mainly contributes to the back-to-back topology at $|\Delta\phi_{e,E_{\mathrm{T}}^{\mathrm{miss}}}| \approx \pi$. The multijet background is contributing equally to the back-to-back and collinear topology. A slight mismodeling can be observed for values $|\Delta\phi_{e,E_{\mathrm{T}}^{\mathrm{miss}}}| < 0.5$. The cause for the mismodeling is the not considered $E_{\mathrm{T}}^{\mathrm{miss}}$ dependency of the fake efficiencies in that region. Very good agreement can be observed for the jet term of the $E_{\mathrm{T}}^{\mathrm{miss}}$ calculation, the number of jets in the event, ϕ and rapidity of the leading jet. Some mismodeling is observed for the p_{T} distribution of the leading jet. However, the W MC is generated using POWHEG, which is not expected to model all jet properties.

9.3.2.4 Systematic Uncertainty of the Multijet Background

The largest uncertainty on the multijet background arises from the uncertainty on the fake efficiencies ϵ_F. The cuts which define the multijet control region in which the fake efficiencies are determined are varied to study the systematic uncertainty. A systematic uncertainty arises from the real electron dilution which is corrected by MC. This contribution is increased to study the systematic uncertainty by removing the veto on Z/γ^* events. The real electron dilution in the control region is normalized to the integrated luminosity of the data. The measured integrated luminosity has an uncertainty of 5%. The dilution is varied up and down by this uncertainty to study the effect on the electron fake efficiency. A systematic uncertainty on the event topology can be obtained by calculating the fake efficiency without the requirement of an additional jet which enhances the di-jet topology (requirement of an additional jet in the event). An uncertainty also arises from the remaining $E_{\mathrm{T}}^{\mathrm{miss}}$ dependency which is studied by varying the $E_{\mathrm{T}}^{\mathrm{miss}}$ region in which the fake efficiencies are obtained to $E_{\mathrm{T}}^{\mathrm{miss}} < 20\,\mathrm{GeV}$ and $20\,\mathrm{GeV} < E_{\mathrm{T}}^{\mathrm{miss}} < 55\,\mathrm{GeV}$. The propagation of the $E_{\mathrm{T}}^{\mathrm{miss}}$ and jet systematics has been studied and was found to be negligible. An additional systematic uncertainty arises from not fully taking into account the p_{T} and η dependency. The effect on the background is studied by not averaging the fake efficiency binned in $|\Delta\phi_{e,E_{\mathrm{T}}^{\mathrm{miss}}}| - p_{\mathrm{T}}$ and $|\Delta\phi_{e,E_{\mathrm{T}}^{\mathrm{miss}}}| - |\eta|$ but applying only one of them. Figure 9.17 shows the effect of these variations on the calculated fake efficiencies binned in p_{T}, η,

[7] Jets can also contain neutrinos from heavy-flavor decays, but this is not expected to be the dominant background in this case.

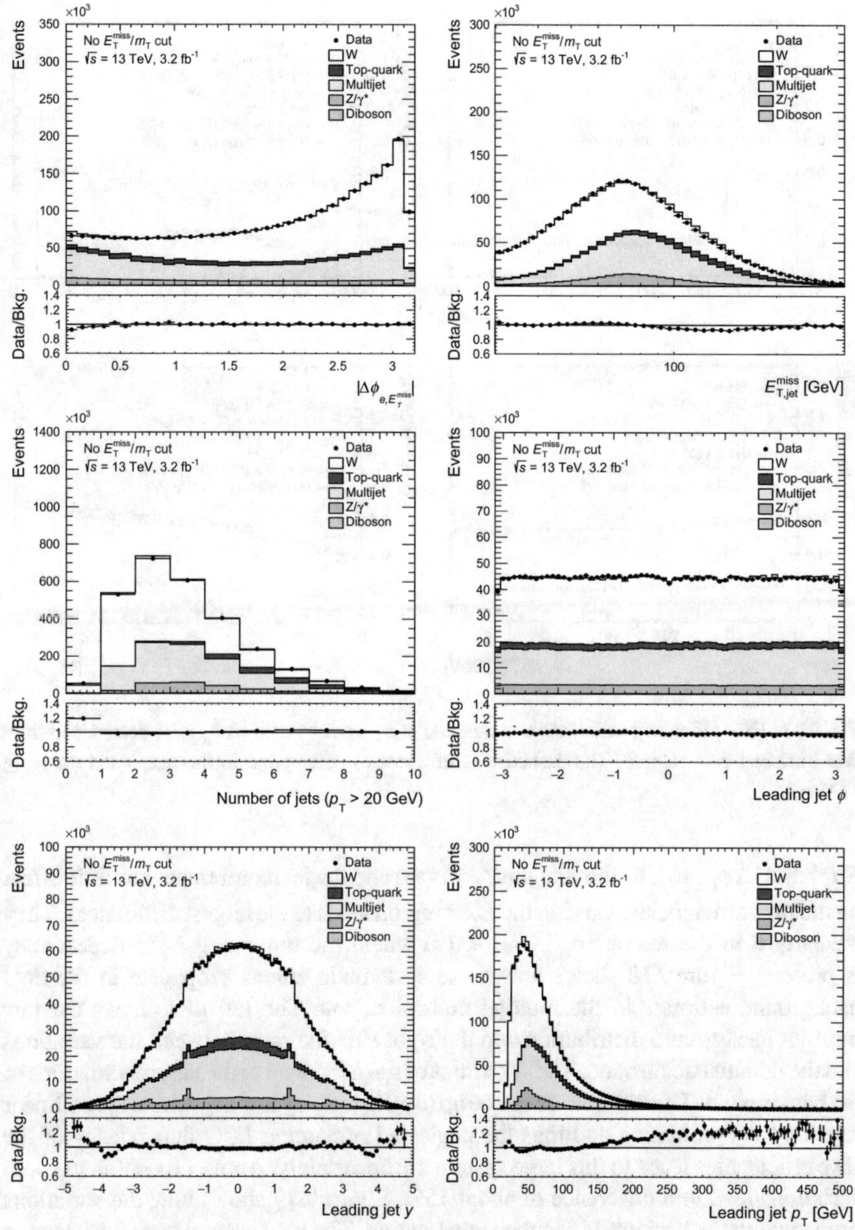

Fig. 9.16 Shown are the distributions for $|\Delta\phi_{e,E_T^{miss}}|$, $E_{T,jet}^{miss}$, the number of jets with $p_T > 20$ GeV, and ϕ, y, p_T of the leading jet for events satisfying all selection criteria for the electron multijet control region (E_T^{miss} and m_T cut released). The distributions are compared to the stacked sum of all expected backgrounds

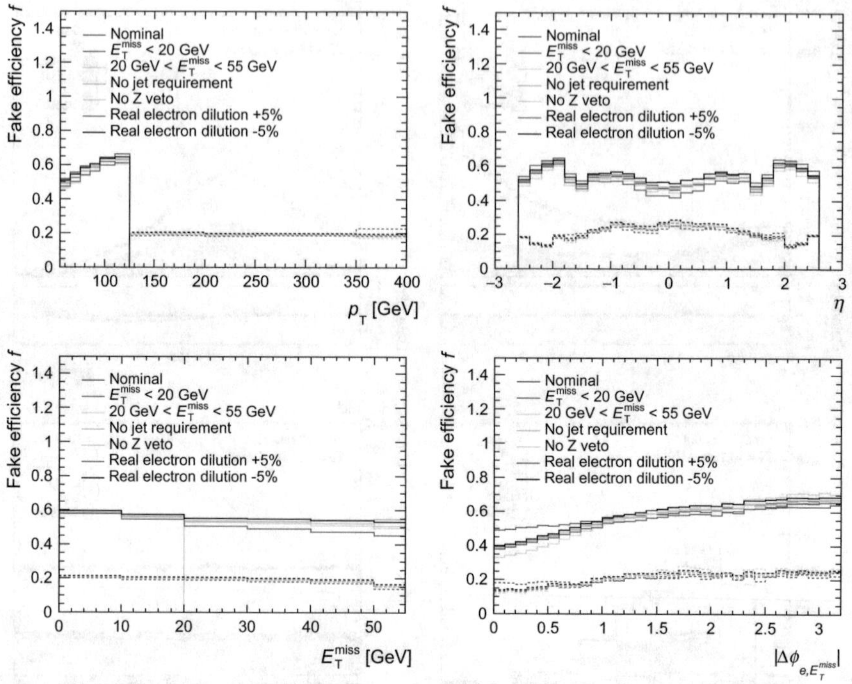

Fig. 9.17 Fake efficiency variations as a function of p_T, η, E_T^{miss} and $|\Delta\phi_{e,E_T^{miss}}|$ determined from fake enriched data sample. The dashed fake efficiencies correspond to the ones used for $p_T \geq$ 125 GeV

E_T^{miss} and $|\Delta\phi_{e,E_T^{miss}}|$. Removing the Z/γ^* veto and the jet requirement has little effect on the fake efficiencies. Varying the E_T^{miss} region leads to the largest differences. They are largest in the region $|\Delta\phi_{e,E_T^{miss}}| < 1.0$ where the remaining E_T^{miss} dependency is present. Figure 9.18 shows how these systematic effects propagate to the final background estimate in the multijet control region. The left plot shows the raw multijet background distribution and the right side the ratio between the variations and the default background. The variations of the E_T^{miss} lead to the largest variations of the background. The differences can be up to 50% for very low m_T, where the collinear topology is dominating and thus the neglected remaining E_T^{miss} dependency of the fake efficiencies leads to this large systematic uncertainty. Above 60 GeV in m_T both variations lead to a difference of about 15%. Figure 9.19 shows how the variations propagate to the background in the signal region. The background variations lead to differences up to 20% for $m_T = 110$ GeV. Larger effects are observed at very high m_T, but here the statistical uncertainty of the background is already very large. Taking the variations in the signal region into account, a conservative 25% uncertainty is estimated on the background yield. The uncertainty is taken to be constant over the whole m_T range.

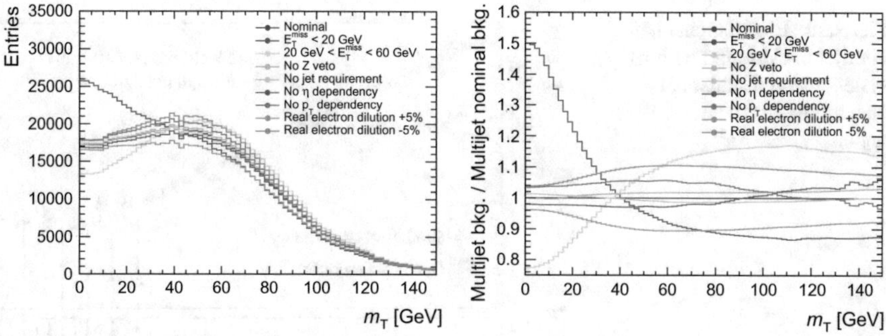

Fig. 9.18 Multijet background variations for the different estimated fake efficiency variations in the multijet control region

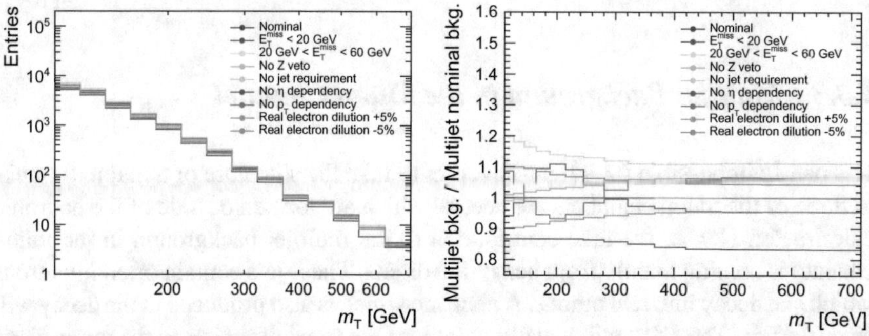

Fig. 9.19 Multijet background estimates for the fake efficiency variations in the signal region

9.3.2.5 Summary

The multijet background was calculated using the matrix method. Figure 9.20 shows the transverse mass spectrum of the background estimate and its systematic and statistical uncertainties. The small bump around $m_T \approx 1.2$ TeV has been studied and it was found that it is a consequence of the low statistics in that region. In some bins at high m_T a negative background expectation is predicted. This unphysical result can occur in the case of low statistics bins. Events which pass the signal selection contribute, according to Eq. 9.3, to the background yield with a negative sign. In case of low statistics, it can happen that all events passing the loosened selection also pass the signal selection. In that case a negative background estimation is predicted. Hence, the estimate of the matrix method becomes unstable in the regime of low statistics since for an accurate estimate both samples N_T and N_L in a bin need to have sufficient statistics. Therefore the background will be extrapolated to obtain an estimate in the high m_T region. The extrapolation will be discussed in Sect. 9.4.

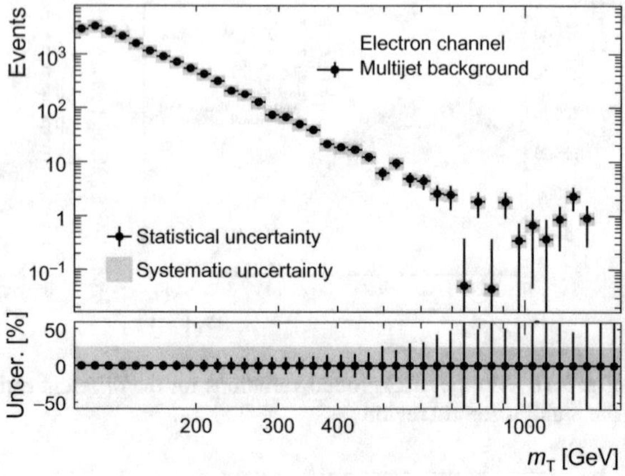

Fig. 9.20 Electron channel multijet background with its systematic and statistical uncertainty in the signal region

9.3.3 *Multijet Background in the Muon Channel*

It is nearly impossible for a light-flavor jet to fake the signature of a muon as tracks in three of the muon chambers are needed which are located outside of the hadronic calorimeter. Hence, the fake contribution to the multijet background in the muon channel is coming mainly from heavy-flavor jets. These jets contain often b-hadrons which can decay into real muons. A neutrino which is also produced in the decay will lead to E_T^{miss}. The E_T^{miss} will usually point into the same direction as the muon since the b-hadrons are highly boosted at the LHC energies and the decay products are therefore collimated. This background will rise for low p_T and low E_T^{miss} values. Thus, the region at low E_T^{miss} can, as in the electron channel, be used as a control region in which the multijet background is enhanced. The overall contribution of the multijet background to the total background is expected to be very small due to the lower cross section of processes including heavy-flavor jets. The multijet background is therefore of much lower importance in the muon channel than in the electron channel.

Muons from jets are usually rejected by requiring the muon to be isolated. Hence, the loosened selection is defined by removing the isolation requirement.

9.3.3.1 Measurement of the Real Efficiency

The real efficiency $r = N_{\text{tight}}^{\text{real}}/N_{\text{loose}}^{\text{real}}$ is, like for the electron case, measured from the W MC. Efficiency corrections are applied to the MC. Muons from jets are discarded by placing a $\Delta R < 0.2$ requirement. The real efficiency binned in p_T (left hand side) and η (right hand side) is shown in Fig. 9.21. The real efficiency is always above 98.8% which corresponds approximately to the targeted isolation efficiency of at least 99%. The efficiency rises with p_T to nearly 100%. No strong dependence of the muon real efficiency on η is observed. The real efficiencies are therefore only binned in p_T.

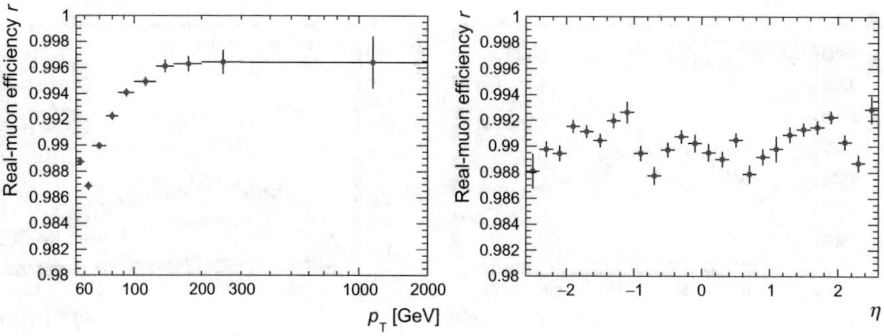

Fig. 9.21 Real-muon efficiency r as a function of p_{T} and η determined from W MC samples

9.3.3.2 Measurement of the Fake Efficiency

The fake efficiency ϵ_F cannot be reliably calculated with simulation and therefore needs to be measured in a fake enriched control region from data. The control region is defined in a similar way as for the electron channel. The multijet background is distributed mainly at low values of $E_{\mathrm{T}}^{\mathrm{miss}}$. Hence, the $E_{\mathrm{T}}^{\mathrm{miss}}$ cut of the analysis is inverted by requiring $E_{\mathrm{T}}^{\mathrm{miss}} < 55$ GeV, to define this fake enriched control region. The m_{T} cut is released to further increase this background contribution. The remaining dilution from real muons in this control region is mainly coming from the W and Z/γ^* processes. The dilution from Z/γ^* can be reduced by applying a veto on all events in which two muon candidates with $p_{\mathrm{T}} > 20$ GeV fulfill the standard selection and have an invariant mass which lies in a window of $66\,\mathrm{GeV} < m_{\mu\mu} < 116\,\mathrm{GeV}$. Heavy-flavor jet tagging relies on the fact that the tracks are misplaced with respect to the vertex since b-hadrons usually have a longer time of flight. This results in a larger d_0 significance. A requirement on the d_0 significance of $|d_0|/\sigma_{d_0} > 1.5$ is placed to further enrich the sample with heavy-flavor jets. The $E_{\mathrm{T}}^{\mathrm{miss}}$ originating from the neutrino from the b-meson decay is usually pointing in the same direction as the muon. Hence, $|\Delta\phi_{e,E_{\mathrm{T}}^{\mathrm{miss}}}|$ is required to be smaller than 0.5. The remaining dilution from real muons is corrected using MC. The dominant background arises from the di-jet heavy-flavor topology. To further increase this topology it is required that a jet with $p_{\mathrm{T}} > 40$ GeV, not overlapping with the muon candidate ($\Delta R_{\mu,jet} > 0.2$), is found in the event. All additional cuts are the same as in the signal selection.

Figure 9.22 shows the real muon dilution in the multijet control region binned in $E_{\mathrm{T}}^{\mathrm{miss}}$. The loosened selection is shown on the left hand side and the signal selection on the right hand side. The dilution is about 12% in the loosened selection and about 63% in the signal selection. The real muon dilution corrected fake efficiency is shown as a function of muon p_{T} (left) and $E_{\mathrm{T}}^{\mathrm{miss}}$ (right) is shown in Fig. 9.23. The fake efficiency for muons is about 6% for a p_{T} of 55 GeV and then rises up to 17% for $p_{\mathrm{T}} > 100$ GeV. No clear trend, like in the electron channel, is observed for the fake efficiency binned in $E_{\mathrm{T}}^{\mathrm{miss}}$. The fake efficiency can therefore be applied in the

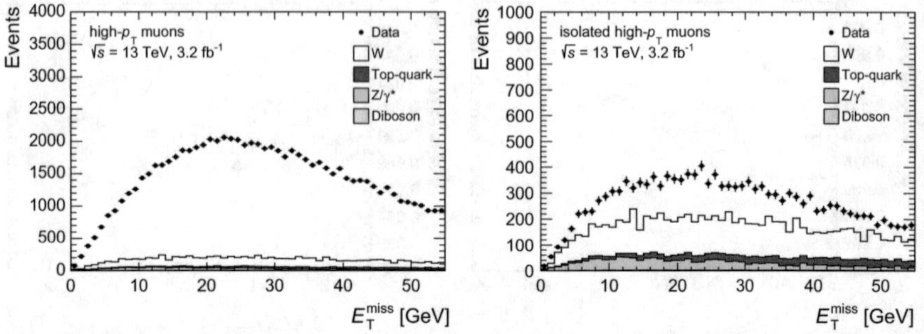

Fig. 9.22 Distribution of data with its statistical uncertainties, binned in E_T^{miss}, in the fake enriched control region used to measure the muon fake efficiency. The left side shows the denominator N_{loose}^{fake} and the right side the numerator N_{tight}^{fake}. The sources of real muon dilution are added on top of each other

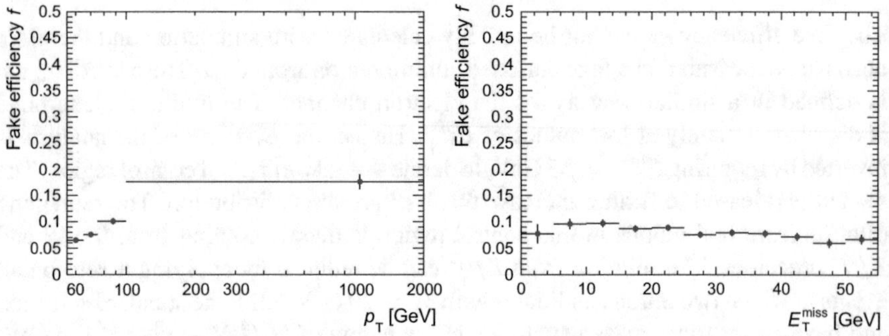

Fig. 9.23 Muon fake efficiency as a function of muon p_T and E_T^{miss}, determined from the fake enriched data sample

signal region. A potential remaining dependency will be addressed with a systematic uncertainty.

9.3.3.3 Validation of the Multijet Background

The validity of the estimated multijet background can be studied in a control region in which this background is enhanced. The E_T^{miss} and m_T cuts are released in order to significantly increase the background contribution from multijet events. Given that the multijet contribution in the muon channel is very small, additional cuts of $|\Delta\phi_{\mu,E_T^{miss}}| < 0.5$ and $|d_0|/\sigma_{d_0} > 2$ are applied to further increase the background. Figure 9.24 shows the η, ϕ and E_T^{miss} distributions in this control region. The data in the central η region is, besides a few dips, approximately flat and is dropping towards higher values. The dip in the region $\eta \approx 0$ is caused by a gap for cables which are connected to the calorimeters and the inner detector. The efficiency is

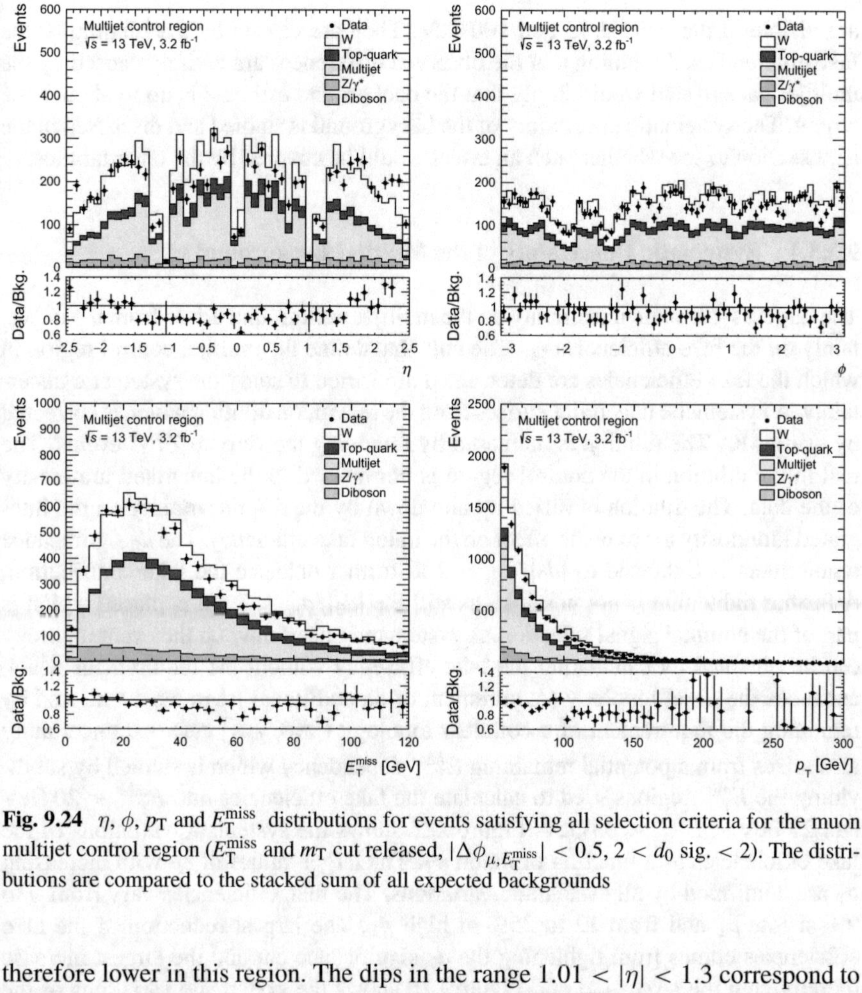

Fig. 9.24 η, ϕ, p_{T} and $E_{\mathrm{T}}^{\mathrm{miss}}$, distributions for events satisfying all selection criteria for the muon multijet control region ($E_{\mathrm{T}}^{\mathrm{miss}}$ and m_{T} cut released, $|\Delta\phi_{\mu,E_{\mathrm{T}}^{\mathrm{miss}}}| < 0.5$, $2 < d_0$ sig. < 2). The distributions are compared to the stacked sum of all expected backgrounds

therefore lower in this region. The dips in the range $1.01 < |\eta| < 1.3$ correspond to the regions which are vetoed in the high-p_{T} muon identification criteria. The contribution of the multijet background is in the control region about 50%. Data and expected background agree well in the region $|\eta| > 1.5$, an offset of about 20% is observed in the region $|\eta| < 1.5$. The ϕ spectrum is approximately flat. Some dips are observed in the region $\phi \approx -1$ and $\phi \approx -2$ which are caused by the structure on which the ATLAS experiment is placed. Here muon chambers are missing and leading therefore to a lower efficiency in this region. Data and expected background agree within 15%. The lower left plot shows the $E_{\mathrm{T}}^{\mathrm{miss}}$ spectrum which has a maximum at around 25 GeV. The multijet background contributes up to 60% of the total background. Good agreement between data and expected background can be observed for low values of $E_{\mathrm{T}}^{\mathrm{miss}}$. Some differences of up to about 20% are observed in the region $E_{\mathrm{T}}^{\mathrm{miss}} > 30$ GeV. The right plot shows the p_{T} spectrum which is steeply falling towards higher values of p_{T}. A similar level of multijet contribution and agreement with data as in the $E_{\mathrm{T}}^{\mathrm{miss}}$ spectrum is also observed in the p_{T} spectrum. Some steps are visible in the ratio at 75 and 100 GeV. They are caused by the binning of the fake efficiencies. Assuming that the observed differences are entirely caused by the multijet background would imply that the background estimate is up to about 40% wrong. The systematic uncertainty of the background is studied and discussed in the next section to see whether such an effect would be covered by the uncertainties.

9.3.3.4 Systematic Uncertainty of the Multijet Background

The largest systematic uncertainty on the multijet background arises from the uncertainty on the fake efficiencies ϵ_F. The cuts that define the multijet control region in which the fake efficiencies are determined are varied to study the systematic uncertainty. A systematic uncertainty arises from the real muon dilution which is corrected by using MC. The dilution is increased by removing the veto on Z/γ^* events. The real muon dilution in the control region is normalized to the integrated luminosity of the data. The dilution is varied up and down by the 5% uncertainty on the integrated luminosity to study the effect on the muon fake efficiency. The d_0 significance requirement is tightened to $|d_0|/\sigma_{d_0} > 2$ to further enhance the fake contribution. A further tightening is not possible, as still the $|d_0|/\sigma_{d_0} < 3$ cut is placed which is part of the nominal signal selection. A systematic uncertainty on the event topology can be obtained by calculating the fake efficiency without the requirement which enhances the di-jet topology (requirement of an additional jet in the event) and by removing the requirement of a collinear topology ($|\Delta\phi_{e,E_{\mathrm{T}}^{\mathrm{miss}}}|$ cut). An uncertainty also arises from a potential remaining $E_{\mathrm{T}}^{\mathrm{miss}}$ dependency which is studied by subdividing the $E_{\mathrm{T}}^{\mathrm{miss}}$ regions used to calculate the fake efficiencies into $E_{\mathrm{T}}^{\mathrm{miss}} < 20$ GeV and 20 GeV $< E_{\mathrm{T}}^{\mathrm{miss}} < 55$ GeV. Figure 9.25 shows the systematic variations of the fake efficiencies as a function of muon p_{T}. The larger values of ϵ_F with increasing p_{T} are confirmed by all systematic variations. The fake efficiencies vary from 6 to 9% at low p_{T} and from 12 to 25% at high p_{T}. The largest reduction of the fake efficiencies comes from tightening the d_0 significance cut and the largest increase by removing the $|\Delta\phi_{e,E_{\mathrm{T}}^{\mathrm{miss}}}|$ cut. Figure 9.26 shows the systematic variations of the multijet background as a function of transverse mass m_{T} in the signal region. The left hand side shows the total background and the right hand side the ratio of the variations with the nominal background. The background is strongly falling towards higher values of m_{T}. All variations show a very similar shape. The largest systematic variation leads to differences on the yield of the background of about 43% at low m_{T} and up to about 58% at higher m_{T}. Taking the variations in the signal region into account, a conservative 60% uncertainty is estimated on the background yield. The uncertainty is assumed to be flat in m_{T}. The uncertainty covers also the differences which were observed in the multijet control region in Sect. 9.3.3.3.

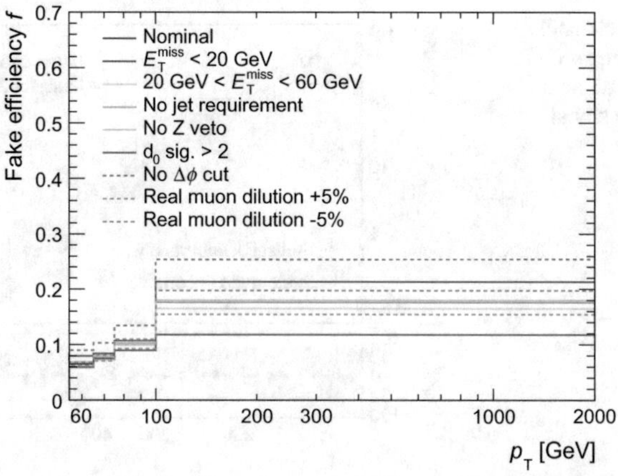

Fig. 9.25 Fake efficiency variations as a function of p_T determined from fake enriched data sample

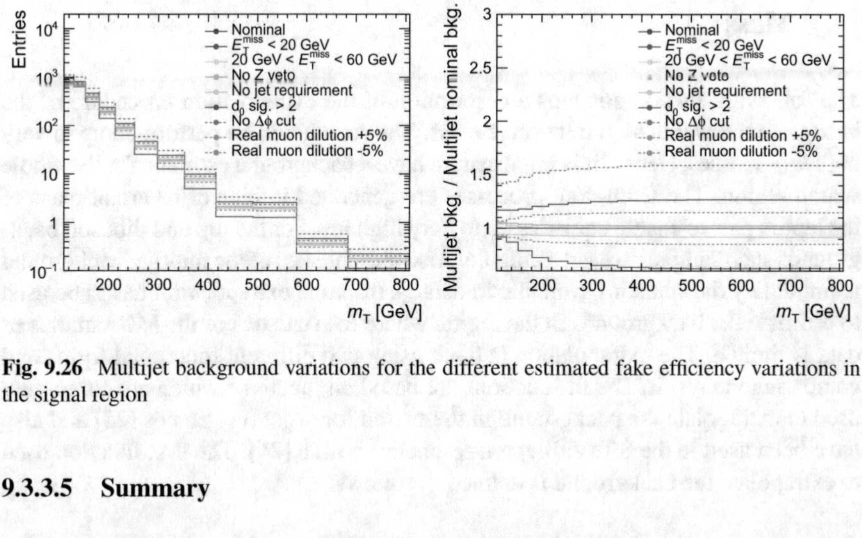

Fig. 9.26 Multijet background variations for the different estimated fake efficiency variations in the signal region

9.3.3.5 Summary

The multijet background was calculated using the matrix method. Figure 9.27 shows the transverse mass spectrum of the background estimate and its systematic and statistical uncertainties. The systematic uncertainty was estimated to be 60% on the background yield. The same negative predictions as in the electron channel are observed at high m_T due to the same reasons. The background can be extrapolated from the low and medium m_T range towards higher values of m_T to obtain a prediction in the region with low statistics. The methodology how this extrapolation is performed is discussed in the next section.

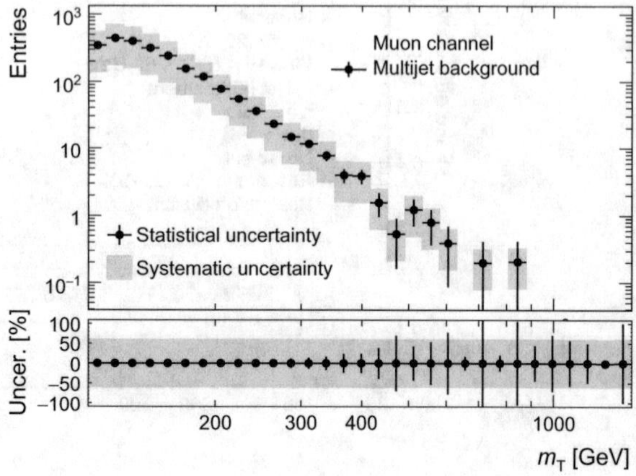

Fig. 9.27 Muon channel multijet background with its systematic and statistical uncertainty in the signal region

9.4 Extrapolation of the Background to High Transverse Mass

The following section contains a description of the extrapolation procedure of the background towards high transverse mass. The search will be performed up to very high $m_{W'}$ masses. Hence, it is important to have a background estimate for the whole search region. The Drell–Yan processes are generated in bins of invariant mass of the lepton pair to ensure statistics up to very high m_T. For the top and diboson backgrounds such samples are not available. Also the statistics of the multijet background is limited by the statistics available in data. A fit-based extrapolation has to be used to estimate the backgrounds in the region where the statistics of the MC samples or data is limited. The extrapolation is done using two different functional forms and comparing the result. The fit functions are based on functions which are commonly used to extrapolate the background in the search for di-jet resonances [25] and also have been used in the 8 TeV dilepton resonance search [26]. The first function used to extrapolate the background is defined as follows:

$$f(m_T) = e^{-a} m_T^b m_T^{c \log(m_T)}. \tag{9.4}$$

The second function is a modified power law function:

$$f(m_T) = \frac{a}{(m_T + b)^c}. \tag{9.5}$$

Several fits are performed with both functions and varying start and end point of the fit range. The fit with the best $\chi^2/N.d.o.f$ of all fits is taken as central value for the extrapolation and the envelope of all fits as systematic uncertainty for the extrapolation. The statistical uncertainty of the fit parameters was found to be negligible.

Table 9.5 Range and increment for the starting and end point of the fit range for the background extrapolation

Background channel	Top-quark		Diboson		Multijet	
	Electron	Muon	Electron	Muon	Electron	Muon
m_{T}^{min} range [GeV]	140–200	140–200	120–240	120–240	140–240	140–240
$\Delta m_{\mathrm{T}}^{min}$ [GeV]	20	20	20	20	20	20
m_{T}^{max} range [GeV]	600–900	600–900	500–700	800–900	800–1000	500–700
$\Delta m_{\mathrm{T}}^{max}$ [GeV]	25	25	25	25	50	50

Table 9.5 shows the range for the starting point m_{T}^{min} and end point m_{T}^{max}. The start and end points have been varied in steps of $\Delta m_{\mathrm{T}}^{min}$ and $\Delta m_{\mathrm{T}}^{max}$, respectively. The parameters were chosen in a way that they lead to a reasonable description of the background and a reasonable systematic uncertainty coming from the fits. The extrapolated backgrounds are used in all cases for $m_{\mathrm{T}} > 600$ GeV.

All fits used for the extrapolation of the three backgrounds are shown in Fig. 9.28. The top-quark and diboson backgrounds are of the same size in the electron and muon channel at low m_{T}. This is different for the multijet background, which is about one order of magnitude larger in the electron channel. All backgrounds fall steeply over several orders of magnitude towards very high m_{T}. The relative background uncertainty on the background yield is above 100% starting from about 2 TeV (4 TeV), 1 TeV (3 TeV), 2 TeV (2.5 TeV), for the top-quark, diboson, and multijet background in the electron (muon) channel, respectively.

9.5 Systematic Uncertainties

This section lists and discusses all systematic uncertainties which affect this analysis. The systematic uncertainties can be subdivided into the experimental uncertainties for the selected electrons, muons, the $E_{\mathrm{T}}^{\mathrm{miss}}$ value, luminosity, and the backgrounds, and uncertainties from theoretical predictions. First, all experimental sources are discussed. Thereafter follows a discussion of the sources for the theoretical uncertainties. The chapter ends with a discussion of all sources and their relative size on the total background estimate.

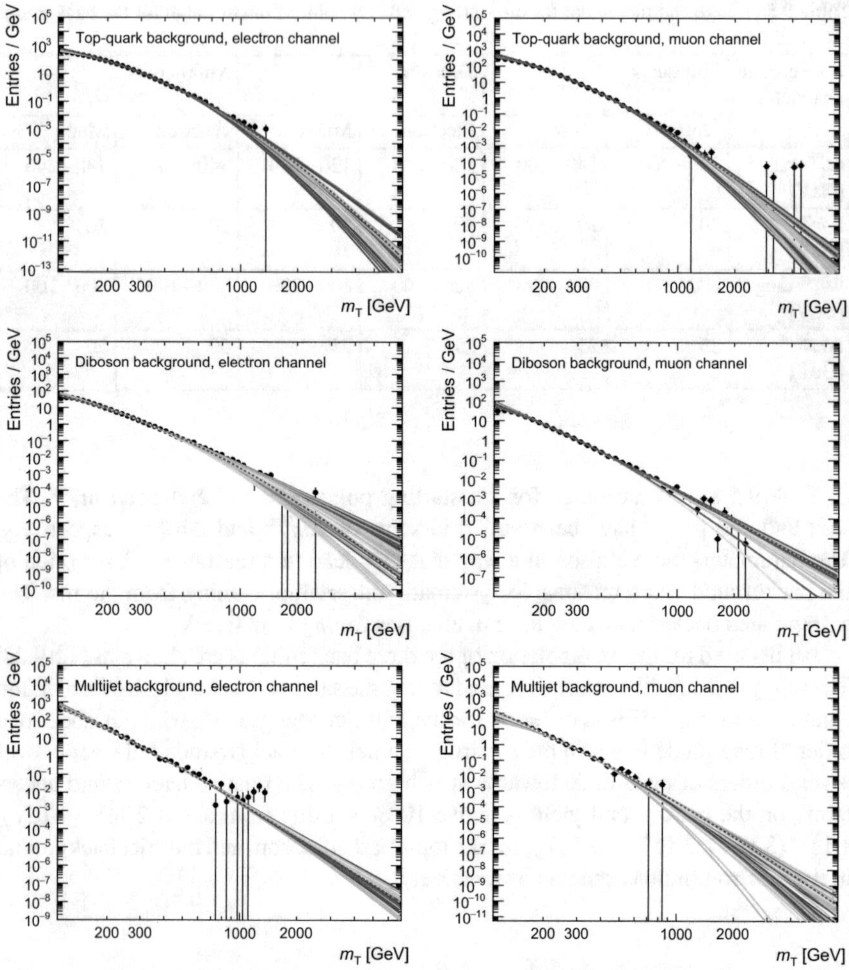

Fig. 9.28 All fits performed to extrapolate the top-quark (top), diboson (middle), and multijet (bottom) backgrounds are shown on the left side for the electron channel and on the right side for the muon channel

9.5.1 Experimental Uncertainties

9.5.1.1 Electron Uncertainties

Electron efficiencies The efficiency corrections provided by the ATLAS electron performance group come with systematic uncertainties [27]. The uncertainties are obtained by varying the tag and probe selection, e.g., identification of the tag electron or the window around the Z-peak, or varying the background model. They are available separately for the reconstruction, identification, trigger and isolation efficiency correction. The full methodology of the tag and probe method and the different systematic sources can be found in Ref. [27]. However, the identification efficiency is only measured precisely up to a p_T of 150 GeV. The effect of extrapolating to higher values of p_T was found to be 2.5% and was estimated by extrapolating the shower shapes to high p_T and by varying the shower shapes for which differences are observed between data and MC. A similar study was performed for the isolation efficiency corrections where the effect was found to be 2% for $p_T > 150$ GeV.

Electron resolution Differences between MC and data in the electron energy resolution are handled by smearing the electron energies in MC. The ATLAS electron performance group provides uncertainties for the smearing. The full correlation model for this uncertainty consists of several nuisance parameters where all sources of uncertainties have been decorrelated in η-bins. A simplified correlation model is used in this analysis which provides one nuisance parameter for the energy resolution. In this simplified model all the effects are considered fully correlated in η and they are summed in quadrature. With this simplification the total effect is usually increased. The full methodology is documented in Ref. [28].

Electron energy scale Corrections for the energy scale of the electrons are applied to data. The effect of varying the respective uncertainties of the corrections, which are provided by the ATLAS electron performance group, up and down is checked in order to determine the systematic uncertainty. This is done in MC, given the higher statistics available. The full correlation model for the corrections consists of 60 nuisance parameters for which several effects have been decorrelated in eta-bins. A simplified correlation model is used in this analysis which provides one nuisance parameter for the energy scale. In this simplified model all the effects are considered fully correlated in η and are summed in quadrature which increases the total effect. The full methodology is documented in Ref. [28].

9.5.1.2 Muon Uncertainties

Muon efficiencies The muon efficiency corrections are provided by the ATLAS muon performance group [23] and have been obtained using the tag and probe method on $Z \rightarrow \mu\mu$ and $J/\psi \rightarrow \mu\mu$ decays in data. Systematic uncertainties have been derived from variations of the tag and probe selection and background subtraction following the methodology documented in Ref. [23]. These uncertainties have been propagated to the signal region. The uncertainties are available separately for the reconstruction, isolation and trigger efficiency correction. The effect of extrapolating the efficiencies to high p_T was studied and a systematic uncertainty assigned, corresponding to the magnitude of the drop in the muon reconstruction and selection efficiency with increasing p_T that is predicted by MC. For the isolation efficiency an additional uncertainty of 5% is assigned for $p_T > 500$ GeV.

Muon momentum resolution The muon momentum corrections are provided by the ATLAS muon performance group [23] and have been obtained by fitting certain correction constants to match the invariant mass distribution in $Z \rightarrow \mu\mu$ and $J/\psi \rightarrow$

$\mu\mu$ decays in MC to that observed in data. The dependence of the muon momentum on the fit parameters is given by a model in which each parameter is associated to a certain source of potential data/MC disagreement. Systematic uncertainties are derived from variations of the fit procedure, the background parameterization, and the muon spectrometer alignment.

9.5.1.3 Jet Uncertainties

Jet energy scale The jet energy scale and resolution uncertainties enter the analysis through the $E_{\mathrm{T}}^{\mathrm{miss}}$ calculation. The uncertainties for the jet energy scale and resolution are provided by the ATLAS jet performance group [29, 30]. A reduced set of uncertainties with three nuisance parameters is chosen for the jet energy scale. This reduced set of nuisance parameters simplifies the correlations between the different sources of the jet energy scale uncertainty. Four scenarios of correlation models are provided. The final result of an analysis using the reduced set must not depend on a specific choice of correlation model. The jet energy scale uncertainty has been tested for all scenarios and no difference was found.

Jet energy resolution The jet energy resolution agrees between data and MC within the estimated uncertainty [31]. Hence, no resolution correction is applied to MC. However, the uncertainties on the resolution are propagated to the $E_{\mathrm{T}}^{\mathrm{miss}}$ calculation. All jet uncertainties are treated as fully correlated between the electron and muon channel.

9.5.1.4 $E_{\mathrm{T}}^{\mathrm{miss}}$ Uncertainties

The systematic uncertainties related to $E_{\mathrm{T}}^{\mathrm{miss}}$ come from both the calculation of the contribution of tracks not belonging to any physics objects in the $E_{\mathrm{T}}^{\mathrm{miss}}$ calculation and directly from the measurements of the physics objects. The jet, electron and muon energy/momentum uncertainties are affecting the $E_{\mathrm{T}}^{\mathrm{miss}}$ calculation in this way. The uncertainties for the $E_{\mathrm{T}}^{\mathrm{miss}}$ scale and resolution arising are provided by the ATLAS $E_{\mathrm{T}}^{\mathrm{miss}}$ performance group [32]. They enter the analysis through the track-based soft term in the $E_{\mathrm{T}}^{\mathrm{miss}}$ calculation. The uncertainties were studied by quantifying the agreement of the balance between soft term and muon p_{T} in $Z \rightarrow \mu\mu$ events. The uncertainties are decomposed into resolution components which are longitudinal and transverse to the p_{T} of the muons and to a component for the overall scale of the $E_{\mathrm{T}}^{\mathrm{miss}}$ value. All $E_{\mathrm{T}}^{\mathrm{miss}}$ uncertainties are treated as fully correlated between the electron and muon channel.

9.5.1.5 Multijet Background Uncertainty

The multijet background is estimated in both channels by the matrix method. Uncertainties on the measured fake efficiencies are propagated to the final background to study the uncertainty on the background estimate in the signal region. The estimation of the multijet uncertainties is described in detail in the corresponding sections (Sect. 9.3.2.4 for the electron channel and Sect. 9.3.3.4 for the muon channel). The uncertainty is found to be 25% in the electron channel and 60% in the muon channel.

9.5.1.6 Extrapolation Uncertainties

The top-quark, diboson and multijet background are extrapolated using a variety of fits. The envelope of all fits is taken as systematic uncertainty on the extrapolation. The results and methods are discussed in detail in Sect. 9.4.

9.5.1.7 Luminosity Uncertainty

The uncertainty on the integrated luminosity is 5% [33], affecting the normalization of all simulated MC samples. It is derived from a preliminary calibration of the luminosity scale using a pair of $x - y$ beam separation scans performed in June 2015. A more detailed description of the measurement method is given in Sect. 5.8 and Ref. [34].

9.5.2 Theoretical Uncertainties

9.5.2.1 PDF Uncertainties

The PDF uncertainties have been studied for the leading W background and the Z/γ^* background. The uncertainty has been estimated using VRAP with the CT14NNLO PDF error set. A single combined PDF uncertainty can be calculated using the full set of 56 eigenvectors. This uncertainty was found to be around 4% for $m_{\ell\nu} < 500$ GeV rising up to 40% at $m_{\ell\nu} = 6$ TeV. However, the 56 eigenvectors were found to have a very different m_T dependence and combining them into one single uncertainty might over-constrain the uncertainty at very high m_{T}. Instead a reduced set of seven eigenvectors with a similar mass dependence is used. The reduced set was provided by the authors of the CT14 PDF set and obtained using MP4LHC [35, 36].

As the central value of the prediction using the NNPDF3.0 set does not lie inside the 90% CL uncertainty of the CT14 PDF set, an additional uncertainty for the arbitrary choice of the central PDF was added. It enlarges the CT14 PDF uncertainty such that the central value of NNPDF3.0 is covered when adding it in quadrature. The same incompatibility was observed for the HERAPDF2.0 PDF set. However, the HERAPDF2.0 set does not use all available high-x data and was therefore not considered.

9.5.2.2 α_s Uncertainty

An uncertainty arises from the uncertainty on α_s which affects the VRAP calculations. The uncertainty on α_s was estimated to be ± 0.003 which corresponds to twice the uncertainty recommended by the PDF4LHC group [36]. The recommended uncertainty is given on the value of $\alpha_S(Q^2)$ at $Q^2 = m_Z^2$. The enlargement of the recommended uncertainty is justified by the much higher invariant masses this analysis aims at.

9.5.2.3 Electroweak Uncertainty

When calculating corrections for the higher order EW effects it is not known how to combine them with higher order QCD effects. An additive and a multiplicative approach has been studied. The additive approach is currently the default and the difference to the multiplicative approach used as an uncertainty.

9.5.2.4 Renormalization and Factorization Scale Uncertainty

An uncertainty on the theory corrections calculated with VRAP arises from the specific choice of the renormalization scale μ_R and the factorization scale μ_F. The calculations were repeated while varying μ_R and μ_F simultaneously up and down by a factor two. The resulting difference is used as an uncertainty.

9.5.2.5 Top-Quark and Diboson Background Uncertainty

An uncertainty arises from the cross section to which the top and diboson samples are normalized to. The predicted $t\bar{t}$ production cross section is $\sigma_{t\bar{t}} = 832^{+20}_{-29}(scale) \pm 35(PDF + \alpha_S) \pm 23(mass)$ pb (see Sect. 9.1.2). The tW background is normalized to a cross section of $\sigma_{tW} = 71.7 \pm 3.8$ pb. The uncertainty on the single top cross section is typically in the order of 4% [17] and therefore smaller than for the other top processes. Given these numbers, the normalization uncertainty is estimated to be 6%. A further uncertainty can arise from the modeling of the top background in the signal region. It is known that the top-p_T spectrum is not well modeled at very high p_T. This can have a sizable impact on the analysis. A data driven $t\bar{t}$ control region was defined to study a potential mismodeling. No such mismodeling has been observed, hence no further uncertainty is added. The study is documented in Appendix D.

Diboson processes have typically an uncertainty on the order of 5–10% [37]. A conservative normalization uncertainty of 10% is assigned to the diboson background.

9.5.3 Summary

The size of the systematic uncertainties on the total background is discussed in the following. Most sources have symmetric uncertainties and the upwards variation leads to the same result as a downward variation. In case of energy/momentum scale or resolution uncertainties, upwards variations usually have a larger effect due to the steeply falling background which causes more events to migrate from a lower bin to a higher bin. For all uncertainties, the upward variation has been performed, and the total uncertainty is taken to be the symmetrized effect. In the following a positive systematic uncertainty means that an upward variation of the source increases the background while a negative uncertainty means that the variation leads to a decrease in background. Previous analyses have shown that the final result of the search does not highly depend on the systematic uncertainties. This is especially true in the region of high transverse masses where the search is performed in a nearly background free region which is limited by statistical uncertainties. Studies have shown that for lower W' masses of 150 GeV the final result of this analysis is not affected by systematic uncertainties which change the background yield by less than 3% over the whole transverse mass range. Uncertainties which change the background yield by less than 3% everywhere will therefore be neglected.

Figure 9.29 shows the lepton related systematic uncertainties. The left plot shows the electron and the right plot the muon uncertainties. The electron energy scale uncertainty is about 3% at low m_{T}, rising up to 6% at high m_{T}. The uncertainty is therefore considered as it leads to a sizable contribution at low m_{T}. The electron energy resolution uncertainty is well below 1% over the whole m_{T} range and is therefore neglected. The same statement holds for the electron trigger efficiency correction uncertainty. The identification and isolation efficiency correction uncertainties are well below 1% at low m_{T}, rising to about 3 and 2% due to the high-p_{T} extrapolation uncertainties. Nevertheless, both uncertainties are neglected since the contribution becomes only sizable at around 500 GeV where the statistical uncertainty of the data is already large. In the muon channel, the muon ID resolution uncertainty gives the largest contribution at very high transverse mass. The uncertainty is well below 1% at low m_{T} and then rising up to 24% at 7 TeV. The muon MS resolution uncertainty is

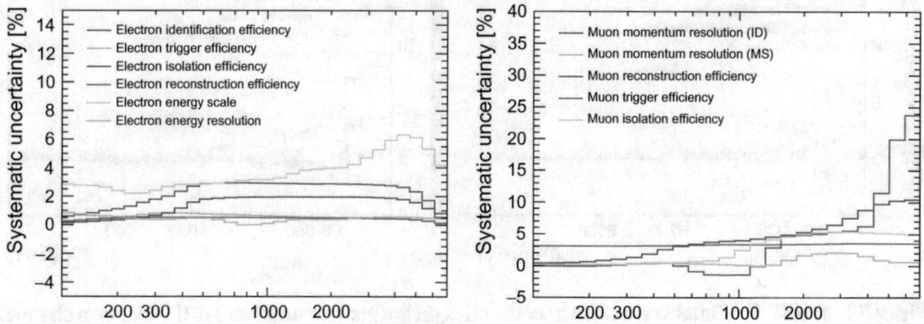

Fig. 9.29 Lepton related systematic uncertainties on the background yield in the electron channel (left) and muon channel (right)

in the region of 3% at high m_T. It could therefore potentially be neglected. However, the experimental uncertainties are also propagated to the modeled W' signal, where a resolution uncertainty can change the width of the signal. It can therefore lead to non negligible effects on the signal modeling and hence will not be neglected. Another sizable uncertainty is arising from the muon reconstruction efficiency correction which is rising up to 10% at high m_T. The uncertainty on the isolation efficiency correction is small at low m_T and rising up to 6% for higher m_T. The uncertainty on the muon trigger efficiency correction is about 3% over the whole m_T range. Both uncertainties are therefore considered.

Figure 9.30 shows the jet and E_T^{miss} related systematic uncertainties which behave very similar in the electron and muon channel. The largest uncertainty here is the jet energy resolution which is highest at low m_T and then falling towards higher m_T. The transverse energy of the jets in the event plays an important role in the E_T^{miss} calculation. Varying the energy can change the E_T^{miss} value and as a consequence change the decision whether an event passes the E_T^{miss} cut. The jet energy resolution plays therefore an important role in the low m_T region. The uncertainty is about 7 and 8% at 110 GeV for the electron and muon channel, respectively. The higher the m_T and therefore the E_T^{miss} of the event, the less important is the contribution from jets since the electron or muon will dominate the E_T^{miss} calculation. The uncertainty is hence becoming smaller towards higher m_T. The E_T^{miss} related uncertainties show a similar shape due to the same reason. They are about 3%, falling towards higher m_T. The jet energy scale uncertainties show a similar behavior. The second nuisance parameter is negligible. The other two nuisance parameters are yielding an uncertainty of 2–4% at low transverse mass. The size of the two nuisance parameters is different in the two channels. The first nuisance parameter is larger in the muon channel while in the electron channel the third is larger. The reason for the behavior are statistical fluctuations of the MC which cause these differences. The uncertainties are therefore neglected, as they only play a subleading role at low m_T and are affected by statistical fluctuations.

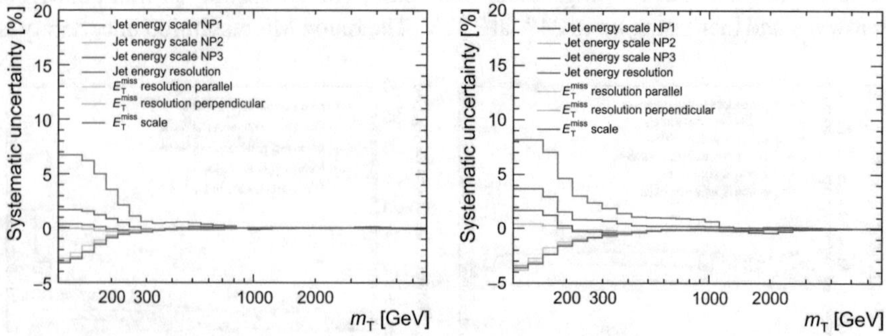

Fig. 9.30 Jet/E_T^{miss} related systematic uncertainties on the background yield in the electron channel (left) and muon channel (right)

Figure 9.31 shows the uncertainties arising from the normalization and extrapolation from the subleading backgrounds. The normalization uncertainty on the top and diboson backgrounds is negligible in both channels. The uncertainty of the multijet in the electron channel can reach up to 3% in the low m_T range and reaches up to 11% at high m_T. It is therefore considered, in contrast to the muon channel where this uncertainty is negligible. The top extrapolation uncertainty has its maximum around 800 GeV and decreases towards higher m_T as the background is steeply falling. The diboson extrapolation is in the electron channel the largest uncertainty at high m_T. The background plays only a minor contribution in the low m_T range. The extrapolation is performed with a fit over several TeV and it is therefore clear that at very high transverse masses, the uncertainty on the background prediction is large. The uncertainty is large enough to also have a sizable effect on the total background uncertainty. However, the uncertainty is about 100% only for $m_T > 4.5$ TeV. It is therefore only very large in a region where the background is close to zero and these uncertainties do not affect the result. The uncertainty is much smaller in the muon channel. The uncertainty is estimated by taking the envelope of all performed fits. This approach is very conservative and therefore preferred, but leads also to instabilities in the uncertainty estimation, as outliers will significantly enlarge the uncertainty. However, this only affects the regions at very high m_T which are nearly background free. The multijet extrapolation in the electron channel reaches up to 200% at 7 TeV. In the muon channel this uncertainty is negligible as it is well below 1% everywhere due to the low contribution of the background.

Figure 9.32 shows all theory related systematic uncertainties. These include the PDF, α_s, scale uncertainties and the uncertainties on the electroweak corrections. The largest uncertainty is coming from the PDF choice, originating from the differences observed between the central PDF and NNPDF3.0. The uncertainty is small at low m_T values but rises up to about 38% in the electron channel at about 4 TeV. The uncertainty is much smaller in the muon channel. This is due to the worse resolution in the muon channel which leads to migration from events at low m_T, where the PDF uncertainties are small, to higher m_T. The high m_T region is therefore also populated by events from lower m_T values and has therefore a much smaller uncertainty. The uncertainty is getting smaller again towards higher transverse masses in the electron

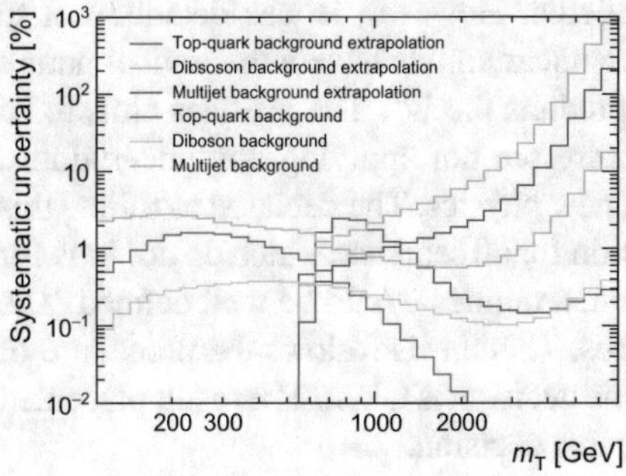

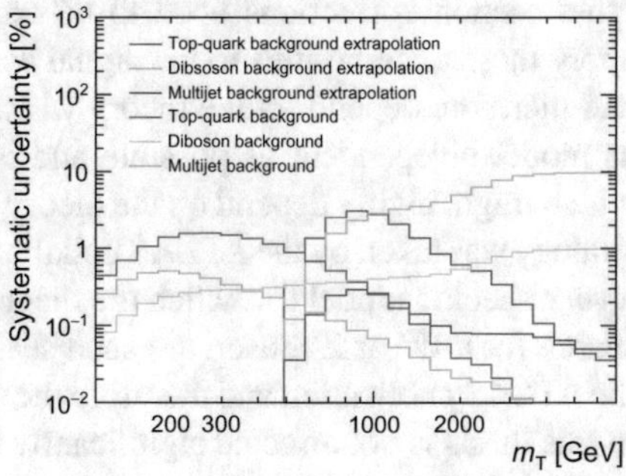

Fig. 9.31 Background related systematic uncertainties on the background yield in the electron channel (left) and muon channel (right)

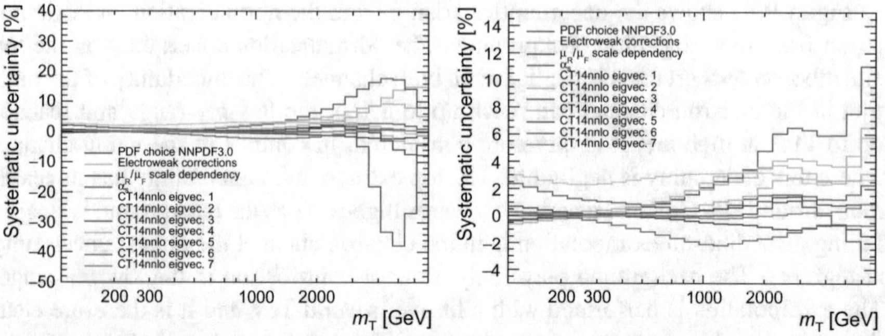

Fig. 9.32 Theory related systematic uncertainties on the background yield in the electron channel (left) and muon channel (right)

channel. This is due to the multijet background, which is one of the subleading backgrounds up to about 4 TeV, where it represents around 10% of the total background. The contribution rises then to about 90% at 7 TeV. Since this background is estimated from data, no theory uncertainties need to be applied to it and the uncertainties play a smaller role with respect to the total background. The different eigenvectors of the CT14 PDF have, as anticipated, very different mass dependencies. The uncertainty from the eigenvectors two to four increases with m_{T}, while the first eigenvector leads to a nearly constant uncertainty of about 3% over the whole mass range. All first four eigenvectors have a sizable contribution and are therefore considered. The eigenvectors five to seven lead to a negligible uncertainty and are neglected. The uncertainty on the electroweak corrections is rising with m_{T} and is of the order of 10% in the electron channel at high m_{T} while it is slightly smaller in the muon channel. The variation of the renormalization and factorization scale leads to a negligible uncertainty. The α_s uncertainty is of the order of 3% at low m_{T} and does not vary much for higher m_{T} values. The chosen uncertainty of ± 0.003 is, as already discussed, only applicable at very high m_{T}. A choice which would aim at the low m_{T} region would lead to much smaller uncertainties. As a consequence this uncertainty is neglected.

The theory uncertainties would in principle apply also to the signal, as the same cross section corrections need to be considered. However, it was decided not to apply these uncertainties to the signal as the uncertainties have a theoretical nature and therefore depend on the theory which predicts the W'. The analysis aims to be as model independent as possible and is therefore not applying any uncertainties which might highly depend on the theory of new physics. The decision to follow this strategy was taken by the ATLAS collaboration for all searches which do not look for a very specific model for which the theory uncertainties would be well defined. The search for a W' or Z' boson are such searches. This thesis follows the procedure of the ATLAS publication and therefore the same decision is taken. The result presented in this thesis is not affected significantly by this decision.

Table 9.6 summarizes again all systematic sources and states whether they are applied or neglected. The last column states whether an uncertainty is fully correlated

Table 9.6 Summary of the correlations for the uncertainties. Entries with "yes" share a nuisance parameter and are treated as correlated. Systematic uncertainties which do not apply to a channel are marked with "n/a"

	Channel		
	$W' \rightarrow e\nu$	$W' \rightarrow \mu\nu$	Correlated between channels
Electron energy scale	Applied	n/a	–
Electron energy resolution	Neglected	n/a	–
Muon momentum resolution (ID)	n/a	Applied	–
Muon momentum resolution (MS)	n/a	Applied	–
Jet energy resolution	Applied	Applied	Yes
Jet energy scale NP1	Neglected	Neglected	–
Jet energy scale NP2	Neglected	Neglected	–
Jet energy scale NP3	Neglected	Neglected	–
$E_{\mathrm{T}}^{\mathrm{miss}}$ resolution parallel	Applied	Applied	Yes
$E_{\mathrm{T}}^{\mathrm{miss}}$ resolution perpendicular	Applied	Applied	Yes
$E_{\mathrm{T}}^{\mathrm{miss}}$ scale	Applied	Applied	Yes
Lepton trigger efficiency	Neglected	Applied	–
Lepton reconstruction efficiency	Neglected	Applied	–
Electron identification efficiency	Neglected	n/a	–
Lepton isolation efficiency	Neglected	Applied	–
Electroweak background normalization	Neglected	Neglected	–
Electroweak background extrapolation	Applied	Neglected	–
Multijet background normalization	Applied	Neglected	–
Multijet background extrapolation	Applied	Neglected	–
PDF choice NNPDF3.0	Applied	Applied	Yes
CT14nnlo eigvec. 1	Applied	Applied	Yes
CT14nnlo eigvec. 2	Applied	Applied	Yes
CT14nnlo eigvec. 3	Applied	Applied	Yes
CT14nnlo eigvec. 4	Applied	Applied	Yes
CT14nnlo eigvec. 5	Neglected	Neglected	–
CT14nnlo eigvec. 6	Neglected	Neglected	–
CT14nnlo eigvec. 7	Neglected	Neglected	–
μ_R/μ_F scale dependency	Neglected	Neglected	–
α_s	Neglected	Neglected	–
Electroweak corrections	Applied	Applied	Yes
Luminosity	Applied	Applied	Yes

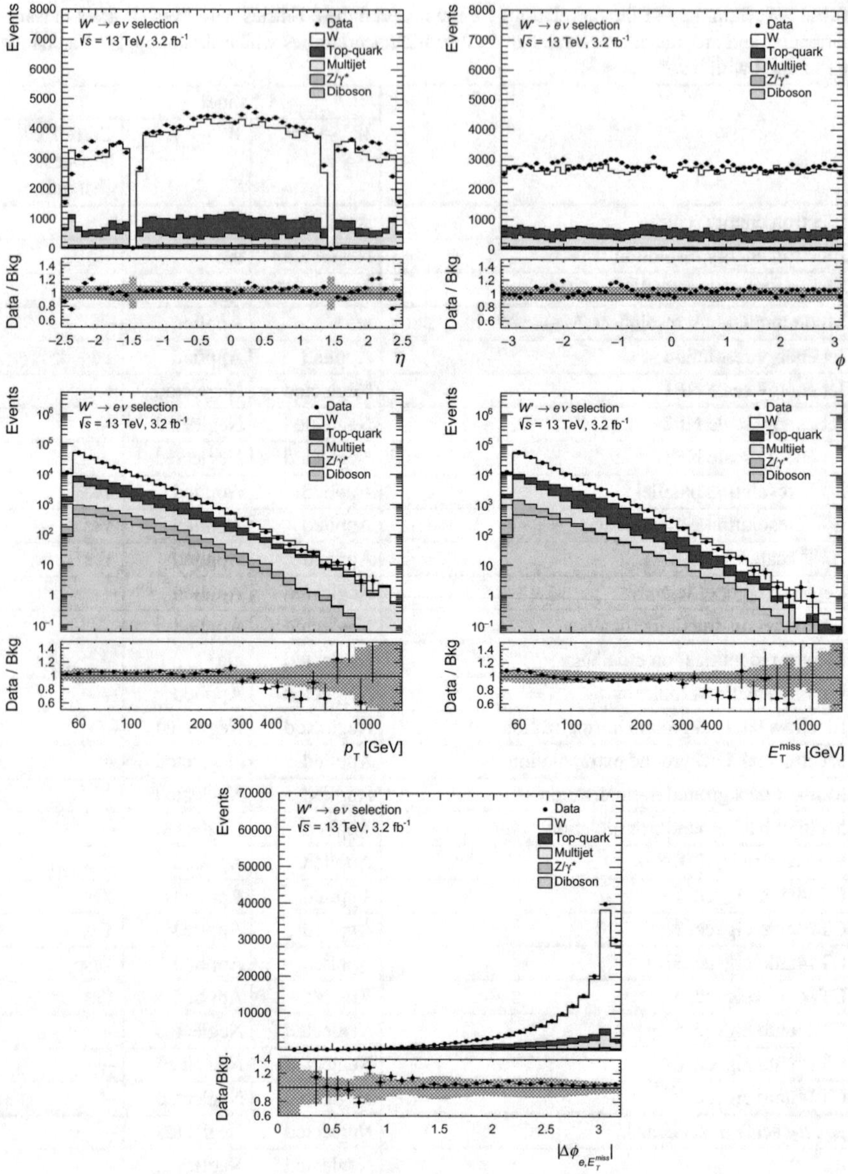

Fig. 9.33 Electron η (top left), ϕ (top right), p_{T} (middle left), $E_{\mathrm{T}}^{\mathrm{miss}}$ (middle right) and $|\Delta\phi_{e,E_{\mathrm{T}}^{\mathrm{miss}}}|$ (bottom) distributions after the event selection in the electron channel

between both channels. Top-quark and diboson background have been summarized as electroweak background. The luminosity uncertainty has not been shown in the Figs. 9.29, 9.30, 9.31 and 9.32. It is 5% and applies to all MC backgrounds.

9.6 Comparison of Background with Data

In the following section the kinematic properties of the events passing the signal selection are discussed. Figure 9.33 shows the η, ϕ, p_T, $E_\mathrm{T}^\mathrm{miss}$ and $|\Delta\phi_{e,E_\mathrm{T}^\mathrm{miss}}|$ distribution of the events passing the $W' \rightarrow e\nu$ selection. The η distribution of the selected electron candidates has its maximum at $\eta = 0$ and is slightly falling towards higher values of $|\eta|$. The region $1.37 < |\eta| < 1.52$ corresponds to the transition region between the calorimeter central region and the calorimeter endcap region. A few events are still falling into this region, since the cut is placed on the η position of the energy deposition in the second layer and here the best estimate for η is shown, which also includes track information. The dominant background is coming from the SM W-boson. The second largest background in the central region is the top background followed by the multijet background. In the more forward region, the multijet background is the subleading background. The ratio between data and background expectation is shown in the lower panel. The systematic uncertainty is shown as a gray hashed band. The systematic uncertainty is of the order of 10%. In the central region, data and background expectation agree well within the systematic uncertainty band. Some differences can be seen in the regions $|\eta| > 2.0$. The same differences between data and background were already observed in the multijet control region and are coming from a too coarse modeling of the electron identification corrections in that region. The selected electron candidates are equally distributed in ϕ. Some small structures in data can be observed which are not modeled in the background expectation. These structures are again coming from regions with slightly different identification efficiencies. The identification efficiency corrections are averaged over ϕ and thus these effects are not well modeled for the background. However, the overall agreement between data and background expectation is described well within the systematic uncertainties. The p_T spectrum of the selected electron candidates is strongly falling over several orders of magnitude. The candidate with the highest p_T is observed around 1 TeV. The multijet background is becoming the leading background towards higher p_T values. The fake efficiency is very stable up to high p_T, hence a higher rate of jets faking the electron signature is not the cause for this behavior. The reason is the underlying p_T spectrum of the jets which is much harder than the p_T spectrum of single electrons. The agreement between data and background expectation is well within the uncertainties up to a p_T of 300 GeV. In the region 400 GeV $< p_\mathrm{T} <$ 700 GeV a slight deficit of the data is observed. The region $p_\mathrm{T} >$ 400 GeV has been studied in more detail to exclude that the deficit of data is coming from background mismodeling. The study is documented in Appendix E. The $E_\mathrm{T}^\mathrm{miss}$ spectrum of the selected events is, like the p_T spectrum, strongly falling over several orders of magnitude. The highest $E_\mathrm{T}^\mathrm{miss}$ events are observed around 1 TeV. The multijet background never plays a dominant role in contrast to the p_T spectrum. The data agrees within the systematic uncertainty with the background expectation. Some small deficits can be again observed in the region 400 GeV $< E_\mathrm{T}^\mathrm{miss} <$ 800 GeV. The ratio between data and background expectation is around 1.1 at 55 GeV, falling to around 0.95 for $E_\mathrm{T}^\mathrm{miss} >$ 120 GeV. The $|\Delta\phi_{e,E_\mathrm{T}^\mathrm{miss}}|$ distribution shows a maximum at around 3 and is falling towards lower opening angles. Nearly no events are observed for $|\Delta\phi_{e,E_T^{miss}}| < 1$. The maximum around $|\Delta\phi_{e,E_T^{miss}}| = 3$ corresponds to the back-to-back topology which is dominating the high transverse mass region. The agreement for $|\Delta\phi_{e,E_T^{miss}}|$ is similar to what was observed for the previous distributions.

Figure 9.34 shows the same distributions for the $W' \rightarrow \mu\nu$ selection. The muon η distribution shows several dips which are corresponding to the regions with not well aligned chambers which are vetoed in the high-p_T selection. The muon η spectrum is in general a bit lower in the central region than in the endcap region due to the lower trigger efficiency in the barrel region for muons. Especially in the region $\eta \approx 0$, where a gap for cables which are connected to the liquid argon calorimeter is leading to a much lower efficiency. The dominating background is the SM W background, followed by the top background and the Z/γ^* background. The Z/γ^* background becomes larger in the forward region and leads to a substantial contribution. This is different to the electron channel where the Z/γ^* background is very small, even in the endcap region. A Z/γ^* event enters the selection if E_T^{miss} is faked. This happens if a muon or electron is outside of the detector acceptance. If an electron is outside of the region covered by the tracking system then it will be most likely still be reconstructed as a jet in the forward calorimeters. A muon which leaves the region of the tracking system is already lost which explains the differences between both channels. The data are in general about 10% above the expected background. The difference is covered by the systematic uncertainty which is of the same order. The same difference between data and expected background is observed in the ϕ spectrum. The spectrum shows some dips which are corresponding to the regions in which the muon spectrometer efficiency is lower. Eight dips can be observed with a distance of about 0.8 which corresponds to the coils of the toroidal magnet. Less muon chambers are installed in this region and therefore the efficiency is reduced. The two larger dips at around $\phi \approx -2$ and $\phi \approx -1$ correspond to the support structure on which the ATLAS detector is placed. The muon p_T spectrum falls, like the corresponding electron spectrum, over several orders of magnitude. The muon candidate with the highest p_T was found around 1.2 TeV. The data are, like in the η and ϕ distributions, about 10% higher than the expected background. The ratio between data and expectation is flat in p_T. The largest deviation is found in a bin at around 300 GeV which is found to be about 3.5σ above the neighboring bins. This behavior is expected to come from a single statistical fluctuation, as no larger deviation is observed in the neighboring bins. The E_T^{miss} spectrum of the muon selection is strongly falling towards higher values of E_T^{miss} and reaches out up to 1.2 TeV in data. The data are about 15% higher than the expected background at low E_T^{miss} values. The differences are covered by the systematic uncertainty. The difference between data and expectation is becoming smaller with higher E_T^{miss} values. Data and expected background agree very well for $E_T^{miss} > 120$ GeV. This behavior leads to the suspicion that an E_T^{miss}-mismodeling is causing the observed differences between data and expected background in the muon channel. However, the same mismodeling should be visible in the electron channel as the systematic sources are correlated between both channels. The shape of the ratio is very similar to the shape observed in the electron channel, which is an indication that the difference between data and expected background are indeed caused by E_T^{miss}

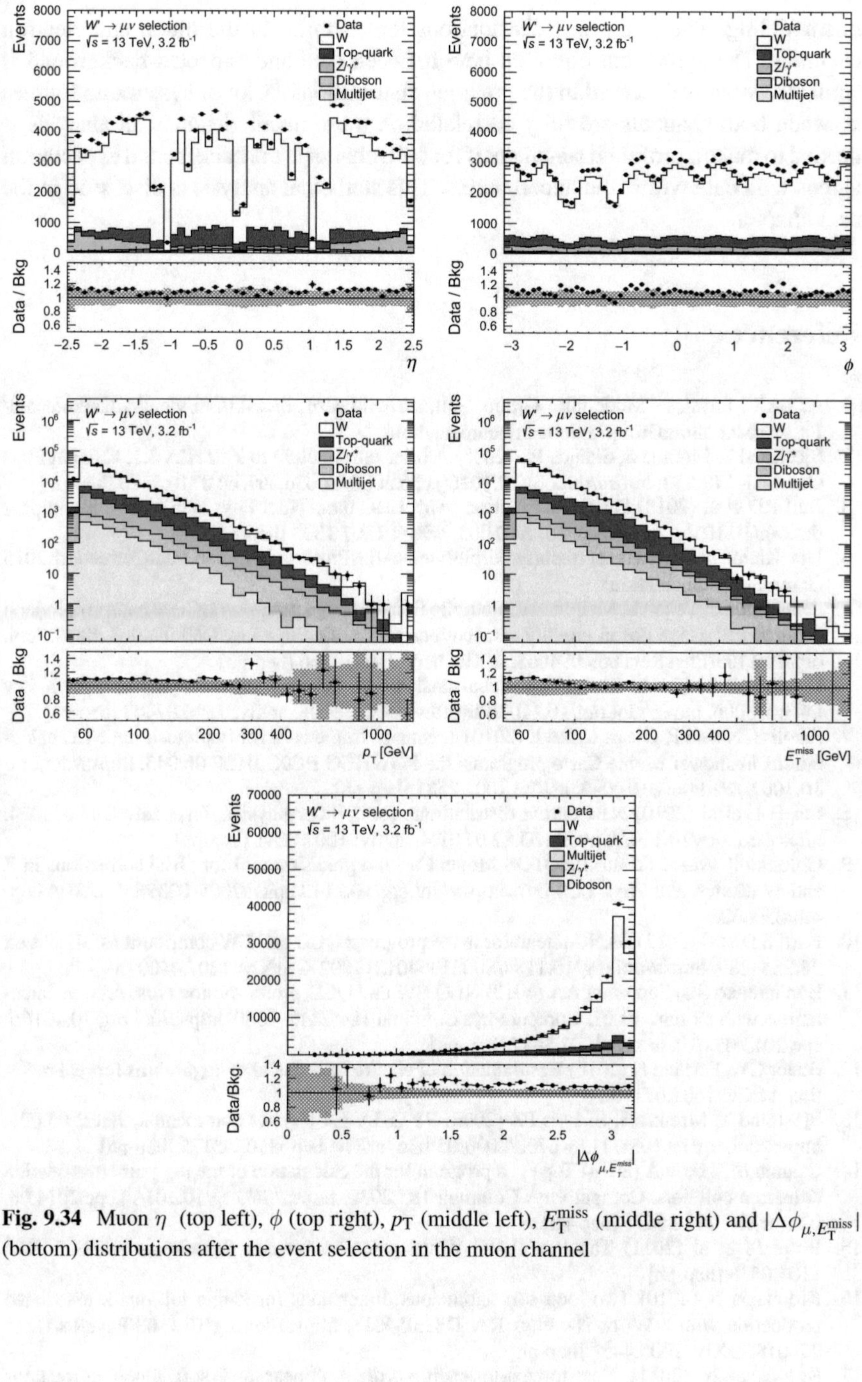

Fig. 9.34 Muon η (top left), ϕ (top right), p_T (middle left), E_T^{miss} (middle right) and $|\Delta\phi_{\mu,E_T^{miss}}|$ (bottom) distributions after the event selection in the muon channel

mismodeling. The $|\Delta\phi_{\mu,E_{\mathrm{T}}^{\mathrm{miss}}}|$ distributions look similar in the muon and electron channels. The agreement observed here between data and expected background is similar to what is observed in the previous distributions. A lot of systematic sources between both channels are fully correlated. A more detailed statistical analysis is needed to make a profound statement if for both channels the background expectation agrees with data within the uncertainties. This statistical analysis is discussed in the next chapter.

References

1. ATLAS Physics Modeling Group. https://twiki.cern.ch/twiki/bin/view/AtlasProtected/PhysicsModellingGroup (Internal documentation)
2. Sjostrand T, Mrenna S, Skands PZ (2008) A brief introduction to PYTHIA 8.1. Comput Phys Commun 178:852. https://doi.org/10.1016/j.cpc.2008.01.036, arXiv: 0710.3820 [hep-ph]
3. Ball RD et al (2013) Parton distributions with LHC data. Nucl Phys B867:244–289. https://doi.org/10.1016/j.nuclphysb.2012.10.003, arXiv: 1207.1303 [hep-ph]
4. Uta Klein. https://twiki.cern.ch/twiki/bin/view/AtlasProtected/HigherOrderCorrections2015 (Internal documentation)
5. Anastasiou C, Dixon L, Melnikov K, Petriello F (2004) High precision QCD at hadron colliders: electroweak gauge boson rapidity distributions at NNLO. Phys Rev D69:094008. https://doi.org/10.1103/PhysRevD.69.094008, arXiv: hep-ph/0312266 [hep-ph]
6. Dulat S et al (2016) The CT14 global analysis of quantum chromodynamics. Phys Rev D93:033006. https://doi.org/10.1103/PhysRevD.93.033006, arXiv: 1506.07443 [hep-ph]
7. Alioli S, Nason P, Oleari C, Re E (2010) A general framework for implementing NLO calculations in shower Monte Carlo programs: the POWHEG BOX. JHEP 06:043. https://doi.org/10.1007/JHEP06(2010)043, arXiv: 1002.2581 [hep-ph]
8. Lai H-L et al (2010) New parton distributions for collider physics. Phys Rev D82:074024. https://doi.org/10.1103/PhysRevD.82.074024, arXiv: 1007.2241 [hep-ph]
9. Golonka P, Was Z (2006) PHOTOS Monte Carlo: a precision tool for QED corrections in Z and W decays. Eur Phys J C45:97. https://doi.org/10.1140/epjc/s2005-02396-4, arXiv: hep--ph/0506026
10. Bardin D et al (2012) SANC integrator in the progress: QCD and EW contributions. JETP Lett 96:285–289. https://doi.org/10.1134/S002136401217002X, arXiv: 1207.4400 [hep-ph]
11. Bondarenko SG, Sapronov AA (2013) NLO EW and QCD proton-proton cross section calculations with mcsanc-v1.01. Comput Phys Commun 184:2343–2350. https://doi.org/10.1016/j.cpc.2013.05.010, arXiv: 1301.3687 [hep-ph]
12. Bauer CW, Ferland N (2016) Resummation of electroweak Sudakov logarithms for real radiation. arXiv: 1601.07190
13. Sjostrand T, Mrenna S, Skands PZ (2006) PYTHIA 6.4 physics and manual. JHEP 05:026. https://doi.org/10.1088/1126-6708/2006/05/026, arXiv: hep-ph/0603175 [hep-ph]
14. Czakon M, Mitov A (2014) Top++: a program for the calculation of the top-pair cross-section at hadron colliders. Comput Phys Commun 185:2930. https://doi.org/10.1016/j.cpc.2014.06.021, arXiv: 1112.5675 [hep-ph]
15. Botje M et al (2011) The PDF4LHC Working Group Interim Recommendations. arXiv: 1101.0538 [hep-ph]
16. Kidonakis N (2010) Two-loop soft anomalous dimensions for single top quark associated production with a W- or H-. Phys Rev D82:054018. https://doi.org/10.1103/PhysRevD.82.054018, arXiv: 1005.4451 [hep-ph]
17. Kidonakis N (2011) Next-to-next-to-leading-order collinear and soft gluon corrections for t-channel single top quark production. Phys Rev D83:091503. https://doi.org/10.1103/PhysRevD.83.091503, arXiv: 1103.2792 [hep-ph]

18. ATLAS Physics Modeling Group. https://twiki.cern.ch/twiki/bin/view/LHCPhysics/TtbarNNLO (Internal documentation)
19. ATLAS Physics Modeling Group. https://twiki.cern.ch/twiki/bin/view/AtlasProtected/MC15SingleTopSamplesPMG (Internal documentation)
20. Gleisberg T et al (2009) Event generation with SHERPA 1.1. JHEP 02:007. https://doi.org/10.1088/1126-6708/2009/02/007, arXiv: 0811.4622 [hep-ph]
21. ATLAS Luminosity Working Group. https://twiki.cern.ch/twiki/bin/view/~AtlasPublic/LuminosityPublicResults
22. ATLAS Isolation Forum. https://twiki.cern.ch/twiki/bin/view/AtlasProtected/IsolationForum (Internal documentation)
23. Aad G et al (2016) Muon reconstruction performance of the ATLAS detector in proton-proton collision data at ps = 13 TeV. Eur Phys J C76.5:292. https://doi.org/10.1140/epjc/s10052-016-4120-y, arXiv: 1603.05598 [hep-ex]
24. Cacciari M, Salam GP, Soyez G (2008) The anti-k(t) jet clustering algorithm. JHEP 04:063. https://doi.org/10.1088/1126-6708/2008/04/063, arXiv: 0802.1189 [hep-ph]
25. Aad G et al (2015) Search for new phenomena in the dijet mass distribution using p – p collision data at ps = 8 TeV with the ATLAS detector. Phys Rev D91.5:052007. https://doi.org/10.1103/PhysRevD.91.052007, arXiv: 1407.1376 [hep-ex]
26. Aad G et al (2014) Search for high-mass dilepton resonances in pp collisions at ps = 8 TeV with the ATLAS detector. Phys Rev D90.5:052005. https://doi.org/10.1103/PhysRevD.90.052005, arXiv: 1405.4123 [hep-ex]
27. Electron efficiency measurements with the ATLAS detector using the 2015 LHC proton-proton collision data (2016). Technical report ATLAS-CONF-2016-024. Geneva: CERN. https://cds.cern.ch/record/2157687
28. ATLAS Collaboration (2014) Electron and photon energy calibration with the ATLAS detector using LHC run 1 data. Eur Phys J C74.10:3071. https://doi.org/10.1140/epjc/s10052-014-3071-4, arXiv: 1407.5063 [hep-ex]
29. ATLAS Collaboration (2015). Jet calibration and systematic uncertainties for jets reconstructed in the ATLAS detector at sqrt(s) = 13 TeV. ATL-PHYS-PUB-2015-015. https://cds.cern.ch/record/2037613
30. ATLAS Collaboration (2015). A method for the construction of strongly reduced representations of ATLAS experimental uncertainties and the application thereof to the jet energy scale. ATL-PHYS-PUB-2015-014. https://cds.cern.ch/record/2037436
31. ATLAS Collaboration (2015). Jet calibration and systematic uncertainties for jets reconstructed in the ATLAS detector at ps = 13 TeV. Technical report ATL-PHYS-PUB-2015-015. Geneva: CERN. https://cds.cern.ch/record/2037613
32. ATLAS Collaboration (2015). Expected performance of missing transverse momentum reconstruction for the ATLAS detector at sqrt(s) = 13 TeV. ATL-PHYS-PUB-2015-023. https://cds.cern.ch/record/2037700
33. ATLAS Physics Modeling Group. https://twiki.cern.ch/twiki/bin/view/Atlas/LuminosityForPhysics (Internal documentation)
34. Aad G et al (2013) Improved luminosity determination in pp collisions at sqrt(s) = 7 TeV using the ATLAS detector at the LHC. Eur Phys J C73.8:2518. https://doi.org/10.1140/epjc/s10052-013-2518-3, arXiv: 1302.4393 [hep-ex]
35. Gao J, Nadolsky P (2014) A meta-analysis of parton distribution functions. JHEP 07:035. https://doi.org/10.1007/JHEP07(2014)035, arXiv: 1401.0013 [hep-ph]
36. Butterworth J et al (2016) PDF4LHC recommendations for LHC Run II. J Phys G43:023001. https://doi.org/10.1088/0954-3899/43/2/023001, arXiv: 1510.03865 [hep-ph]
37. Butterworth J et al (2010) Single Boson and Diboson production cross sections in pp collisions at sqrts = 7 TeV. Technical report ATL-COM-PHYS-2010-695. Geneva: CERN. https://cds.cern.ch/record/1287902

Chapter 10
Statistical Interpretation

In the following chapter first the basics of the statistical framework are introduced in Sect. 10.1 and the observed transverse mass spectrum is discussed in terms of the ratio of the number of observed events and the number of expected events in Sect. 10.2. The quantification of a potential observed excess is discussed in Sect. 10.3 and a limit on the cross section of a W' times the branching ratio for the decay into electron or muon is calculated in Sect. 10.4.

10.1 Statistical Framework

The data are described statistically following a multi-bin counting experiment approach. This approach has the advantage that the shape of the signal in the transverse mass distribution is taken into account. The number of observed events in each m_{T} bin of each channel is described by a Poisson distributed stochastic variable. The expectation value λ_{kl} in bin l of channel k is written as the sum of the signal[1] and background contributions,

$$\lambda_{kl}(\sigma B, \vec{\theta}) = s_{kl}(\sigma B, \vec{\theta}) + b_{kl}(\vec{\theta}), \tag{10.1}$$

where the nuisance parameters $\vec{\theta}$ describe the effect of systematic uncertainties. The signal cross section multiplied with the branching ratio into the electron or muon final state σB is the parameter of interest in the statistical analysis. The number of observed events in bin l of channel k is denoted as n_{kl}. The likelihood is built by multiplying the Poisson probabilities for each bin in each channel:

[1]In the following description of the statistical analysis, the W' mass is assumed to be specified. The whole statistical analysis is repeated for each candidate W' mass.

M. Zinser, *Search for New Heavy Charged Bosons and Measurement of High-Mass Drell-Yan Production in Proton-Proton Collisions*, Springer Theses,
https://doi.org/10.1007/978-3-030-00650-1_10

$$\mathcal{L}(\sigma B, \vec{\theta}) = P(\vec{n}|\sigma B, \vec{\theta}) = \prod_{k=1}^{N_{\text{chan}}} \prod_{l=1}^{N_{\text{bin}}} \frac{\lambda_{kl}(\sigma B, \vec{\theta})^{n_{kl}}\, e^{-\lambda_{kl}(\sigma B, \vec{\theta})}}{n_{kl}!}. \quad (10.2)$$

Here, the set of all individual-bin observations n_{kl} is denoted as $\vec{n}$. The number of channels N_{chan} is equal to 1 when the electron and muon channels are analyzed individually and 2 for their combination.

The number of expected signal events in bin l of channel k can be written as

$$s_{kl}(\sigma B, \vec{\theta}) = \overline{s_{kl}}(\sigma B) \left(1 + \sum_{i=1}^{N_{\text{sys}}} \theta_i \frac{(\delta s_{kl})_i}{\overline{s_{kl}}}. \right) \quad (10.3)$$

It has a central value of

$$\overline{s_{kl}}(\sigma B) = L_{\text{int}}\, \sigma B\, A_k\, \varepsilon_{kl}. \quad (10.4)$$

Here, L_{int} is the integrated luminosity of the data and A_k is the product of acceptance and efficiency for signal events in channel k to be triggered, passing the event selection, and having a reconstructed transverse mass within the limits of the transverse mass histogram used for the statistical analysis ($110 < m_{\text{T}} < 7000\,\text{GeV}$). This quantity is shown as function of the W' mass in Fig. 9.4. Furthermore, ε_{kl} takes into account the shape of the W' signal. It is defined as the fraction of the events in the signal transverse mass histogram in channel k that fall into bin l. The quantity $(\delta s_{kl})_i/\overline{s_{kl}}$ in Eq. (10.3) is the relative shift in s_{kl} induced by a 1σ variation of the nuisance parameter θ_i associated with a systematic uncertainty i. The corresponding equation for the number of background events is

$$b_{kl}(\vec{\theta}) = \overline{b_{kl}} \left(1 + \sum_{i=1}^{N_{\text{sys}}} \theta_i \frac{(\delta b_{kl})_i}{\overline{b_{kl}}} \right), \quad (10.5)$$

where $\overline{b_{kl}}$ is the central value of b_{kl} and $(\delta b_{kl})_i/\overline{b_{kl}}$ is the corresponding relative shift of a systematic uncertainty i (shown in Figs. 9.29–9.32). Correlations between signal uncertainties and background uncertainties and between channels are properly accounted for, since the same set of systematic nuisance parameters affects s_{kl} and b_{kl}. Some nuisance parameters affect only either the electron or the muon channel. In this case the corresponding relative shift in the other channel is zero. Equations (10.3)–(10.5), contain all inputs needed for the statistical analysis. These are the integrated luminosity L_{int}, the acceptances A_k and signal shapes ε_{kl}, the background estimates $\overline{b_{kl}}$, and the signal and background systematic variations $(\delta s_{kl})_i/\overline{s_{kl}}$ and $(\delta b_{kl})_i/\overline{b_{kl}}$. The evaluation of all these inputs is described in Sect. 9.

10.2 Results

The p_T and E_T^{miss} distributions which were discussed in Sect. 9.6 showed no obvious deviation of the data from the expected background. Figure 10.1 shows the resulting transverse mass spectrum in the range 110 GeVto 7 TeV in the electron channel (left) and the muon channel (right). The bin width is constant in $\log(m_T)$. A reasonable number of bins (62 bins in the electron channel and 50 bins muon channel) were chosen by hand since the result does not strongly depend on the binning. The muon channel has less bins due to the worse m_T resolution. The transverse mass spectrum is strongly falling over several orders of magnitudes. In both channels events with a transverse mass above 1 TeV are observed. In the electron channel (muon channel) the event with the highest transverse mass is found at $m_T = 1.95$ TeV ($m_T = 2.51$ TeV). Example W' signals with a mass of 2, 3 and 4 TeV are shown on top of the background. The Jacobian peak of the W' signal is much wider in the muon channel due to the worse p_T resolution. The background expectation is in general in good agreement with the data. No strong deviation can be observed. A small excess is observed in the muon channel at a transverse mass of around 1.5 TeV, where four events fall into one bin. The panel in the middle shows the ratio between data and expected background. The gray shaded band shows the systematic uncertainty on the background. At a transverse mass of 110 GeV data are about 7% above the expected background in the electron channel and about 12% in the muon channel. The agreement gets slightly better towards higher values of m_T.

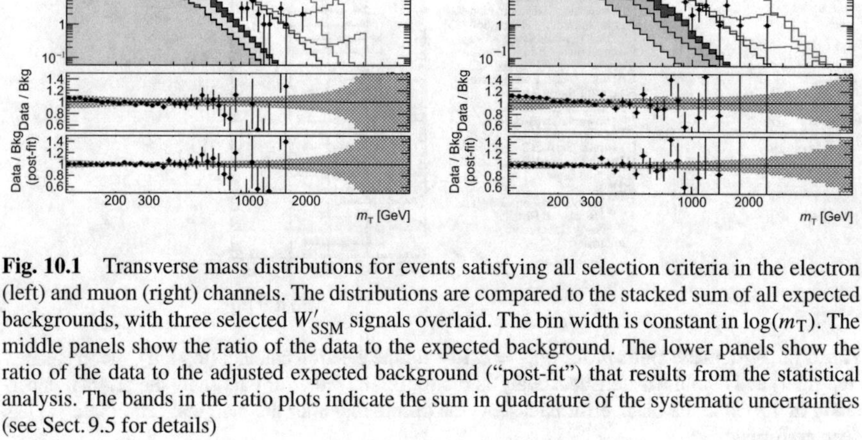

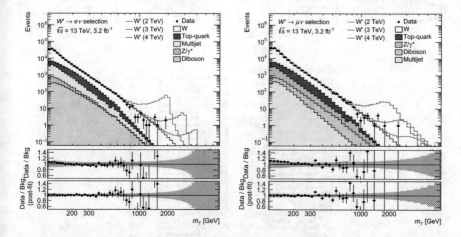

Fig. 10.1 Transverse mass distributions for events satisfying all selection criteria in the electron (left) and muon (right) channels. The distributions are compared to the stacked sum of all expected backgrounds, with three selected W'_{SSM} signals overlaid. The bin width is constant in $\log(m_T)$. The middle panels show the ratio of the data to the expected background. The lower panels show the ratio of the data to the adjusted expected background ("post-fit") that results from the statistical analysis. The bands in the ratio plots indicate the sum in quadrature of the systematic uncertainties (see Sect. 9.5 for details)

The bottom panel shows the ratio between data and expected background after applying the shifts of the nuisance parameters θ_i to the background ("post-fit"). The values of θ_i were obtained by computing the marginalization integral (see Eq. 10.15 and the description in Sect. 10.4.1 for more details) with the Markov Chain Monte Carlo technique using the Bayesian Analysis Toolkit (BAT) [1]. The systematic uncertainties might be constrained by the data after the marginalization integral has been performed. A Gaussian distribution of the shifts θ_i was assumed and a W' mass of 2 TeV was used for the marginalization. Changing to a different W' mass or not assuming any signal does not affect the result. Figure 10.2 shows the parameters θ_i for each nuisance parameter. The red error bar shows the initial uncertainty of 1σ and the black error bar the uncertainty after the marginalization integral has been performed. The largest shifts in the combined analysis are introduced for the

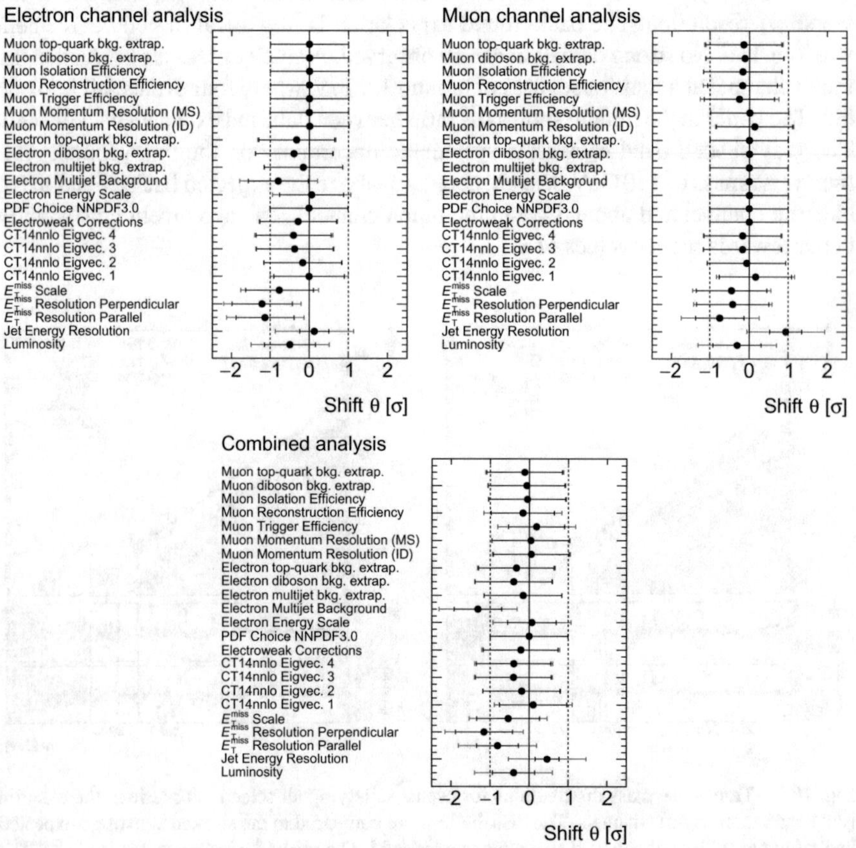

Fig. 10.2 Systematic shifts θ_i for $m_{W'} = 2\,\text{TeV}$ for the electron channel (top left), muon channel (top right) and combined statistical analysis (bottom). The red error bars show the original uncertainty of 1σ while the black error bars show the uncertainty after the marginalization integral has been performed

E_T^{miss} uncertainties and the electron channel multijet background. All three E_T^{miss} uncertainties are shifted down in the combined analysis by about 0.6 to 1.2σ. These nuisance parameters can correct for the shape in the ratio which is observed in both channels in the region $110 < m_T < 300\,\text{GeV}$. The jet energy resolution leads to a similar shape change and is shifted up by 0.4σ. The jet energy resolution uncertainties and the E_T^{miss} uncertainties have opposite effects on the background yield. Hence, the shifts lead to a higher background contribution at low m_T. A slight offset between both channels is left after correcting the slope at low m_T. The electron multijet background is shifted by about -1.3σ to correct for the offset between electron channel and muon channel. The luminosity is finally shifted by -0.4σ to correct for the remaining offset of both channels. The shifts of all other nuisance parameters do not lead to a significant change of the background expectation. The same behavior is observed for the statistical analysis of the single channels. Very similar shifts of the nuisance parameters for the luminosity, jet energy resolution, E_T^{miss} uncertainties and electron channel multijet background are observed.

The background expectation and data are in a very good agreement after applying the shifts of the combined analysis (see lower panel in Fig. 10.1). The systematic uncertainty is significantly reduced in the low m_T region in which the data are able to constrain the nuisance parameters. The high m_T region is unaffected by the data and therefore no uncertainty reduction is observed. The strongest uncertainty reduction is observed for the jet energy resolution. This might be an indication that the uncertainty is overestimated, since not always a significant pull is introduced.

10.3 Search for a New Physics Signal

10.3.1 Likelihood-Ratio Test

To test for excesses in data, a log-likelihood ratio test is carried out using RooStats [2]. A potential excess can be quantified by calculating the probability that the background fluctuates creating a signal-like excess equal or larger than what is observed. This probability is called p-value and usually denoted as p_0. The p-value p_0 is computed by defining a test statistic

$$q_0 = \begin{cases} 0 & \text{for } \hat{\mu} < 0, \\ -2\ln\left[\frac{\mathcal{L}(\vec{n}|0,\vec{\theta}_0)}{\mathcal{L}(\vec{n}|\hat{\mu},\vec{\theta})}\right] & \text{for } \hat{\mu} \geqslant 0, \end{cases} \tag{10.6}$$

where $\hat{\mu}$ is a signal strength parameter and defined as $\hat{\mu} = \sigma B/\sigma_{SSM} B$. The more important an excess is, the larger is the difference between both likelihoods. For a given dataset, $\mathcal{L}(\vec{n}|\hat{\mu},\vec{\theta})$ is always larger or equal to $\mathcal{L}(\vec{n}|0,\vec{\theta}_0)$ and therefore the ratio of the likelihoods always smaller or equal to one. The test statistic q_0 is set to 0 for values of $\hat{\mu} < 0$. This is justified by the fact that a potential signal should have a positive cross section value. This definition will lead to a δ-peak in the distribution of the test

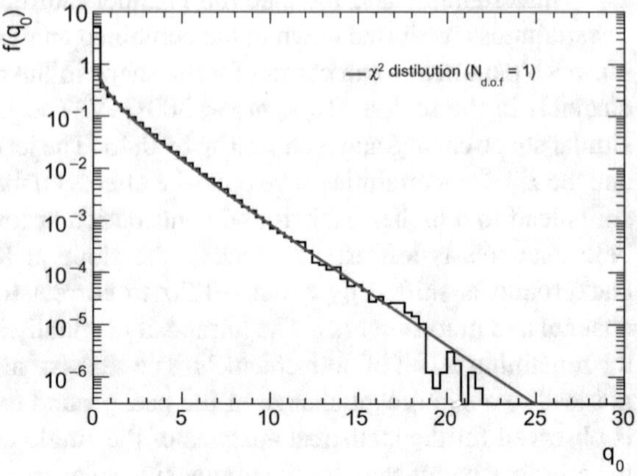

Fig. 10.3 Probability density function of the q_0 test statistic for background-only pseudo-experiments. The red curve shows the expected distribution for the asymptotic approximation

statistic at $q_0 = 0$. It represents a 50% occurrence that a background-only dataset will have a downward fluctuation. A probability density function for q_0 can be obtained from background-only pseudo-experiments. The pseudo-experiments are generated using the observed nuisance parameter values $\vec{\theta}_0^{obs}$. The observed nuisance parameter values are determined by RooStats using a maximum likelihood fit using the Minuit package [3]. An additional Gaussian probability density function is introduced for each nuisance parameter to avoid large shifts of single nuisance parameters. The nuisance parameter shifts $\vec{\theta}_0^{obs}$ determined by RooStats are very similar to the shifts shown in Fig. 10.2 which were determined with a slightly different method using BAT. The number of events in a bin is following a Poisson probability. From these pseudo-experiments a probability density function $f(q_0|0, \vec{\theta}_0^{obs})$ can be built. Figure 10.3 shows f for the muon channel obtained from 10,0000 pseudo-experiments. A peak at $q_0 = 0$ can be observed from the cases in which the background has a downward fluctuation. This peak has been ignored when normalizing the distribution of pseudo-experiments to unity.

The probability p_0 corresponding to a given experimental observation q_0^{obs} is evaluated as follows:

$$p_0 = P(q_0 \geqslant q_0^{obs}|\text{background-only}) = \int_{q_0^{obs}}^{\infty} f(q_0|0, \vec{\theta}_0^{obs})dq_0. \tag{10.7}$$

The p-values p_0 are usually translated into a scale of significance (z) in terms of Gaussian standard deviations:

$$z = \Phi^{-1}(1 - p_0) \tag{10.8}$$

Here, Φ^{-1} is the inverse of the cumulative distribution of the standard Gaussian probability density function. A discovery of a new particle and therefore the rejection of the background-only hypothesis is usually announced with a significance of at least 5σ ($z = 5$). This corresponds to a probability of $p = 2.87 \times 10^{-7}$.

10.3.2 Asymptotic Approximation

According to Wilks' theorem [4], the distribution of q_0 for background-only datasets follows in the asymptotic limit a χ^2 distribution for one degree of freedom. This behavior is illustrated in Fig. 10.3, where the solid red line corresponds to a χ^2 distribution for one degree of freedom. The probability p_0 can then be calculated directly from q_0 [5]:

$$p_0 = 1 - \Phi(\sqrt{q_0}). \tag{10.9}$$

The significance is then given by

$$z = \sqrt{q_0}. \tag{10.10}$$

This largely reduces the amount of computing time and computing resources required.

10.3.3 Look-Elsewhere Effect

The p-value calculated so far gives the probability that for a given W' pole mass the background fluctuates, creating a signal-like excess equal or larger than what is observed. It ignores the fact that W' masses over a large range in m_{T} have been tested and is therefore also called "local" p-value. The more masses are tested, the more likely it is that for a certain mass a small p-value is observed. This effect is called "look-elsewhere effect". It can be taken into account by calculating a "global" p-value which is the probability of measuring a local p-value somewhere in the background transverse mass spectrum which is at least as significant as the one observed in data. Calculating the global p-value requires a knowledge of the fluctuations inherent in purely background processes. The statistical behavior of the background can be modeled by observing excesses in an ensemble of pseudo-experiments generated under the background-only hypothesis. The distribution of the largest local significance found in each independent pseudo-experiment represents the ability of the background to exhibit false signals. From this distribution the global p-value can be calculated:

$$p_{global} = P(z_{local} \geq z_0^{obs} | \text{background-only}) \frac{1}{N_{pe}} = \int_{z_0^{obs}}^{\infty} n(z_0) dz_0, \tag{10.11}$$

where N_{pe} is the number of pseudo-experiments.

Figure 10.4 shows the global p-value expressed in Gaussian significance as a function of the local significance calculated from 10,0000, pseudo-experiments. Negative values of global significance indicate cases which are less signal-like than expected from the background-only case. The global p-value is typically about 1σ lower than the local p-value. The electron and muon channel have a very similar mapping.

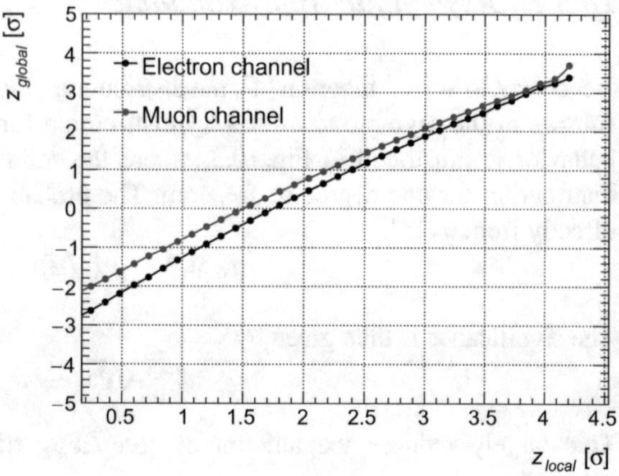

Fig. 10.4 Mapping from local to global significance in the electron and muon channel

More signal-like fluctuations are expected for the electron channel due to its superior electron energy resolution which makes it slightly more affected by single bin fluctuations. No mapping for the combination of both channels has been calculated, as the mapping will be similar to the electron channel due to the better electron energy resolution. Hence, the electron channel mapping was used for the combined channel.

10.3.4 Results

Figure 10.5 shows the observed p-values for the electron channel (top left), muon channel (top right) and the combination of both (bottom). The local significance is shown as a gray dashed line and the global significance as red dashed line. The p-values were obtained using the procedure described in Sect. 10.3.1 with the background estimation as discussed in Sect. 9.3 and the systematic uncertainties described in Sect. 9.5. Signal templates have been generated, following the procedure discussed in Sect. 9.1.1, in steps of 50 GeVfor pole masses ranging from 150 GeV to 3.5 TeV, thus leading to 67 signal templates in total.

The largest excesses in the electron channel are about 1.4σ local at $m_{W'} =$ 600 GeV and $m_{W'} = 2$ TeV. They correspond to about three bins around 600 GeV which are above the background expectation and two events which are observed near 2 TeV. The largest excess in the muon channel is about 1.8σ local at $m_{W'} = 350$ GeV. It corresponds to a single bin which is above the background expectation by about three times its statistical uncertainty. A broad excess of about 1σ is observed for masses above 1.5 TeV. It corresponds to the region $m_T > 1$ TeV where several bins are above the background expectation. Especially one bin at about 1.5 TeV in m_T in which four events are found. The excess does not disappear for higher $m_{W'}$, like in

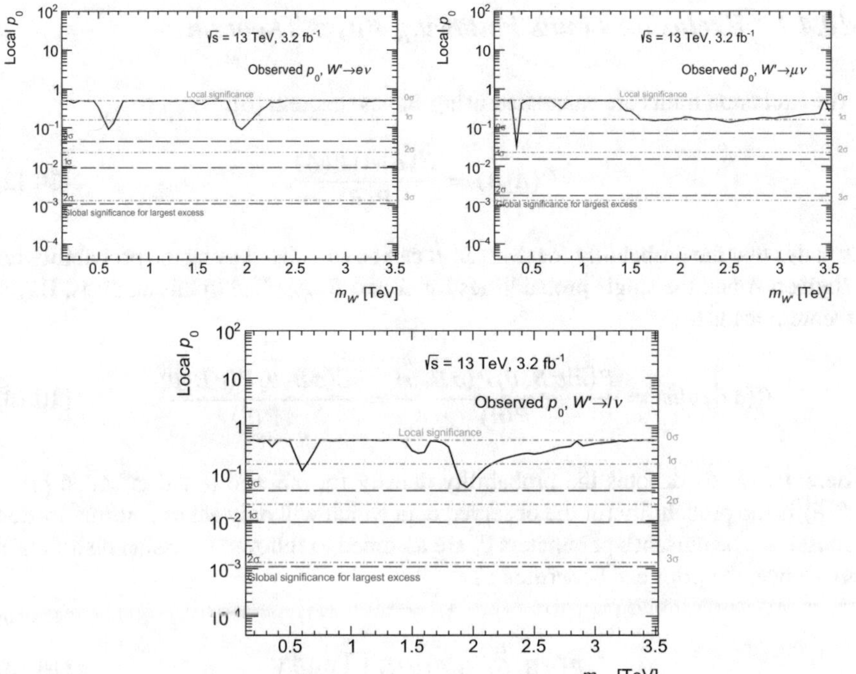

Fig. 10.5 The local significance derived from ideal W'_{SSM} signals between 0.15 and 3.5 TeV. Accompanying local significance levels are shown as gray dashed lines and global significance levels as red dashed lines. Shown are results for the electron (upper left), muon (upper right), and the combined channel (lower)

the electron channel, due to the worse m_{T} resolution in the muon channel. Signals with higher masses will due to the resolution also contribute to the lower m_{T} regions. The excesses at 350 and 600 GeV are reduced in the combination as they only appear in one of the channels. The excess at higher masses gets slightly stronger, rising up to 1.6σ local, since an excess in that region in both channels is observed. The global significance never exceeds one sigma. The observed data are therefore consistent with the Standard Model hypothesis.

10.4 Exclusion Limits

The statistical analysis performed showed no significant excess of data with respect to the expected SM background. Therefore exclusion limits will be calculated for the SSM W' model.

10.4.1 Exclusion Limits Following Bayes Theorem

The exclusion limits are calculated using Bayes theorem [6]:

$$P(A|B) = \frac{P(B|A)P(A)}{P(B)}. \tag{10.12}$$

It states that the probability for A given B can be calculated using the probability for B given A and the single probabilities for A and B. Applied to this analysis, Bayes theorem leads to

$$P(\sigma B, \vec{\theta}|\vec{n}) = \frac{P(\vec{n}|\sigma B, \vec{\theta})P(\sigma B, \vec{\theta})}{P(\vec{n})} = \frac{\mathcal{L}(\sigma B, \vec{\theta})P(\sigma B, \vec{\theta})}{P(\vec{n})}. \tag{10.13}$$

Here $P(\sigma B, \vec{\theta})$ denotes the probability density for σB and $\vec{\theta}$ and is called prior. $P(\vec{n})$ is the probability for the observed data which will only act as a normalization constant. The nuisance parameters θ_i are assumed to follow a Gaussian distribution Φ. Hence, the prior can be written as

$$P(\sigma B, \vec{\theta}) = P(\sigma B) \prod_{i=1}^{N_{\text{sys}}} \Phi(\theta_i), \tag{10.14}$$

where Φ denotes the probability density function for the standard normal distribution and $P(\sigma B)$ is the prior for the signal cross section times branching ratio. The choice of a proper prior $P(\sigma B)$ for the signal can be very controversial. In principle the prior knowledge from earlier W' searches can be used to strengthen the resulting limits. However, a conservative choice is made by assuming a "flat" prior, i.e. zero for $\sigma B < 0$ and constant for $\sigma B \geq 0$. This is a common choice in high-energy physics experiments and makes the results comparable between experiments. The prior is absorbed into a normalization constant together with $P(\vec{n})$. The explicit expression for the probability for σB takes the form (for $\sigma B \geq 0$)

$$P(\sigma B|\vec{n}) = \int P(\sigma B, \vec{\theta}|\vec{n})\, d\vec{\theta} = N \int \prod_{k=1}^{N_{\text{chan}}} \prod_{l=1}^{N_{\text{bin}}} \frac{\lambda_{kl}(\sigma, \vec{\theta})^{n_{kl}}\, e^{-\lambda_{kl}(\sigma,\vec{\theta})}}{n_{kl}!} \prod_{i=1}^{N_{\text{sys}}} \Phi(\theta_i)\, d\vec{\theta}, \tag{10.15}$$

where N is the normalization constant which absorbs the prior $P(\sigma B)$ and the probability for the data $P(\vec{n})$. It is determined by the condition

$$\int_0^\infty P(\sigma B|\vec{n})\, d\sigma = 1\,. \tag{10.16}$$

The integral over the nuisance parameters (called marginalization integral) is performed in BAT using Markov Chain MC. The number of channels N_{chan} is equal to 1 when the electron and muon channels are analyzed individually and 2 for their combination.

The upper limit σB_{limit} on the cross section times branching ratio can be computed by

$$\int_{\sigma B_{\text{limit}}}^{\infty} P(\sigma B|\vec{n})\, d\sigma = \delta\,, \tag{10.17}$$

i.e. it is the signal cross section times branching ratio for which the probability that the true value of σB is equal or larger than σB_{limit} given the data $\vec{n}$ is δ. The confidence level[2] is defined as $1-\delta$. Exclusions at 95% CL are commonly used in high energy physics experiments.

The observed limit σB_{limit}, given the data, is calculated using the procedure described above. The expected limit σB_{limit}, given the background-only hypothesis, is calculated using a sample of background-only pseudo-experiments. One pseudo-experiment is performed by generating sample values for all nuisance parameters θ_i according to their Gaussian priors. Thereafter the number of events in a bin l of channel k are sampled following a Poisson probability with the expectation value $b_{kl}(\vec{\theta})$ for the generated sample values of the nuisance parameters $\vec{\theta}$. The resulting m_{T} distribution is treated as actual data and a limit σB_{limit} is calculated following the above procedure. After a suitable number of pseudo-experiments, the expected limit is extracted by taking the median of the limit distribution. One and two sigma bands are extracted by taking the 68 and 95% quantiles of the limit distribution. These bands do not have any specific meaning in the Bayesian paradigm. They visualize how the limit could change by a one or two sigma fluctuation of the nuisance parameters and data.

10.4.2 Results

Figure 10.6 shows the resulting 95% CL limits on the cross section times branching ratio σB_{limit} for the electron channel (top left), muon channel (top right) and the combination of both (bottom). The limits were obtained using the SSM W' signal shapes with the background estimation as discussed in Sect. 9.3 and the systematic uncertainties described in Sect. 9.5. Signal templates have been generated, following the procedure discussed in Sect. 9.1.1, every 50 GeV in the pole mass range from 150 GeV to 6 TeV, leading to 118 signal templates in total. For each pole mass, a total of 1000 pseudo-experiments have been carried out to estimate the expected limit and the 68% and 95% uncertainty at each point. Tables with the observed and expected limits can be found in Appendix F. The expected limit is indicated as a dashed black line, which is surrounded by the 68% (95%) uncertainty band drawn in green (yellow). The observed limit is shown as a solid black line. For a given pole

[2]The confidence level is in the Bayesian paradigm often also called *credibility level*. To avoid confusion throughout this thesis the more common term *confidence level* is used.

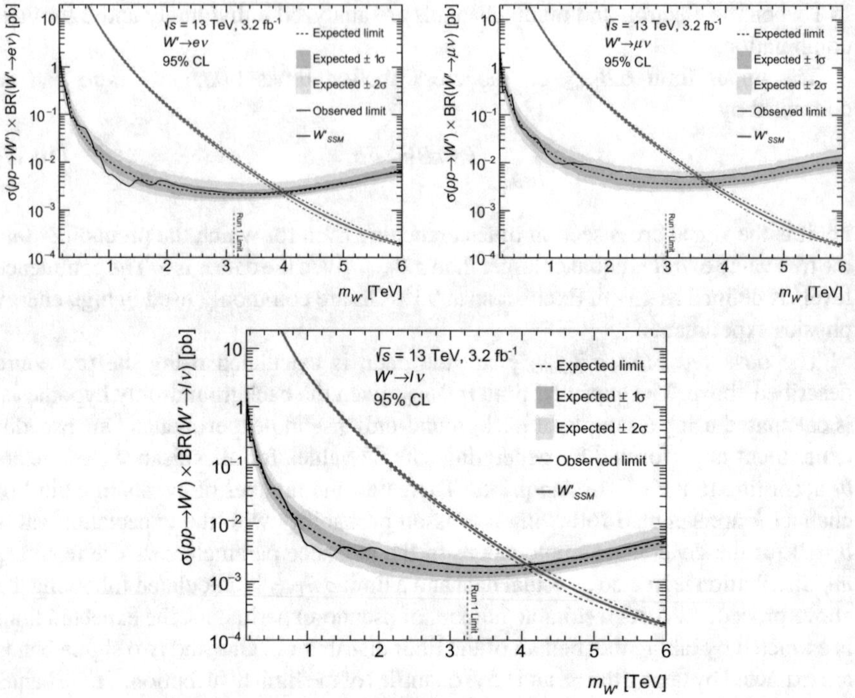

Fig. 10.6 The calculated 95% CL limits on the cross section times branching ratio are shown as a function of the pole mass of the SSM W'. The results are shown for the electron channel (top), the muon channel (middle) and the combination of both (bottom). The expected limit is indicated as a dashed black line, which is surrounded by the 68% (95%) uncertainty band drawn in green (yellow). The observed limit is shown as a solid black line. For a given pole mass, all values of σB which are above the value of the observed limit are excluded with 95% CL. The theory prediction for the SSM W' model is shown at NNLO in QCD as a red solid line surrounded by its uncertainty. Masses for the SSM W' boson below the intersection between observed limit and theory prediction are excluded with 95% CL

mass, all values of σB which are above the value of the observed limit are excluded with 95% CL. The curve of the expected limit is strongly falling over about three orders of magnitude up to pole masses of 3.5 TeV. The expected limit then bends up again due to the lower acceptance times efficiency for the higher pole masses (as shown in Fig. 9.4). The curve of the observed limit for the electron channel and muon channel agrees well with the expected limit and always lies within the 2σ uncertainty band. Slight excesses are observed in the electron channel at $m_{W'} = 600\,\text{GeV}$ and in the muon channel at $m_{W'} = 350\,\text{GeV}$ which were already discussed in Sect. 10.3.4. A deficit is observed in both channels around 1 TeV. The deficit is stronger in the electron channel where it is almost outside of the 2σ band. The deficits correspond to deficits in data in the region $m_{\text{T}} \approx 1$ TeV in Fig. 10.1. For higher pole masses the observed

Table 10.1 Expected and observed 95% CL lower limits on the W'_{SSM} mass in the electron and muon channel and their combination

Decay	$m_{W'}$ limit [TeV]	
	Expected	Observed
$W' \to e\nu$	3.99	3.96
$W' \to \mu\nu$	3.72	3.56
$W' \to \ell\nu$	4.18	4.07

limit is again above the expected limit due to the slight excess observed at high m_{T}. Whereas the excesses at $m_{W'} = 600\,\text{GeV}$ and $m_{W'} = 350\,\text{GeV}$ are smaller in the combined limit, the latter described deficits/excesses are becoming more pronounced as they appear in both channels in a similar range. The deficit at 1 TeV is slightly beyond the 2σ band. The limit in the electron channel is in general stronger than the limit in the muon channel due to its superior signal efficiency and resolution.

The theory prediction for the SSM W' model is shown as a red solid line. The cross section for each pole mass was obtained with PYTHIA at LO. A NNLO QCD theory correction is applied to the theory prediction. It was obtained by calculating the yield in the range $110\,\text{GeV} < m_{\text{T}} < 7\,\text{TeV}$ for each signal template at LO and after applying the QCD theory correction described in Sect. 9.1.1. The ratio of these two yields defines the theory correction and fully takes the shape and normalization differences between LO and NNLO into account. Uncertainties on σB from the PDF, α_s and scale are shown as a red-dashed line. They were determined as described in Sect. 9.5. Masses for the SSM W' boson below the intersection between observed limit and theory prediction are excluded with 95% CL. Table 10.1 lists the observed and expected mass limits on the SSM W' for the electron, muon and combined channel. The expected mass limit in the electron channel is 3.99 TeV and therefore well in agreement with the observed limit of 3.96 TeV. The expected mass limit in the muon channel is 3.72 TeV, which is slightly higher than the observed limit of 3.56 TeV. The difference in expected and observed limit is caused by the excess at high m_{T}. This excess does affect the muon channel in a stronger way as, due to the worse resolution compared to the electron channel, also signals at higher masses contribute significantly at lower m_{T} values. The expected muon channel limit is 270 GeV lower than the expected electron channel limit due to the lower signal efficiency and worse resolution. The expected combined limit is with 4.18 TeV about 200 GeV stronger than the electron channel limit. The observed combined limit is 4.07 TeV. These limits on $m_{W'}$ are about 800 GeV stronger than all previous published results using data taken at $\sqrt{s} = 8\,\text{TeV}$ [7, 8].

10.4.3 Comparison to Other Analyses

Exclusion limits with 95% CL obtained by other analyses are summarized in Table 10.2 and discussed in the following.

Table 10.2 Expected and observed 95% CL lower limits on the W'_{SSM} mass in the electron channel, the muon channel, and their combination

Analysis by	$\sqrt{s}$ [TeV]	L_{int}[fb^{-1}]	$W' \to e\nu$ limit [TeV]		$W' \to \mu\nu$ limit [TeV]		$W' \to \ell\nu$ limit [TeV]	
			Expected	Observed	Expected	Observed	Expected	Observed
ATLAS [7]	8	20.3	3.13	3.13	2.97	2.97	3.17	3.24
CMS [8]	8	19.7	3.18	3.22	3.09	2.99	3.26	3.28
ATLAS [9]	13	3.2	3.99	3.96	3.72	3.56	4.18	4.07
CMS [10]	13	2.2	3.8	3.8	3.8	4.0	4.2	4.4
ATLAS [11]	13	13.3	4.59	4.64	4.33	4.19	4.77	4.74

As already discussed in Sect. 2.3.3, the analyses performed by ATLAS and CMS at $\sqrt{s} = 8$ TeV, using the full 2012 data set, excluded SSM W' masses below 3.24 and 3.28 TeV, respectively. The obtained combined exclusion limit by the analysis of the 2015 data set is 4.07 TeV, i.e., about 800 GeV stronger. Figure 7.1 shows the ratio of the exclusion limit on the cross section and the predicted SSM cross section as a function of the W' pole mass. Exclusion limits from this analysis and the ATLAS analysis using 8 TeV data are shown. It can be seen that the presented analysis substantially improves the cross section limits also at low masses up to 500 GeV. This is due to the shape-based limit setting in this analysis which is more sensitive in this region than the single-bin approach used in the previous analysis. In addition triggers with a lower p_T threshold were used in the electron channel. In the intermediate mass range between 500 GeV and 2.5 TeV both limits are comparable due to the higher integrated luminosity of the data set used for the 8 TeV analysis. From 2.5 TeV onwards, this analysis provides more stringent limits due to the higher center of mass energy (Fig. 10.7).

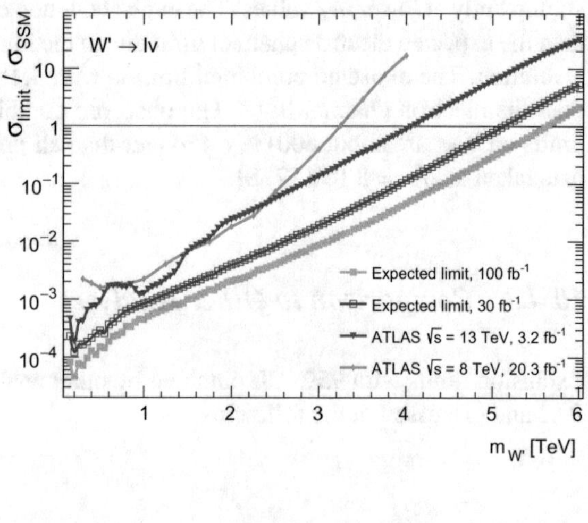

Fig. 10.7 Normalised cross-section limits ($\sigma_{\text{limit}}/\sigma_{\text{SSM}}$) for W' bosons as a function of mass for the measurement of a previous ATLAS analysis [7], for this measurement, and the expected exclusion limit for 30 and 100 fb^{-1}. The cross-section calculations assume the W' has the same couplings as the SM W boson. The region above each curve is excluded at 95% CL

CMS has also performed an preliminary analysis of the 2015 data set studying W' masses above 1 TeV. Due to a problem with the solenoid magnet they only collected 2.2 fb^{-1} of data. Nevertheless, they derived a stronger exclusion limit of 4.4 TeV. Comparing the results of the individual channels shows that the expected and observed limits in the electron channel are with 3.8 TeV about 200 GeV weaker than the result of this analysis. They obtain an efficiency times acceptance of 75% for a W' boson with a mass of 3 TeV which is comparable but slightly lower than the efficiency times acceptance obtained by this analysis (80%). Hence, the lower CMS limit in the electron channel results from the slightly lower acceptance times efficiency and the smaller amount of integrated luminosity collected. However, in the muon channel, despite the smaller data set, CMS obtains an expected limit of 3.8 TeV and an observed limit of 4.0 TeV. Their expected exclusion limit in the muon channel is therefore equal to the electron channel. For this analysis the expected exclusion limit is about 300 GeV weaker in the muon channel. The difference is coming from the muon efficiency of both detectors. While CMS obtains for $m_{W'} = 3$ TeV the same acceptance times efficiency of 75% as in the electron channel, this analysis only reaches about 51%. The difference is directly coming from the differences of the detectors. While the CMS muon spectrometer has a very high trigger efficiency (above 90% in most regions), the ATLAS muon spectrometer has only about 70–80% trigger efficiency. The very stringent quality criteria in the ATLAS analysis and the veto on some muon chambers reduces further the efficiency and leads to the observed differences. Regardless of these differences, the expected muon limits of both experiments are with 3.72 TeV and 3.8 TeV very similar but while CMS observes a small deficit which is resulting in a stronger limit of 4.0 TeV, in this analysis a small excess is observed resulting in a weaker limit of 3.56 TeV. This difference translates also to the combined limit. The expected limits are again very similar with 4.17 TeV and 4.2 TeV, while the observed limits differ with 4.07 and 4.4 TeV, respectively.

Very recently, the ATLAS collaboration published a preliminary result for the ICHEP summer conference using 13.3 fb^{-1} of data collected in the year 2015 and 2016 at $\sqrt{s} = 13$ TeV. With the larger integrated luminosity, masses below 4.74 TeV are excluded.

10.4.4 Impact on Other BSM Models

In the following the obtained results shall be briefly discussed in the light of other BSM models.

In the ATLAS and CMS analysis of the 2012 data also dark matter models have been considered (see Sect. 2.3.3). It was found that the signal models which gave the strongest limits were assuming constructive interference which was, as later pointed out, violating the electroweak gauge invariance [12]. Therefore these models have not been considered any longer. The effective field theory models have been replaced by simplified models in which the dark matter production is mediated by a Z' boson [13]. The sensitivity of this analysis for models in which a W is produced by initial state radiation and models in which the W is produced in association with the dark matter pair has been studied and was found to be very small. These models where as a consequence not considered. Limits have been set on these models in an analyses where the W decays hadronically [14]. The limits this analysis would yield are expected to be similar but most likely would be a bit worse.

The W^* model which is also discussed in Sect. 2.3.3 has a slightly different signal shape than the SSM W'. In the analysis of the 8 TeV data it lead nevertheless to very similar limits of $m_{W^*} > 3.21$ TeV. The limits on this model for this analysis are hence expected to be also in the 4 TeV range.

The right handed W'_R should have the same signal shape as the SSM W'. The limits on cross section times branching ratio should therefore be directly translatable into limits on the cross section of these bosons as long as the handedness of the couplings to quarks and leptons are equal [15].

The SUSY model which was introduced in Sect. 2.3.3 was only brought up recently by the authors and has so far never been considered in any of the W' searches. It is thus not possible to make any claim about the sensitivity.

References

1. Caldwell A, Kollar D, Kroninger K (2009) BAT: the bayesian analysis Toolkit'. Comput Phys Commun 180: 2197–2209 (2009). https://doi.org/10.1016/j.cpc.2009.06.026, arXiv: 0808.2552 [physics.data-an]
2. Moneta L et al (2010) The RooStats project. PoS(ACAT2010)057. http://pos.sissa.it/archive/conferences/093/057/ACAT2010_057.pdf
3. Lazzaro A, Moneta L (2010) MINUIT package parallelization and applications using the RooFit package. J Phys Conf Ser 219:042044. https://doi.org/10.1088/1742-6596/219/4/042044
4. Wilks SS (1938) The large-sample distribution of the likelihood ratio for testing composite hypotheses. Ann Math Stat 9(1):60–62. https://doi.org/10.1214/aoms/1177732360
5. Cowan G et al (2011) Asymptotic formulae for likelihood-based tests of new physics. Eur Phys J C71: 1554. [Erratum: Eur Phys J C73:2501 (2013)]. https://doi.org/10.1140/epjc/s10052-011-1554-0,10.1140/epjc/s10052-013-2501-z, arXiv: 1007.1727 [physics.data-an]
6. Cowan G (1998) Statistical data analysis. Oxford University Press, Oxford
7. Aad G et al (2014) Search for new particles in events with one lepton and missing transverse momentum in pp collisions at ps = 8 TeV with the ATLAS detector. JHEP 09: 037 (2014). https://doi.org/10.1007/JHEP09(2014)037, arXiv: 1407.7494 [hep-ex]
8. Khachatryan V et al (2015) Search for physics beyond the standard model in final states with a lepton and missing transverse energy in proton-proton collisions at sqrt(s) = 8 TeV'. Phys Rev D91.9: 092005. https://doi.org/10.1103/PhysRevD.91.092005, arXiv: 1408.2745 [hep-ex]
9. Aaboud M et al (2016) Search for new resonances in events with one lepton and missing transverse momentum in pp collisions at ps = 13 TeV with the ATLAS detector. arXiv: 1606.03977 [hep-ex]
10. Search for SSM W' production, in the lepton+MET final state at a center-of-mass energy of 13 TeV. Technical report CMS-PAS-EXO-15-006. Geneva: CERN (2015). https://cds.cern.ch/record/2114864
11. Search for new resonances decaying to a charged lepton and a neutrino in pp collisions at ps = 13 TeV with the ATLAS detector. Technical report, ATLAS-CONF-2016-061. Geneva: CERN (2016). https://cds.cern.ch/record/2206177

12. Haisch U, Re E (2015) Simplified dark matter top-quark interactions at the LHC. JHEP 06: 078. https://doi.org/10.1007/JHEP06(2015)078, arXiv: 1503.00691 [hep-ph]
13. Abercrombie D et al (2015) Dark matter benchmark models for early LHC run-2 searches: report of the ATLAS/CMS dark matter forum. In: Boveia A et al (eds). arXiv: 1507.00966 [hep-ex]
14. Search for dark matter produced in association with a hadronically decaying vector boson in pp collisions at ps = 13 TeV with the ATLAS detector at the LHC. Technical report, ATLAS-CONF-2015-080. Geneva: CERN (2015). https://cds.cern.ch/record/2114852
15. Haber HE (1984) Signals of new W's And Z's'. In: Proceedings of the workshop on electroweak symmetry breaking, Berkley, USA, 3–22 June 1984. http://www-public.slac.stanford.edu/sciDoc/docMeta.aspx?slacPubNumber=SLAC-PUB-3456

Chapter 11
Conclusion and Outlook

A search for a new heavy charged gauge boson, a so-called W', has been performed in the final state of an electron or muon and missing transverse momentum. Those new gauge bosons are predicted by some theories extending the Standard Model gauge group to solve some of its conceptual and experimental problems. The analyzed data set was recorded by the ATLAS experiment during proton-proton collisions at a center of mass energy of $\sqrt{s} = 13$ TeV and corresponds to an integrated luminosity of $3.2\,\text{fb}^{-1}$.

The electron and muon transverse mass spectrum has been measured and transverse masses up to about 2 TeV have been observed. The expected amount of Standard Model background has been estimated using Monte Carlo simulations and data-driven methods. The main contribution to the background is arising from high-p_T electrons and muons produced by the decay of an off-shell W boson. Further backgrounds are arising from processes including a top- or antitop-quark, from jets misidentified as leptons, and from diboson processes, where pairs of W and/or Z bosons were produced.

The Standard Model expectation has been compared to data. Possible deviations have been quantified in terms of local and global significances with a likelihood ratio test. The largest excesses observed are around 1.4 and 1.8σ local in the electron and muon channel, respectively. When combining both channels, an excess of 1.6σ at around $m_{W'} = 2$ TeV is observed. The global significance of these excesses is well below 1σ and hence the data are compatible with the Standard Model only hypothesis.

As a consequence, exclusion limits have been set on the mass of a Sequential Standard Model W' and masses below 3.56 and 3.96 TeV are excluded with 95% confidence level (CL) in the electron and muon channel, respectively. A combination of both channels leads to an improved exclusion limit of $m_{W'} > 4.07$ TeV. The obtained exclusion limit is about 800 GeV stronger than the exclusion limit obtained

M. Zinser, *Search for New Heavy Charged Bosons and Measurement of High-Mass Drell-Yan Production in Proton-Proton Collisions*, Springer Theses,
https://doi.org/10.1007/978-3-030-00650-1_11

from ATLAS and CMS analyses using the 2012 data set at $\sqrt{s} = 8$ TeV. Also the cross section limit at low masses improved partially by about an order of magnitude.

Future analyses could be improved by trying to recover some of the efficiency losses in the muon channel. In the presented analysis, muons which fall into regions in which the alignment of the muon chambers is not well enough understood have been vetoed. These regions will be further studied and are expected to be included in future analyses. It could be furthermore studied if the very stringent requirements on the muon quality could be relaxed without increasing the risk of fake muons imitating a signal event and without worsening the momentum resolution too much. In general a cut on the balance between the transverse momentum of the lepton and the missing transverse momentum could be introduced as events coming from a decay of a heavy resonance are expected to be well balanced. This has briefly been studied while developing the analysis methods, but was not applied as no improvement at high transverse masses was observed. However, this cut could be revisited to test if intermediate transverse mass ranges could profit. The backgrounds arising from top- and/or anti-top quarks and diboson production had to be extrapolated due to limited statistics in the simulated Monte Carlo samples. A fit was performed to obtain an estimate at very high transverse masses. A more robust estimation can be achieved by producing the simulated samples of these processes in bins of transverse mass. A first test using the SHERPA generator led to promising results. Besides the systematic uncertainties arising from the background extrapolation, theory uncertainties arising from the knowledge of the parton distribution functions (PDFs) are the dominating uncertainties at high transverse masses. The presented analysis is relatively insensitive to systematic uncertainties. If more data are collected the tail at high transverse masses will be further populated and the knowledge of these PDFs will become important. An improved understanding can be obtained by precisely measuring Standard Model processes at very high masses.

Figure 10.7 in the previous section shows the ratio of the exclusion limit on the cross section and the predicted SSM cross section as a function of the W' pole mass. Exclusion limits (95% CL) from this analysis and an earlier ATLAS analysis are shown. In addition, also expected limits are shown for data sets of 30 (expected until end of 2016) and 100 fb^{-1} (expected until end of 2017). The expected limits have been calculated by using the Standard Model background calculated for this analysis as pseudo-data and scaling it according to the luminosity. For 30 fb^{-1} W' masses are expected to be excluded around 5.15 TeV and for 100 fb^{-1} around 5.6 TeV.[1] It can already be seen that the exclusion limit between the data set expected for 2016 and 2017 shows a smaller increase in mass limit than the expected 2016 result with respect to the 2015 result. The exclusion limits will hence increase only slowly with more data. The total expected luminosity delivered by the LHC in its lifetime is 3000 fb^{-1}. For this amount of data W' masses are expected to be excluded up to

[1] Author's comment: When preparing this thesis for the Springer publication (July 2018), the mentioned searches have been performed by the ATLAS collaboration, giving unfortunately as well no sign for a signal. A search using 36 fb^{-1} of data excluded with 95% C.L. W' masses below 5.1 TeV [1] and a search using 80 fb^{-1}, masses below 5.6 TeV [2]. These results are in good agreement with the expected exclusion limits when writing this thesis.

around 7 TeV. Plans are being made for a proton-proton collider located at CERN, operated at a center of mass energy of 100 TeV. This collider would significantly increase the reach in W' masses. With $1000\,\mathrm{fb}^{-1}$ of collected data at such a collider SSM W' bosons with masses up to 31.6 TeV [3] could be discovered and excluded up to masses of 35 TeV [4].

References

1. Aaboud M et al (2018) Search for a new heavy gauge boson resonance decaying into a lepton and missing transverse momentum in 36 fb^{-1} of pp collisions at $\sqrt{s} = 13$ TeV with the ATLAS experiment. Eur Phys J C78.5:401. https://doi.org/10.1140/epjc/s10052-018-5877-y, arXiv:1706.04786 [hep-ex]
2. Search for a new heavy gauge boson resonance decaying into a lepton and missing transverse momentum in 79.8 fb^{-1} of pp collisions at $\sqrt{s} = 13$ TeV with the ATLAS experiment. Technical report ATLAS-CONF-2018-017. Geneva: CERN (2018). https://cds.cern.ch/record/2621303
3. Rizzo TG (2014) Exploring new gauge bosons at a 100 TeV collider. Phys Rev D89.9:095022. https://doi.org/10.1103/PhysRevD.89.095022, arXiv:1403.5465 [hep-ph]
4. T Golling et al. Physics at a 100 TeV pp collider: beyond the Standard Model phenomena. Technical report. CERN-TH-2016-111 (2016). arXiv:1606.00947. https://cds.cern.ch/record/2158187

Part IV
High-Mass Drell-Yan Cross Section Measurement at $\sqrt{s} = 8$ TeV

Chapter 12
Motivation

For tests of the Standard Model and measurements of its parameters, precise predictions of the processes at the LHC are needed. To obtain a high level of accuracy for these predictions, a very good understanding of the structure of the proton is essential. In this context, the knowledge of the parton distribution functions of the proton plays a key role. For example, the PDF uncertainties can be one of the largest uncertainties on the predictions of the Higgs production cross sections [1]. It was already seen in the analysis discussed in the previous chapters that at very high transverse masses the PDF uncertainties are the dominating theoretical uncertainties. This is also true for searches for new heavy neutral gauge bosons (commonly called Z'). These uncertainties are due to the not well constrained high-x region in the PDFs. The measurement of the high-mass Drell–Yan cross section gives access to the high-x quark and antiquark information. Measuring at high invariant masses and high dilepton rapidities $y_{\ell\ell}$ gives access to even higher values of x, as discussed in Sect. 2.2.3 and visible in Fig. 2.8. The measurement of this process at high invariant mass can furthermore be sensitive to electroweak corrections which have not yet been constrained [2].

The $\gamma\gamma$ initiated dilepton pair production via the photon induced process, introduced in Sect. 2.2.4, has a significant contribution at high invariant masses. The photon PDF of the proton is only weakly constrained. It is an important background for example to searches in the dilepton invariant mass spectrum and a measurement of this process is therefore of importance. A measurement of the pseudorapidity separation of the two leptons $\Delta\eta_{\ell\ell}$ gives a possibility to separate the photon induced contribution from the Drell–Yan production. The Drell–Yan process dominates due to its large s-channel contribution at low values of absolute pseudorapidity separation while the photon induced process especially contributes at high values due to the t-channel contribution.

M. Zinser, *Search for New Heavy Charged Bosons and Measurement of High-Mass Drell-Yan Production in Proton-Proton Collisions*, Springer Theses,
https://doi.org/10.1007/978-3-030-00650-1_12

In the following chapters an analysis is presented measuring the Drell–Yan cross section $pp \rightarrow Z/\gamma^* + X \rightarrow \ell^+\ell^- + X$ ($\ell = e, \mu$) and photon induced cross section in the range 116 GeV $< m_{\ell\ell} <$ 1500 GeV. Three cross section measurements are performed, one single-differential measurement as a function of invariant mass $m_{\ell\ell}$, and two double-differential measurements. The first double-differential measurement is performed as a function of invariant mass $m_{\ell\ell}$ and absolute dilepton rapidity $|y_{\ell\ell}|$ and a second measurement as a function of invariant mass $m_{\ell\ell}$ and absolute pseudorapidity separation $|\Delta\eta_{\ell\ell}|$. The data used for this measurement was collected by the ATLAS experiment during 2012 at $\sqrt{s} = 8$ TeV and corresponds to an integrated luminosity of 20.3 fb^{-1}.

The measurement as a function of absolute rapidity will measure the rapidity spectrum up to $|y_{\ell\ell}| = 2.4$. This measurement will therefore probe x values as low as about $x \approx 10^{-3}$ going up to $x \approx 1$. The dominant part of the data is distributed in the region $x = 10^{-2}$ to $x = 10^{-1}$ (see Appendix G).

Figure 12.1 shows the measured cross section in the final state of an electron and positron performed by the ATLAS collaboration [3] using data with an integrated luminosity of 4.9 fb^{-1}, collected at a center of mass energy of 7 TeV. The measurement was performed single-differential as a function of invariant mass and only in the e^+e^--channel. A comparison to various PDFs is shown in the lower panels and shows a systematic offset for all PDFs. Additional measurements, especially of the muon channel, will give further insight whether these differences are originating from a systematic effect or not. A measurement of both channels is an important cross check and the uncertainty on the measurement can be reduced by combining these.

The following chapters will concentrate on the measurement of the electron channel cross section which I already started to work on in my master thesis [5]. Since then several improvements have been made in the context of this thesis. Besides adding the measurement as a function of $|\Delta\eta_{\ell\ell}|$ also the uncertainties of the measurement have been significantly reduced. The cuts of the analysis have partially been refined to further reduce background processes, a lot of systematic checks have been performed, and additional sources of systematic uncertainties have been studied. In the meantime also a cross section measurement of the muon channel has been performed in the context of another thesis [6]. It will be used in this work to calculate combined cross sections with reduced uncertainties. These combined cross sections have been compared to different theory predictions and interpreted in terms of sensitivity to PDFs. The results are published in JHEP [7].

The analysis is structured as follows. In Chap. 13 first briefly the analysis strategy is introduced. Chapter 14 describes the analysis of the electron channel. In Sect. 14.1 the Monte Carlo samples are discussed, followed by the description of the event and object selection in Sect. 14.2. Section 14.3 describes the estimation of the multijet and W+jets background which is arising from jets faking the electron signature. The chapter closes with a comparison of the selected data with the expected signal and background in Sect. 14.4. In Chap. 15 the methodology of the cross section measurement is explained. In Sect. 15.1 first the resolution of the observables and thereafter the chosen binning and its purity is discussed. The unfolding method is

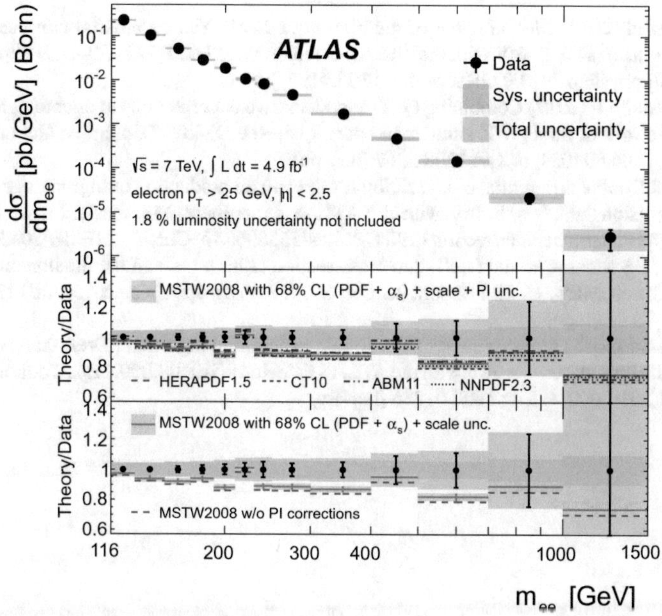

Fig. 12.1 Fiducial Drell–Yan cross section at $\sqrt{s} = 7\,\text{TeV}$, binned in invariant mass of the electron pair in the range 116 to 1500 GeV. Shown is the measured fiducial cross section with its statistical uncertainty. The green bands show the systematic and total uncertainty, excluding the 1.8% luminosity uncertainty. Different theory predictions, calculated at NNLO with FEWZ [4] using different PDFs are shown. The predictions include corrections for photon induced processes and W/Z radiation. Figure taken from Ref. [3]

discussed in Sect. 15.2. The systematic uncertainties on the cross section are studied in Sect. 15.3 and the unfolded cross section with their uncertainties are afterwards shown in Sect. 15.4. Following is Chap. 16 with a brief discussion of the muon channel cross section measurement which is needed as an input for the following Chap. 17. Here first the combination of the electron and muon channel is described in Sect. 17.1. The combined cross sections are afterwards compared to theory predictions in Sect. 17.2. An interpretation of the obtained results in terms of sensitivity to PDFs is discussed in Sect. 17.3. The analysis discussion closes with a conclusion and an outlook in Chap. 18.

References

1. Andersen JR et al (2013) In: Heinemeyer S et al (eds) Handbook of LHC Higgs cross sections: 3. Higgs properties. https://doi.org/10.5170/CERN-2013-004, arXiv: 1307.1347 [hep-ph]
2. Boughezal R, Li Y, Petriello F (2014) Disentangling radiative corrections using the high-mass Drell–Yan process at the LHC. Phys Rev D89.3:034030. https://doi.org/10.1103/PhysRevD.89.034030, arXiv: 1312.3972 [hep-ph]

3. Aad G et al (2013) Measurement of the high-mass Drell–Yan differential cross-section in pp collisions at $\sqrt(s) = 7$ TeV with the ATLAS detector. Phys Lett B725:223–242. https://doi.org/10.1016/j.physletb.2013.07.049, arXiv: 1305.4192 [hep-ex]
4. Li Y, Petriello F (2012) Combining QCD and electroweak corrections to dilepton production in the framework of the FEWZ simulation code. Phys Rev D86:094034. https://doi.org/10.1103/PhysRevD.86.094034, arXiv: 1208.5967 [hep-ph]
5. Zinser M Double differential cross section for Drell–Yan production of high-mass $e + e^-$-pairs in *pp* collisions at $\sqrt{s} = 8$ TeV with the ATLAS experiment. MA thesis. Mainz U., 07 Aug 2013. http://inspirehep.net/record/1296478/files/553896852_CERN-THESIS-2013-258.pdf
6. Hickling RS Measuring the Drell–Yan cross section at high mass in the dimuon channel. PhD thesis. Queen Mary, U. of London, 12 Sep 2014. http://inspirehep.net/record/1429540/files/fulltext_WiZAbu.pdf
7. Aad G et al (2016) Measurement of the double-differential high-mass Drell–Yan cross section in pp collisions at $\sqrt{s} = 8$ TeV with the ATLAS detector. JHEP 08:009. https://doi.org/10.1007/JHEP08(2016)009, arXiv: 1606.01736 [hep-ex]

Chapter 13
Analysis Strategy

Aim of the analysis is to measure the cross section of $\ell^+\ell^-$-pairs ($\ell = e, \mu$) from neutral current Drell-Yan production and photon induced production and to measure the invariant mass $m_{\ell\ell}$ dependence, the absolute rapidity $|y_{\ell\ell}|$ dependence, and dependence on the absolute pseudorapidity separation of the two leptons $|\Delta\eta_{\ell\ell}|$. The Drell-Yan process typically leaves a very clean signature of two leptons with high transverse momenta which give a high invariant mass. Figure 13.1 shows an event display of the Drell-Yan dielectron event with the highest invariant mass. The event contains two electron candidates originating from the same vertex with very high transverse momenta of $p_T = 588$ GeV and $p_T = 584$ GeV. The green towers depict the energy deposits in the EM calorimeter. The two electron candidates are back-to-back in the transverse plane and have an invariant mass of $m_{ee} = 1526$ GeV. Given this signature, the general strategy of the analysis is to select events with two high-p_T lepton candidates giving a high invariant mass.

The background for this processes is typically very low. The leading background originates from $t\bar{t}$ production. The top- and antitop-quarks dominantly decay via $t \rightarrow Wb$, therefore leading to two W bosons plus additional b-quarks. The W bosons can further decay into leptons leading to a final state containing two leptons. A further background originates from multijet and W+jets events where either at least two jets (multijet) or one jet (W+jets) is misidentified as lepton. This background has a much higher contribution in the electron channel, since the probability for a jet to fake a muon signature is very low. The multijet background is, like the photon induced process, dominantly a t-channel process. It is therefore more forward and the jets are measured at higher $|\eta|$. A precise estimation of this background in the electron channel is therefore crucial since both processes have very similar kinematics. Finally, a background arises from the diboson processes WW, WZ and ZZ which can also lead to two or more leptons. MC simulation reliably predicts all backgrounds containing at least two real leptons and will be used to estimate these.

M. Zinser, *Search for New Heavy Charged Bosons and Measurement of High-Mass Drell-Yan Production in Proton-Proton Collisions*, Springer Theses,
https://doi.org/10.1007/978-3-030-00650-1_13

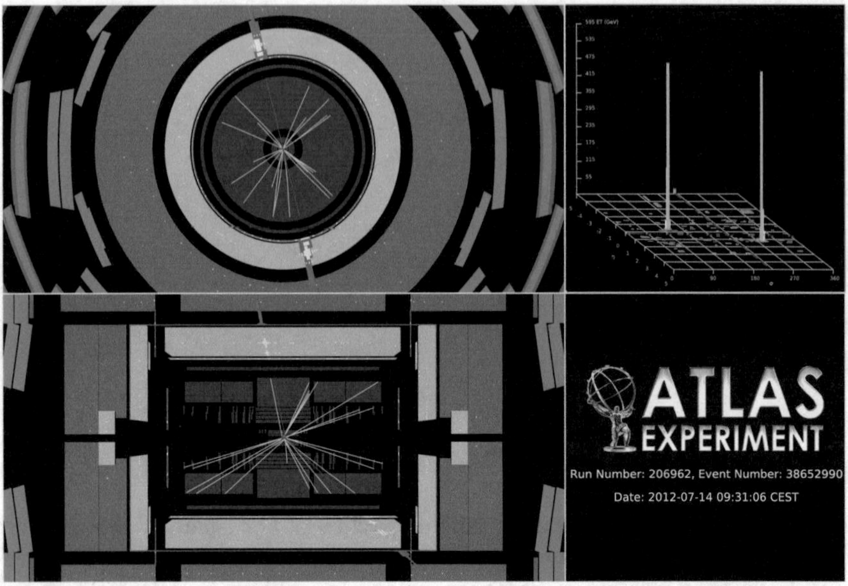

Fig. 13.1 Event with the electron pair that has the highest invariant mass in the 2012 data set is shown ($m_{ee} = 1526$ GeV). On the upper left, the r-ϕ-plane and on the lower left, the r-η-plane of the detector is shown. On the upper right, the energy deposition in the electromagnetic calorimeter is shown in the ϕ-η-plane. Tracks with $p_T > 0.5$ GeV are shown and colored depending on the vertex they originate from. The event display was made using ATLANTIS [1]

Backgrounds containing one or more misidentified jets need to be estimated from data as MC simulation in general fails to accurately describe the probability that a lepton signature is faked.

The backgrounds are subtracted from the data to obtain the pure Drell-Yan and photon induced spectrum. The sum of the Drell-Yan and photon induced event yields are then corrected for detector effects using a MC based unfolding technique. The electron and muon channel are measured in slightly different phase spaces due to the different coverage of the calorimeters and the muon spectrometer. Both event yields are therefore further extrapolated to a common fiducial phase space with an acceptance correction. The corrected yield in the common fiducial phase space is finally divided by the integrated luminosity of the data to obtain the sum of the Drell-Yan and photon induced cross section. The estimation of the systematic uncertainties is of special importance for this analysis as they are, especially at lower invariant mass, a limiting factor for the precision of the extracted cross sections. At higher invariant mass, the cross section measurement is limited by the available statistics in data. Finally, the cross section for both channels is combined using a statistical procedure to further reduce the statistical and systematic uncertainties by exploiting the fact that some systematic uncertainties are correlated between both channels.

The combined measured cross section can then be compared to state of the art theory predictions and sensitivity studies can be performed to show the impact of the measurement on the uncertainties of modern PDFs.

Reference

1. Atlantis. http://atlantis.web.cern.ch/atlantis/

Chapter 14
Analysis of the Electron Channel

The following chapter describes the analysis of the electron channel. First, all MC samples used for the analysis are discussed in Sect. 14.1. The data used and the event and electron selection criteria are discussed in Sect. 14.2. The determination and validation of the multijet and W+jets background is discussed in Sect. 14.3 and selected data are compared to the background and signal expectation in Sect. 14.4.

14.1 Monte Carlo Simulations

The following section contains a description of the Monte Carlo samples used in this analysis. The first part describes the signal simulations and a second part the simulations for background processes. All MC samples used in this analysis have been centrally provided by the ATLAS collaboration [1].

14.1.1 Simulation of Signal Processes

14.1.1.1 Drell-Yan Process

The matrix element of the hard scattering process for the Drell-Yan process ($pp \rightarrow Z/\gamma^* + X \rightarrow ee + X$) is generated with the CT10 PDF set [2] at NLO in QCD using POWHEG [3]. PYTHIA 8.170 [4] is used for the modeling of the parton shower, hadronization and particle decays and QED FSR is simulated using Photos [5]. The cross section for Z/γ^* production is strongly falling with higher invariant masses. Very large statistics would be needed to sufficiently populate the tail at high invariant masses. Therefore, the signal is produced in 15 slices in invariant mass m_{ee} to save

M. Zinser, *Search for New Heavy Charged Bosons and Measurement of High-Mass Drell-Yan Production in Proton-Proton Collisions*, Springer Theses,
https://doi.org/10.1007/978-3-030-00650-1_14

computing time. The slices start from 120 GeV $< m_{ee} <$ 180 GeV and reach up to $m_{ee} >$ 3000 GeV. The number of generated events for each sample reaches from 5, 000, 000 down to 100, 000 corresponding to an integrated luminosity of 508 fb^{-1} to 1.9×10^8 fb^{-1}. Additionally, three inclusive samples are generated over the whole mass range using the same MC setup. The three samples are filtered at the generation stage for events containing exactly two, one or no electron with $p_T >$ 10 GeV and $|\eta| < 2.8$. The efficiency for the filters is 55.65%, 31.47%, 12.89% for the two, one or no electron sample, respectively. The three inclusive samples have been generated with 50 million, 10 million and 3 million events, corresponding to an integrated luminosity of 81 fb^{-1}, 29 fb^{-1} and 21 fb^{-1}. The cross section times branching ratio σB predicted by POWHEG for the sum of the three inclusive samples is 1.109 nb. Events generated with an invariant mass of $m_{ee} >$ 120 GeV are rejected to avoid overlap between the inclusive samples and the mass-binned samples. Detailed information about the MC samples can be found in the appendix in Table H.1.

Figure 14.1 shows the resulting invariant mass spectrum for the POWHEG Z/γ^* process. The colored lines show the individual mass bins and the black line shows the resulting sum of all samples scaled up by a factor of two for easier visibility. The inclusive samples filtered for exactly two, one, and no electrons in the required phase space are respectively shown in blue, red, and yellow. At around 91 GeV the Breit-Wigner resonance of the Z-boson is visible. The cross section is steeply falling towards higher masses. The samples provide sufficient statistics up to several TeV.

For systematic checks an alternative Drell-Yan MC sample is simulated using the same PDF with the MC@NLO 4.09 [6, 7] generator interfaced with HERWIG++ [8]. Here MC@NLO is used to generate the matrix element and HERWIG++ to model the parton shower, hadronization, particle decays and QED FSR. The samples are

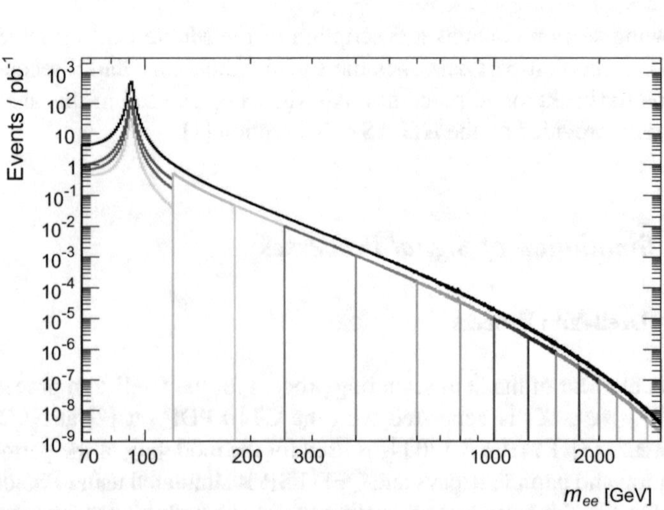

Fig. 14.1 Generated invariant mass m_{ee} of the POWHEG Z/γ^* MC samples. The colored lines show the different mass slices and the black line the sum of all samples, scaled up by a factor of two for easier visibility

again generated in slices of invariant mass to have sufficient statistics up to several TeV. Detailed information about the MC samples can be found in the appendix in Table H.2.

The calculation of the matrix element of the hard scattering process is done in the default POWHEG sample at NLO in QCD. Theory correction factors are derived to correct for differences between the NLO calculation and calculations at higher order in QCD. These correction factors are obtained by a polynomial fit to the ratio of the Z/γ^* cross section as a function of m_{ee} predicted by POWHEG and at NNLO, calculated using FEWZ 3.1 [9–11]. The FEWZ calculation also includes NLO EW corrections beyond QED FSR. For the NNLO cross section calculation the MSTW2008NNLO PDF set [12] is used. The renormalization (μ_R) and factorization scales (μ_F) are set equal to the m_{ee} at which the cross section is calculated. An additional small correction arises from single boson production in which the final-state charged lepton radiates a real W or Z boson. This was estimated using Madgraph 5 [13], following the prescription outlined in reference [14]. The corrections are given as a function of m_{ee} and were provided by the ATLAS collaboration [15]. The resulting correction factors are 1.025, 1.023 and 1.018 at 200 GeV, 1 TeV and 1.5 TeV, respectively.

14.1.1.2 Photon Induced Process

The photon induced process ($\gamma\gamma \rightarrow ee$) has been generated at LO using PYTHIA 8.170 and the MRST2004qed PDF set [16]. PYTHIA also modeled the parton shower, hadronization and particle decays and QED FSR is simulated using Photos. The cross section of the photon induced process is steeply falling towards higher invariant masses. Hence, the same strategy as for the Drell-Yan samples has been used and several mass-binned samples have been produced. In total five mass-binned samples have been used starting from 60 GeV $< m_{ee} <$ 200 GeV reaching up to $m_{ee} >$ 2500 GeV. The number of events generated for each sample reaches from 500, 000 down to 100, 000 corresponding to an integrated luminosity of 185 fb^{-1} to 4.4×10^7 fb^{-1}. The cross section predicted by PYTHIA for the first mass-binned sample is 2.69 pb and therefore about three orders of magnitude smaller than the Drell-Yan process. However, away from the Z-resonance, the relative contribution from the photon induced process becomes sizable compared to Drell-Yan. The cross section of the photon induced process predicted by PYTHIA in the region 600 GeV $< m_{ee} <$ 1500 GeV is 3.5×10^{-3} pb and therefore about 16% of the cross section of the Drell-Yan process in the same region (21.5×10^{-3} pb). Detailed information about the MC samples can be found in the appendix in Table H.3.

Figure 14.2 shows the resulting invariant mass spectrum for the photon induced process generated by PYTHIA. The colored lines show the individual mass bins and the black line shows the resulting sum of all samples scaled up by a factor of two for easier visibility. The cross section is steeply falling towards higher masses. The drop of the cross section is lower than the drop observed for the Drell-Yan process. MRST2004qed was the only available photons PDF by the time the MC sample was produced. The PDF does not contain a full set of eigenvectors representing the PDF

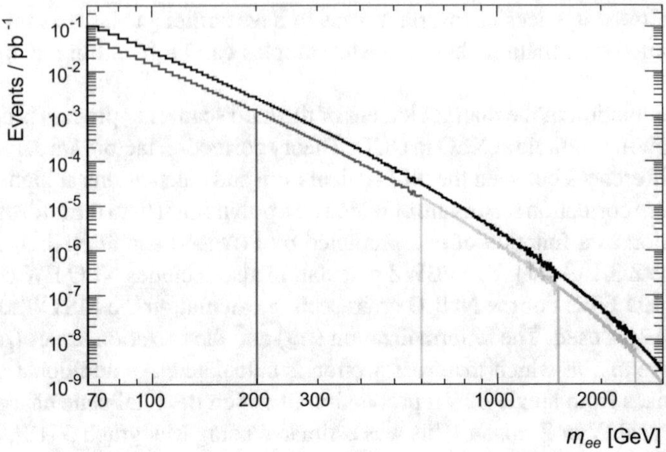

Fig. 14.2 Generated invariant mass m_{ee} of the invariant mass-binned PYTHIA photon induced MC samples. The colored lines show the different mass slices and the black line the sum of all samples, scaled up by a factor of two for easier visibility

uncertainty. Only two possible options were provided, a PDF using constituent quark masses in the proton and a PDF using current quark masses. For the production of the MC samples the latter was used. This leads to a higher cross section since lighter quarks are more likely to radiate photons. The cross section of the MC sample is therefore scaled down by a factor of 0.7 in order to match NLO calculations from the SANC group [17, 18].

14.1.2 Simulation of Background Processes

14.1.2.1 Top-Quark Processes

Top- and antitop-quarks dominantly decay into b- and $\bar{b}$-quarks under emission of a W boson. These W bosons can then further decay into leptons. Two different processes of top- and antitop-quark production are considered. The dominant process is the production of top-antitop pairs ($pp \rightarrow t\bar{t} + X \rightarrow W^+bW^-\bar{b} + X$). The sub-dominant process is the production of a single top- or antitop-quark in association with a W boson.

The matrix element of the hard scattering process has been generated at NLO in QCD using POWHEG with the CT10 PDF set. PYTHIA 6.427.2 [19] is used for the modeling of the parton shower, hadronization and particle decays and QED FSR is simulated using Photos. For systematic checks a second sample has been generated with MC@NLO 4.06 using the same PDF and interfaced with HERWIG 6.520 [20] for the modeling of the parton shower, hadronization, particle decays and QED FSR.

The MC samples are normalized to a cross section of $\sigma_{t\bar{t}} = 253^{+13}_{-15}$ pb as calculated with the Top++2.0 program [21] at NNLO in QCD, including soft-gluon resummation to next-to-next-to-leading-log order, and assuming a top-quark mass of 172.5 GeV. PDF and α_s uncertainties on the $t\bar{t}$ cross section are calculated using the PDF4LHC [22] prescription[1] and are added in quadrature to the scale uncertainty. Varying the top-quark mass by ± 1 GeV leads to an additional systematic uncertainty of $+8$ pb and -7 pb, which is also added in quadrature. The produced samples are all filtered at the generation stage for events in which at least one of the W bosons decays into an electron, muon or tau. The predicted cross section agrees within the assigned uncertainties with the measured cross section [23].

Single top production in association with a W boson can also lead up to two leptons. The MC sample is generated using the same configuration as the default $t\bar{t}$-sample and is normalized to a cross section of $\sigma_{tW} = 22.4 \pm 1.5$ pb [24]. All cross sections have been calculated by the ATLAS top group [25] or are taken from the given references. Detailed information about the samples can be found in the appendix in Table H.4.

14.1.2.2 Diboson Processes

Further important backgrounds are due to diboson production (WW, WZ, and ZZ). These W- and Z-bosons can then decay into electrons leading to two or more electrons. The diboson processes are generated at LO with HERWIG 6.520, using the CTEQ6L1 PDF [26]. The samples were filtered for decays with at least one charged lepton. Since the diboson spectrum is strongly falling with invariant mass, two additional mass binned samples were produced. Here only events were generated in which the decay leads to at least two electrons with an invariant mass in a certain window. If there were more than two electrons, the pair with the highest invariant mass is chosen. The inclusive sample is used up to an invariant mass of 400 GeV, a second sample from 400 GeV to 1000 GeV and a third sample above 1000 GeV. The diboson cross sections for pp collisions are known up to NLO. The WZ and ZZ cross section values used are 20.3 ± 0.8 pb and 7.2 ± 0.3 pb respectively, as calculated at NLO with MCFM [27, 28] and the CT10 PDF. The WW cross section is assumed to be 70.4 ± 7 pb, derived by scaling the MCFM value of 58.7 pb by a factor of 1.20 ± 0.12. This scale factor and its uncertainty correspond to an approximate mean of the two scale factors for WW production with zero and one extra jet, as discussed in reference [29]. The cross section is in agreement with the recent ATLAS measurement of the WW cross section at $\sqrt{s} = 8$ TeV, which yields a value of $71.1 \pm 1.1\,(\text{stat})\,^{+5.7}_{-5.0}\,(\text{sys}) \pm 1.4$ pb [30]. These cross sections were used to normalize the samples to get a better description of the processes. All cross section

[1] The PDF4LHC prescription for calculating the PDF uncertainties is to take the envelope of the uncertainties from the MSTW2008 68% CL NNLO, CT10 NNLO and NNPDF2.3 5f FFN PDF sets.

calculations have been provided by the ATLAS collaboration [1]. Detailed information about the samples used can be found in the appendix in Table H.5.

14.1.2.3 $Z/\gamma^* \rightarrow \tau\tau$ Process

The Drell-Yan process can also lead to a pair of τ-leptons which can further decay into electrons via $\tau \rightarrow \nu_\tau \nu_e e$. Both τ-leptons have to decay in this way to produce a pair of electrons. This reduces the contribution from the process by 97%, as only about 3% of the τ-pairs decay into two electrons. Due to the three-body decay of the τ lepton, the resulting electron has a much lower transverse momentum and the resulting invariant mass of the two electrons is therefore much smaller than the invariant mass of the initial τ-pair. The contribution from this process has been studied using the expectation from a MC sample and found to be negligible (<0.1%) at the high invariant masses which are studied in this analysis.

14.1.2.4 *W* Process

The decay of a W boson can lead only to one electron. The background will therefore not be estimated with MC. However, MC samples for this process are needed for studies of the multijet and W+jets background. The process was generated with POWHEG using the CT10 PDF. The modeling of the parton showers and hadronization is done afterwards by PYTHIA. Two samples are used, one for the process $W^+ \rightarrow e^+\nu_e$ and the other one for $W^- \rightarrow e^-\bar{\nu}_e$. The W cross section for pp collisions is known up to NNLO [31]. These cross sections were used to normalize the samples to get a better description of the process. Details about the samples and NNLO cross section values used can be found in the appendix in Table H.6.

14.2 Data and Selection Criteria

The following section contains a description of the data set which is used in the analysis and all selection criteria applied to the events and electrons.

14.2.1 Data

The data used in this analysis was delivered by the LHC at $\sqrt{s} = 8$ TeV and recorded by the ATLAS experiment. The data taking period was from April 2012 to December 2012 and the recorded data set corresponds to a total integrated luminosity of 21.3 fb^{-1}. The collisions were performed with a 50 ns spacing of the proton bunches in the LHC. Figure 14.3 shows the sum of the integrated luminosity delivered by the LHC (green), recorded by ATLAS (yellow) and ready for physics analyses (blue) for the data taking period in the year 2012.

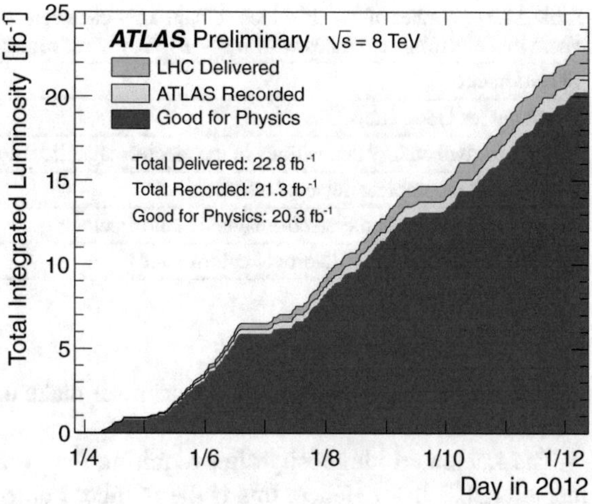

Fig. 14.3 Sum of integrated luminosity delivered by the LHC by day is shown in green for data taking in 2012. The sum of the from ATLAS recorded integrated luminosity is shown in yellow. Figure taken from reference [32]

14.2.2 Event Selection

The data has been preselected in order to reduce the amount of data to be analyzed and the required amount of disk space. Only events which contain at least two reconstructed electron candidates with $p_T > 18$ GeV are used. The 2012 data set is divided into the periods A to L (see Sect. 5.7.1 for more details on the structure of the data). All events used for this analysis have to be in a luminosity block which is listed in the Good Runs List.[2] The events have to fulfill a trigger[3] which requires at least two energy depositions in the electromagnetic calorimeter which have $E_T > 35$ GeV and $E_T > 25$ GeV. For these energy depositions, requirements on the shape of the shower and the leakage into the hadronic calorimeter are imposed. No cuts on the track of the electron candidates are imposed by this trigger. This is the trigger with the lowest available p_T thresholds which has simultaneously the least requirements on the energy deposition. The efficiency of the trigger is about 99% for electron candidates with a transverse momentum 5 GeV above the threshold. It rises to 99.8% and higher for electrons with $p_T > 60$ GeV. In addition, events are discarded in which a noise burst was observed in the electromagnetic or hadronic calorimeter. Such a noise burst could fake energy depositions and would make an accurate energy measurement impossible. It might be sometimes necessary to restart the trigger system during data taking. During such a restart events might not have the complete detector information and are therefore also rejected. Table 14.1 shows the number of events remaining after each of these cleaning cuts. The requirement of the trigger reduces the number of events strongly to a subset of events. Events with incomplete detector

---

[2]The Good Runs List used in this analysis is: data12_8TeV.periodAllYear_DetStatus-v61-pro14-02_DQDefects-00-01-00_PHYS_StandardGRL_All_Good.xml.

[3]EF_g35_loose_g25_loose.

Table 14.1 Number of events which remain after each selection cut. Preselected data were used, where two electron candidates with $p_T > 18$ GeV were required

Selection cut	Number of events
Event passes Good Runs List	368,648,710
Trigger for two energy depositions in the electromagnetic calorimeter	40,873,695
Events with incomplete detector information	40,873,669
Veto on noise burst in the electromagnetic calorimeter	40,783,645
Veto on noise burst in the hadronic calorimeter	40,783,644

information or noise bursts in the calorimeter make up only a very small fraction of the total events.

The integrated luminosity after requiring the events to pass all quality requirements is 20.3 fb^{-1}. Hence, this is the number quoted as the integrated luminosity for the data set. The sum of the integrated luminosity ready for physics analysis is shown in blue in Fig. 14.3.

14.2.3 Electron Selection

In the selected events, pairs of electron candidates have to be found. Therefore several selection criteria are applied to the single electrons and the pairs. These selection criteria are chosen in such a way that they reduce background from other physics processes by obtaining at the same time a high signal efficiency. Each pair of electron candidates consists of a leading and a subleading candidate, where the leading candidate is the one with higher p_T and the subleading the one with lower p_T.

All electrons are considered which are reconstructed by a reconstruction algorithm which first searches for an energy deposition in the electromagnetic calorimeter and then searches for a track matching this energy deposition. A more detailed description of the electron reconstruction is given in Sect. 6.2.1. The electron candidates have to be detected in the central detector region of $|\eta| < 2.47$, in order to have tracking information available. The tracking detectors have a coverage up to $|\eta| = 2.5$, the region of $|\eta| < 2.47$ is chosen to ensure that the shower is contained in the region $|\eta| < 2.5$. In addition, electron candidates which are in the transition region $1.37 < |\eta| < 1.52$ between the barrel and endcap electromagnetic calorimeter are rejected, as these candidates have a worse energy resolution. The η information for the restriction of the electron candidates is chosen to be the η information from the electromagnetic shower, as energy resolution is the motivation for these restrictions. With an object-quality check it is ensured, that the electron is measured in a region where the electromagnetic calorimeter was working properly at that time. This excludes electron candidates in regions where for instance some electronic device was broken or problems with the high-voltage supply occurred. The p_T threshold for the electron candidates is chosen to be 5 GeV above the threshold of the trigger, i.e., $p_T > 40$ GeV for the leading electron candidate and $p_T > 30$ GeV for the subleading electron candidate. The cut is chosen to ensure that no threshold effects affect the trigger efficiency and that the trigger is fully efficient. To reduce background from misidentified objects, both electron candidates are first required to fulfill the cut-based *medium* electron identification, described in Sect. 6.2.2. Electrons from Drell-Yan production are expected to be well isolated from other energy depositions not associated with the lepton. The requirement of calorimeter isolation is a very efficient way to reduce background from jets. The leading electron candidate is therefore required to fulfill $\sum E_T(\Delta R = 0.4) < 0.007 \cdot E_T + 5$ GeV and the subleading electron candidate $\sum E_T(\Delta R = 0.4) < 0.022 \cdot E_T + 6$ GeV [33]. The quantity $\sum E_T(\Delta R = 0.4)$ is the sum of the energy deposition in a cone with size $\Delta R = 0.4$ around the electron candidate. The cut value on the isolation is less strict for the subleading candidate, since it has most likely less p_T because it radiated Bremsstrahlung, which leads to a worse calorimeter isolation. The functions are chosen in such a way that the cut has an efficiency of 99%. Finally, a requirement of $|\Delta\eta_{ee}| < 3.5$ is imposed on the absolute pseudorapidity separation of the leptons. Fully hadronic processes, like multijet production, are dominating at large opening angles and the imposed cut reduces the background contribution from these processes. No further requirements are made on the charge of the electron candidates, since for very high transverse momentum the charge identification efficiency gets worse mainly due to Bremsstrahlung and due to the limited momentum resolution of the electron in the tracking detector. For example, for an electron with $p_T = 1$ TeV, the efficiency to reconstruct the correct charge decreases to 95% [34]. It is also very difficult to measure the charge identification efficiency for high p_T, and thus derived efficiency corrections would come with large systematic uncertainties. The pairs are required to have an invariant mass of $m_{ee} > 66$ GeV. If there is more than one pair in one event, all combinations are considered. This is the case only in less than one per mille of the events. Table 14.2 shows the number of events with at least two electrons remaining after each selection cut. A detailed table with all event yields for all backgrounds can be found in Appendix I.

The left plot in Fig. 14.4 shows the selection efficiency times acceptance (fraction of all generated events which pass the selection) of the signal selection for the Drell-Yan and photon induced process. The acceptance times efficiency was calculated using the corresponding MC samples and is binned in the invariant mass of the electron pair. In the range of the Z-resonance from 66 GeV to 116 GeV, the selection efficiency times acceptance for the Drell-Yan process is only on the order of 19%. This is due to the large p_T thresholds for the two electrons. The measurement of this analysis starts at 116 GeV, where the selection efficiency times acceptance is about 30% and then rises with invariant mass up to 65%. The selection efficiency times acceptance for the photon induced process is lower throughout the whole mass range. It is only about 9% at 116 GeV and then rises up to about 20%. The photon induced process is mainly a t-channel process and therefore yields more forward leptons than Drell-Yan which are at the same time distributed at lower values of p_T. Imposing the high p_T cuts has therefore a much larger impact on the acceptance. On the right side

Table 14.2 Number of events with at least two electrons remaining after each selection cut

Selection cut	Number of events
After event selection	40,783,644
At least two objects reconstructed as electron candidates by a specific algorithm	39,328,689
At least two electrons with $\|\eta\| < 2.47$, which are not in the transition region $1.37 < \|\eta\| < 1.52$	37,796,480
At least two electrons fulfilling the object quality check	37,717,667
Leading electron $p_T > 40$ GeV, subleading electron $p_T > 30$ GeV	19,647,642
At least two electrons fulfilling the *medium* identification	4,619,892
At least two electrons fulfilling the isolation requirements	4,573,716
At least one electron pair has $\|\Delta\eta_{ee}\| < 3.5$	4,573,047
At least one electron pair has $m_{ee} > 66$ GeV	4,551,899
At least one electron pair has $m_{ee} > 116$ GeV	124,648

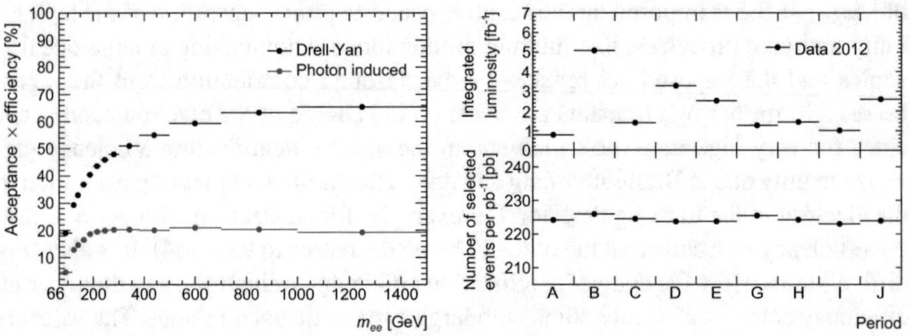

Fig. 14.4 The selection efficiency times acceptance of the signal selection for the Drell-Yan and photon induced process is shown on the left side. The efficiency was calculated using the corresponding MC samples. The right plot shows in the upper half the amount of integrated luminosity for each period. The yield of Z-candidates per pb^{-1} over the different periods of data taking is shown in the lower half

of Fig. 14.4, the yield of the selected candidates per pb^{-1} is shown, as well as the integrated luminosity for different data taking periods. The yield is constant over all data periods, as expected if there are no time dependent efficiency losses.

14.3 Background Determination

This section describes how the background for the analysis is estimated. For most background processes the MC samples which are introduced in Sect. 14.1 are used. An additional background arises from falsely identified leptons. This background is not well described in MC and has to be measured in data. This background is mainly arising from multijet events in which two jets are misidentified and are passing the electron selection. Another important contribution is the W+jets background, where one real electron is originating from the decay of the W and an additionally produced jet is misidentified and passing the electron selection. The same method to estimate this background as for the $\wp$ search was used (see Sect. 9.3). It is first briefly discussed again and extended to the two-lepton final state. The groundwork for the results of the following background estimation has been performed in my master thesis [35]. The results have been partially further refined and validated in the context of the presented thesis. The main results are in the following repeated to be able to present the analysis in a coherent way.

14.3.1 Matrix Method

The multijet and W+jets background in this analysis is estimated using the same method as in Sect. 9.3. The idea of the method is to loosen some of the identification criteria for electrons and to measure the efficiency for these looser objects to pass the signal selection (also denoted as "tight" selection). The efficiency gives a handle on the contribution from misidentified electrons in the signal selection. It is defined for real electrons ($r = N^{\text{real}}_{\text{tight}}/N^{\text{real}}_{\text{loose}}$) and fake electrons ($f = N^{\text{fake}}_{\text{tight}}/N^{\text{fake}}_{\text{loose}}$) separately. The same definitions for r and f as in Eq. 9.1 are used. Equation 9.2 can be extended by two additional dimensions to describe the background for a two-electron selection:

$$\begin{pmatrix} N_{TT} \\ N_{TL} \\ N_{LT} \\ N_{LL} \end{pmatrix} = M \begin{pmatrix} N_{RR} \\ N_{RF} \\ N_{FR} \\ N_{FF} \end{pmatrix} \tag{14.1}$$

$$M = \begin{pmatrix} r_1 r_2 & r_1 f_2 & f_1 r_2 & f_1 f_2 \\ r_1(1-r_2) & r_1(1-f_2) & f_1(1-r_2) & f_1(1-f_2) \\ (1-r_1)r_2 & (1-r_1)f_2 & (1-f_1)r_2 & (1-f_1)f_2 \\ (1-r_1)(1-r_2) & (1-r_1)(1-f_2) & (1-f_1)(1-r_2) & (1-f_1)(1-f_2) \end{pmatrix} \tag{14.2}$$

The indices R and F refer again to real and fake electrons, while the indices T and L denote whether an object in the loosened selection passes the signal selection or not. The first index of the number of events in a given bin N_{xy} refers always to the leading candidate while the second index refers to the subleading candidate. Similarly f_1 and r_1 denote the efficiencies for the leading object while f_2 and r_2 those for the subleading object. The isolation criteria for leading and subleading electron candidates differ and therefore also r and f. The number of events in the signal selection N_{TT} consists of the following contributions:

$$N_{TT} = r_1 r_2 N_{RR} + r_1 f_2 N_{RF} + f_1 r_2 N_{FR} + f_1 f_2 N_{FF}. \tag{14.3}$$

Interesting is here only the part originating from events containing at least one fake object:

$$\begin{aligned} N_{TT}^{e+jet} &= r_1 f_2 N_{RF} + f_1 r_2 N_{FR} \\ N_{TT}^{multijet} &= f_1 f_2 N_{FF} \\ N_{TT}^{fake} = N_{TT}^{e+jet\&multijet} &= r_1 f_2 N_{RF} + f_1 r_2 N_{FR} + f_1 f_2 N_{FF}. \end{aligned} \tag{14.4}$$

The truth quantities N_{RF}, N_{FR} and N_{FF} can again be replaced by calculating M^{-1}. The number of events in the signal selection containing one or two fake objects is then given by:

$$\begin{aligned} N_{TT}^{e+jet\&multijet} =& \alpha[r_1 f_2(f_1-1)(1-r_2) + f_1 r_2 (r_1-1)(1-f_2) + f_1 f_2 (1-r_1)(1-r_2)] N_{TT} \\ &+ \alpha f_2 r_2 [r_1(1-f_1) + f_1(1-r_1) + f_1(r_1-1)] N_{TL} \\ &+ \alpha f_1 r_1 [f_2(1-r_2) + r_2(1-f_2) + f_2(r_2-1)] N_{LT} \\ &- \alpha f_1 f_2 r_1 r_2 N_{LL} \end{aligned} \tag{14.5}$$

where

$$\alpha \equiv \frac{1}{(r_1-f_1)(r_2-f_2)}. \tag{14.6}$$

With this equation the number of background events in any given bin can be calculated and therefore any distribution of the background can be predicted. The fake efficiencies and real efficiencies depend in general on kinematic properties like p_T or η of the electron. They can therefore be binned in these variables to take these dependencies into account. The background will in this case be calculated on an event by event basis. In this analysis the loosened selection is given by the *loose* cut-based identification level (see Sect. 6.2.2). The cut on the difference in η of the track measured in the inner detector and the energy deposition measured in the electromagnetic calorimeter has been removed from the *loose* identification level to further increase the fake contribution. This loosened selection is slightly stricter than the requirement imposed by the trigger. The signal selection (tight selection) is given by the *medium* cut-based identification level plus the respective isolation requirements for the leading and subleading object. All other requirements which are imposed in the signal selection are also imposed on the loosened selection ($|\eta|$ restriction, object quality etc.).

14.3.1.1 Systematic Variations of the Matrix Method

Variation 1

The efficiency for real electrons in the loosened selection to pass the signal selection is usually very high. Hence, to simplify Eq. 14.5, the approximation $r_1 = r_2 = 1$ can be made. This assumes that every real electron in the loosened selection passes also the signal selection. Equation 14.5 then simplifies to

$$N_{TT}^{e+jet\&multijet} = F_2 N_{TL} + F_1 N_{LT} - F_1 F_2 N_{LL}, \tag{14.7}$$

where

$$F_i \equiv \frac{f_i}{1-f_i} = \frac{N_{\text{tight}}^{\text{fake}}/N_{\text{loose}}^{\text{fake}}}{1-N_{\text{tight}}^{\text{fake}}/N_{\text{loose}}^{\text{fake}}} = \frac{N_{\text{tight}}^{\text{fake}}}{N_{\text{loose}}^{\text{fake}}-N_{\text{tight}}^{\text{fake}}}. \tag{14.8}$$

The quantity F_i is called fake factor. The following expression is valid, since the signal (tight) selection is a subset of the loosened selection:

$$N_{\text{loose}}^{\text{fake}} - N_{\text{tight}}^{\text{fake}} = N_{\text{fail tight}}^{\text{fake}}. \tag{14.9}$$

The fake factor then simplifies to

$$F_i^{FT} = \frac{N_{\text{tight}}^{\text{fake}}}{N_{\text{fail tight}}^{\text{fake}}}. \tag{14.10}$$

The events falling into the category $N_{\text{fail tight}}^{\text{fake}}$ passed the loosened selection but fail the signal selection (*medium* identification or isolation requirement). When assuming $r_1 = r_2 = 1$, entries which accounted for real electron contributions in the selection L (pass loosened selection but fail signal selection), simplify to zero. Real electrons contributing to the multijet and W+jets background still have to be accounted for. The calculation of the background will therefore be performed on MC samples of the dielectron processes instead of using data. The obtained contribution to the background from these real electron processes can then be subtracted from the background obtained from data.

Variation 2

The selection $N_{\text{fail tight}}^{\text{fake}}$ contains contamination from real electrons since it is possible for a real electron to fail the *medium* identification or the isolation requirement. Some cuts of the *medium* identification are particularly able to separate real electrons from jet events. Jets faking electrons contain often neutral pions, decaying to collimated photons which lead to an electromagnetic energy deposition, and accompanying charged pions which leave a track in the detector. The tracks and the electromagnetic energy deposition are therefore not perfectly aligned and a cut on the track match between the η values of the electron track and the electron energy deposition is a very efficiency way of reducing such backgrounds. However, for the $N_{\text{fail tight}}^{\text{fake}}$ category, a clean fake sample is required. A cleaner set of fake objects, and therefore smaller corrections from MC resulting in a more stable method, can be obtained by modifying the definitions of the fake factor

$$F_i^{FTM} \equiv \frac{N_{\text{tight}}^{\text{fake}}}{N_{\text{fail track match}}^{\text{fake}}}, \tag{14.11}$$

where events in the category $N^{\text{fake}}_{\text{fail track match}}$ fail the track matching cut of the *medium* identification criteria, which is requiring the absolute difference of the track and cluster η position to be less than 0.005. The definition of L (for N_{LL}, N_{LT}, N_{TL}) has to also change from "pass loosened selection but fail signal selection" to "pass loosened selection but fail track matching cut" if the fake factor F_i^{FTM} is applied.

14.3.2 Measurement of the Real Efficiency

The real efficiency $r = N^{\text{real}}_{\text{tight}}/N^{\text{real}}_{\text{loose}}$ is measured from MC, since it is usually well modeled in the simulation and given the higher statistics available. The usual efficiency corrections are applied to account for small differences of the trigger, identification and isolation efficiencies between data and MC. Hence, the real efficiency measured from MC is effectively matched to the real efficiency which would be measured in data. The Drell-Yan background MC provides a large sample of real electrons which can be used to measure r_i. The electrons are required to be reconstructed within a cone of $\Delta R < 0.2$ around the generated electrons to avoid dilution from misidentified jets. Figure 14.5 shows the real efficiencies binned in p_{T} for four different detector regions: $|\eta| < 1.37$, $1.52 < |\eta| < 2.01$, $2.01 < |\eta| < 2.37$ and $|\eta| > 2.37$.

The left plot shows the real efficiency for the leading electron r_1 starting from $p_{\text{T}} = 40$ GeV and the right plot shows the real efficiency of the subleading electron r_2 starting from $p_{\text{T}} = 30$ GeV. Both efficiencies behave very similar, the leading efficiency being slightly higher. The real efficiency in the barrel region of the detector ($|\eta| < 1.37$) rises as a function of p_{T} from 93% to 96.5%. In the endcap regions, the efficiency in general is lower. The real efficiency in the region which is still covered by the TRT ($1.52 < |\eta| < 2.01$) rises from 90% to 96.5%. In region $2.01 < |\eta| < 2.37$, the real efficiency ranges from 91.5% to 97% and in the region $|\eta| > 2.37$ from 96% to 98%.

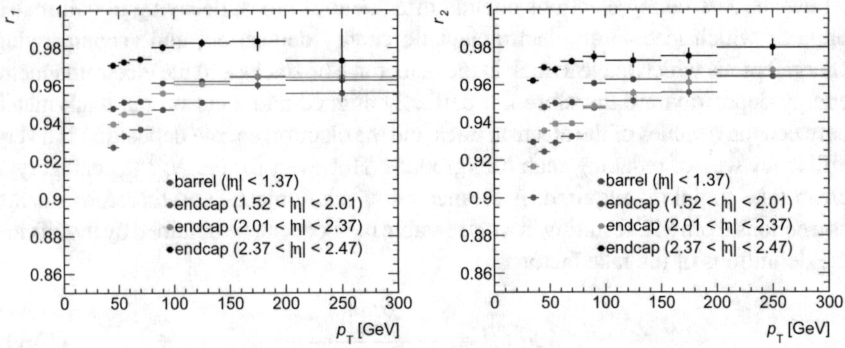

Fig. 14.5 Real electron efficiency with its statistical uncertainty, determined from Drell-Yan MC and binned in p_{T} separately for the barrel and three endcap regions. For leading electrons the efficiency is shown on the left side and for subleading electrons on the right side

14.3.3 Measurement of the Fake Efficiency

The fake efficiency $f = N^{\text{fake}}_{\text{tight}}/N^{\text{fake}}_{\text{loose}}$ and fake factor $F_i^{FTM} = \frac{N^{\text{fake}}_{\text{tight}}}{N^{\text{fake}}_{\text{fail track match}}}$ cannot be reliably calculated with simulation and therefore need to be measured from data. Three different methods are performed which aim to obtain a jet enriched control region in which the fake efficiencies and fake factors can be calculated. Two of the methods implement a tag and probe like procedure. The idea is to tag di-jet events by requiring one object to fail a certain identification level and to probe a second object which is assumed to be also a jet. The method is performed with two different types of triggers, with single jet triggers and with the signal trigger. The methods slightly differ due to requirements on the tag object imposed by the trigger choice. A second method is studying single objects in a data sample collected with single jet triggers. All methods reduce dilution from real electrons (mainly from the Drell-Yan and W process) by imposing additional requirements. This results overall in three different measured fake efficiencies. The methods to measure these fake efficiencies have been developed and studied in detail during my master thesis [35]. They are therefore not discussed again in detail here. Appendix J contains a detailed description of the methods. Figure 14.6 shows the fake efficiencies for the leading electron on the left side and the subleading electron on the right side. The fake efficiencies are separated into the same detector regions in η as the real efficiencies. The results from all three methods are shown. The fake efficiencies vary in the central barrel region ($|\eta| < 1.37$) from about 5% to 8% and 6% to 7% for the leading and subleading electron, respectively. The leading fake efficiency is slightly falling towards higher p_{T} while the subleading fake efficiency stays flat. The falling behavior is more pronounced in the endcap regions. The fake efficiencies become slightly larger towards higher $|\eta|$ values as it becomes more difficult to discriminate electrons from jets. The three methods are generally in good agreement. Some differences can be observed in the last two endcap bins. The fake efficiencies from the reverse tag and probe method with the electron trigger predicts higher fake efficiencies for higher p_{T}. The fake factors F_i^{FTM} compare in a similar way and are shown in the appendix in Fig. J.1.

14.3.4 Comparison of All Methods

Three different fake efficiencies have been presented. Each of the methods comes with its own advantages and disadvantages. While the default method (single object method using single jet triggers) provides statistics up to very high p_{T}, it does not correct for real electron dilution[4] which enters the fake enriched control region. The reverse tag and probe method aims to measure the fake efficiencies in a di-jet

[4] The dilution is here, due to the lower identification requirement, much lower than in the previous analysis in part III.

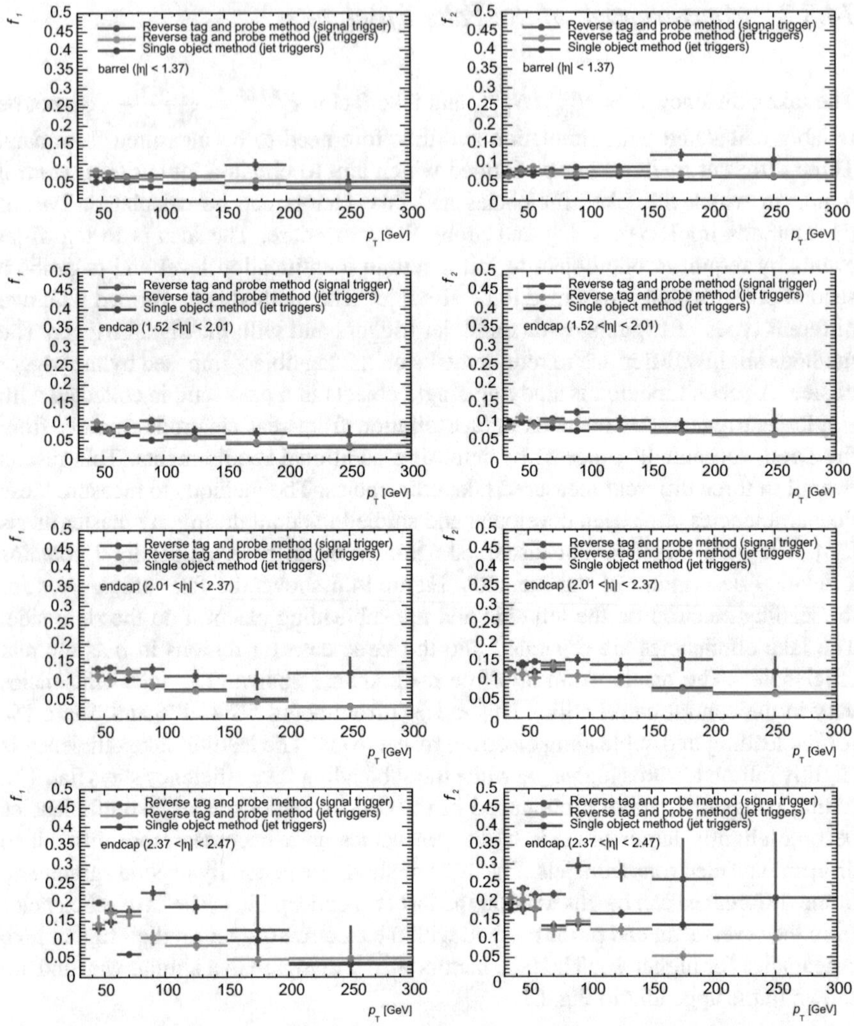

Fig. 14.6 Comparison of the fake efficiencies f_i with their statistical uncertainties, calculated with the three different methods (reverse tag and probe with signal trigger, reverse tag and probe with single jet triggers and single object method with single jet triggers). The upper row shows the fake efficiencies for the barrel region ($|\eta| < 1.37$). The corresponding fake efficiencies for the endcap regions ($1.52 < |\eta| < 2.01$, $2.01 < |\eta| < 2.37$ and $2.37 < |\eta| < 2.47$) are shown from the second to the fourth row. The fake efficiencies for the leading object are shown on the left side and for the subleading object on the right side

enriched control region and provides further possibilities to suppress real electron dilution (i.e. same charge of tag and probe). However, also this method does not correct the remaining real electron dilution in the fake enriched control region. The same method has therefore been repeated with the signal trigger which makes it easier to correct for the dilution while having a slightly less stringent requirement on the tag

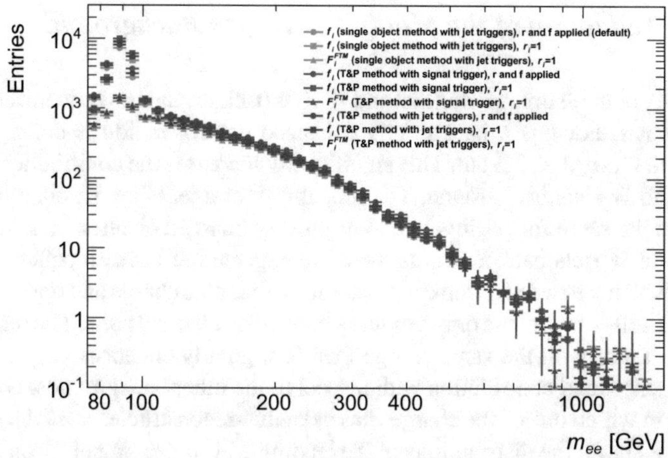

Fig. 14.7 Nine fake background estimates using the three different methods for the background determination and three different methods for measuring the fake efficiencies and fake factors. The marker color represents the method used for the determination of the fake efficiencies or fake factors and the marker symbol represents the method used for the determination of the background

object due to the trigger identification requirements. All methods have advantages and disadvantages and they are therefore all considered to study the impact of the systematic uncertainty from the fake efficiencies on the background yield.

Also three different variations of the matrix method have been performed. The original method relies on measuring the real efficiency r. A different way of correcting for the contribution has been studied by setting r equal to one and using direct MC predictions to correct for electrons failing the signal selection. These corrections are further reduced when restricting the loosened selection to a subset of events in which the track matching cut of the *medium* identification is failed. All three methods to measure the fake efficiencies have been used with all three variations of the matrix method leading to in total nine different background predictions. Figure 14.7 shows the multijet and W+jets background prediction for the invariant mass spectrum of all nine different variations.

Large differences between the predictions can be observed in the region $m_{ee} <$ 116 GeV, where all methods show a maximum at around 91 GeV. The Z-resonance leads in this region to a very large dilution from real electrons. The N_{TT} selection is therefore dominating the selection, leading to very large corrections to the total background. The method in which the loosened selection is required to fail the track/cluster matching is assumed to be least sensitive to these corrections. This assumption is confirmed by the observed background shape. All three methods using this variation are predicting a yield which is about one order of magnitude smaller than the other methods. However, also these methods predict a small peak at around 91 GeV. All methods agree well in the region in which this measurement is performed ($m_{ee} >$ 116 GeV). The background is smoothly falling over four orders of magnitude towards an invariant mass of 1.5 TeV.

14.3.5 Validation of the Multijet & W+jets Background

The validity of the estimated multijet and W+jets background has been further studied for this thesis. For this a background enhanced control region is defined by first releasing the $|\Delta\eta_{ee}| < 3.5$ cut. This significantly increases the contribution from the multijet and W+jets background. The probability of a jet being misidentified as an electron is the same for positively and negatively charged electron candidates. The multijet and W+jets background is therefore expected to have an equal amount of events in which the two electron candidates have the same charge and opposite charge while the Drell-Yan process only produces oppositely charged pairs. The requirement of having a pair with the same charge therefore greatly enhances the multijet and W+jets background contribution with respect to the other backgrounds. Only Drell-Yan pairs in which one of the charges has been mis-reconstructed pass the selection. Figure 14.8 shows the distributions of the leading and subleading electron p_{T} in this fake enriched selection for pairs with $m_{ee} > 116$ GeV and $m_{ee} > 300$ GeV. The data are compared to the sum of the expected background and signal.

The multijet and W+jets background is dominating at low p_{T}. The contribution is here in the region $m_{ee} > 300$ GeV close to 100%. The Z/γ^* from the Drell-Yan process is produced mainly at rest with its p_{T} being close to zero. The Z/γ^* is therefore mainly decaying with a back-to-back topology in the transverse plane. The momen-

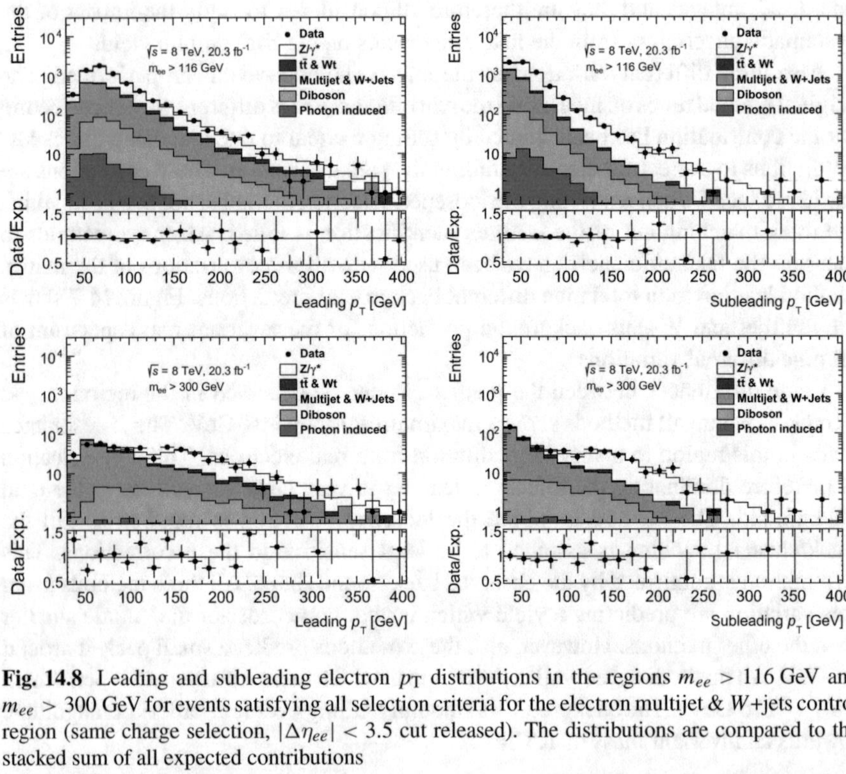

Fig. 14.8 Leading and subleading electron p_{T} distributions in the regions $m_{ee} > 116$ GeV and $m_{ee} > 300$ GeV for events satisfying all selection criteria for the electron multijet & W+jets control region (same charge selection, $|\Delta\eta_{ee}| < 3.5$ cut released). The distributions are compared to the stacked sum of all expected contributions

tum of both electrons is in this case about $m_{ee}/2$ due to momentum conservation. This leads to the observed p_{T} spectra of electron candidates from the Drell-Yan process which starts to contribute from about 60 GeV on for $m_{ee} > 116$ GeV and from about 150 GeV on for $m_{ee} > 300$ GeV. The data are in good agreement with the expected background and signal yield. Figure 14.9 shows the measured observables $|y_{ee}|$ and $|\Delta\eta_{ee}|$ in the multijet and W+jets background enriched selection for the regions $m_{ee} > 116$ GeV and the region $m_{ee} > 300$ GeV.

The total relative contribution of the multijet and W+jets background for both observables is 44% in the region $m_{ee} > 116$ GeV and 61% in the region $m_{ee} >$ 300 GeV, i.e., the relative contribution is rising with invariant mass. The reason for this rise is that the multijet invariant mass spectrum reaches up to much higher invariant masses than the Drell-Yan spectrum. It becomes therefore more and more important at higher invariant masses. This is still true when considering that in the same charge selection, the contribution from Drell-Yan is also amplified by the rising charge misidentification rate at higher masses due to the straighter tracks from high-p_{T} electrons. The multijet and W+jets background has the largest contribution at low values of $|y_{ee}|$ and is slowly falling towards higher values of rapidity. The same kinematic behavior is expected for the Drell-Yan process. However, the charge misidentification rate is larger at large values of electron $|\eta|$ due to the lower magnetic

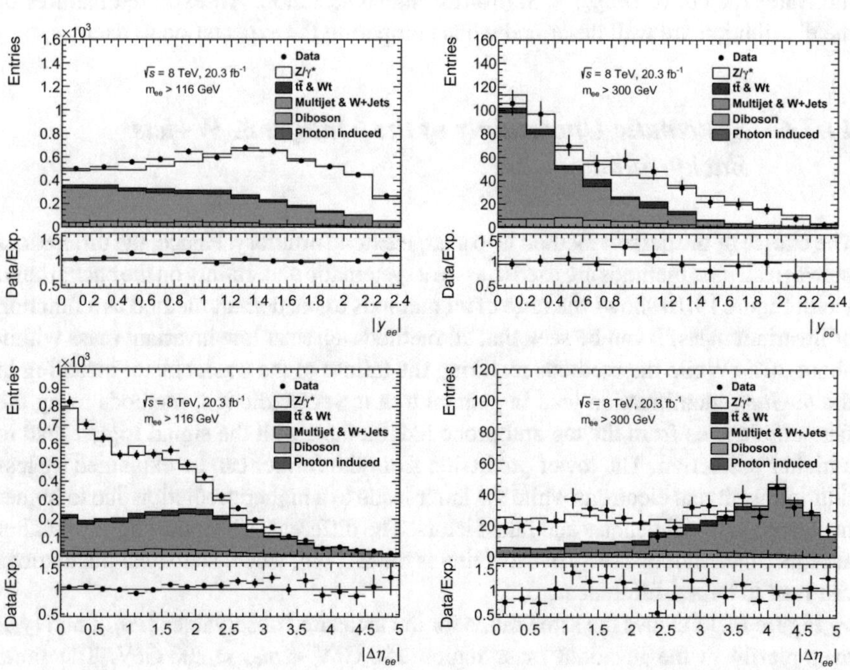

Fig. 14.9 Distributions of the absolute rapidity $|y_{ee}|$ and of the absolute pseudorapidity separation $|\Delta\eta_{ee}|$ in the region $m_{ee} > 116$ GeV (left side) and $m_{ee} > 300$ GeV (right side) for events satisfying all selection criteria for the electron multijet & W+jets control region (same charge selection, $|\Delta\eta_{ee}| < 3.5$ cut released). The distributions are compared to the stacked sum of all expected contributions

field strength at high $|\eta|$ and a worse track reconstruction due to the coverage of the TRT until $|\eta| = 2.0$, which leads to a different behavior of the Drell-Yan process in the same charge selection. Both electron candidates are therefore expected to be more at higher values of $|\eta|$ which leads to a shift of the Drell-Yan contribution to higher values of rapidity. Both processes are nicely separated due to this different kinematic behavior. The general agreement between data and expected yield is very good for the $|y_{ee}|$ distributions.

The Drell-Yan contribution in the $|\Delta\eta_{ee}|$ spectrum is slowly falling towards higher values of $|\Delta\eta_{ee}|$. The multijet and W+jets background is reaching its maximum at higher values of $|\Delta\eta_{ee}|$. The multijet production is dominantly a t-channel process and therefore leading to a higher contribution in the forward direction, i.e., leading to larger values of $|\Delta\eta_{ee}|$. When comparing the distributions for $m_{ee} > 116$ GeV and $m_{ee} > 300$ GeV it can be seen that for higher invariant masses the multijet and W+jets contribution moves to higher values of $|\Delta\eta_{ee}|$. A high invariant mass can either be reached by high p_{T} of the objects or by a large opening angle $|\Delta\eta_{ee}|$. Figure 14.8 shows that the multijet and W+jets background is dominantly distributed at low values of p_{T}. A high invariant mass can therefore only be reached if both objects have a large opening angle $|\Delta\eta_{ee}|$. The background from fake electrons starts to be close to 100% of the total expected events for $|\Delta\eta_{ee}| > 3.5$. This a posteriori motivates the cut of $|\Delta\eta_{ee}| < 3.5$ for the signal selection. All expected features of the distribution are well described when comparing the expectation to data.

14.3.6 Systematic Uncertainty of the Multijet & W+jets Background

The choice of the default method is to a large extend arbitrary. Hence, the differences between all nine methods are used to asses a systematic uncertainty on the background yield. Figure 14.10 shows the ratio of all methods to the default method as a function of invariant mass. It can be seen that all methods agree at low invariant mass within about 20%. While the methods requiring the failure of the track/cluster matching in the *medium* identification lead in general to a lower prediction, methods using the fake efficiencies from the tag and probe like method with the signal trigger lead to a higher prediction. The lower prediction from the former can be explained by less dilution from real electrons while the latter leads to a higher prediction due to higher measured fake efficiencies and fake factors. The differences become slightly smaller at higher invariant masses. An exception is the last bin, where statistical fluctuations can lead to larger differences.

Figure 14.11 shows the same ratios for the measured observables $|\Delta\eta_{ee}|$ and $|y_{ee}|$ exemplarily in the invariant mass region 150 GeV $< m_{ee} <$ 200 GeV. The same trends as in the invariant mass spectrum are here observed concerning the prediction of the different methods. The methods agree within about 15% at low values of $|y_{ee}|$. The agreement gets worse when going towards higher rapidities, where also the background is smaller (as discussed in the previous Sect. 14.3.5). At low values of $|\Delta\eta_{ee}|$, where again the background is smaller, differences of about 35% are

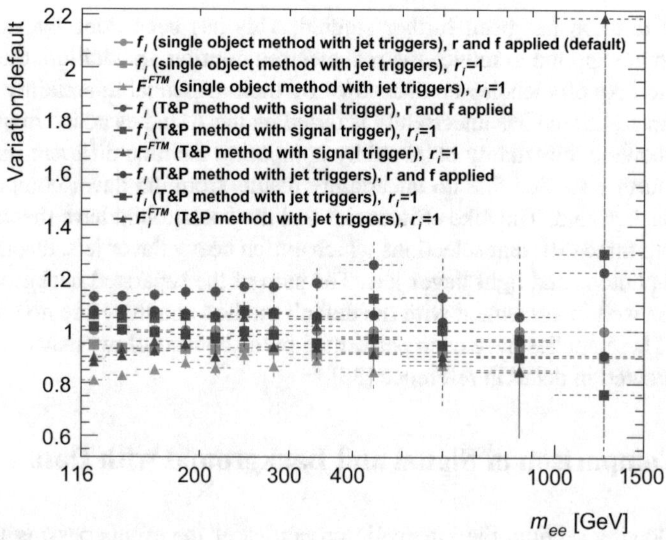

Fig. 14.10 Ratio of the final background estimate of all method variations to the default method

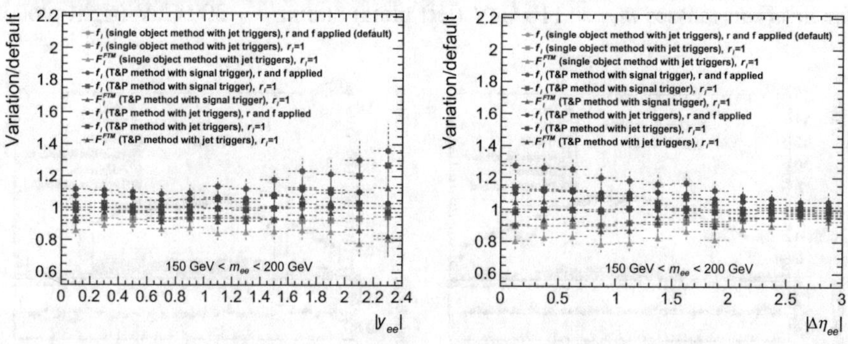

Fig. 14.11 Ratio of the final background estimate of all method variations to the default method as a function of $|y_{ee}|$ (left plot) and $|\Delta\eta_{ee}|$ (right plot) in an invariant mass window of 150 GeV < m_{ee} < 200 GeV

observed while the agreement is getting better towards higher values of $|\Delta\eta_{ee}|$, where the background becomes large. Hence, the variations are smaller and therefore the methods more predictive in regions where the background yield is large. All other invariant mass regions of the measurement are shown in Appendix K. The systematic uncertainty coming from the different variations of the method is calculated in each bin by taking the difference between the maximum and minimum prediction and dividing it by two. These results represent an improved uncertainty estimation when compared to [35], where a flat 20% uncertainty was assessed.

As for the default method to measure the fake efficiencies (single object method using single jet triggers) no real electron dilution is corrected, the impact of the remaining dilution has been further studied. This has been done by varying the cuts which are applied to reduce the real electron dilution. In addition, the effect of varying the fake efficiencies up and down by their statistical uncertainty has been studied. An additional 5% uncertainty is based on these studies added in quadrature to the systematic uncertainty obtained by comparing the nine different methods. It has been further studied that no uncertainty results from the flavor composition of the fake background. The fake efficiencies and the background have therefore been divided into three different selections which enrich heavy flavor jets, electrons from converted photons and light flavor jets. The sum of the separated backgrounds was found to be well in agreement with the default method and therefore no uncertainty is added. These studies have been performed in the context of my master thesis and are documented in detail in reference [35].

14.4 Comparison of Signal and Background with Data

In the following section, the kinematic properties of the events passing the signal selection are discussed. Figure 14.12 shows the η distributions (upper plots) and the p_{T} distributions (lower plots) of the electron candidates passing the signal selection for invariant masses $m_{ee} > 116$ GeV (left plots) and $m_{ee} > 300$ GeV (right plots).

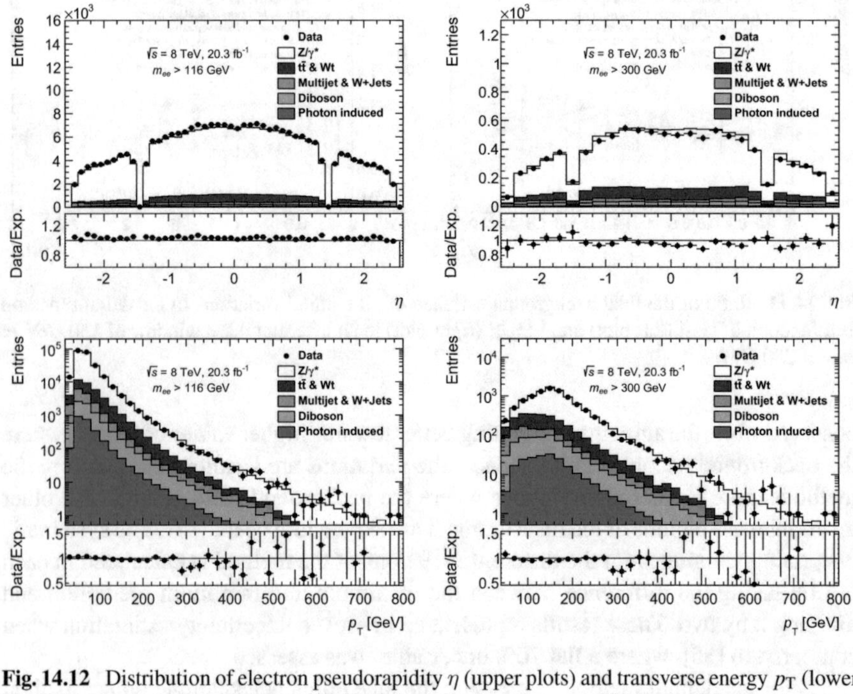

Fig. 14.12 Distribution of electron pseudorapidity η (upper plots) and transverse energy p_{T} (lower plots) for invariant masses $m_{ee} > 116$ GeV (left plots), and $m_{ee} > 300$ GeV (right plots), shown for data (solid points) and expectation (stacked histogram) after the complete selection. The lower panels show the ratio of data to the expectation

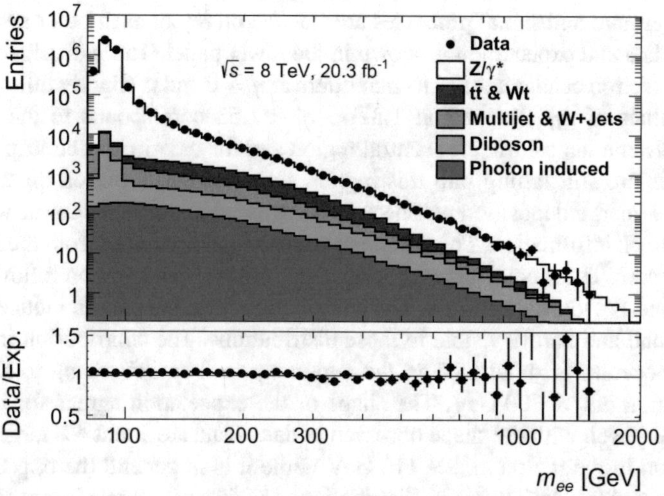

Fig. 14.13 The invariant mass (m_{ee}) distribution after event selection for the electron selection, shown for data (solid points) compared to the expectation (stacked histogram). The lower panels show the ratio of data to the expectation. The binning is chosen to be constant in $\log(m_{ee})$

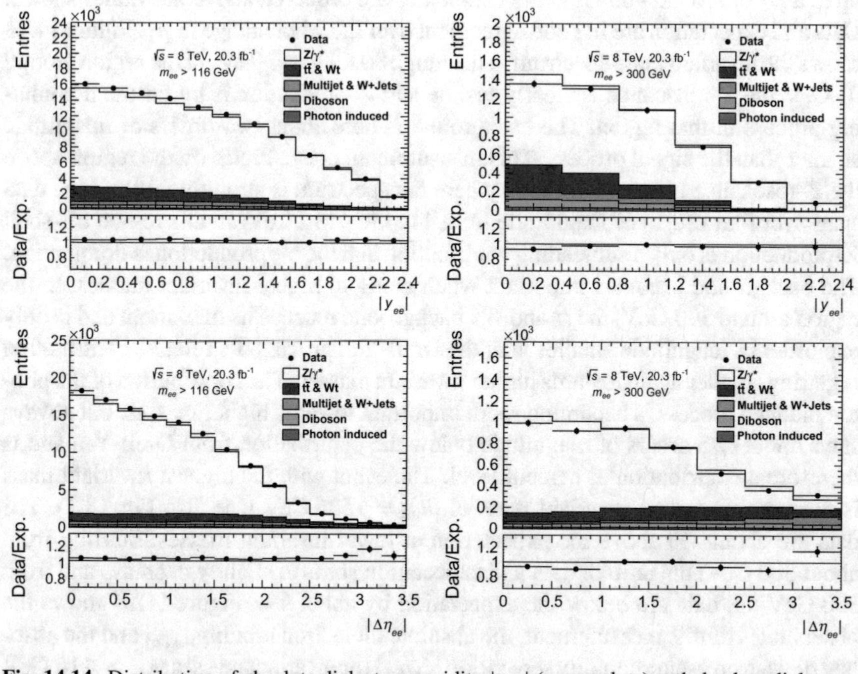

Fig. 14.14 Distribution of absolute dielectron rapidity $|y_{ee}|$ (upper plots) and absolute dielectron pseudorapidity separation $|\Delta\eta_{ee}|$ (lower plots) for invariant mass $m_{ee} > 116$ GeV (left plots), and $m_{ee} > 300$ GeV (right plots), shown for data (solid points) and expectation (stacked histogram) after the complete selection. The lower panels show the ratio of data to the expectation

The background and signal processes are stacked on top of each other and the ratio between data and expectation is shown in the lower panel. The η distribution of the selected electron candidates has its maximum at $\eta = 0$ and is slightly falling towards higher values of $|\eta|$. The region $1.37 < |\eta| < 1.52$ corresponds to the transition region between the calorimeter central region and the calorimeter endcap region. A few events are still falling into this region, since the cut is placed on the η position of the energy deposition and here the best estimate for η is shown, which also includes track information. The dominant contribution is coming from the Drell-Yan signal process. The second largest background is the top background followed by the multijet and W+jets background. The contribution from the photon induced process is very small and hardly visible in these distributions. The contribution from background processes is about 15% in the region $m_{ee} > 116$ GeV rising to about 25% in the region $m_{ee} > 300$ GeV. The shape of the expectation agrees for both mass regions very well with the shape observed in data. Data are about 4% larger than the expectation in the region $m_{ee} > 116$ GeV while it is in general the opposite in the region $m_{ee} > 300$ GeV. In the p_{T} distributions, the data have a maximum around half of the invariant mass threshold. The p_{T} spectra of the selected electrons is strongly falling over five orders of magnitude for the region $m_{ee} > 116$ GeV and three orders of magnitude for the region $m_{ee} > 300$ GeV. Both distributions show candidates up to a p_{T} of around 800 GeV. No candidates are observed above the values shown. Data and expectation are in good agreement over the whole range in p_{T}. Figure 14.13 shows the invariant mass spectrum in the range 80 GeV to 2 TeV. In the region around 91 GeV the Z-resonance is clearly visible and Z-production is by far the dominating process in that region. The background is here about two orders of magnitude smaller than the signal process. This measurement concentrates on the region above the Z-resonance ($m_{ee} > 116$ GeV) where the spectrum is strongly falling. The data span over four orders of magnitude from 116 GeV to 2 TeV. In this region off-shell Z-production is only a subleading contribution and the γ^*-production is dominating. The background becomes important when going to higher invariant masses. In the region around 300 GeV, the $t\bar{t}$ and Wt background reaches its maximum and is only one order of magnitude smaller than the Drell-Yan process. Its relative contribution is getting smaller again towards higher invariant masses. The contribution of the photon induced process is becoming more important towards higher masses, but always stays about two orders of magnitude below the contribution from Drell-Yan and is therefore a contribution at percent level. The event with the highest invariant mass is reconstructed at an invariant mass of $m_{ee} = 1526$ GeV (see also Fig. 13.1). The data are about 4% above the expectation at lower invariant masses. Starting from about 300 GeV, the ratio of data and expectation starts to slightly decrease and from 350 GeV on, data are below the expectation by about 4%. Figure 14.14 shows the observables of this measurement, the absolute dielectron rapidity $|y_{ee}|$ and the absolute dielectron pseudorapidity separation $|\Delta\eta_{ee}|$ for invariant masses $m_{ee} > 116$ GeV (left plots) and $m_{ee} > 300$ GeV (right plots). The data are slowly falling towards higher values of $|y_{ee}|$ and $|\Delta\eta_{ee}|$. The dominant background is, as already seen in all other distributions, the $t\bar{t}$ and tW background, followed by the multijet and W+jets background. The backgrounds contribute largest at low rapidities, where also the

Drell-Yan process has its largest contribution. The opposite behavior can be seen in the pseudorapidity separation, where the backgrounds have their highest contribution towards higher values, especially at higher masses, while the Drell-Yan process mainly contributes at low values of $|\Delta\eta_{ee}|$. When comparing data and expectation, the same features can be observed as for all other distributions. The shape generally agrees very well. In the region $m_{ee} > 116$ GeV, data are slightly above expectation while the opposite behavior is observed in the region $m_{ee} > 300$ GeV. It seems that the difference at higher masses is most prominent in the first rapidity bin. Only the absolute rapidity and pseudorapidity separation is shown. It has been checked that the distributions are symmetric for positive and negative values.

References

1. ATLAS Physics Modeling Group. https://twiki.cern.ch/twiki/bin/view/AtlasProtected/PhysicsModellingGroup (Internal documentation)
2. Lai H-L et al (2010) New parton distributions for collider physics. Phys Rev D 82:074024. https://doi.org/10.1103/PhysRevD.82.074024, arXiv:1007.2241 [hep-ph]
3. Alioli S, Nason P, Oleari C, Re E (2010) A general framework for implementing NLO calculations in shower Monte Carlo programs: the POWHEG BOX. In: JHEP 06, p 043. https://doi.org/10.1007/JHEP06(2010)043, arXiv: 002.2581 [hep-ph]
4. Sjöstrand T, Mrenna S, Skands PZ (2008) A brief introduction to PYTHIA 8.1. Comput Phys Commun 178:852. https://doi.org/10.1016/j.cpc.2008.01.036, arXiv:0710.3820 [hep-ph]
5. Golonka P, Was Z (2006) PHOTOS Monte Carlo: a precision tool for QED corrections in Z and W decays. Eur Phys J C45:97. https://doi.org/10.1140/epjc/s2005-02396-4. arXiv:hep-ph/0506026
6. Frixione S, Webber BR (2002) Matching NLO QCD computations and parton shower simulations. In: JHEP 06, p 029. https://doi.org/10.1088/1126-6708/2002/06/029, arXiv:hep-ph/0204244 [hep-ph]
7. Frixione S, Nason P, Webber BR (2003) Matching NLO QCD and parton showers in heavy avor production. In: JHEP 0308, p 007. https://doi.org/10.1088/1126-6708/2003/08/007, arXiv:hep-ph/0305252 [hep-ph]
8. Bahr M et al (2008) Herwig++ physics and manual. Eur Phys J C 58:639. https://doi.org/10.1140/epjc/s10052-008-0798-9, arXiv:0803.0883 [hep-ph]
9. Li Y, Petriello F (2012) Combining QCD and electroweak corrections to dilepton production in the framework of the FEWZ simulation code. Phys Rev D 86:094034. https://doi.org/10.1103/PhysRevD.86.094034, arXiv:1208.5967 [hep-ph]
10. Melnikov K, Petriello F (2006) Electroweak gauge boson production at hadron colliders through $\mathcal{O}(\alpha_s^2)$'. Phys Rev D 74:114017. https://doi.org/10.1103/PhysRevD.74.114017, arXiv:hep-ph/0609070 [hep-ph]
11. Gavin R et al (2011) FEWZ 2.0: a code for hadronic Z production at next-to-next-to-leading order. Comput Phys Commun 182:2388. https://doi.org/10.1016/j.cpc.2011.06.008, arXiv:1011.3540 [hep-ph]
12. Martin AD et al (2009) Parton distributions for the LHC. Eur Phys J C 63:189. https://doi.org/10.1140/epjc/s10052-009-1072-5, arXiv:0901.0002 [hep-ph]
13. Alwall J et al (2011) MadGraph 5 : going beyond. JHEP 06:128. https://doi.org/10.1007/JHEP06(2011)128, arXiv:1106.0522 [hep-ph]
14. Baur U (2007) Weak boson emission in hadron collider processes. Phys Rev D 75:013005. https://doi.org/10.1103/PhysRevD.75.013005, arXiv:hep-ph/0611241 [hep-ph]

15. Uta Klein. https://twiki.cern.ch/twiki/bin/view/AtlasProtected/Zprime2012Kfactors (Internal documentation)
16. Martin AD et al (2005) Parton distributions incorporating QED contributions. Eur Phys J C 39:155. https://doi.org/10.1140/epjc/s2004-02088-7, arXiv:hep-ph/0411040 [hep-ph]
17. Bardin D et al (2012) SANC integrator in the progress: QCD and EW contributions. JETP Lett 96:285. https://doi.org/10.1134/S002136401217002X, arXiv:1207.4400 [hep-ph]
18. Bondarenko SG, Sapronov AA (2013) NLO EW and QCD proton-proton cross section calculations with mcsanc-v1.01. Comput Phys Commun 184:2343. https://doi.org/10.1016/j.cpc.2013.05.010, arXiv:1301.3687 [hep-ph]
19. Sjostrand T, Mrenna S, Skands PZ (2006) PYTHIA 6.4 physics and manual. JHEP 05:026. https://doi.org/10.1088/1126-6708/2006/05/026, arXiv:hep-ph/0603175 [hep-ph]
20. Corcella G et al (2001) HERWIG 6: an event generator for hadron emission reactions with interfering gluons (including supersymmetric processes). JHEP 01:010. https://doi.org/10.1088/1126-6708/2001/01/010, arXiv:hep-ph/0011363 [hep-ph]
21. Czakon M, Mitov A (2014) Top++: a program for the calculation of the top-pair cross-section at hadron colliders. Comput Phys Commun 185:2930. https://doi.org/10.1016/j.cpc.2014.06.021, arXiv:1112.5675 [hep-ph]
22. Botje M et al (2011) The PDF4LHC working group interim recommendations. arXiv:1101.0538 [hep-ph]
23. Aad G et al (2014) Measurement of the $t\bar{t}$ production cross-section using e_{μ} events with b -tagged jets in pp collisions at $\sqrt{s}$ = 7 and 8 TeV with the ATLAS detector. Eur Phys J C 74(10):3109. https://doi.org/10.1140/epjc/s10052-014-3109-7, arXiv:1406.5375 [hep-ex]
24. Kidonakis N (2010) Two-loop soft anomalous dimensions for single top quark associated production with a W- or H-. Phys Rev D 82:054018. https://doi.org/10.1103/PhysRevD.82.054018, arXiv:1005.4451 [hep-ph]
25. ATLAS Top Group. https://twiki.cern.ch/twiki/bin/view/AtlasProtected/Top2011MCCrossSectionReference (Internal documentation)
26. Pumplin J et al (2002) New generation of parton distributions with uncertainties from global QCD analysis. JHEP 0207:012. https://doi.org/10.1088/1126-6708/2002/07/012, arXiv:hep-ph/0201195 [hep-ph]
27. Campbell JM, Ellis RK (1999) An update on vector boson pair production at hadron colliders. Phys Rev D 60:113006. https://doi.org/10.1103/PhysRevD.60.113006, arXiv:hep-ph/9905386 [hep-ph]
28. Campbell JM, Ellis RK, Williams C (2011) Vector boson pair production at the LHC. JHEP 07:018. https://doi.org/10.1007/JHEP07(2011)018, arXiv:1105.0020 [hep-ph]
29. Aad G et al (2015) Observation and measurement of Higgs boson decays to WW_* with the ATLAS detector. Phys Rev D 92(1):012006. https://doi.org/10.1103/PhysRevD.92.012006, arXiv:1412.2641 [hep-ex]
30. Aad G et al (2016) Measurement of total and differential W^+W- production cross sections in proton-proton collisions at $\sqrt{s}$ = 8 TeV with the ATLAS detector and limits on anomalous triple-gauge-boson couplings. arXiv:1603.01702 [hep-ex]
31. Butterworth J et al (2010) Single Boson and Diboson Production Cross Sections in pp Collisions at $sqrts = 7TeV$. Technical report ATL-COM-PHYS-2010-695. Geneva: CERN, 2010. https://cds.cern.ch/record/1287902
32. ATLAS Luminosity Working Group. https://twiki.cern.ch/twiki/bin/view/~AtlasPublic/LuminosityPublicResults
33. Aad G et al Search for high-mass dilepton resonances in pp collisions at $\sqrt{s}$ = 8 TeV with the ATLAS detector. Phys Rev D 90(5):052005. https://doi.org/10.1103/PhysRevD.90.052005, arXiv: 1405.4123 [hep-ex]
34. Aad G et al (2008) The ATLAS experiment at the CERN large hadron collider. JINST 3:S08003. https://doi.org/10.1088/1748-0221/3/08/S08003.
35. Zinser M (2013) Double differential cross section for Drell-Yan production of high-mass e^+e^- pairs in pp collisions at $\sqrt{s}$ = 8 TeV with the ATLAS experiment. MA thesis. Mainz U. http://inspirehep.net/record/1296478/files/553896852_CERN-THESIS-2013-258.pdf. Accessed 08 July 2013

Chapter 15
Electron Channel Cross Section Measurement

The following chapter describes the methodology of the cross section measurement. First the binnings of the measurement are discussed in Sect. 15.1 and the unfolding procedure is described in Sect. 15.2. The systematic uncertainties on the cross section are afterwards discussed in Sect. 15.3 and finally the results of the electron channel cross section measurement are presented and briefly discussed in Sect. 15.4.

15.1 Resolution and Binning

A sensible binning has to be chosen for the measurement of the differential cross section. It is important to choose a binning which is coarse enough to have sufficient statistics in every bin. The binning has in addition to be coarser than the detector resolution of the measured observable. Bin migration effects become otherwise large and it becomes difficult to extract the cross section without having large uncertainties from the unfolding procedure. If on the other hand the binning is too coarse, then information about the shape of the measured spectra is lost and therefore the physics value of the measurement is decreased. Hence, the resolution of the measured quantities is studied to define a lower range for the bin width.

Figure 15.1 shows the resolution of the invariant mass m_{ee}, the absolute rapidity $|y_{ee}|$, and the absolute pseudorapidity separation $|\Delta\eta_{ee}|$. The resolution was determined by taking the difference between the measured quantity and the generated Born level truth quantity. In case of the invariant mass, the relative resolution is shown. The Born level truth definition does not include losses due to the QED final state radiation and Bremsstrahlung. It was chosen since the main result will be cross sections at Born level and it is therefore the relevant definition to study the lower range for the bin width. The resulting distribution of the difference between the mea-

M. Zinser, *Search for New Heavy Charged Bosons and Measurement of High-Mass Drell-Yan Production in Proton-Proton Collisions*, Springer Theses,
https://doi.org/10.1007/978-3-030-00650-1_15

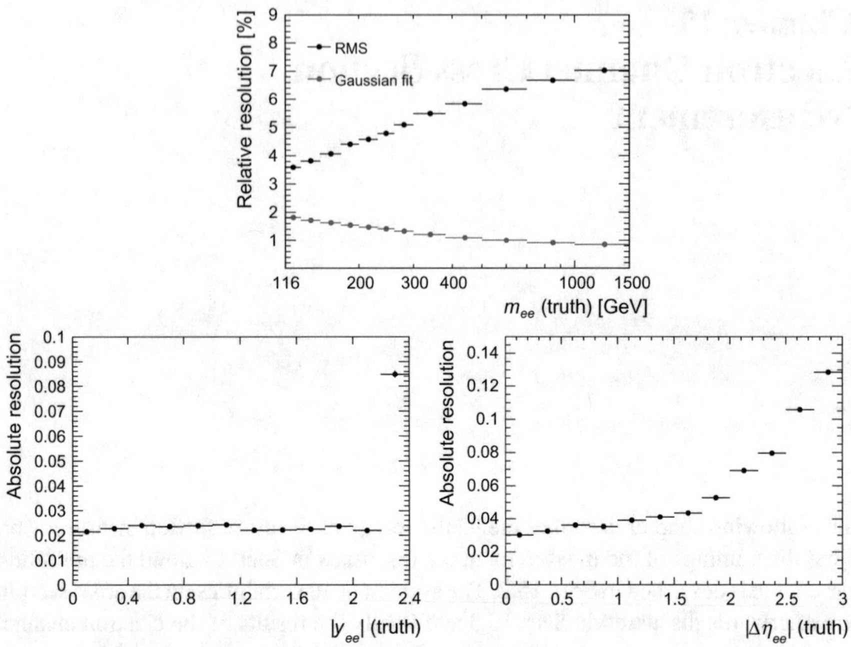

Fig. 15.1 Resolution of the invariant mass m_{ee} (top), absolute rapidity $|y_{ee}|$ (bottom left) and absolute pseudorapidity separation $|\Delta\eta_{ee}|$ (bottom right). The resolution was determined with respect to the Born level using the Drell-Yan simulation

sured value and the Born level value is therefore non-Gaussian as it has tails from large radiation. The RMS of the distribution is taken as resolution.

The invariant mass resolution at $m_{ee} = 116$ GeV is about 3.5%. The pseudorapidity of the electron candidates can be measured very precisely. Therefore only the energy resolution of the calorimeter is a limiting factor for the invariant mass resolution. The relative energy resolution at high energies gets better (see Eq. 5.6) and consequently also the invariant mass resolution is expected to get better towards higher masses. However, the invariant mass resolution is getting worse up to 7% at 1.5 TeV. The reason is the definition of the resolution with respect to the invariant mass at Born level. Radiation at large angles is not included in the measurement of the electron energy. The radiation of Bremsstrahlung is getting more likely towards higher energies and therefore the improving energy resolution is canceled by the higher probability of Bremsstrahlung, leading to the observed increasing resolution. As an example is for the invariant mass also shown the resolution when taking the Gaussian core of the relative resolution (red dots). In a single bin the resolution is not expected to follow a Gaussian distribution, as the effect of radiation will lead to a tail at the side where a lower invariant mass was reconstructed. To extract the width of the distribution, a Gaussian fit was performed to the side in which a higher invariant mass was measured. This side is nearly not affected by QED final state radiation.[1] Here the expected behavior coming from the calorimeter energy resolution can be observed. The resolution is 1.8% at 116 GeV and slightly falling towards 0.8% at 1.5 TeV.

The resolution of the absolute rapidity is about 0.02 at $|y_{ee}| = 0.0$, staying flat except for the last bin where the resolution jumps to 0.085. One of the electrons has to have $|\eta| > 2.2$ to build a pair which has an absolute rapidity above $|y_{ee}| = 2.2$ and therefore to fall into the last bin. This is close to the cut value of $|\eta| < 2.47$. The resolution is only studied for events which pass the signal selection. This leads to a bias of the resolution, as less events are considered in which the electron was reconstructed with a higher absolute rapidity. These events will most likely not enter as they are not passing the $|\eta| < 2.47$ requirement. This leads in the last bin to a bias for events which are generated with a higher absolute rapidity then reconstructed. The resolution therefore jumps up.

The η resolution is in general very good, since the inner tracking detector has a good performance and the η of the electromagnetic energy deposition and track are required to match well by the identification criteria. This translates also to a good $|\Delta\eta_{ee}|$ resolution which can be seen in the lower right plot. The resolution was determined inclusively for all invariant masses and is therefore dominated by the low invariant masses. Here the cross section for very high $|\Delta\eta_{ee}|$ is very low. Therefore more events which were mismeasured enter the bins at large $|\Delta\eta_{ee}|$. This effect leads to the observed increase in resolution.

The resolution of all measured quantities is very good and not a limiting factor on the bin width. The binning was therefore chosen in a way to have a reasonable statistical uncertainty in data. The binning for the one dimensional invariant mass measurement was chosen to be:

$$m_{ee} = [116, 130, 150, 175, 200, 230, 260, 300, 380, 500, 700, 1000, 1500]\ \text{GeV}.$$

Figure 15.2 shows in the top plot the purity of the one dimensional binning. The purity is defined as the fraction of simulated events which are reconstructed in a given m_{ee} bin and the simulated events generated in the same m_{ee} bin. The generated invariant mass is taken at Born level. The purity in the first bin is about 84%, rising to 89% in the third bin. The purity then varies around 88% as the increasing bin width roughly cancels with the improving resolution. The resolution jumps at 300 GeV to about 94% due to the wider bin width and then increases constantly to about 98% in the last bin.

The two dimensional binning for the $|y_{ee}|$ measurement was chosen in the same way, i.e., to have a reasonable amount of statistics in each bin:

$$\begin{aligned} &m_{ee} = [116, 150, 200, 300, 500, 1500]\ \text{GeV} \times |y_{ee}| \\ &m_{ee} < 300\ \text{GeV} : |y_{ee}| = [0.0, 0.2, 0.4, 0.6, 0.8, 1.0, 1.2, 1.4, 1.6, 1.8, 2.0, 2.2, 2.4] \\ &m_{ee} \geq 300\ \text{GeV} : |y_{ee}| = [0.0, 0.4, 0.8, 1.2, 1.6, 2.0, 2.4]. \end{aligned}$$

[1] The side of higher measured invariant mass is only affected if a larger invariant mass was measured and at the same time QED final state radiation and Bremsstrahlung leads to a lower invariant mass.

In the first three invariant mass bins a constant bin width of $\Delta|y_{ee}| = 0.2$ is chosen which is doubled for the last two invariant mass bins. The top plot in Fig. 15.2 also shows the purity of the two dimensional invariant mass binning. The mass bins are quite wide to have enough statistics for a two dimensional distribution, the purity is therefore quite high. It rises from 93% up to 98.5%. The bottom left plot in Fig. 15.2 shows the purity binned in $|y_{ee}|$. The purity ranges from about 88% at low invariant masses to about 98% in the last invariant mass bin. The purity starts high and is slightly lower for values around ≈ 1 before it rises again.

The two dimensional binning for the $|\Delta\eta_{ee}|$ measurement is chosen to be:

$$m_{ee} = [116, 150, 200, 300, 500, 1500]\ \text{GeV} \times |\Delta\eta_{ee}|$$
$$m_{ee} < 300\ \text{GeV} : |\Delta\eta_{ee}| = [0.0, 0.25, 0.5, 0.75, 1.0, 1.25, 1.5, 1.75, 2.0, 2.25, 2.5, 2.75, 3.0]$$
$$m_{ee} \geq 300\ \text{GeV} : |\Delta\eta_{ee}| = [0.0, 0.5, 1.0, 1.5, 2.0, 2.5, 3.0].$$

In the first three invariant mass bins a constant bin width of $\Delta|\Delta\eta_{ee}| = 0.25$ is chosen which is doubled for the last two invariant mass bins. The bottom right plot in Fig. 15.2 shows the purity binned in $|\Delta\eta_{ee}|$. Except for the first invariant mass bin, the purity always lies above 89% and no strong dependency is observed. In the first invariant mass bin, the purity drops in the last two bins down to 48–74%. The

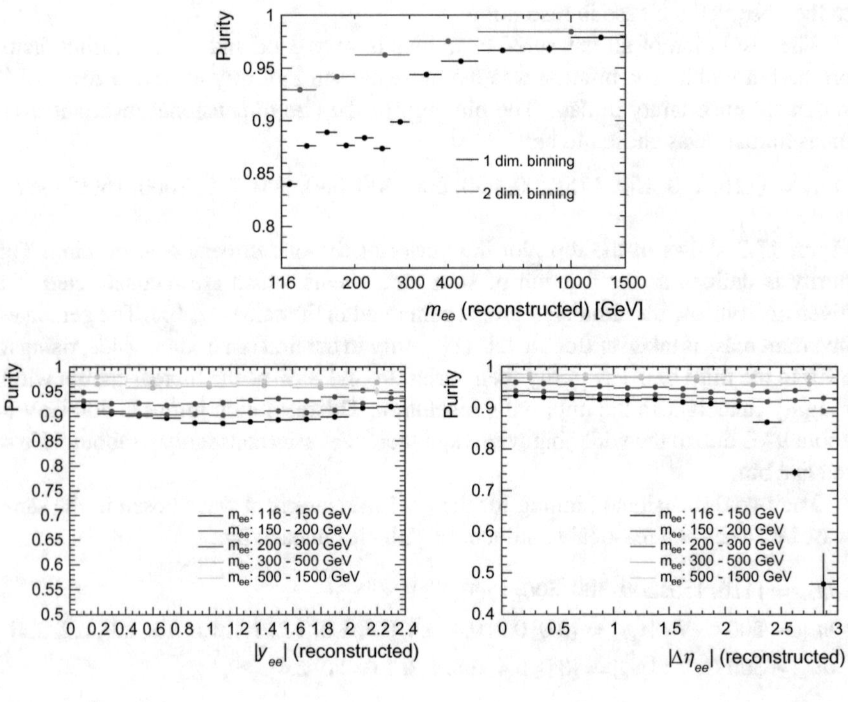

Fig. 15.2 Fraction of events for which the reconstructed mass and the true mass (Born level) fall in the same bin in the Drell-Yan simulation

drop here is expected as the resolution is getting worse in this region. The observed purities confirm again that the resolution at large $|\Delta\eta_{ee}|$ was rising mainly due to mismeasured events at low mass.

15.2 Unfolding

15.2.1 Differential Cross Section

To determine a differential cross section, the measured signal spectra have to be unfolded. In this analysis, the single-differential cross section as a function of the invariant mass m_{ee}, the double-differential cross section as a function of the invariant mass m_{ee} and the absolute rapidity $|y_{ee}|$, and the double-differential cross section as a function of the invariant mass m_{ee} and absolute pseudorapidity separation $|\Delta\eta_{ee}|$ are calculated by a simple bin-by-bin unfolding technique. A bin-by-bin unfolding technique is assumed to be sufficient since the chosen binning has a high purity (see previous section) and thus bin-migration effects are small. The double-differential cross section as a function of dilepton mass and rapidity in a fiducial phase space is calculated as:

$$\left(\frac{d\sigma}{dm_{ee}\,d|y_{ee}|}\right)_i = \frac{N_{data,i} - N_{bkg,i}}{\mathcal{L}_{int}\,\mathcal{A}_i\,\mathcal{E}_i\,\Delta m_{ee,i}\,\Delta|y_{ee}|_i}. \tag{15.1}$$

$N_{data,i}$ is the number of selected events and $N_{bkg,i}$ the number of estimated background events in a given bin i. To unfold the cross section for efficiency and acceptance effects, bin-by-bin correction factors $\mathcal{E}_i$ and $\mathcal{A}_i$ are used, respectively. Finally, to get the cross section, the unfolded number of signal events have to be divided by the integrated luminosity of the dataset $\mathcal{L}_{int}$ and the width of the bins $\Delta m_{ee,i}$ and $\Delta|y_{ee}|_i$. The double-differential cross section as a function of mass and $|\Delta\eta_{ee}|$ and the single-differential measurement as a function of invariant mass are defined accordingly.

15.2.2 Efficiency and Acceptance

The number of selected events has to be corrected, since due to inefficiencies of the detector, not every inside the acceptance produced Drell-Yan and photon induced event is measured. This efficiency correction can be determined from the signal simulation and can, for a bin i, be derived with the following formula:

$$\mathcal{E}_i = \frac{N^{\text{sim}}_{\text{sel},i}}{N^{\text{sim}}_{\text{gen},\Sigma,i}}. \tag{15.2}$$

$N^{sim}_{sel,i}$ is the number of selected signal events simulated on detector level. This number has been derived in a phase space Σ which is defined by the signal selection:

$$|\eta| < 2.47 \quad \text{excl. } 1.37 < |\eta| < 1.52, \quad p_{\mathrm{T}}^{leading} > 40 \text{ GeV}, \quad p_{\mathrm{T}}^{subleading} > 30 \text{ GeV}, \quad |\Delta\eta_{ee}| < 3.5.$$

$N^{sim}_{gen,\Sigma,i}$ is the number of generated events in the phase space Σ. The efficiency covers also the effect of bin migration, since for the numerator, the event is not required to be generated and reconstructed in the same bin. Figure 15.3 shows on the left side the efficiency of the one dimensional m_{ee} binning for the Drell-Yan process, the photon induced process and the combined efficiency. The latter was determined by using the photon induced and Drell-Yan simulation and weighting them to their cross section accordingly. The combined efficiency starts at $m_{ee} = 116$ GeV at around 70% and then rises up to 82%. The rising behavior is due to the *medium* identification efficiency. At higher invariant mass both electrons have on average higher energy. The cut values of the cut-based *medium* identification are only binned up to a transverse energy of 80 GeV. They afterwards stay constant while the electromagnetic showers become more and more narrow for higher energies. This leads to a rising identification efficiency for higher invariant masses m_{ee}. The combined efficiency is dominated due to the high cross section by the Drell-Yan process. Hence, the efficiency for the Drell-Yan process only closely follows the combined efficiency. Also the efficiency of the photon induced process follows closely the combined efficiency, despite at lower masses where its efficiency is up to about 7% higher. In Fig. 15.4, the combined efficiency binned in absolute dielectron rapidity is shown on the left side for all five invariant mass bins. In all bins the efficiency is higher at low rapidity values and then slightly drops towards higher values. At higher rapidities, the two electrons are more likely to be at higher $|\eta|$ and therefore measured in the endcaps of the electromagnetic calorimeter. This leads to a falling behavior with $|y_{ee}|$, since in the endcaps there is more material between beam axis and electromagnetic calorimeter and thus the identification becomes more problematic. Figure 15.5 shows the combined efficiency for the dielectron pseudorapidity separation in all five invariant mass bins. The same behavior as for the rapidity can be observed due to the same reason. At higher pseudorapidity separation, the electrons are more likely to be in the endcaps of the electromagnetic calorimeter which leads to the falling behavior. In the last bin of the first invariant mass bin, the efficiency jumps from about 65% to about 86%. This is assumed to be due to a statistical fluctuation in the MC since the Drell-Yan and photon induced MC do have poor statistics in that region.

The efficiency correction, as already discussed, is defined for a given phase space Σ. The calculated cross section is thereby only valid in this phase space. To give a more convenient result, which is more independent from the detector geometry, a phase space extrapolation can be made via an acceptance correction. The acceptance correction can also be determined from the Drell-Yan and photon induced simulation and is given by:

$$\mathcal{A}_i = \frac{N^{\mathrm{sim}}_{\mathrm{gen},\Sigma,i}}{N^{\mathrm{sim}}_{\mathrm{gen},\Omega,i}}, \tag{15.3}$$

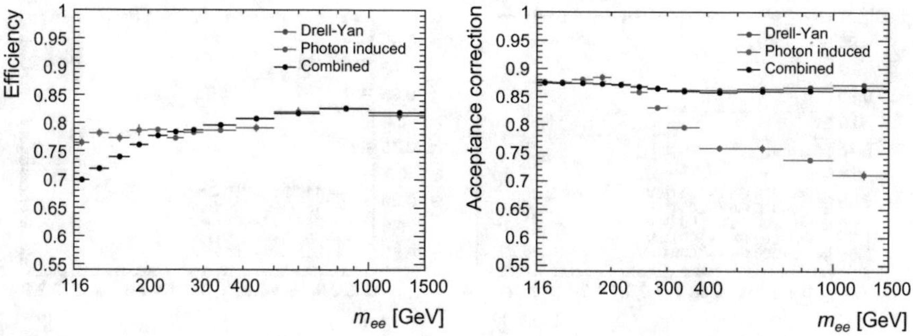

Fig. 15.3 Efficiency $\mathcal{E}$ (left) and the acceptance correction $\mathcal{A}$ (right) for the one dimensional m_{ee} binning. Efficiency and acceptance correction were determined on Born level using the Drell-Yan and photon induced simulation

where $N^{sim}_{gen,\Omega,i}$ is the number of generated events in a fiducial phase space Ω to which the cross section shall be extrapolated. For this analysis Ω is chosen to be:

$$|\eta| < 2.5, \quad p_{\mathrm{T}}^{leading} > 40 \text{ GeV}, \quad p_{\mathrm{T}}^{subleading} > 30 \text{ GeV}$$

This includes the extrapolations from $|\Delta\eta_{ee}| < 3.5$ to infinity, over the transition region $1.37 < |\eta| < 1.52$ to have a continuous interval, and the extrapolation from $|\eta| < 2.47$ up to $|\eta| < 2.5$ due to simplicity. A correction up to higher $|\eta|$ and smaller p_{T} would have, mainly due to the chosen PDF, a stronger model dependency and thus would introduce larger theoretical uncertainties. The phase space Ω will be later also used for the muon cross section which makes it possible to compare both cross sections and to perform a combination. The acceptance correction binned in invariant mass can be seen for the separate processed and their combination on the right side in Fig. 15.3. The combined acceptance correction is rather constant at around 86–87.5%. The slight drop at higher invariant masses is caused by the extrapolation from $|\Delta\eta_{ee}| < 3.5$ to infinity since at higher mass, the electrons are more likely to have a larger pseudorapidity separation. The acceptance correction of the Drell-Yan process is, like for the efficiency, closely following the combined acceptance correction. The acceptance correction for the photon induced process slightly drops towards higher invariant masses. This is due to the t-channel contribution which leads to more forward electrons with low p_{T} values. The high invariant mass is therefore mainly generated via a large pseudorapidity separation and therefore higher invariant masses are more affected by the extrapolation from $|\Delta\eta_{ee}| < 3.5$ to infinity.

Figure 15.4 shows on the right side the acceptance correction for the five invariant mass bins binned in rapidity of the dielectron system. For low rapidities, and therefore low boosts along the z-axis, both electrons decay mainly into the central region of the detector and are thus not so much affected by the acceptance extrapolations. The acceptance correction ranges therefore from 86% to 99%, getting lower towards higher invariant masses, as here the extrapolation from $|\Delta\eta_{ee}| < 3.5$ to infinity play

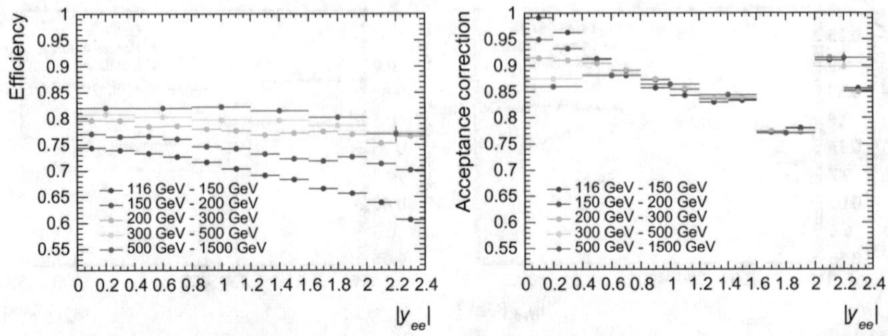

Fig. 15.4 Efficiency $\mathcal{E}$ (left) and the acceptance correction $\mathcal{A}$ (right) in the five invariant mass bins for the $|y_{ee}|$ binning. Efficiency and acceptance correction were determined on Born level using the Drell-Yan and photon induced simulation

a bigger role. When going to higher $|y_{ee}|$, it is more and more likely for one of the two electrons to be in the transition region. This results in a minimum acceptance correction of about 77% for all invariant mass bins at around $|y_{ee}| = 1.8$. For absolute rapidities above 2.0, both electrons are dominantly above the transition region between central region and the endcap of the electromagnetic calorimeter. Hence, they are mainly affected by the extrapolation up to $|\eta| = 2.5$ and $|\Delta\eta_{ee}|$ to infinity. The acceptance correction is in the last bin slightly dropping again due to the extrapolation to $|\eta| = 2.5$.

Figure 15.5 shows on the right side the acceptance correction for the five invariant mass bins binned in the absolute pseudorapidity separation of the two electrons. Here the extrapolation from $|\Delta\eta_{ee}| < 3.5$ to infinity does not play a role. The acceptance correction in $|\Delta\eta_{ee}|$ does therefore not show any strong mass dependency. The acceptance correction is slightly dropping up to $|\Delta\eta_{ee}| = 1.0$. In this region both electrons are mainly in the central region of the detector and it becomes more likely for one to be inside of the transition region of the electromagnetic calorimeter. Above $|\Delta\eta_{ee}| = 1.0$, the more forward electron is most likely above the transition region while it becomes with rising pseudorapidity separation more likely that the second lepton is in the transition region or the first lepton above $|\eta| = 2.47$. The acceptance correction is therefore slightly dropping again.

15.2.3 Correction Factor C_{DY}

The efficiency and acceptance corrections can be combined to a common correction factor:

$$C_{\mathrm{DY},i} = \mathcal{A}_i \mathcal{E}_i = \frac{N^{\mathrm{sim}}_{\mathrm{sel},i}}{N^{\mathrm{sim}}_{\mathrm{gen},\Omega,i}}. \tag{15.4}$$

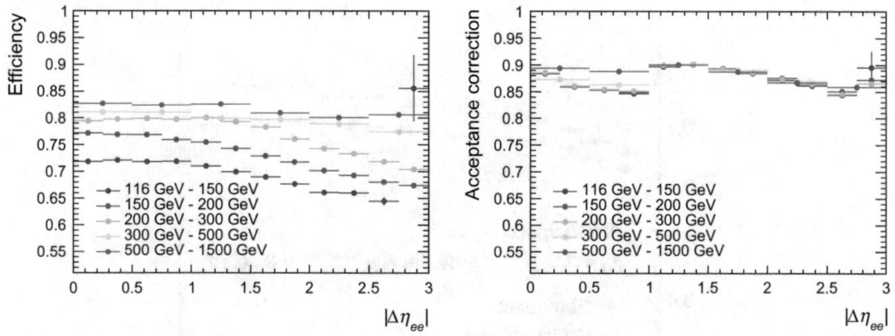

Fig. 15.5 Efficiency $\mathcal{E}$ (left) and the acceptance correction $\mathcal{A}$ (right) in the five invariant mass bins for the $|\Delta\eta_{ee}|$ binning. Efficiency and acceptance correction were determined on Born level using the Drell-Yan and photon induced simulation

The correction factor C_{DY} is affected by the limited statistics of the signal sample used to calculate it. For a perfect detector resolution, the statistical uncertainty of C_{DY} would be the uncertainty of a binomial distribution, since in one bin $N_{\mathrm{sel}}^{\mathrm{sim}}$ is a subset of $N_{\mathrm{gen},\Omega}^{\mathrm{sim}}$. Due to finite resolution, migration between bins occurs and thus $N_{\mathrm{gen},\Omega}^{\mathrm{sim}}$ does not any longer completely contain $N_{\mathrm{sel}}^{\mathrm{sim}}$. Assuming an uncertainty of a Gaussian distribution would however be too conservative and would lead to a too large uncertainty. Due to the rather small amount of migration there is still a large correlation between numerator and denominator. To get the correct statistical uncertainty, the calculation of C_{DY} can be split into uncorrelated samples:

$$C_{\mathrm{DY}} = \frac{N_{\mathrm{sel}}^{\mathrm{sim}}}{N_{\mathrm{gen},\Omega}^{\mathrm{sim}}} = \frac{N_{\mathrm{stay}} + N_{\mathrm{come}}}{N_{\mathrm{stay}} + N_{\mathrm{leave}}}, \tag{15.5}$$

where N_{stay} is the number of events generated and reconstructed in a certain bin, $N_{\mathrm{come}} = N_{\mathrm{sel}}^{\mathrm{sim}} - N_{\mathrm{stay}}$ are the events reconstructed in a certain bin, but generated elsewhere, and $N_{\mathrm{leave}} = N_{\mathrm{gen},\Omega}^{\mathrm{sim}} - N_{\mathrm{stay}}$ are the events generated in a certain bin, but migrating out or failing the selection cuts. Following reference [1], the uncertainty on C_{DY} can then be expressed as:

$$\begin{aligned}(\Delta C_{DY})^2 =& \frac{(N_{\mathrm{gen},\Omega}^{\mathrm{sim}} - N_{\mathrm{sel}}^{\mathrm{sim}})^2}{(N_{\mathrm{gen},\Omega}^{\mathrm{sim}})^4}(\Delta N_{\mathrm{stay}})^2 + \frac{1}{(N_{\mathrm{gen},\Omega}^{\mathrm{sim}})^2}(\Delta N_{\mathrm{come}})^2 \\ &+ \frac{(N_{\mathrm{sel}}^{\mathrm{sim}})^2}{(N_{\mathrm{gen},\Omega}^{\mathrm{sim}})^4}(\Delta N_{\mathrm{leave}})^2.\end{aligned} \tag{15.6}$$

The correction factor C_{DY} is shown binned in invariant mass in Fig. 15.6. The C_{DY} corrections are given at Born level and dressed level. For the results at dressed level, the leptons after QED FSR are recombined with radiated photons within a cone of $\Delta R = 0.1$. The cross section results will be provided for both definitions. The

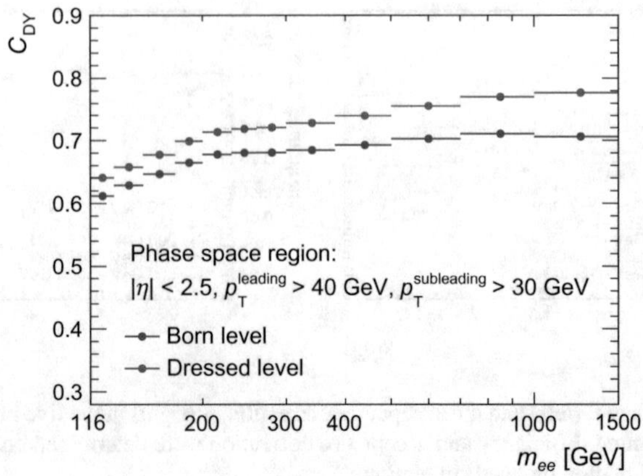

Fig. 15.6 Central values for C_{DY} from Drell-Yan and photon induced MC binned in invariant mass at Born and dressed level with statistical uncertainties

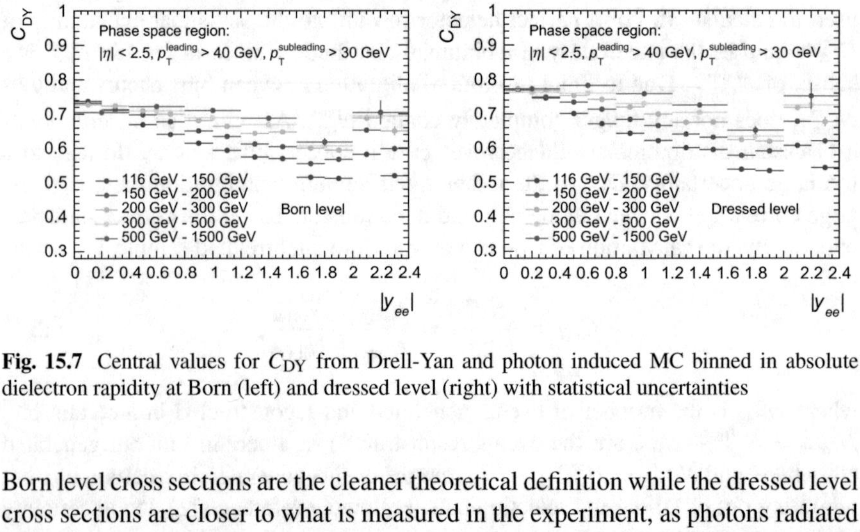

Fig. 15.7 Central values for C_{DY} from Drell-Yan and photon induced MC binned in absolute dielectron rapidity at Born (left) and dressed level (right) with statistical uncertainties

Born level cross sections are the cleaner theoretical definition while the dressed level cross sections are closer to what is measured in the experiment, as photons radiated with a small angle to the electron cannot be resolved by the calorimeter. The corrections for the dressed level are therefore in general smaller. The difference between both definitions gets larger towards higher invariant masses, as the probability for Bremsstrahlung is getting larger. The C_{DY} factor combines the effects of both, the acceptance extrapolation and the efficiency correction. The dependencies are therefore the same as discussed before. Figures 15.7 and 15.8 show the C_{DY} factors binned in absolute dielectron rapidity and absolute pseudorapidity separation for the born (left) and dressed (right) level.

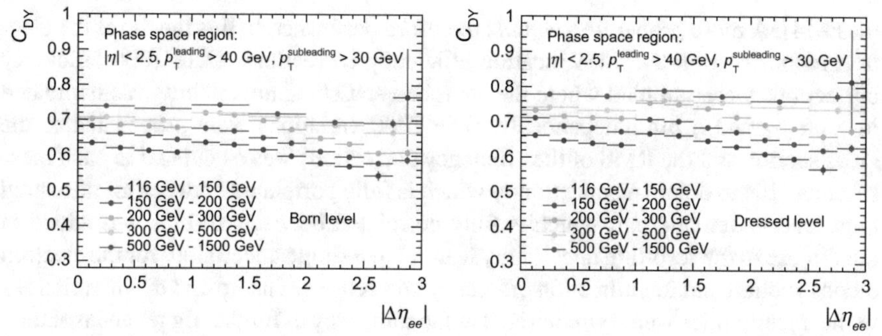

Fig. 15.8 Central values for C_{DY} from Drell-Yan and photon induced MC binned in absolute pseudorapidity separation at Born (left) and dressed level (right) with statistical uncertainties

15.3 Systematic Uncertainties

Several systematic uncertainties have to be considered for this measurement. They are in the following discussed and summarized in Figs. 15.13, 15.14, and 15.15.

Trigger and isolation efficiency The p_{T} dependent trigger and isolation efficiency corrections provided by the ATLAS group [2] come with systematic and statistical uncertainties. They are correcting for differences of the trigger and isolation efficiency in data and simulation. Efficiency corrections enter only the numerator of C_{DY} and therefore propagate directly to the cross section measurement. The systematic uncertainties are obtained by varying the tag and probe selection, e.g., identification of the tag electron or window of the Z-peak or varying the background model. These uncertainties are propagated to the cross section by recomputing C_{DY} with the efficiency corrections varied coherently up and down by their full systematic and statistical uncertainty. The statistical and systematic component are varied separately. The up and down variations are symmetrized by taking the larger of the two variations.

Electron reconstruction and identification efficiency The p_{T} and η dependent reconstruction and identification efficiency corrections provided by the ATLAS electron performance group [3] come with systematic uncertainties. The uncertainties were determined in a similar way as for the trigger and isolation efficiency corrections. The statistical uncertainty of the efficiency corrections is uncorrelated between all bins in p_{T} and η, as they are measured on orthogonal data sets. Varying the efficiency correction by the statistical uncertainty in all bins simultaneously up and down and propagating the effect to the cross section gives a result which is still partially correlated between the measurement bins as the same η/p_{T} bin can contribute to several measurement bins. This procedure in general overestimates the statistical uncertainty. It was performed for the trigger and isolation efficiency where these effects are small. However, the identification and reconstruction efficiency uncertainties were one of the largest uncertainties on the measurement described in my master thesis [4]. A more proper treatment is therefore performed in this thesis for the electron reconstruction and identification efficiency corrections. Here 1000 efficiency corrections were sampled where the Gaussian statistical uncertainty was fluctuated in each p_{T} and η bin independently. The 1000 variations were propagated to the cross section and the RMS of the efficiency corrections was calculated in each measurement bin to obtain an uncertainty which is fully correlated between all measured bins. The systematic part which is fully correlated between all bins was added in quadrature to the resulting uncertainty leading to a single uncertainty for the electron reconstruction and identification efficiency corrections. The up and down variations of the uncertainties were symmetrized in the same way as for the trigger and isolation efficiency corrections.

Electron energy scale Corrections for the energy scale of the electrons are applied to data [5]. The effect of varying all their respective uncertainties up and down is checked in order to determine the systematic uncertainty. This can in principle be done on data or on MC, given the higher statistics available it is preferable to check the effect of the up/down variation on C_{DY} in MC. The reconstructed energy is varied in the simulation, according to the systematic uncertainties which are given for the energy corrections. Correcting the energy of the electrons leads to different invariant masses and thereby to bin migration in invariant mass or rapidity. This can distort the shape of the reconstructed spectra and thus lead to differences in C_{DY}. The uncertainty of the energy scale is subdivided into 14 systematic uncertainty components and a statistical uncertainty. The statistical and a systematic component are arising from the method with which the energy scale corrections are extracted (*Zee*). The position of the Z-resonance in the invariant mass distribution of the electron pairs in data is calibrated by matching it to the position in MC. The uncertainty is dominantly driven by uncertainties on the background estimation in the electron selection, which is used to select candidates in the Z-region. The statistical uncertainty was found to be below the per mille level and as such is neglected. Three of the systematic uncertainties are due to the limited knowledge of the material in the inner detector (*MatID*), the cryostat of the liquid argon calorimeter (*MatCryo*), and the passive material of the calorimeter itself (*MatCalo*). Another four additional sources of uncertainties arise from the limited knowledge of the internal liquid argon calorimeter geometry itself (*LArCalib*, *LARUnconvCalib*, *LArElecUnconv*, *LArElecCalib*). The energy scale is reevaluated using a Monte Carlo sample where the amount of material in the detector part was varied according to its systematic uncertainty. Differences in the energy scale are then quoted as systematic uncertainty of the material. Additionally, there is an uncertainty due to the knowledge of the energy scale in the presampler detector (*PS*), which is used to correct for energy lost upstream of the active electromagnetic calorimeter. A similar uncertainty arises from the energy scale in the first and second layer of the calorimeter (*S12*). Two sources of uncertainty are assigned to the intrinsic accuracy of the electromagnetic shower development simulation (*G4*, *Pedestal*), by varying physics modeling options in GEANT4. Finally, two uncertainties arise from the electronic gain of the signals in the first and second layer of the calorimeter (*L1Gain*, *L2Gain*). The uncertainty from the gain in the second layer is typically the largest at high masses, followed by the uncertainty from the material knowledge of the liquid argon calorimeter and the method to extract the energy scale corrections. All uncertainties are symmetric but do not lead to symmetric effects in C_{DY}, since varying the energy scale up has a larger effect on a strongly falling spectrum, due to larger bin migrations. Because of this asymmetry, not the maximum deviation from the up and down variations is used as systematic uncertainty on C_{DY}, but the average of the up and down variation. This is not the most conservative treatment, but the asymmetries of the uncertainties are small and this approach is less affected by statistical fluctuations. For simplicity sometimes a single uncertainty is quoted which is calculated by adding all up and down variations in quadrature and taking the larger uncertainty in each bin.

Energy smearing The smearing of the energy in the simulation, to correct for a better energy resolution modeled in MC than observed in data, has a systematic uncertainty [5]. Varying the smearing within its uncertainties can, like for the energy scale, distort the reconstructed invariant mass spectrum and thus cause differences in C_{DY}. The uncertainty of the energy resolution is subdivided into seven systematic uncertainty components. A single source of uncertainty arises from the method with which the corrections are obtained (*ZSmearing*). After correcting the energy scale, the MC is corrected in such a way that the width of the Z-resonance is corresponding to the width in data. Four of the systematic uncertainties are due to the limited knowledge of the material in the inner detector (*MaterialID*), the cryostat of the liquid argon calorimeter (*MaterialCryo*), the inner detector support material in the transition region between barrel and endcap (*MaterialGap*), and the passive material of the calorimeter itself (*MaterialCalo*). Finally, additional uncertainties arise from deriving the resolution corrections for different pile-up conditions (*PileUp*) and from test beam measurements of the sampling term of the energy resolution parameterization[2] (*SamplingTerm*). The uncertainties are symmetrized in the same way as the electron energy scale uncertainties. The uncertainties are typically much smaller. Also here sometimes a single uncertainty is quoted for simplicity. It is obtained in the same way as for the electron energy scale.

Electroweak background An uncertainty is arising from the normalization of the electroweak backgrounds ($t\bar{t}$ & Wt and diboson). The $t\bar{t}$ MC sample is normalized to a cross section of $\sigma_{t\bar{t}} = 253^{+15}_{-17}$ pb for a top-quark mass of 175.5 GeV. The single-top background in association with a W boson has a cross section of $\sigma_{Wt} = 22.4 \pm 1.5$ pb. Given that the Wt contribution is about 10% compared to the $t\bar{t}$ cross section, an overall normalization uncertainty of 6% is estimated on the background including top-quarks. A more detailed description of the systematic uncertainties can be found in Sect. 14.1.2. The top background is the largest background in most of the phase space. Therefore further studies have been added for this thesis to check whether the top background agrees within the assigned normalization uncertainties. The number of b-jets and the $E_{\mathrm{T}}^{\mathrm{miss}}$ distributions, which are dominated by the top background

---

[2] $\frac{\sigma}{E} = \frac{a}{\sqrt{E}} \oplus \frac{b}{E} \oplus c$, where a, b and c are η-dependent parameters; a is the sampling term, b is the noise term, and c is the constant term.

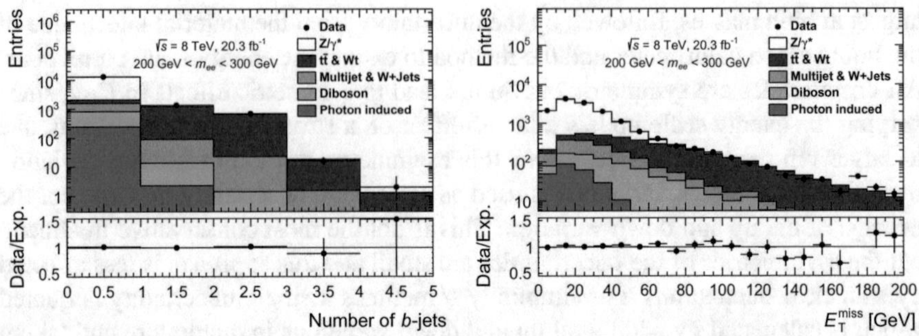

Fig. 15.9 Distributions of the number of b-jets and $E_{\mathrm{T}}^{\mathrm{miss}}$ in the region 200 GeV $< m_{ee} <$ 300 GeV after the final high-mass Drell-Yan signal selection. Shown for data (solid points) and expectation (stacked histogram). The lower panels show the ratio of data to the expectation. The first bin for the number of b-jets distribution corresponds to 0 observed b-jets whereas the second bin corresponds to one observed b-jet (and so on)

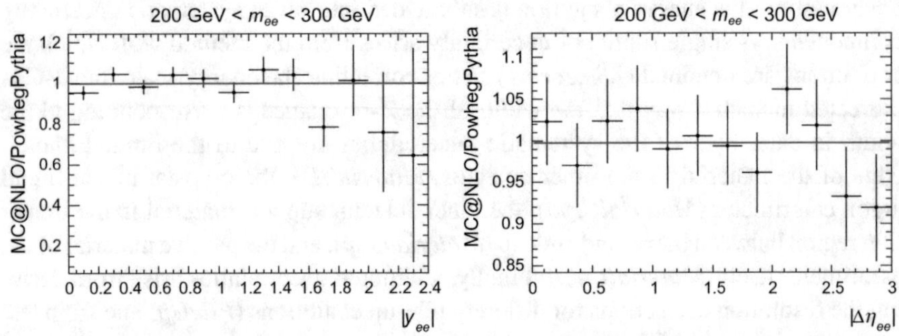

Fig. 15.10 Ratio between the $t\bar{t}$ background simulated by POWHEG and MC@NLO binned in $|y_{ee}|$ (left) and $|\Delta\eta_{ee}|$ (right) for the range 200 GeV $< m_{ee} <$ 300 GeV. A linear fit has been performed to the ratio (red line)

for $N_{b-jet} > 0$ and large $E_{\mathrm{T}}^{\mathrm{miss}}$, are shown in Fig. 15.9 for the region 200 GeV $< m_{ee} <$ 300 GeV. Good agreement of the top background within the normalization uncertainty is observed. The distributions for all other invariant mass bins of the measurement are shown in Appendix L.

The default POWHEG $t\bar{t}$ MC sample has been compared to a MC sample generated with MC@NLO to check for uncertainties on the modeling of the measured observables. The ratios of the two MC samples for the two dimensional observables in the range 200 GeV $< m_{ee} <$ 300 GeV is shown in Fig. 15.10. No systematic uncertainty is added since the differences are within the statistical fluctuations which are propagated to the measurement. The ratios of the two MC samples for the two dimensional observables for all other invariant mass bins of the measurement are shown in Appendix L.

Finally, a further check has been performed [6] by selecting oppositely charged electron-muon pairs and comparing expectations for the invariant mass spectrum and distributions of the measured observables with data. This selection strongly suppresses the Drell-Yan process and leads to a sample with an about 80% contribution from top-quark production. The remaining events are mainly due to diboson production. No systematic differences are found and therefore no further uncertainties are assigned. The normalization uncertainties on the diboson background are listed in Sect. 14.1.2.2. They are 4, 4.2 and 10% on the WZ, ZZ and WW process, respectively. The uncertainties of the backgrounds are combined to a single uncertainty by adding them in quadrature. The statistical and systematic part are given separately.

Multijet & W+jets background The determination of the systematic uncertainty of the multijet and W+jets background is described in Sect. 14.3.6. The dominant contribution is coming from the differences in the nine performed methods to calculate the background. Systematic and statistical uncertainty are quoted separately.

Drell-Yan and photon induced MC statistic The uncertainty is arising from the limited number of simulated events for the signal MC when determining the C_{DY} factor. This uncertainty is calculated following Eq. 15.6 and treated as fully uncorrelated between all measurement bins.

Drell-Yan theory uncertainty Varying the Drell-Yan NNLO theory corrections for the C_{DY} calculation within their uncertainties affects both, numerator and denominator. Thus the variation cancels in large parts. The resulting uncertainty is below the per mille level and will not be considered in the following.

Photon induced uncertainty As the photon induced processes has not been studied at length, any MC simulation attempting to describe these will have a high level of uncertainty. A 40% MC cross section uncertainty is calculated as the difference between photon induced calculations using a current quark model and a constituent quark model [7]. As the photon induced contribution is small and the uncertainties affect again both, numerator and denominator, the effect this uncertainty has on the cross section is found to be below the per mille level and as such is neglected.

Monte Carlo modeling uncertainty To account for any MC model dependences when calculating the C_{DY} factor used for the unfolding, an additional Drell-Yan MC is compared to the nominal Drell-Yan sample. The alternative MC@NLO sample used has a different matrix element calculation, parton shower model and FSR model. This uncertainty was not studied for the measurement described in my master thesis as the alternative MC was not available at that time. The ratio of C_{DY} at Born level between using POWHEG or MC@NLO is shown in Fig. 15.11 for two invariant mass bins of the measurement as a function of absolute rapidity. A linear fit to the ratio has been performed and is also shown (red line). No systematic uncertainty is added since the differences are within the statistical fluctuations which are propagated to the measurement. The same ratios for all other measurement bins are shown in Appendix M for born and dressed level separately.

Unfolding uncertainty An uncertainty can arise from the chosen unfolding method. This uncertainty was not studied for the measurement described in my master thesis. For this thesis, the difference between bin-by-bin unfolding via C_{DY} and using an

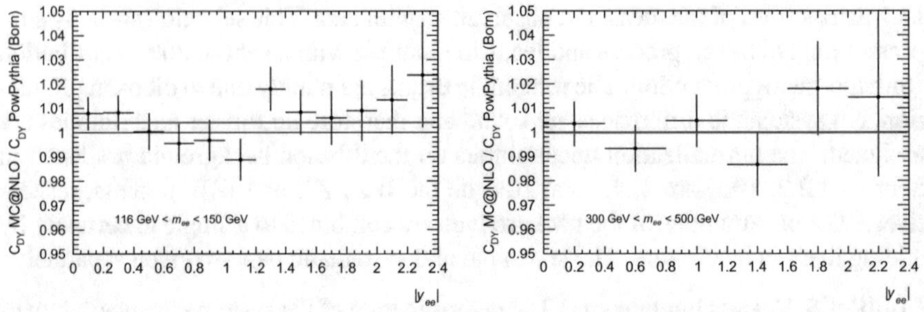

Fig. 15.11 Ratio between C_{DY} calculated with MC@NLO and C_{DY} calculated with POWHEG interfaced with PYTHIA at Born level. A linear fit has been performed to the ratio (red line)

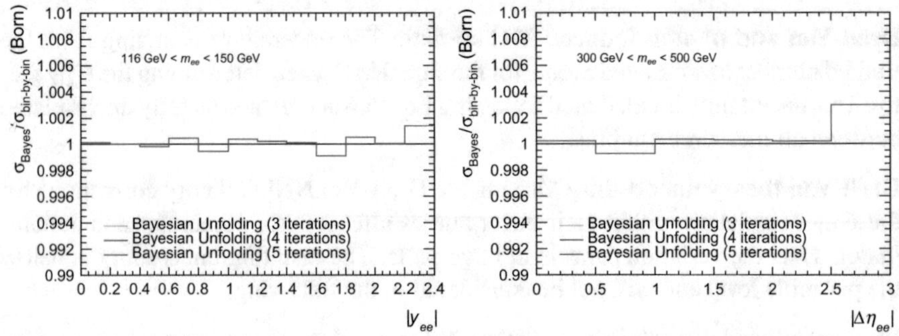

Fig. 15.12 Ratio between the final cross section unfolded with the standard bin-by-bin unfolding method and a Bayesian unfolding method

Bayesian unfolding method [8], implemented in the RooUnfold package [9], has been investigated. Figure 15.12 shows the ratio between the Born level cross section unfolded with bin-by-bin unfolding to the unfolded cross section unfolded with Bayesian unfolding. Shown is the difference for 3, 4 and 5 iterations of the Bayesian unfolding. No uncertainty is added as the observed differences between both methods are negligible when compared to the statistical uncertainty of the cross section and no systematic trends are observed. The same ratios for all other measurement bins are shown in Appendix M.

15.3.1 Summary

Figure 15.13 shows the resulting statistical and systematic uncertainties on the unfolded single-differential cross section. Also the total correlated systematic (total syst.) and total systematic uncertainty (total syst.+stat.) is shown and calculated by adding all correlated and uncorrelated systematic sources in quadrature. At low invariant mass the cross section measurement is dominated by the systematic uncer-

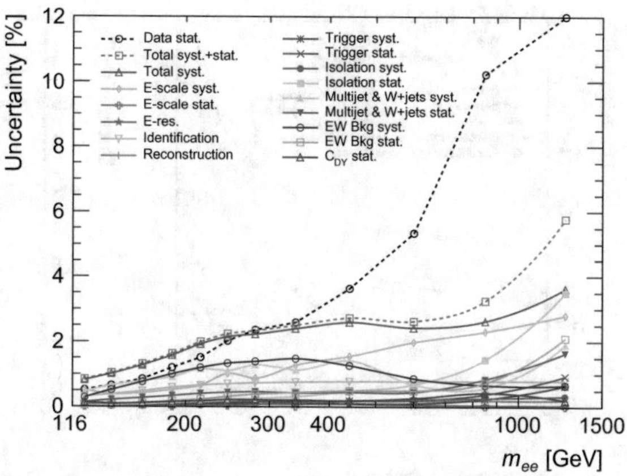

Fig. 15.13 The relative size of the systematic and statistical uncertainties on the single-differential cross section measurement. Points that are drawn at the maximum are off-scale

tainty which is below 1%. The largest contributions are coming from the multijet & W+jets background, the electroweak backgrounds and the electron energy scale. The uncertainties on the electron energy scale and energy resolution are combined into a single uncertainty each. A detailed breakdown of all uncertainty sources can be found in the tables in Appendix O. The electroweak background uncertainty and the multijet & W+jets background uncertainty is each rising up to about 1.5% between 300 GeV and 400 GeV as the relative contribution of the background becomes larger. The total systematic uncertainty in this region is about 2.5% and therefore similar to the statistical uncertainty on data. For higher invariant masses, the measurement uncertainties are dominated by the statistical uncertainty on data. Besides the last bin, the systematic uncertainty is only slowly rising. The total systematic uncertainty in the last bin is about 6%. Here, the uncorrelated component of the systematic uncertainty has a sizable contribution, mainly coming form the uncorrelated statistical uncertainty on the isolation efficiency correction.

Figure 15.14 shows the uncertainties in all five invariant mass bins for the measurement as a function of absolute rapidity. The same dependency of the systematic uncertainties as for the invariant mass is observed across the invariant mass bins. Here, the statistical uncertainty of the data is dominating in all measurement bins. The total systematic uncertainty is 0.8% in the first invariant mass bin at low absolute rapidities while the statistical uncertainty on data is already about 1.2%. For low values of absolute rapidity, the dominating source of systematic uncertainty is coming from the electroweak background uncertainty, it is falling towards higher values since the relative contribution of the background becomes smaller. At high values of absolute rapidity, the multijet & W+jets background is leading to the largest uncertainty.

Figure 15.15 shows the uncertainty in all five invariant mass bins for the measurement as a function of absolute pseudorapidity separation. The same dependency of the systematic uncertainties as for the invariant mass is observed across the invariant mass bins. Also here, the statistical uncertainty on data is larger or equal to the total

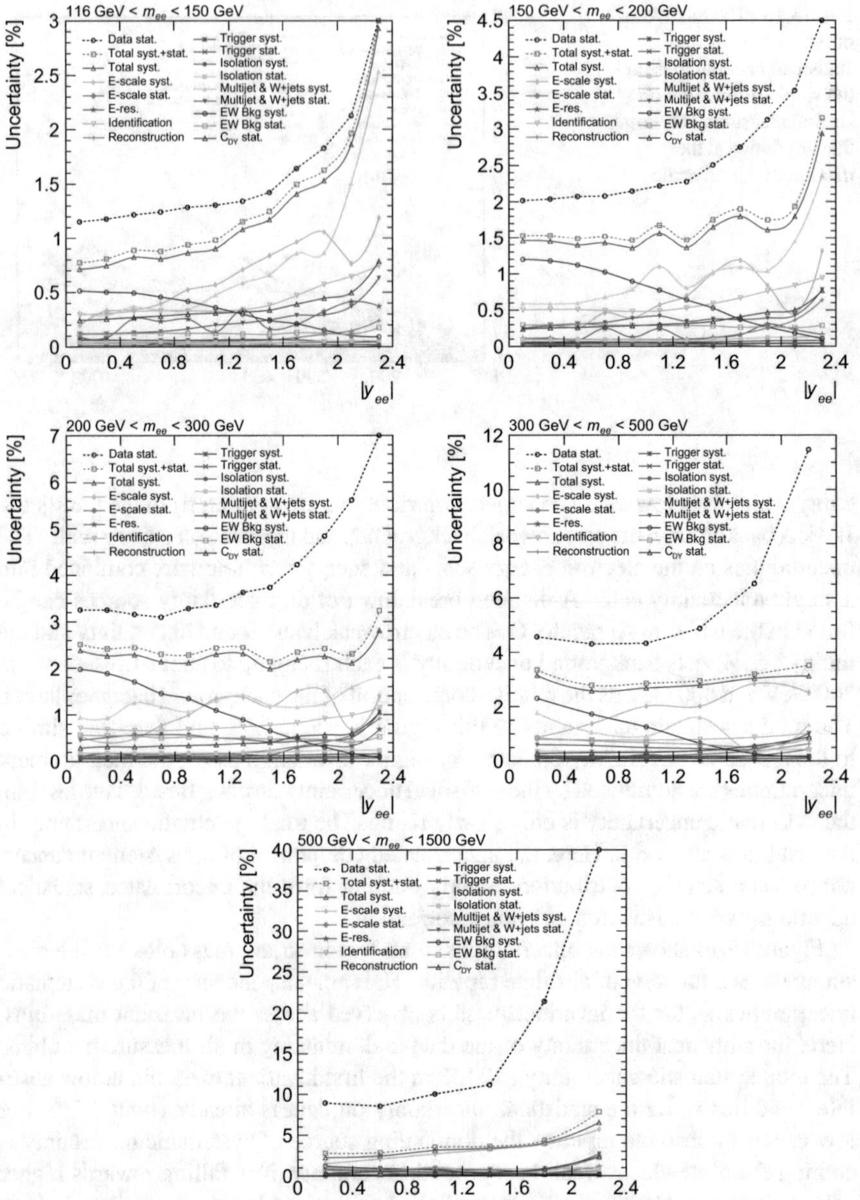

Fig. 15.14 The relative size of the systematic and statistical uncertainties on the double-differential cross section measurement as a function of m_{ee} and $|y_{ee}|$. Points that are drawn at the maximum are off-scale

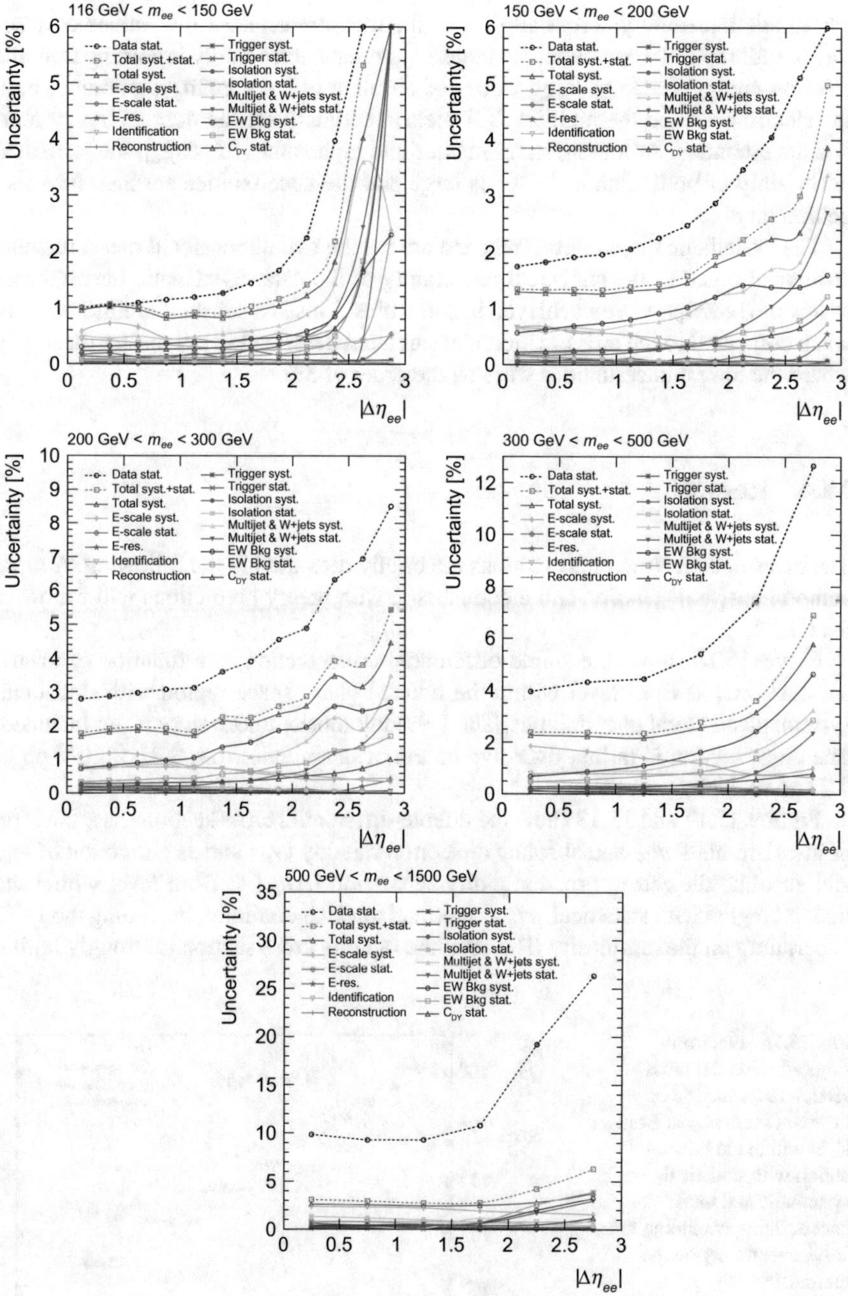

Fig. 15.15 The relative size of the systematic and statistical uncertainties on the double-differential cross section measurement as a function of m_{ee} and $|\Delta\eta_{ee}|$. Points that are drawn at the maximum are off-scale

systematic uncertainty across all bins of the measurement. At low values of absolute pseudorapidity separation, the largest systematic uncertainty is coming from the electron energy scale. At large values of absolute pseudorapidity separation, both the electroweak and the multijet & W+jets uncertainty are the dominating systematic uncertainties. At low invariant masses and high values of $|\Delta\eta_{ee}|$, the statistical uncertainty of both, data and MC, is large and the uncertainties are therefore also getting large.

The systematic uncertainties reached are for the two dimensional measurements smaller or equal to the statistical uncertainty of the data. Systematic uncertainties as low as 0.8% have been achieved in some bins. This is a substantial improvement when compared to the cross section measurement described in my master thesis [4], where the lowest uncertainties were on the order of 3%.

15.4 Results

The measured electron cross sections are briefly discussed in the following. A more comprehensive discussion and a comparison with theory predictions will follow in Sect. 17.2.

Figure 15.16 shows the single-differential cross section as a function of invariant mass m_{ee} at Born level within the fiducial phase space region with statistical, systematic and total uncertainties. The 1.9% luminosity uncertainty is not included. The cross section is falling over five orders of magnitude from 2.31×10^{-1} pb to 3.23×10^{-6} pb.

Figures 15.17 and 15.18 show the double-differential cross sections as a function of invariant mass m_{ee} and absolute dielectron rapidity $|y_{ee}|$ and as a function of m_{ee} and absolute dielectron pseudorapidity separation $|\Delta\eta_{ee}|$ at Born level within the fiducial region with statistical, systematic and total uncertainties, excluding the 1.9% uncertainty on the luminosity. The absolute rapidity cross section is strongly falling

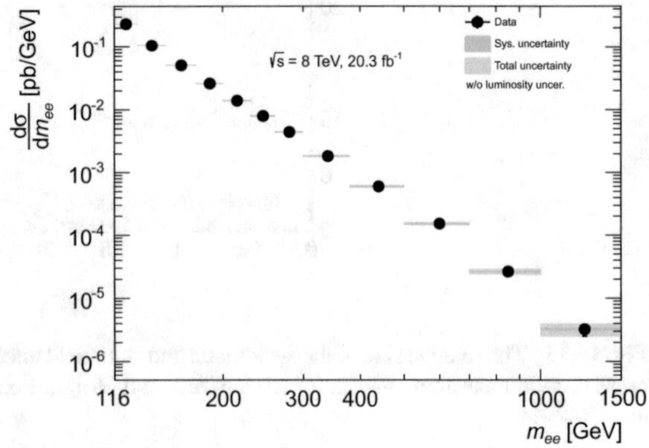

Fig. 15.16 Electron single-differential cross section as a function of invariant mass m_{ee} at Born level within the fiducial region with statistical, systematic and total uncertainties, excluding the 1.9% uncertainty on the luminosity

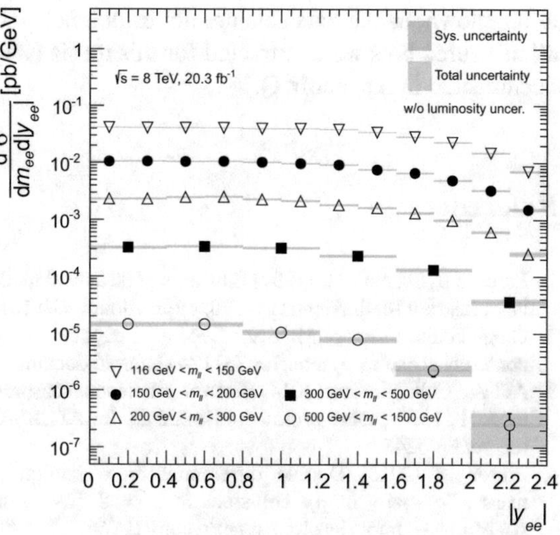

Fig. 15.17 Electron double-differential cross section as a function of invariant mass m_{ee} and absolute dielectron rapidity $|y_{ee}|$ at Born level within the fiducial region with statistical, systematic and total uncertainties, excluding the 1.9% uncertainty on the luminosity

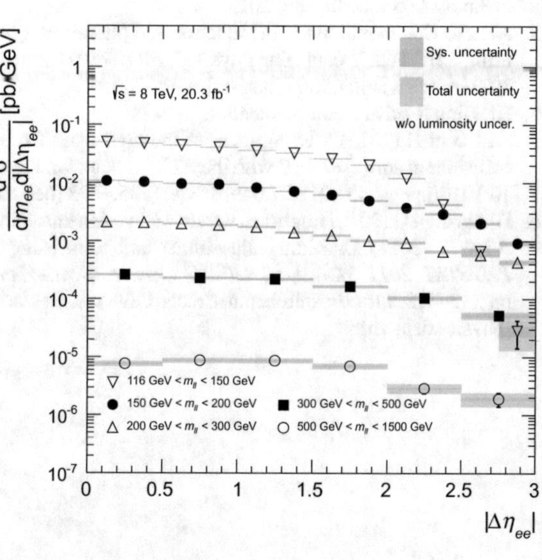

Fig. 15.18 Electron double-differential cross section as a function of invariant mass m_{ee} and absolute dielectron pseudorapidity separation $|\Delta\eta_{ee}|$ at Born level within the fiducial region with statistical, systematic and total uncertainties, excluding the 1.9% uncertainty on the luminosity

with invariant mass and slowly falling towards higher values of absolute rapidity. It is spanning five orders of magnitude from 4.15×10^{-2} pb to 2.51×10^{-7} pb. The same behavior is observed for the cross section as a function of $|\Delta\eta_{ee}|$ which is ranging from 4.99×10^{-2} pb to 1.90×10^{-6} pb. Detailed tables of the single-differential and double-differential cross sections with all systematic uncertainties can be found in Appendix P. Also given for each bin is a factor $k_{dressed}$ which is the ratio of C_{DY} at Born level to the C_{DY} at dressed level. The dressed level cross section can be obtained by multiplying the Born level cross section with this factor. To use all three measurements at the same time, the statistical correlations between them need to be known, as the data sets are not orthogonal. The statistical correlations across all measured bins were extracted for this thesis using the bootstrap method and are documented in Appendix Q.

References

1. Laforge B, Schoeffel L (1997) Elements of statistical methods in high-energy physics analyses. Nucl Instrum Meth A394:115–120. https://doi.org/10.1016/S0168-9002(97)00649-9
2. Oleg Fedin, Victor Maleev, Evgeny Sedykh, Victor Solovyev. https://indico.cern.ch/conferenceDisplay.py?confId=251134 (Internal documentation)
3. ATLAS Collaboration (2014) Electron efficiency measurements with the ATLAS detector using the 2012 LHC proton-proton collision data. In: ATLAS-CONF-2014-032. https://cds.cern.ch/record/1706245
4. Zinser M (2013) Double differential cross section for Drell-Yan production of high-mass e^+e^--pairs in pp collisions at $\sqrt{s}$= 8 TeV with the ATLAS experiment. MA thesis. Mainz U. http://inspirehep.net/record/1296478/files/553896852$_$CERN-THESIS-2013-258.pdf. Accessed 08 July 2013
5. ATLAS Collaboration (2014) Electron and photon energy calibration with the ATLAS detector using LHC Run 1 data. Eur Phys J C 74(10):3071. https://doi.org/10.1140/epjc/s10052-014-3071-4, arXiv:1407.5063 [hep-ex]
6. Hickling R private communication
7. Aad G et al (2013) Measurement of the high-mass Drell–Yan differential cross-section in pp collisions at sqrt(s)=7 TeV with the ATLAS detector. Phys Lett B 725:223–242. https://doi.org/10.1016/j.physletb.2013.07.049, arXiv:1305.4192 [hep-ex]
8. D'Agostini G (2010) Improved iterative Bayesian unfolding. arXiv:1010.0632 [physics.data-an]
9. Adye T (2011) Unfolding algorithms and tests using RooUnfold. In: *Proceedings of the PHYSTAT 2011 Workshop, CERN, Geneva, Switzerland, January 2011, CERN-2011-006*, pp 313–318. http://inspirehep.net/record/898599/files/arXiv:1105.1160.pdf, arXiv:1105.1160 [physics.data-an]

Chapter 16
Muon Channel Cross Section

The measurement of the muon channel cross section was not performed by myself and was added after the time of my master thesis. The analysis is documented in detail in the following thesis [1] and in the publication [2]. Nonetheless, the measurement of the cross section will be briefly discussed in the following, since the muon channel cross section is used as input in the following chapters.

The analysis strategy is very similar to the strategy of the electron channel analysis. Drell–Yan and photon induced signal MC samples have been generated in slices of invariant mass, using the same MC setup as in Sect. 14.1. The same higher order corrections are applied to the Drell–Yan sample. Also the MC setup for the background processes is identical to the setup used in the electron channel analysis.

The analysis is performed using the same 2012 data set. Events are required to pass the same quality requirements as described in Sect. 14.2.2. The largest fraction of the data sample is collected by a trigger which requires a muon with a transverse momentum above 36 GeV. A supplementary trigger requires a muon with a transverse momentum above 24 GeV but also imposes a track isolation requirement. The track isolation for muons is defined by building the scalar sum of the transverse momenta $\sum p_{\mathrm{T}}$ of tracks surrounding the muon candidate and dividing it by the transverse momentum of the muon. Isolation criteria provide a good discriminant against multijet background arising from semileptonic decays of heavy-flavor quarks. The muons trigged by the low-threshold trigger are required to satisfy $\sum p_{\mathrm{T}}(\Delta R = 0.2)/p_{\mathrm{T}} < 0.12$.

The muon candidates are reconstructed by matching tracks measured in the muon spectrometer to tracks reconstructed in the inner detector. All candidates have to satisfy $|\eta| < 2.4$ and must pass in addition the *medium* identification criteria which is documented in reference [3]. It is based on the number of hits in the inner detector and muon spectrometer as well as on the significance of the charge/momentum ratio imbalance between the muon spectrometer and inner detector measurements. An isolation cut of $\sum p_{\mathrm{T}}(\Delta R = 0.2)/p_{\mathrm{T}} < 0.1$ is applied. This cut is slightly tighter than

M. Zinser, *Search for New Heavy Charged Bosons and Measurement of High-Mass Drell-Yan Production in Proton-Proton Collisions*, Springer Theses,
https://doi.org/10.1007/978-3-030-00650-1_16

the requirement imposed by the low-threshold trigger. Background from cosmic-ray muons is removed by requiring the longitudinal impact parameter to the primary interaction vertex, z_0, to be less than 10 mm. The primary interaction vertex is defined as the vertex with the largest $\sum p_{\mathrm{T}}^2$ of all tracks associated to it. Events which contain two oppositely charged muons, where the leading muon fulfills $p_{\mathrm{T}} > 40\,\mathrm{GeV}$ and the subleading muon $p_{\mathrm{T}} > 30\,\mathrm{GeV}$ are selected. The transverse momentum requirements are imposed in order to be in the same phase space as in the electron channel measurement.

The multijet and W+jets background, which is largely arising from heavy-flavor b- and c-quark decays, is estimated using a data-driven technique. The, so called, *ABCD* method is based on four orthogonal control regions. The region A is the standard signal selection in which the background needs to be known. The regions B, C are background enriched by inverting the isolation requirement (B) or inverting the muon-pair charge requirement (C). For the region D, both requirements are inverted at the same time. In each control region contaminations from signal, top-quark, and diboson background is subtracted using MC simulations. The $|y_{\mu\mu}|$ and $|\Delta\eta_{\mu\mu}|$ shape of the background in each $m_{\mu\mu}$ region is obtained from region D. The shape of the multijet background is normalized to the yield of multijet events in the signal region. It is obtained using the constraint that the yield ratio of opposite-charge to same-charge muon pairs is identical in the isolated and non-isolated regions. The contribution from the multijet background ranges from 0.1 to 1% and is therefore much smaller than in the electron channel.

Figure 16.1 shows the invariant mass distribution $m_{\mu\mu}$ after the final muon channel event selection. The distribution looks very similar to the electron channel distribution. The data are, like in the electron channel, at lower invariant masses about 4% above the expectation, while good agreement is seen at higher invariant masses.

The muon channel spectra are unfolded using the same bin-by-bin unfolding method and the same binning as for the electron channel. The unfolding includes the acceptance extrapolation to the same phase space region as the electron channel cross

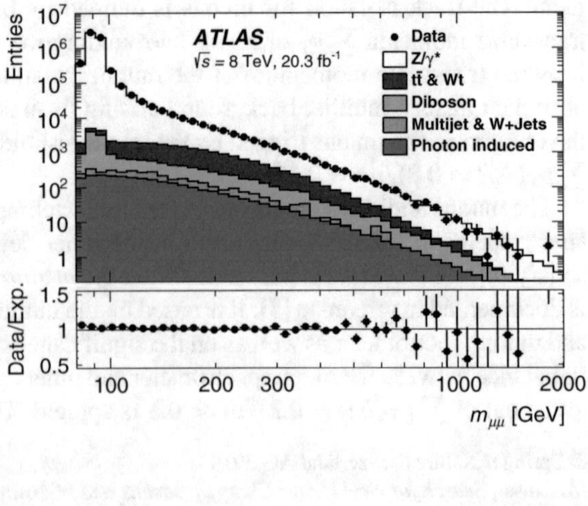

Fig. 16.1 The invariant mass ($m_{\mu\mu}$) distribution after event selection, shown for data (solid points) compared to the expectation (stacked histogram). The lower panels show the ratio of data to the expectation. Figure taken from reference [2]

section was measured in. This extrapolation includes in case of the muon channel just the extrapolation from $|\eta| < 2.4$ to $|\eta| < 2.5$. The unfolding factor C_{DY}, which combines efficiency and acceptance effects, is for the muon channel about 80% and constant in invariant mass.

The systematic uncertainties on the cross section related to the muon are coming from the trigger, reconstruction, isolation and impact parameter efficiencies, as well as the muon momentum scale and resolution. They are all studied using the $Z \to \mu^+\mu^-$ process and a tag and probe method. Of these muon related uncertainties, the largest uncertainty is coming from the reconstruction efficiency corrections and the muon momentum scale calibration. However, the top-quark and diboson background are the dominant sources of uncertainty. The uncertainty is estimated in the same way as for the electron channel and is discussed in Sect. 15.3.

Figure 16.2 shows the single-differential muon channel cross section as a function of invariant mass. The cross section is compared to a NNLO FEZW theory calculation using the CT10 PDF including the contribution from the photon induced process. Theory and cross section agree well within their uncertainties.

Detailed tables with the single- and double-differential muon channel cross sections and with a breakdown of all uncertainties can be found in reference [2].

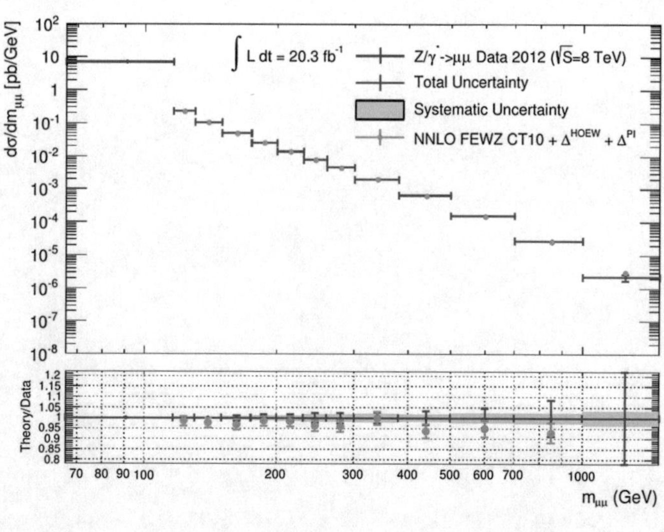

Fig. 16.2 The single-differential Drell–Yan cross section in the muon channel compared to NNLO theory, which includes NLO electroweak corrections (Δ^{HOEW}) and the photon induced contribution (Δ^{PI}). The shaded errors on the data show the systematic uncertainty and the error bars show the total uncertainty. The lower plot shows the ratio of theory to data. The shown experimental uncertainties do not include an overall 1.9% normalisation uncertainty due to the luminosity determination. Figure taken from reference [1]

References

1. Hickling RS (2014) Measuring the Drell–Yan cross section at high mass in the dimuon channel. PhD thesis, Queen Mary University of London. http://inspirehep.net/record/1429540/files/fulltext_WiZAbu.pdf. Accessed 12 Sept 2014
2. Aad G et al (2016) Measurement of the double-differential high-mass Drell–Yan cross section in pp collisions at $\sqrt{s} = 8$ TeV with the ATLAS detector. JHEP 08: 009. https://doi.org/10.1007/JHEP08(2016)009, arXiv: 1606.01736
3. ATLAS Collaboration (2014) Measurement of the muon reconstruction performance of the ATLAS detector using 2011 and 2012 LHC proton-proton collision data. Eur Phys J C74.5: 3130. https://doi.org/10.1140/epjc/s10052-014-3130-x, arXiv: 1407.3935

Chapter 17
Results and Interpretation

17.1 Combination

The electron and muon cross section measurements can be combined to further reduce the statistical and systematic uncertainties on the measurement. The combination is performed using the HERAVERAGER tool [1]. In the following the combination method is briefly introduced and afterwards the obtained results are discussed.

17.1.1 The Combination Method

The combination method is based on a method developed at HERA for the combination of DIS cross section data [2] and is explained in the following. In the simplest case, where no systematic uncertainties affect the measurement, the averaged cross section $\overline{\sigma_i}$ and the absolute uncertainty $\overline{\delta_i}$ are given for a specific bin i by the formula

$$\overline{\sigma_i} = \frac{1}{\overline{\delta_i}^{-2}} \sum_k^{N_{chan}} \frac{\sigma_{i,k}}{\delta_{i,k}^2}, \quad \overline{\delta_i}^{-2} = \sum_k^{N_{chan}} \frac{1}{\delta_{i,k}^2}. \tag{17.1}$$

Here $\sigma_{i,k}$ is the measured cross section in channel k and $\delta_{i,k}^2$ is the absolute statistical uncertainty squared.

Including systematic uncertainties complicates the averaging procedure. The measured cross section value $\sigma_{i,k}$ has then an uncorrelated uncertainty and a systematic uncertainty which is correlated between bins. The former is related to the relative statistical uncertainty on the data $\delta_{i,stat}$, and the relative uncorrelated systematic uncertainty $\delta_{i,unc}$ as $\delta_i^2 = \delta_{i,stat}^2 + \delta_{i,unc}^2$. A χ^2 function, taking into account the systematic uncertainties, can be defined for a single measurement by

M. Zinser, *Search for New Heavy Charged Bosons and Measurement of High-Mass Drell-Yan Production in Proton-Proton Collisions*, Springer Theses,
https://doi.org/10.1007/978-3-030-00650-1_17

$$\chi^2_{tot}(\overline{\sigma}, \theta) = \sum_i^{N_{bin}} \frac{(\overline{\sigma_i} - \sum_j^{N_{sys}} \gamma_{i,j}\, \overline{\sigma_i}\, \theta_j - \sigma_i)^2}{(\delta_{i,stat}\, \sigma_i)^2 + (\delta_{i,unc}\, \overline{\sigma_i})^2} + \sum_j^{N_{sys}} \theta_j^2. \quad (17.2)$$

Here $\gamma_{i,j}$ is the relative uncertainty in bin i of a correlated systematic source j and θ_j the shift of the correlated systematic uncertainty source j. The quantities $\overline{\sigma}$ and θ, without an index i correspond to the set of all bins. A shift of $\theta = 1$ corresponds hereby to a shift of the source by 1σ. The last term accounts for a contribution from the systematic uncertainties to the χ^2_{tot}. If large shifts are introduced, also the contribution to the χ^2_{tot} gets large. The shifts θ_j and the averaged cross section $\overline{\sigma}$ are determined by minimizing the χ^2 function. The minimum is given by the following extremum conditions

$$\frac{\partial \chi^2_{tot}(\overline{\sigma}, \theta)}{\partial \overline{\sigma}} = 0, \quad \frac{\partial \chi^2_{tot}(\overline{\sigma}, \theta)}{\partial \theta} = 0. \quad (17.3)$$

A trivial solution with $\overline{\sigma} = \overline{\sigma}$ and $\overline{\theta} = \vec{0}$ is found if only a single channel is considered. The solution is non-trivial when considering both, the electron channel and the muon channel measurement. The χ^2 function is in this case given by

$$\chi^2_{tot}(\overline{\sigma}, \theta) = \sum_i^{N_{bin}} \sum_k^{N_{chan}} \frac{(\overline{\sigma_i} - \sum_j^{N_{sys}} \gamma_{i,j,k}\overline{\sigma_i}\theta_j - \sigma_{i,k})^2}{(\delta_{i,k,stat}\sigma_{i,k})^2 + (\delta_{i,k,unc}\overline{\sigma_i})^2} + \sum_j^{N_{sys}} \theta_j^2. \quad (17.4)$$

The relative systematic uncertainty $\gamma_{i,j,k}$ is equal to zero if the systematic source j does not apply to the channel k. The minimization is based on an iterative procedure and is described in more detail in reference [1] and in the appendix of reference [2]. The minimization of Eq. 17.4 determines the average cross sections $\overline{\sigma_i}$ and shifts of the systematic nuisance parameters θ_j together with their uncertainties. The minimization introduces correlations among parameters θ_j. The corresponding covariance matrix is diagonalized and re-normalized, such that the average cross sections are represented using independent nuisance parameters with expectation values of zero and standard deviations of unity. The resulting uncertainty on the combined cross section can therefore not directly be related to the input sources. Table 17.1 lists all nuisance parameters for the combination. Each number represents a nuisance parameter. Columns which share a nuisance parameter are treated as correlated between channels. This only applies to the correlated top-quark and diboson background uncertainty. Columns which contain a u are treated as uncorrelated between all bins of the measurement.

17.1.2 Combination Cross Section Results

The combination is performed with the HERAVERAGER tool which uses the procedure mentioned above. It would in principle be possible to combine all measured cross sections at the same time in a single combination. However, the bins between

Table 17.1 Summary of the correlations for the uncertainties. Each number represents a nuisance parameter. Columns with a shared nuisance parameter are treated as correlated between channels, whereas columns containing *u* are treated as uncorrelated between bins

Uncertainty source	Channel	
	ee	$\mu\mu$
Lepton energy (momentum) scale	1–14	15
Lepton energy (momentum) scale (stat.)	u	–
Electron energy resolution	16–22	–
Muon momentum resolution (ID)	–	23
Muon momentum resolution (MS)	–	24
Lepton trigger efficiency	25	26
Lepton trigger efficiency (stat.)	u	–
Lepton reconstruction efficiency	27	28
Lepton reconstruction efficiency (stat.)	–	–
Electron identification efficiency	29	–
Lepton isolation efficiency	30	31
Lepton isolation efficiency (stat.)	u	u
Top-quark background	32	32
Diboson background	33	33
Top-quark and diboson background (stat.)	u	u
Multijet and W+jets background	34	35
Multijet and W+jets background (stat.)	u	u
C_{DY} (stat.)	u	u

the measurements are statistically correlated. Such a combination needs therefore knowledge about the statistical correlations between all measurement. For the electron measurement, this information is provided in Appendix Q but it is not available for the muon measurement. The three measurements are therefore treated separately.

Figure 17.1 shows in the top panel the electron channel (red triangles), the muon channel (blue triangles), and the combined Born level cross section (black dots) as a function of invariant mass $m_{\ell\ell}$. The middle panel shows the ratio of the individual channels to the combination. The error bars on the data points represent the pure statistical uncertainty on data. The systematic uncertainty of the combined cross section is shown as a dark green band and the total uncertainty as a light green band. The luminosity uncertainty of 1.9% is excluded as it affects both measurements in the same way. The lower panel shows the pull for the two individual measurements, which is defined as the single-channel measurement subtracted from the combined result in units of the total uncertainty. Both individual measurements are in good agreement with one another. The single-differential cross section falls rapidly over five orders of magnitude as $m_{\ell\ell}$ increases by about a factor of ten. The minimum χ^2 per degree of freedom, χ^2/dof is found to be $14.2/12 = 1.19$ for the single-differential cross section. This corresponds to a probability of 0.29. The χ^2 value includes the second term in Eq. 17.4, which is coming from the shifts of the correlated systematic uncertainties which had to be applied. The χ^2/dof excluding this term is

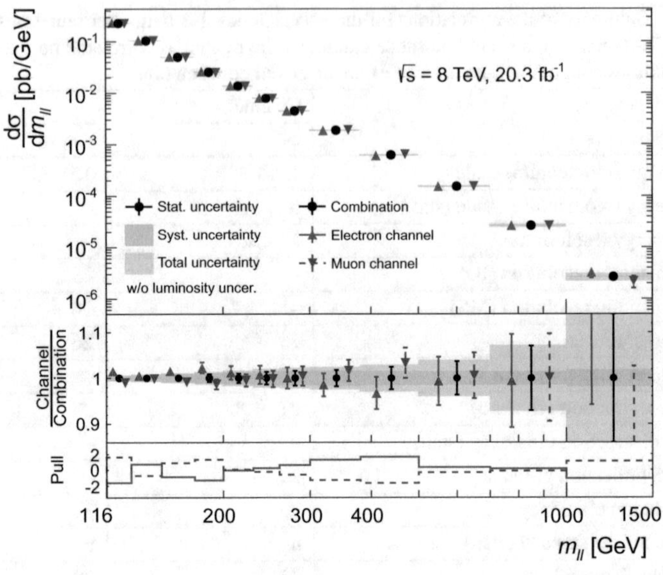

Fig. 17.1 Comparison of the electron (red points), muon (blue points) and combined (black points) single-differential fiducial Born level cross sections as a function of invariant mass $m_{\ell\ell}$. The error bars represent the statistical uncertainty. The inner shaded band represents the systematic uncertainty on the combined cross sections, and the outer shaded band represents the total measurement uncertainty (excluding the luminosity uncertainty). The central panel shows the ratio of each measurement channel to the combined data, and the lower panel shows the pull of the electron (red) and muon (blue) channel measurements with respect to the combined data

found to be $11.9/12 = 0.99$. The χ^2 excluding the systematic contribution represents the pure statistical agreement of both measurements after applying the systematic shifts. No pulls above 2σ are observed in the individual bins. Figure 17.2 shows the systematic shifts θ_j for the different nuisance parameters. The red error bar shows the original uncertainty while the black error bar shows the reduced uncertainty after the combination. The largest shifts for the single-differential are observed for the uncertainty on the method of the electron energy scale extraction ($+0.83\sigma$) and on the muon momentum scale (-0.71σ). All nuisance parameters which are shifted receive an uncertainty reduction. None of the uncertainties is neither drastically pulled nor is its uncertainty drastically reduced. This indicates that none of the uncertainties is either too small or too conservative. The resulting combined cross section has a statistical precision of 0.34% in the first and of 17.05% in the last bin. The systematic uncertainty in the corresponding bins is 0.53 and 2.95%. At low $m_{\ell\ell}$ the combined measurement is dominated by the experimental systematic uncertainties. For $m_{\ell\ell} \gtrsim 400$ GeV the statistical uncertainty of the data dominates the measurement precision. Detailed information on the cross section with a breakdown of all statistical and correlated systematic uncertainties can be found in the Appendix in Table R.1.

Figure 17.3 shows the measured double-differential cross section of the individual channels and their combination as a function of invariant mass $m_{\ell\ell}$ and the absolute

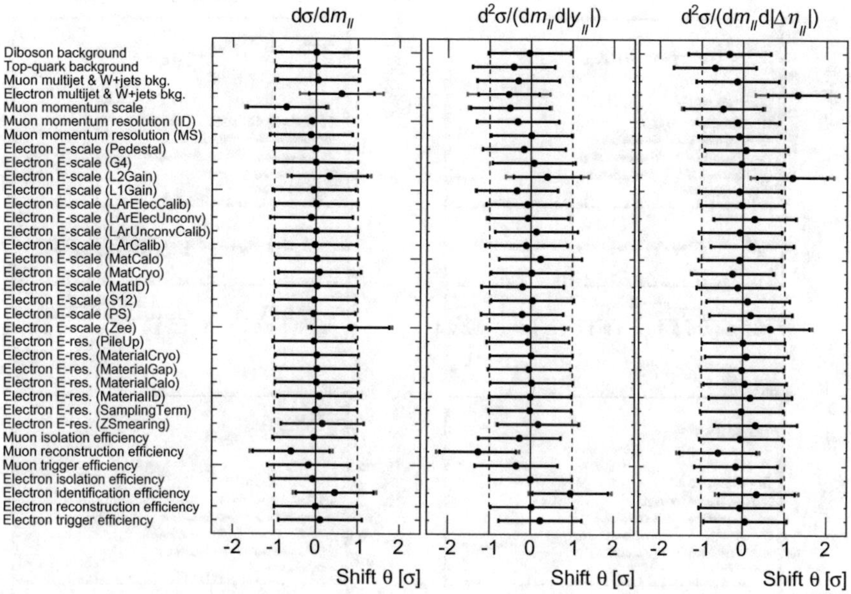

Fig. 17.2 Shifts and uncertainties of the correlated systematic uncertainties from the combination of the cross section measurements. The red error bars show the original uncertainty while the black error bars show the uncertainty after the combination

dilepton rapidity $|y_{\ell\ell}|$. Both individual measurements are in good agreement with one another. The cross sections show a marked narrowing of the rapidity plateau width as $m_{\ell\ell}$ increases. The minimum χ^2 per degree of freedom, χ^2/dof is found to be $53.1/48 = 1.11$ including the contribution from the correlated systematic uncertainties. This corresponds to a probability of 0.28. The χ^2/dof when excluding the systematic part is found to be $48.9/48 = 1.02$. The largest pull of 2.4σ is observed in the first absolute rapidity bin in the range 300 GeV $< m_{\ell\ell} <$ 500 GeV. The largest shifts are observed for the uncertainty on the muon reconstruction efficiency (-1.27σ) and on the electron identification efficiency ($+0.96\sigma$). The shifts go into the same direction as for the single-differential cross section combination, but they are more pronounced. The largest uncertainty reduction is observed for the electron multijet and W+jets background uncertainty, which is reduced by 26% after the combination. The resulting combined cross section has a statistical precision of 0.81% in the first bin of the measurement (116 GeV $< m_{\ell\ell} <$ 150 GeV, $0.0 < |y_{\ell\ell}| < 0.2$) and of 35.7% in the last bin (500 GeV $< m_{\ell\ell} <$ 1500 GeV, $2.0 < |y_{\ell\ell}| < 2.4$). The systematic uncertainty in the corresponding bins is 0.62 and 7.63%. The combined measurement is therefore in each bin dominated by the statistical uncertainty of the data. Detailed information on the cross section with a breakdown of all statistical and correlated systematic uncertainties can be found in the Appendix in Table R.2.

Figure 17.4 shows the measured double-differential cross section of the individual channels and their combination as a function of invariant mass $m_{\ell\ell}$ and the absolute

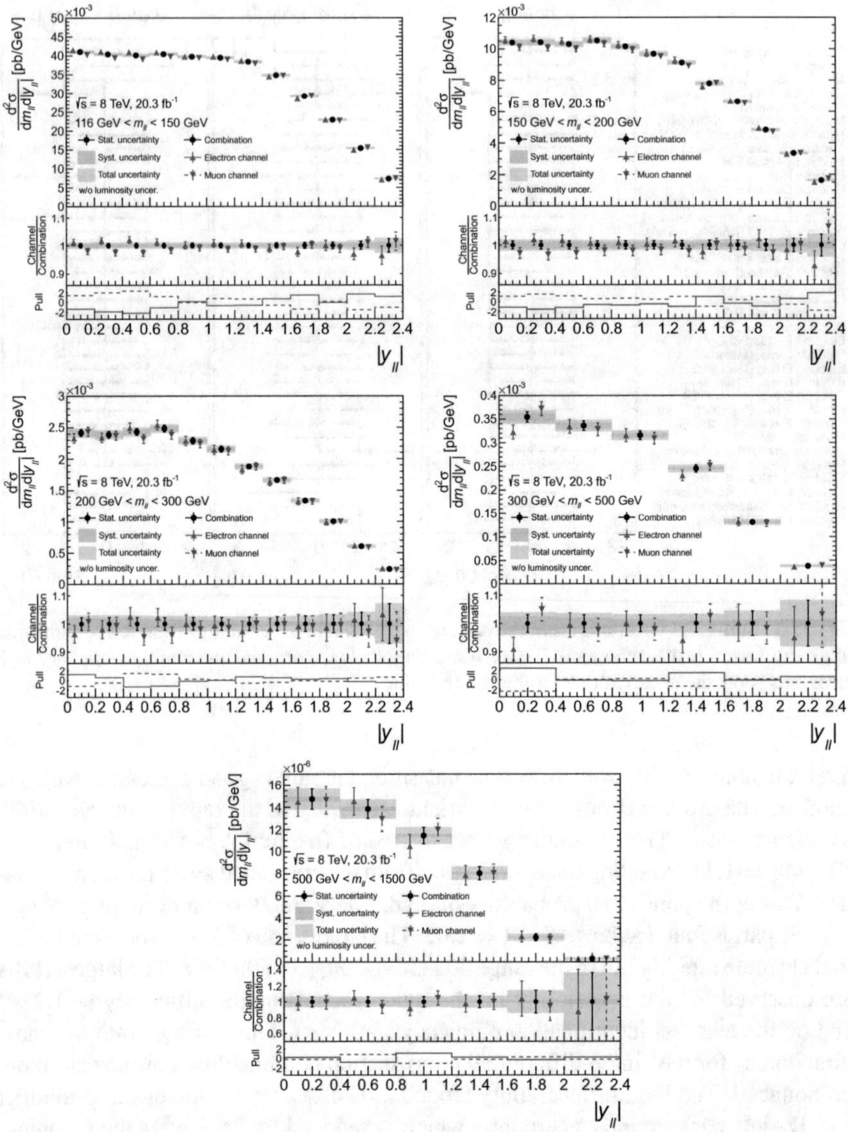

Fig. 17.3 Comparison of the electron (red points), muon (blue points) and combined (black points) fiducial Born level cross sections, differential in invariant mass $m_{\ell\ell}$ and absolute dilepton rapidity $|y_{\ell\ell}|$. The error bars represent the statistical uncertainty. The inner shaded band represents the systematic uncertainty on the combined cross sections, and the outer shaded band represents the total measurement uncertainty (excluding the luminosity uncertainty). The central panel shows the ratio of each measurement channel to the combined data, and the lower panel shows the pull of the electron (red) and muon (blue) channel measurements with respect to the combined data

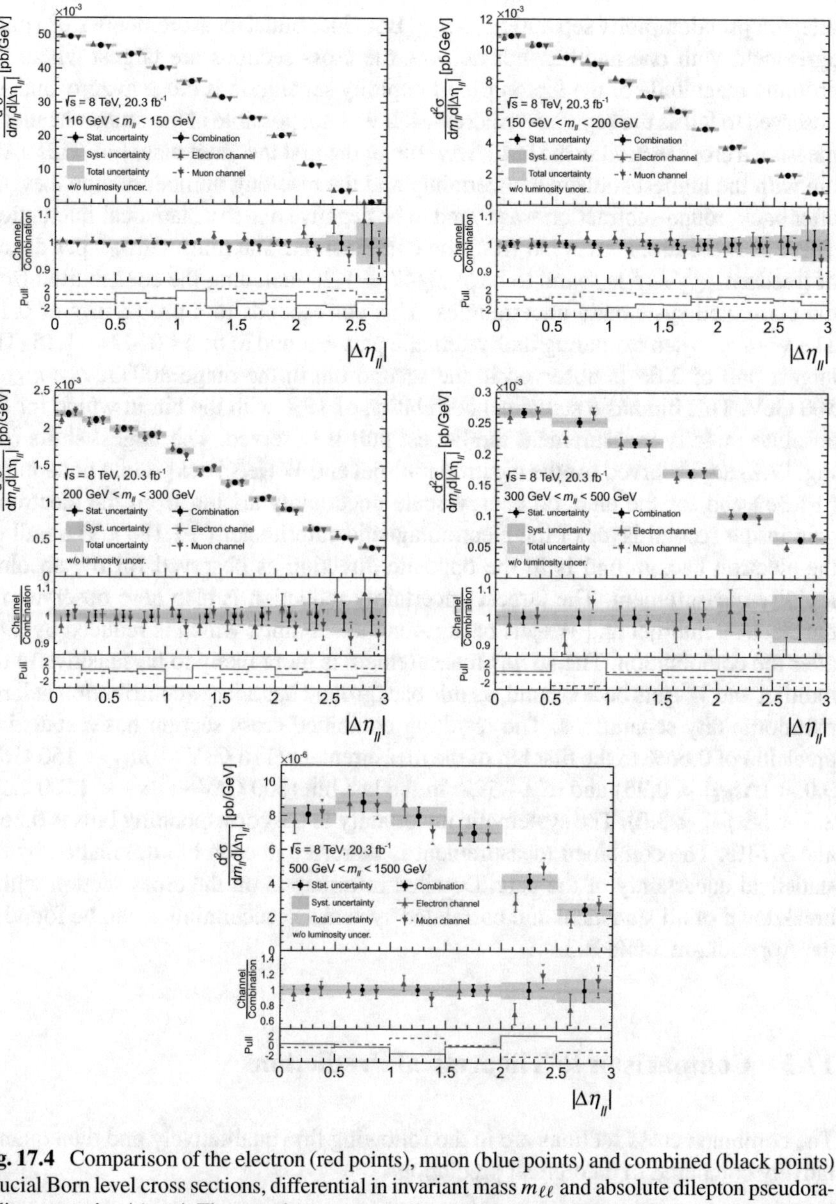

Fig. 17.4 Comparison of the electron (red points), muon (blue points) and combined (black points) fiducial Born level cross sections, differential in invariant mass $m_{\ell\ell}$ and absolute dilepton pseudorapidity separation $|\Delta\eta_{\ell\ell}|$. The error bars represent the statistical uncertainty. The inner shaded band represents the systematic uncertainty on the combined cross sections, and the outer shaded band represents the total measurement uncertainty (excluding the luminosity uncertainty). The central panel shows the ratio of each measurement channel to the combined data, and the lower panel shows the pull of the electron (red) and muon (blue) channel measurements with respect to the combined data

dilepton pseudorapidity separation $|\Delta\eta_{\ell\ell}|$. Both individual measurements are in good agreement with one another. For all $m_{\ell\ell}$, the cross sections are largest where the absolute magnitude of the lepton pseudorapidity separation is close to zero, and are observed to fall as the separation increases. It was not possible in the muon channel to measure a cross section in the last $|\Delta\eta_{\ell\ell}|$ bin of the first invariant mass bin. This is the bin with the highest statistical uncertainty and the resulting number of signal events after background subtraction was found to be negative due to a statistical fluctuation. This bin is therefore excluded from the combination. The minimum χ^2 per degree of freedom, χ^2/dof is found to be $59.3/47 = 1.26$ including the contribution from the correlated systematic uncertainties. This corresponds to a probability of 0.11. The χ^2/dof when excluding the systematic part is found to be $54.0/47 = 1.15$. The largest pull of 2.6σ is observed in the second bin in the range 300 GeV $< m_{\ell\ell} <$ 500 GeV. This bin has a statistical correlation of 19% with the bin in which for the absolute rapidity measurement the largest pull is observed. The largest shifts (see Fig. 17.2) are observed for the electron multijet and W+jets background uncertainty ($+1.3\sigma$) and for the electron energy scale uncertainty arising from the electronic gain in the second layer of the electromagnetic calorimeter ($+1.18\sigma$). The pull on the electron background is in the opposite direction as observed for the absolute rapidity measurement. The largest uncertainty reduction is also here observed for the electron multijet and W+jets background uncertainty, which is reduced by 29% after the combination. The $|\Delta\eta_{\ell\ell}|$ measurement is more likely to be sensitive to the multijet and W+jets background, as this background has a large contribution at large pseudorapidity separations. The resulting combined cross section has a statistical precision of 0.66% in the first bin of the measurement (116 GeV $< m_{\ell\ell} <$ 150 GeV, $0.0 < |\Delta\eta_{\ell\ell}| < 0.25$) and of 14.59% in the last bin (500 GeV $< m_{\ell\ell} <$ 1500 GeV, $2.5 < |\Delta\eta_{\ell\ell}| < 3.0$). The systematic uncertainty in the corresponding bins is 0.56% and 3.74%. The combined measurement is therefore in each bin dominated by the statistical uncertainty of the data. Detailed information on the cross section with a breakdown of all statistical and correlated systematic uncertainties can be found in the Appendix in Table R.3.

17.2 Comparison to Theoretical Predictions

The combined cross sections are in the following first qualitatively and then quantitatively compared to theoretical predictions.

17.2.1 Theoretical Predictions

The combined fiducial cross sections at Born level are compared to NNLO perturbative QCD calculations using various PDFs. All calculations have been performed using the FEWZ 3.1 framework. They include NLO electroweak corrections using the G_μ scheme [3]. The renormalization and factorization scales are set to $\mu_R = \mu_F = m_{\ell\ell}$. The calculations also include the contribution form the photon induced process, $\gamma\gamma \rightarrow \ell\ell$. It is estimated at LO using the photon PDF from the NNPDF2.3qed PDF set [4]. This is a more recent photon PDF than the MRST2004qed PDF which was used for the signal MC. In the extraction of the PDF already LHC measurements have been used, for example also the measurement of the high-mass Drell–Yan cross section at $\sqrt{s} = 7$ TeV, which is shown in Fig. 12.1. Uncertainties have been assigned to the theoretical predictions. They take into account the PDF uncertainties at 68% confidence level and the α_s uncertainty, which is determined by varying α_s by 0.001 with respect to its default value of 0.118. The scale uncertainty is calculated by changing μ_R and μ_F by a factor of two simultaneously and independently.[1] The envelope of all variations is taken as uncertainty. The NNPDF collaboration provides, instead of eigenvectors, a large number of MC replicas for their PDFs. The central value of the PDF is calculated by the mean of all replicas and the uncertainty is defined as the region covering 68% of all MC replicas. The uncertainty on the photon induced contribution is calculated in the same way using the NNPDF2.3qed replicas. The latter uncertainty is rather large, ranging from 62 to 92%. All theoretical predictions and their uncertainties have been provided by the ATLAS collaboration and are listed in detail in the auxiliary material of reference [5].

17.2.2 Comparison to Theoretical Predictions

Theoretical predictions using the MMHT2014 NNLO PDF set [6] are compared to the single-differential cross section at Born level as a function of $m_{\ell\ell}$ in Fig. 17.5. The left plot shows the whole measurement range while the right plot shows a zoomed version in the region 116 GeV $< m_{\ell\ell} <$ 380 GeV. The middle panel shows the ratio of the measured cross section to the MMHT2014 prediction. The red dashed line shows the ratio excluding the contribution from the photon induced process. The MMHT2014 prediction is about 2–3% below the measured cross section until $m_{\ell\ell} = 200$ GeV. The prediction is above the measurement in the region $m_{\ell\ell} > 300$ GeV. The uncertainty of the measurement is in all bins larger than the uncertainty on the measurement. Hence, the measurement should be able to further constrain the theory prediction. The expected contribution from the photon induced process is small at low $m_{\ell\ell}$, rising up to 20% in the last bin. In the regions where the photon induced contribution is large the uncertainty on the photon induced process dominates the total uncertainty band, otherwise the PDF uncertainty is dominant. The change when replacing the MMHT PDF by other NNLO PDFs such as HERAPDF2.0 [7], CT14 [8], ABM12 [9] or NNPDF3.0 [10] is shown in the lower panel. The uncertainty band of the various PDFs is not shown for easier visibility. However, they have been calculated at 68% confidence level and are found to be smaller (ABM12), larger (CT14, NNPDF3.0)

[1] The case in which μ_R is scaled up by two and μ_F at the same time divided by two is by convention not included. The reverse case is also not included.

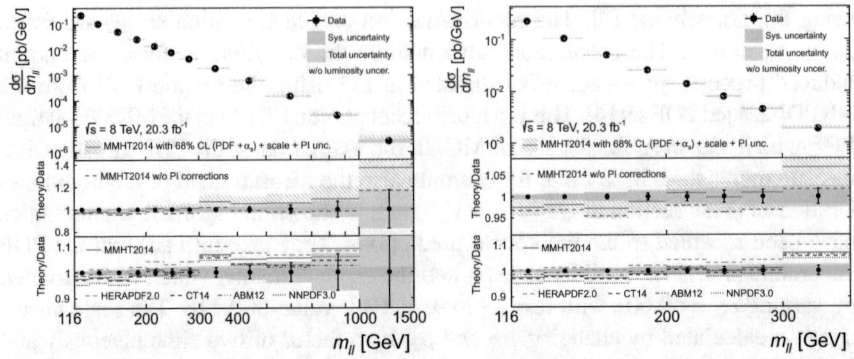

Fig. 17.5 The combined (electron and muon channel) single-differential cross section as a function of invariant mass $m_{\ell\ell}$ at Born level within the fiducial region with statistical, systematic and total uncertainties, excluding the 1.9% uncertainty on the luminosity. Data are compared to combined NNLO QCD and NLO electroweak calculations using the MMHT2014 PDF, where the uncertainty band displays the combined 68% confidence level PDF and α_s variation, the renormalization and factorization scale uncertainties and the uncertainty on the photon induced process. The two ratio panels show the ratio of the calculation with and w/o the photon induced contribution w.r.t. to data (middle panel), as well as the ratio for calculations using different PDFs (bottom panel). On the right, the results are shown for a restricted range of $m_{\ell\ell}$

or even much larger (HERAPDF2.0) than the ones from the MMHT2014 PDF. All PDFs in general agree with the measurement. Some normalization uncertainties are observed for the different predictions, while no large shape differences are observed. NNPDF3.0 shows the least agreement at low mass, while it gets better towards higher mass. The opposite behavior is observed for HERAPDF2.0, where the best agreement is observed at lower masses, getting worse towards higher masses. The spread between all PDF sets is at low mass larger than the uncertainty on the measurement, indicating the sensitivity of the data to the PDFs, and the potential to constrain them.

Figure 17.6 shows the same comparison to theoretical predictions for the double-differential cross section at Born level as a function of $m_{\ell\ell}$ and $|y_{\ell\ell}|$. The same general features can be observed when comparing to the single-differential measurement. The predictions tend to be below the measured cross section at lower masses while at medium and higher masses the agreement gets better. In the central region, the uncertainty on the measurement is always smaller than the uncertainty on the prediction. At high $|y_{\ell\ell}|$, the measurement is partially statistically limited, leading to larger uncertainty when compared to the prediction. All predictions describe the measured cross section across all invariant mass bins reasonably well. The largest differences between the various PDFs can be observed at large dilepton rapidities where HERAPDF2.0 predicts a higher cross section than all other PDFs. The photon induced process contributes up to 15% at low rapidities and high invariant mass. Figure 17.7 shows the same comparison to theoretical predictions for the double-differential cross section at Born level as a function of $m_{\ell\ell}$ and $|\Delta\eta_{\ell\ell}|$. The same general features are observed when comparing to the two other measurements. Some shape differences are observed for the first invariant mass bin at high $|\Delta\eta_{\ell\ell}|$. At low $|\Delta\eta_{\ell\ell}|$, the uncertainty on the measurement is always smaller than the uncertainty on the prediction. At high $|\Delta\eta_{\ell\ell}|$, the measurement is partially statistically limited, leading to larger uncertainty when compared to the prediction. All predictions describe the measured cross section across all invariant mass bins reasonably well. The largest differences between the various PDFs can be observed at small lepton pseudorapidity separation, where HERAPDF2.0 predicts a higher cross section than all other PDFs.

A χ^2 minimization procedure is used to quantify the agreement between the measurement and the various PDFs. The minimization procedure is implemented in the xFitter package [11] and is similar to the procedure described in Sect. 17.1.1. All correlated and uncorrelated experimental uncertainties, the luminosity uncertainty and the theoretical uncertainties are included in the χ^2 minimization. The correlated theoretical uncertainties include the uncertainties on the respective PDF, the photon induced contribution, α_s, and the factorization and renormalization scale. The PDF uncertainties for all the PDF sets except for the photon PDF are further decomposed into the full set of eigenvectors. In case of NNPDF3.0, the replica have been transformed into an eigenvector representation. This has been done by calculating the covariance matrix from the replica and performing a Cholesky decomposition [12] into an eigenvector representation. A single nuisance parameter is used for the photon induced contribution. Also the statistical uncertainties of the theoretical predictions are taken into account. They are at the level of 0.1% for the Drell–Yan calculations and 0.2% for the photon induced calculations. All correlated uncertainties are included as nuisance parameters. Table 17.2 gives the resulting χ^2 values after the minimization.

The χ^2/dof values range from 14.1/12 (ABM12) to 20.0/12 (NNPDF3.0) for the single-differential measurement. For the double-differential measurements, the χ^2 values reach from 51.0/48 for CT14 to 59.3/48 for MMHT2014 ($|y_{\ell\ell}|$) and 53.5/47 for ABM12 to 62.8/47 for MMHT2014 ($|\Delta\eta_{\ell\ell}|$). These values indicate general compatibility between the data and the theory. The overall best agreement is found for ABM12, especially when taking into account the smaller PDF uncertainties when compared to all other PDF sets. The largest χ^2 values are observed for NNPDF3.0 and MMHT2014. The central values of the nuisance parameters may, after the minimization procedure, be shifted from unity and their uncertainties may be reduced.

Table 17.2 The χ^2/dof values for the compatibility of data and theory after the minimization procedure

	$m_{\ell\ell}$	$\|y_{\ell\ell}\|$	$\|\Delta\eta_{\ell\ell}\|$
MMHT2014	18.2/12	59.3/48	62.8/47
CT14	16.0/12	51.0/48	61.3/47
NNPDF3.0	20.0/12	57.6/48	62.1/47
HERAPDF2.0	15.1/12	55.5/48	60.8/47
ABM12	14.1/12	57.9/48	53.5/47

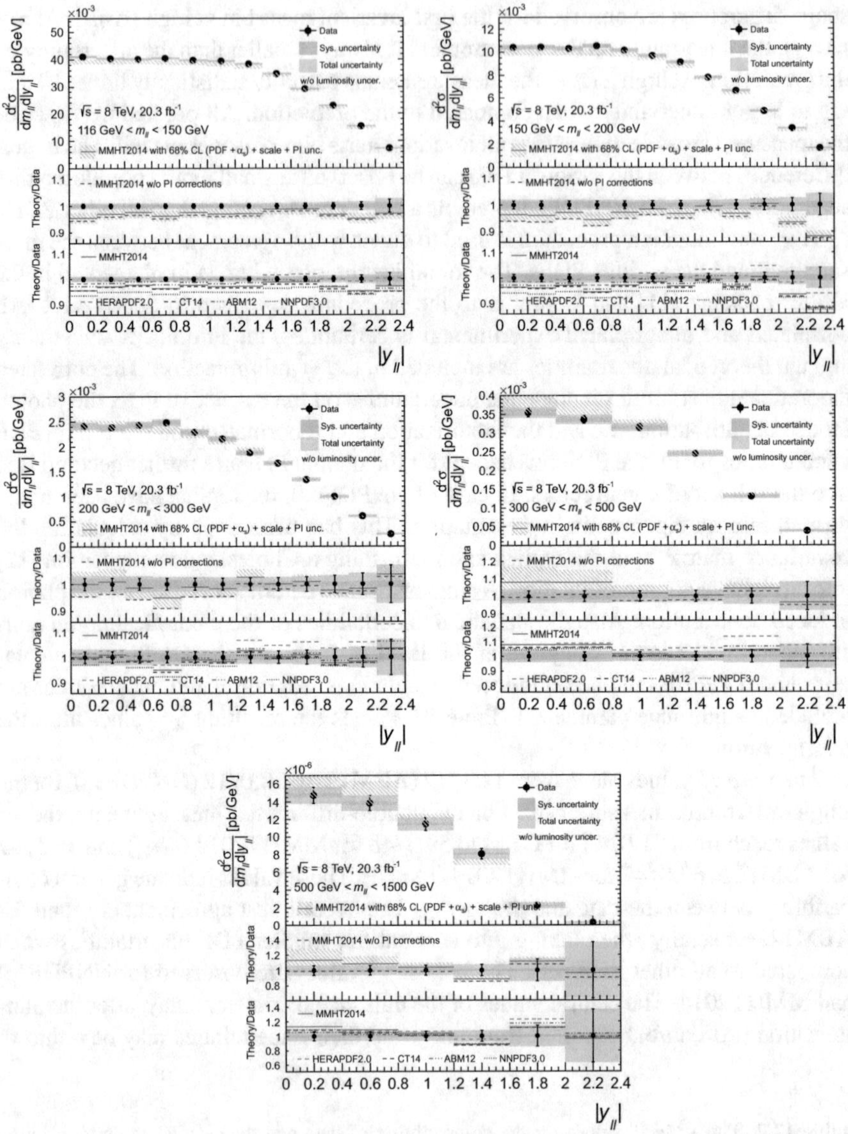

Fig. 17.6 The combined (electron and muon channel) double-differential cross section as a function of invariant mass $m_{\ell\ell}$ and absolute rapidity $|y_{\ell\ell}|$ at the Born level within the fiducial region with statistical, systematic and total uncertainties, excluding the 1.9% uncertainty on the luminosity. Data are compared to combined NNLO QCD and NLO electroweak calculations using the MMHT2014 PDF, where the uncertainty band displays the combined 68% confidence level PDF and α_s variation, the renormalization and factorization scale uncertainties and the uncertainty on the photon induced process. The two ratio panels show the ratio of the calculation with and w/o the photon induced contribution w.r.t. to data (middle panel), as well as the ratio for calculations using different PDFs (bottom panel)

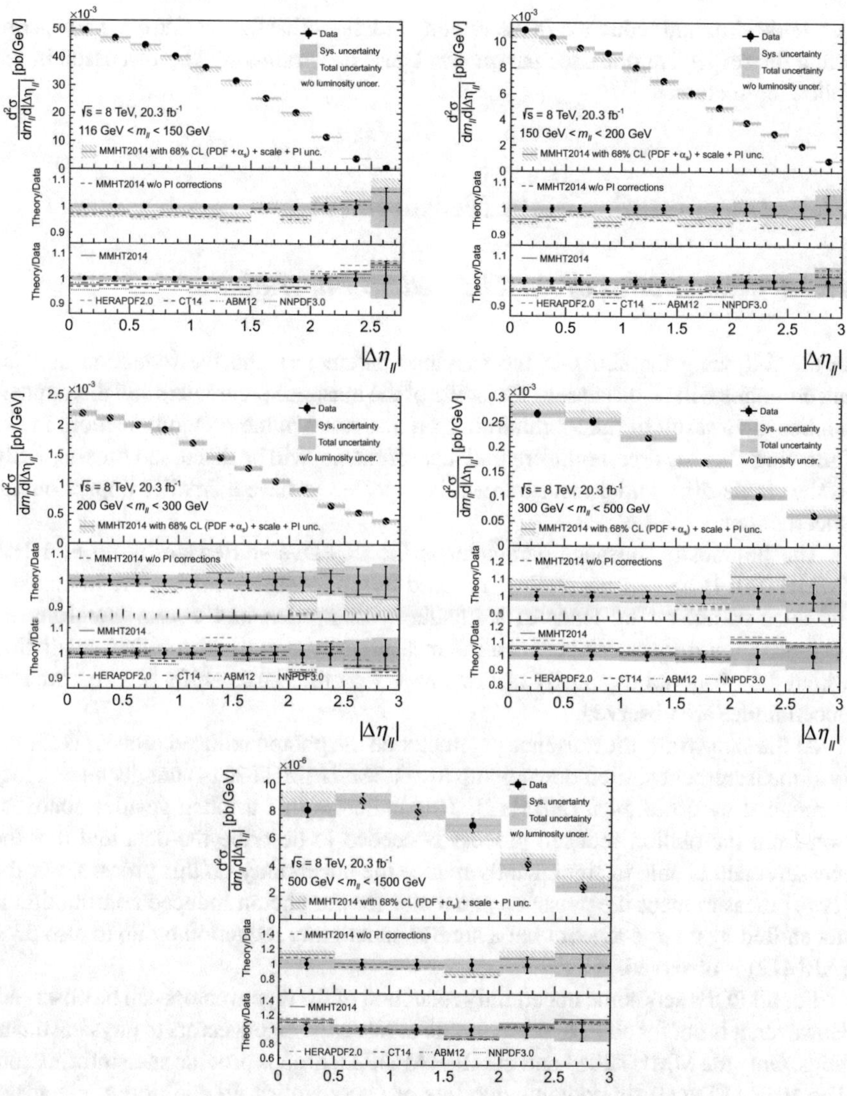

Fig. 17.7 The combined (electron and muon channel) double-differential cross section as a function of invariant mass $m_{\ell\ell}$ and absolute pseudorapidity separation $|\Delta\eta_{\ell\ell}|$ at the Born level within the fiducial region with statistical, systematic and total uncertainties, excluding the 1.9% uncertainty on the luminosity. Data are compared to combined NNLO QCD and NLO electroweak calculations using the MMHT2014 PDF, where the uncertainty band displays the combined 68% confidence level PDF and α_s variation, the renormalization and factorization scale uncertainties and the uncertainty on the photon induced process. The two ratio panels show the ratio of the calculation with and w/o the photon induced contribution w.r.t. to data (middle panel), as well as the ratio for calculations using different PDFs (bottom panel)

A sizable shift and reduction in uncertainty indicates that the measurement can constrain the respective nuisance parameter. These constraints will be discussed in the following section.

17.3 Interpretation of the Measurement

17.3.1 Constraints of the Theoretical Uncertainties

In the following the shifts of the nuisance parameters and the reduction of their uncertainties will be discussed. The shifts of the nuisance parameters and their uncertainties are a result of the χ^2 minimization discussed in the previous section. In the following only the double-differential measurements will be discussed for simplicity as the single-differential measurement is expected to have a smaller impact on the uncertainties.

The luminosity nuisance parameter is for all PDFs shifted up by up to 1.18σ (MMHT2014) and its uncertainty reduced by up to 40%. Shifting the luminosity nuisance parameter up leads to a smaller cross section and covers therefore the normalization differences between the measurement and the theoretical prediction which are observed especially at low mass. No other large shifts of experimental uncertainties are observed.

At the same time, the nuisance parameter on the photon induced process is for the $|y_{\ell\ell}|$ measurement shifted down by up to -1.38σ (MMHT2014) and its uncertainty is reduced by up to 54% (ABM12). This indicates that a much smaller contribution from the photon induced process is needed to describe the data and that the measurement is able to significantly reduce the uncertainty on this process. For the $|\Delta\eta_{\ell\ell}|$ measurement the nuisance parameter for the photon induced contribution is not shifted by a large amount but a similar uncertainty reduction by up to also 53% (ABM12) is observed.

For all PDFs sets some uncertainty reduction of the eigenvectors can be observed. However, it is not for all PDF sets possible to relate the eigenvectors to physical quantities. Only the MMHT2014 and the HERAPDF2.0 groups provide such information. The HERAPDF2.0 set contains two sets of uncertainties. In addition to eigenvectors which correspond to the experimental uncertainties on the input data sets, also uncertainties related to the assumed parameterization of the PDF at the input scale are provided. A significant uncertainty reduction of up to 44% is observed for the PDF variations 1 and 2 which represent the uncertainty on the parameter r_s. The parameter r_s is the ratio of the strange sea quark distribution to the down quark distribution at the input scale. For MMHT2014 an uncertainty reduction of up to 28% is observed for the eigenvector 21. Also this eigenvector is sensitive especially to the sea quark and strange sea quark distribution [6]. These two observations indicate that the data are able to significantly constrain these distributions. For these distributions previous ATLAS data on on-shell W and Z production [13] is already the most constraining data set and an analysis using this data suggests that the strange contribution was underestimated in the past [14]. However, the MMHT2014 PDF set already includes these measurements. The observed sensitivity should therefore exceed the sensitivity of previous measurements.

In addition to the sensitivity of the data to the PDF uncertainties, a large constraint of the scale uncertainty is observed when comparing the theory prediction with the $|\Delta\eta_{\ell\ell}|$ measurements. An uncertainty reduction of up to 62% is achieved for the scale uncertainty when comparing it to all five PDF sets. At lower masses and large $|\Delta\eta_{\ell\ell}|$, the uncertainty of the theory calculations due to the choice of μ_R and μ_F can be as large as 4.5% and therefore can be the dominant uncertainty. This is not observed for the cross sections as a function of absolute rapidity, where the scale uncertainty is small compared to other sources. The scale uncertainties are arising from missing corrections due to QCD contributions beyond NNLO. A sensitivity to this nuisance parameter might be an indication that the measurement is sensitive to these missing corrections. However, an interpretation of this nuisance parameter is difficult. The scale uncertainty was obtained by changing μ_R and μ_F by and arbitrary factor of two. There is, in addition, no underlying true value of μ_R and μ_F. The scale uncertainty is therefore not a real uncertainty and the interpretation in terms of a Gaussian nuisance parameter is not possible, although this was implicitly assumed in the χ^2 minimization. Still, the statement that the uncertainty is, in some regions of phase space, larger than the uncertainties of the measurement is true.

Figures S.1, S.2, S.3, S.4 and S.5 in the appendix provide additional detailed information on the nuisance parameter shifts for all five PDF sets using the $|y_{\ell\ell}|$ measurement. Figures S.6, S.7, S.8, S.9 and S.10 show the same shifts for the $|\Delta\eta_{\ell\ell}|$ measurement. Only the theoretical nuisance parameters and the luminosity uncertainty are shown, as no other large experimental shifts are observed.

17.3.2 Photon PDF Reweighting

The previous section has shown that the measured cross section can significantly constrain the uncertainty on the photon PDF. However, in the χ^2 minimization procedure a single nuisance parameter was used for the uncertainty. In the following a Bayesian reweighting method is used to further quantify the constraining power of the data on the photon PDF. The reweighting method was developed by the NNPDF collaboration and is described in more detail in references [15, 16]. In this approach, the χ^2 between each of the original $N_{rep} = 100$ Monte Carlo replicas of the NNPDF2.3qed PDF and the experimental data is used to assign a weight to each replica. The PDF is then reweighted in a way that a new PDF can be calculated from the weighted replica, which then estimates the result that would be found in a new NNPDF PDF fit which includes this measurement. The theory calculations used for this approach combine the MMHT2014 NNLO PDF set for the quark and gluon PDFs with the NNPDF2.3qed PDF set for the photon PDF. This approach is justified, given the substantial uncertainties that currently affect the photon PDF, and very weak sensitivity of the photon PDF evolution to the DGLAP evolution mixing with quarks and gluons [17]. However, this approach violates the momentum sum-rule which might be a problem if the photon PDF is large.

In practice, 100 χ^2 values are calculated between the measurement and the theoretical predictions using the central value of the MMHT2014 NNLO PDF set and each of the 100 NNPDF2.3qed replicas. The full MMHT2014 uncertainty, decomposed into the eigenvectors is used in the χ^2 minimization, but no uncertainty on the photon PDF. These χ^2 values are calculated for both, the measurement as a function of $|y_{\ell\ell}|$ and $|\Delta\eta_{\ell\ell}|$ with $N_{data} = 48$ and 47 data points, respectively. The χ^2/dof values are ranging from 58.7/48 to 222.8/48 for the $|y_{\ell\ell}|$ cross section and from 65.0/47 to 243.3/47 for the $|\Delta\eta_{\ell\ell}|$. These values are showing that some of the replica are not compatible with the presented measurement. All χ^2 values are listed in the Appendix in Table T.1. No χ^2 values for the single-differential measurement have been calculated as this measurement is expected to have a smaller constraint on the photon PDF.

The weight associated with each replica i is computed in the following by first computing

$$e_i = \frac{1}{2}((N_{data} - 1)\log \chi_i^2 - \chi_i^2), \tag{17.5}$$

where χ_i^2 is the χ^2 value for a replica i. The weights are then given by

$$w_i = \mathcal{N}\exp[e_i - \langle e_i\rangle], \qquad \mathcal{N} = N_{rep}/\sum_{i=1}^{N_{rep}}\exp[e_i - \langle e_i\rangle]. \tag{17.6}$$

where $\langle e_i\rangle = \frac{1}{N_{rep}}\sum_{i=1}^{N_{rep}} e_i$. These formulae can be derived from Bayes' theorem using basic principles. A detailed derivation of these formulae is given in reference [15]. Figure 17.8 shows the resulting weights for the $|y_{\ell\ell}|$ measurement on the left and for the $|\Delta\eta_{\ell\ell}|$ measurement on the right side. For the $|y_{\ell\ell}|$ measurement, 39 of the assigned weights are below 1, 25 below 10^{-1}, and 9 even below 10^{-7}. For the $|\Delta\eta_{\ell\ell}|$ measurement, 28 of the assigned weights are below 1, 19 below 10^{-1}, and 3 below 10^{-7}. The other weights are clustering around $1-2$ and are therefore similarly probable. The weights indicate that the $|y_{\ell\ell}|$ measurement has a larger constraint on the photon PDF than the $|\Delta\eta_{\ell\ell}|$ measurement. The better constraining power of the $|y_{\ell\ell}|$ measurement compared to the $|\Delta\eta_{\ell\ell}|$ measurement is counterintuitive given the expected sensitivity of the measurements. However, in the present case of finite precision, it does make sense, since the $|y_{\ell\ell}|$ measurement has smaller experimental uncertainties than the $|\Delta\eta_{\ell\ell}|$ measurement in the region where the photon induced contribution is large (at central rapidities and large pseudorapidity separation).

From the calculated weights, the effective number of replicas left after the reweighting can be calculated by using the Shannon entropy:

$$N_{eff} = \exp\left\{\frac{1}{N_{rep}}\sum_{i=1}^{N_{rep}} w_i \ln(N_{rep}/w_i)\right\} \tag{17.7}$$

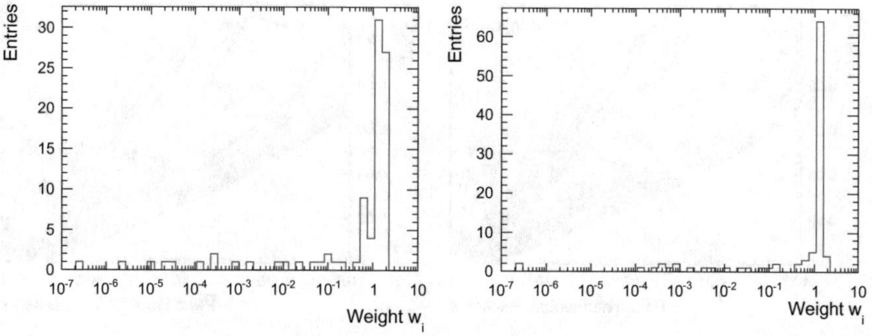

Fig. 17.8 Distribution of the weights w_i calculated from the 100 χ^2 values from the 100 NNPDF2.3qed replica. The left plot shows the weights for the $|y_{\ell\ell}|$ measurement and the right plot shows the weights for the $|\Delta\eta_{\ell\ell}|$ measurement

The Shannon entropy is a measure for the loss of accuracy of the representation of the underlying distribution using the new compared to the old set of replicas. For the $|y_{\ell\ell}|$ measurement a Shannon entropy of 71.3 and for the $|\Delta\eta_{\ell\ell}|$ measurement a Shannon entropy of 78.1 is calculated. These values indicate again that there is some constraining power of the data on the photon PDF, since the number of effective replica is significantly smaller than the initial number of replicas. At the same time, these numbers are still reasonably large, indicating that the reweighting procedure is reliable in the present analysis. If N_{eff} was considerably smaller, the reweighting procedure will no longer be reliable, either because the data contain a lot of information on the PDFs, necessitating a full refitting with more replicas, or because the data are inconsistent with the original PDF set and the data already contained in it. The smaller N_{eff} for the $|y_{\ell\ell}|$ measurement again confirms its larger constraining power.

Once the weights for each replica are calculated, subsequently the resulting PDF set can be unweighted. In the unweighting procedure again a full PDF set with 100 replicas is constructed by keeping only replicas with a reasonably large weight. Replicas with a large weight might be kept twice to construct a full PDF set with 100 replicas again. The unweighting procedure is described in detail in reference [16]. Figure 17.9 shows all 100 replicas of the original PDF set at $Q^2 = 10^4$ GeV^2 as a function of momentum fraction x. The replica of the PDF set have been obtained using LHAPDF6 [18]. Replica which are kept after the unweighting procedure are shown in blue while discarded replica are shown in red. The left plot shows the remaining replica if the $|y_{\ell\ell}|$ measurement is used and the right plot the remaining replica if the $|\Delta\eta_{\ell\ell}|$ measurement is used. These plots show that the NNPDF2.3qed NNLO PDF set is highly asymmetric with a number of very large outliers. It is visible that a significant constraint can be put on the PDF set by discarding the large outliers. All replica which are discarded by the $|\Delta\eta_{\ell\ell}|$ measurement are also discarded by the $|y_{\ell\ell}|$ measurement. Seven replica are only discarded by the $|y_{\ell\ell}|$ measurement. A simultaneous reweighting of both measurements is therefore not expected to increase the sensitivity.

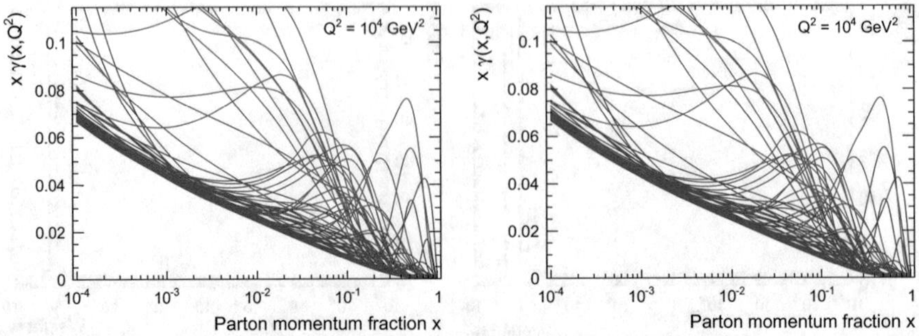

Fig. 17.9 Replicas of the NNPDF2.3Qqed NNLO PDF set before (red) and after the reweighting procedure (blue). The left plot shows the replica after using the $|y_{\ell\ell}|$ measurement and the right plot the replica after using the $|\Delta\eta_{\ell\ell}|$ measurement

From the newly constructed PDF set, the central value and the uncertainties can be calculated. The central value is given by the mean of all replica while the uncertainties have been computed as 68% confidence level intervals[2] around the mean of all replicas, using the same prescription as in reference [4]. This is important in this case since the underlying probability distribution associated to the photon PDF is highly non-gaussian, as seen in Fig. 17.9. Figure 17.10 shows the 68% confidence level interval of the NNPDF2.3qed NNLO photon PDF as a function of momentum fraction x at the input scale $Q^2 = 2$ GeV^2 (left plot) and at $Q^2 = 10^4$ GeV^2 (right plot) before (yellow solid area) and after (grey shaded area) inclusion of the double-differential cross section measurement as a function of $m_{\ell\ell}$ and $|y_{\ell\ell}|$. A significant constraint on the photon PDF uncertainty is visible over the whole range in x. The central value of the reweighted PDF is close to zero at the input scale and on the lower bound of the original PDF uncertainty for a scale of $Q^2 = 10^4$ GeV^2. This supports the nuisance parameter shifts which where obtained in the previous section and indicates that the contribution from the photon induced process is lower than indicated by the NNPDF2.3qed PDF set. Also shown is the MRST2004qed photon PDF in a current quark (blue dashed line) and a constituent quark (blue dotted line) mass scheme. In the current quark mass scheme, the quarks radiate more photons as their mass is lower. This leads to a higher predicted photon PDF. The CT14qed PDF [19] is shown in green with its 68% confidence level band. At a scale of $Q^2 = 10^4$ GeV^2, which is close to the momentum scale at which the measurement is performed, both the MRST2004qed PDF and the CT14qed PDF show a similar behavior by predicting a larger photon PDF at lower x values than the NNPDF2.3qed and the reweighted PDF. Since all PDFs, except MRST2004qed, agree within their uncertainties at the input scale, these differences must come from a different PDF evolution. No conclusive statement can be made whether the MRST2004qed and CT14qed PDF sets or the NNPDF2.3qed set predicts the correct behavior at lower x values. The reweighting

[2] The difference of all 100 replica to the central value is calculated and the 68% confidence level interval is given by the envelope of the 68 closest replica.

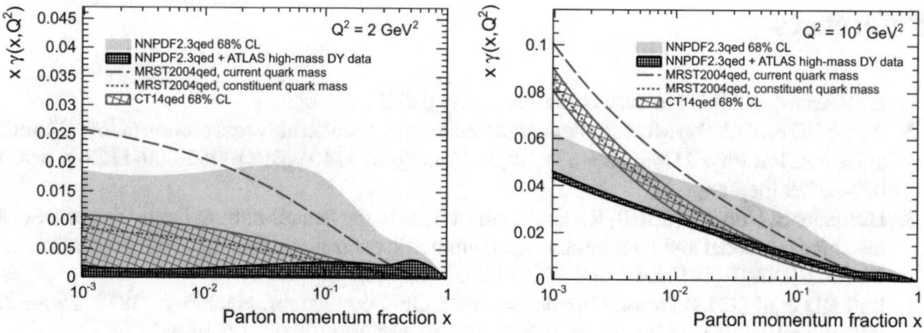

Fig. 17.10 The 68% confidence level interval of the NNPDF2.3qed NNLO photon PDF as a function of momentum fraction x at the input scale $Q^2 = 2$ GeV^2 (left plot) and at $Q^2 = 10^4$ GeV^2 (right plot) before (yellow solid area) and after (grey shaded area) inclusion of the double-differential cross section measurement as a function of invariant mass $m_{\ell\ell}$ and absolute dilepton rapidity $|y_{\ell\ell}|$. Also shown is the MRST2004qed photon PDF in a current quark (blue dashed line) and a constituent quark (blue dotted line) mass scheme, and the 68% confidence level band (green hatched area) for the CT14qed photon PDF

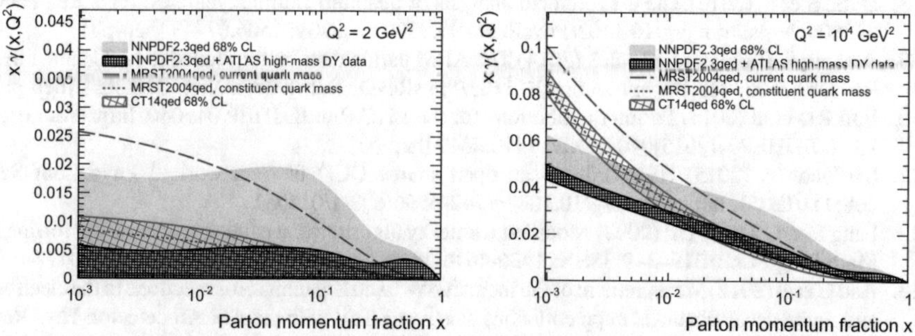

Fig. 17.11 The 68% confidence level interval of the NNPDF2.3qed NNLO photon PDF as a function of momentum fraction x at the input scale $Q^2 = 2$ GeV^2 (left plot) and at $Q^2 = 10^4$ GeV^2 (right plot) before (yellow solid area) and after (grey shaded area) inclusion of the double-differential cross section measurement as a function of invariant mass $m_{\ell\ell}$ and absolute lepton pseudorapidity separation $|\Delta\eta_{\ell\ell}|$. Also shown is the MRST2004qed photon PDF in a current quark (blue dashed line) and a constituent quark (blue dotted line) mass scheme, and the 68% confidence level band (green hatched area) for the CT14qed photon PDF

was performed using the available replica of the NNPDF2.3qed set and can only lead to results which are within the given range of the original PDF set. A full new PDF fit including the presented data is needed to asses the full potential of the measurement.

Finally, Fig. 17.11 shows the same reweighted PDF when using the cross section as a function of $m_{\ell\ell}$ and $|\Delta\eta_{\ell\ell}|$. The observations are here similar, but the constraint on the photon PDF uncertainty is smaller due to the reasons discussed earlier.

References

1. HERAverager-0.0.2. https://wiki-zeuthen.desy.de/HERAverager
2. Aaron FD et al (2009) Measurement of the inclusive ep scattering cross section at low Q2 and x at HERA. Eur Phys J C63: 625–678. https://doi.org/10.1140/epjc/s10052-009-1128-6, arXiv: 0904.0929 [hep-ex]
3. Dittmaier S, Huber M (2010) Radiative corrections to the neutral-current Drell–Yan process in the standard model and its minimal supersymmetric extension. JHEP 01: 060. https://doi.org/10.1007/JHEP01(2010)060, arXiv: 0911.2329 [hep-ph]
4. Ball RD et al (2013) Parton distributions with QED corrections. Nucl Phys B877: 290–320. https://doi.org/10.1016/j.nuclphysb.2013.10.010, arXiv: 1308.0598 [hep-ph]
5. Aad G et al (2016) Measurement of the double-differential high-mass Drell–Yan cross section in pp collisions at $\sqrt{s} = 8$ TeV with the ATLAS detector. JHEP 08: 009. https://doi.org/10.1007/JHEP08(2016)009, arXiv: 1606.01736 [hep-ex]
6. Harland-Lang LA et al (2015) Parton distributions in the LHC era: MMHT 2014 PDFs. Eur Phys J C75.5: 204. https://doi.org/10.1140/epjc/s10052-015-3397-6, arXiv: 1412.3989 [hep-ph]
7. Abramowicz H et al (2015) Combination of measurements of inclusive deep inelastic $e^{\pm}p$ scattering cross sections and QCD analysis of HERA data. Eur Phys J C75.12: 580. https://doi.org/10.1140/epjc/s10052-015-3710-4, arXiv: 1506.06042 [hep-ex]
8. Dulat S et al (2016) The CT14 global analysis of quantum chromodynamics. Phys Rev D93: 033006. https://doi.org/10.1103/PhysRevD.93.033006, arXiv: 1506.07443 [hep-ph]
9. Alekhin S, Blumlein J, Moch S (2014) The ABM parton distributions tuned to LHC data. Phys Rev D89.5: 054028. https://doi.org/10.1103/PhysRevD.89.054028, arXiv: 1310.3059 [hep-ph]
10. Ball RD et al (2015) Parton distributions for the LHC Run II. JHEP 04: 040. https://doi.org/10.1007/JHEP04(2015)040, arXiv: 1410.8849 [hep-ph]
11. Sapronov A (2015) HERAFitter - an open source QCD fit framework. J Phys Conf Ser 608(1):012051. https://doi.org/10.1088/1742-6596/608/1/012051
12. Fang H-r, O'Leary DP (2008) Modified Cholesky algorithms: a catalog with new approaches. Math Prog 115.2: 319–349. ISSN: 1436-4646. https://doi.org/10.1007/s10107-007-0177-6
13. Aad G et al (2012) Measurement of the inclusive $W^{\pm}$ and Z/gamma cross sections in the electron and muon decay channels in pp collisions at $\sqrt{s} = 7$ TeV with the ATLAS detector. Phys Rev D85: 072004. https://doi.org/10.1103/PhysRevD.85.072004, arXiv: 1109.5141 [hep-ex]
14. Aad G et al (2012) Determination of the strange quark density of the proton from ATLAS measurements of the $W \to l\nu$ and $Z \to ll$" cross sections. Phys Rev Lett 109: 012001. https://doi.org/10.1103/PhysRevLett.109.012001, arXiv: 1203.4051 [hep-ex]
15. Ball RD et al (2011) Reweighting NNPDFs: the W lepton asymmetry. Nucl Phys B849: 112–143. [Erratum: Nucl. Phys.B855,927(2012)]. https://doi.org/10.1016/j.nuclphysb.2011.03.017,10.1016/j.nuclphysb.2011.10.024,10.1016/j.nuclphysb.2011.09.011, arXiv: 1012.0836 [hep-ph]
16. Ball RD et al (2012) Reweighting and unweighting of parton distributions and the LHC W lepton asymmetry data. Nucl Phys B855: 608–638. https://doi.org/10.1016/j.nuclphysb.2011.10.018, arXiv: 1108.1758 [hep-ph]
17. Bertone V, Carrazza S, Rojo J (2014) APFEL: A PDF evolution library with QED corrections. Comput Phys Commun 185: 1647–1668. https://doi.org/10.1016/j.cpc.2014.03.007, arXiv: 1310.1394 [hep-ph]
18. Buckley A et al (2015) LHAPDF6: parton density access in the LHC precision era. Eur Phys J C75: 132. https://doi.org/10.1140/epjc/s10052-015-3318-8, arXiv: 1412.7420 [hep-ph]
19. Schmidt C et al (2015) CT14QED PDFs from isolated photon production in deep inelastic scattering. arXiv: 1509.02905 [hep-ph]

Chapter 18
Conclusion and Outlook

A measurements of the double-differential Drell–Yan cross sections for the decay into an electron-positron pair, at a center of mass energy of $\sqrt{s} = 8$ TeV of the colliding protons was performed. The cross section measurements are expected to have sensitivity to the PDFs at high values of the Bjorken-x scaling variable. In particular sensitivity to the PDFs of the antiquarks in the proton is expected, since these are not well constrained at high values of x. The contribution from $\gamma\gamma$ initiated e^+e^--pairs is included in the measured cross section and therefore it also provides sensitivity to the photon part of the PDF. The measurements were performed as a function of the invariant mass and absolute rapidity of the e^+e^--pair and as a function of invariant mass and absolute pseudorapidity separation of the electron and positron. The measurement covered an invariant mass range from $m_{e^+e^-} = 116$ GeV up to $m_{e^+e^-} = 1500$ GeV. The analyzed data set was recorded by the ATLAS experiment in the year 2012 and corresponds to an integrated luminosity of 20.3 fb^{-1}.

The expected amount of e^+e^--pairs produced by Standard Model processes has been estimated using Monte Carlo simulations and data-driven methods. A main part of this work addressed the combination of the measured electron channel cross section with the muon channel cross section and its interpretation in terms of sensitivity to PDFs.

With the combination of the two measurement channels a precision of better than 1% has been achieved in some regions of phase space. The single-differential measurement as a function of invariant mass has systematic uncertainties in the range 0.63–2.95% and a statistical uncertainty of 0.34–17.05%. In the lowest mass bin of the double-differential rapidity measurement, systematic uncertainties in the range 0.62% − 1.46% have been achieved and the statistical uncertainties are in the range 0.81–2.33%. In the highest invariant mass bin the systematic uncertainties rise to 1.89–7.63% and the statistical uncertainties to 6.15–35.7%. The double-differential measurement as a function of pseudorapidity separation has a similar level of accuracy. The measured cross section is compared to several theory predictions using

M. Zinser, *Search for New Heavy Charged Bosons and Measurement of High-Mass Drell-Yan Production in Proton-Proton Collisions*, Springer Theses,
https://doi.org/10.1007/978-3-030-00650-1_18

different PDFs. In general, good agreement is seen for all PDFs, although for most predictions a small offset between data and theory is seen especially in the low mass region. A similar difference was already seen in an analysis performed at $\sqrt{s} = 7$ TeV [1]. The uncertainty of the measurement is smaller than the uncertainty on the theory predictions. The measurement can hence be used to constrain the predictions. It was shown that the measurement can constrain the strange part of the PDFs and that the uncertainty on the photon PDF can be significantly reduced using this measurement.[1] The sensitivity to the latter was further studied using a Bayesian reweighting technique.

No opposite charge requirement had been imposed on the electron channel measurement. The reason is possible charge misidentification mainly due to Bremsstrahlung and due to the limited momentum resolution of the electron in the tracking detector. It is difficult to measure the charge misidentification rate precisely at very high momenta due to limited statistics of data. It would nevertheless be possible to perform this study and to see if the uncertainties on the cross section imposed by this are reasonable. An opposite charge requirement would reduce the background originating from misidentified electrons by a factor of two. This could potentially make the measurement of the electron channel more precise. In addition, different triggers, with lower p_T thresholds, could be tested to further increase the contribution of the photon induced process. However, this would at the same time increase the background from multijet processes and these triggers would have more stringent identification criteria which would lead to a larger uncertainty on the multijet and W+jets background.

Since the year 2015, the LHC collides protons at a center of mass energy of $\sqrt{s} =$ 13 TeV. A measurement using this data would be at a higher center of mass energy, the x values covered by the range 116–1500 GeV would therefore be smaller by a factor of approximately two [4]. Higher values of x could be reached by extending the measurement to higher invariant masses. The gg-luminosity increased with respect to 8 TeV, depending on the mass of the final state, by a factor of about four, whereas the $q\bar{q}$-luminosity only increased approximately by a factor of two [4]. Since $t\bar{t}$ events are mainly produced via gluon-fusion, the cross section increase of this process is approximately two times larger than the cross section increase of the Drell–Yan process. This will double the $t\bar{t}$ background to an amount of about 30% for some ranges of the signal selection. To reduce the amount of $t\bar{t}$ background it might be necessary to impose additional requirements to reject this background (e.g. small $E_{\mathrm{T}}^{\mathrm{miss}}$ or b-jet veto). This will lead to further systematic uncertainties.

[1] Author's comment: When preparing this thesis for the Springer publication (July 2018), the presented measurement has already been used in PDF fits to estimate the photon PDF [2, 3]. It has has been shown that the data is able to constrain the photon PDF and resulted in very good χ^2 values in the PDF fit.

References

1. Aad G et al (2013) Measurement of the high-mass Drell–Yan differential cross-section in pp collisions at $\sqrt(s)$=7 TeV with the ATLAS detector. Phys Lett B725: 223–242. https://doi.org/10.1016/j.physletb.2013.07.049, arXiv: 1305.4192 [hep-ex]
2. Bertone V (2017) The photon PDF from high-mass Drell–Yan data at the LHC. In: Photon 2017: international conference on the structure and the interactions of the photon and 22th international workshop on photon–photon collisions and the international workshop on high energy photon colliders CERN, Switzerland, 22–26 May 2017. arXiv: 1708.00912 [hep-ph]
3. Giuli F (2017) The photon PDF determination within the xFitter framework. In: PoS EPS-HEP2017, p 375. https://doi.org/10.22323/1.314.0375, arXiv: 1708.03623 [hep-ph]
4. Stirling W Private communication

Chapter 19
Summary

The Large Hadron Collider (LHC) at CERN delivers data from proton-proton collisions at an unprecedented center of mass energy and allows for a first look into a new energy regime. Precise predictions of the processes at the LHC are essential to do precise tests of the Standard Model and to search for new physics phenomena in this energy regime. A key role for the predictions of these processes plays the knowledge of the parton distribution functions (PDFs) of the proton.

In this thesis two analyses have been presented using data recorded by the ATLAS experiment. In the first analysis, data recorded at a center of mass energy of proton-proton collisions at 13 TeV has been used to search for a so-called W' boson, a new heavy charged gauge boson. Final states with a charged lepton ($\ell^{\pm} = e^{\pm}, \mu^{\pm}$) and the corresponding (anti-)neutrino have been studied. In the second analysis a double-differential cross section measurement of the process $pp \rightarrow Z/\gamma^* + X \rightarrow \ell^+\ell^- + X$ ($\ell = e, \mu$) at $\sqrt{s} = 8$ TeV has been performed in the invariant mass range of 116–1500 GeV.

The search for a new heavy charged gauge boson at $\sqrt{s} = 13$ TeV has been carried out using data with an integrated luminosity of 3.2 fb^{-1}. These new bosons appear in theories beyond the Standard Model which extend the Standard Model gauge group. The transverse mass spectrum has been measured, in which such a W' boson would be apparent as an excess. Transverse masses up to about 2 TeV have been observed. The expected amount of background from Standard Model processes has been estimated using Monte Carlo simulations and methods based on data. It has been compared to data and possible deviations have been quantified in terms of local and global significances using a likelihood ratio test. Local excesses around 1.4σ and 1.8σ were observed in the electron and muon channel, respectively. Combining both channels lead to an excess of 1.6σ for a W' with a mass of 2 TeV. However, the observed global significance of these excesses is well below 1σ and thus the data are compatible with the Standard Model only hypothesis. As a consequence, limits on the mass of a Sequential Standard Model W', a gauge boson with the same couplings

M. Zinser, *Search for New Heavy Charged Bosons and Measurement of High-Mass Drell-Yan Production in Proton-Proton Collisions*, Springer Theses,
https://doi.org/10.1007/978-3-030-00650-1_19

as the Standard Model W, have been set using a Bayesian approach. Masses below 4.07 TeV have been excluded with 95% confidence level. The obtained exclusion limit is a substantial improvement of the previous limits by around 800 GeV. Using data collected by the LHC at 13 TeV until the end of the year 2017 should, if no excess is observed, improve this limit further to around 5.6 TeV. The complete data set which is expected to be collected in the lifetime of the LHC would increase this limit further to around 7 TeV. A 100 TeV collider, which is still in the conceptual phase, would allow to probe W' masses up to about 35 TeV.

The measurement of the double-differential Drell–Yan cross section at $\sqrt{s} =$ 8 TeV has been performed using data with an integrated luminosity of 20.3 fb^{-1}. In the search for a W', one of the largest uncertainties at high transverse mass was due to the limited knowledge of the parton distribution functions (PDFs) at high Bjorken-x. The cross section measurement is expected to probe exactly this region and to constrain the uncertainties of the PDFs in that region. The measurement includes also $\gamma\gamma$ initiated $\ell^+\ell^-$-pair production from photons inside the proton.

This process is an important contribution at high invariant masses and has so far not been studied in detail. The measurement of the cross section for the decay into an e^+e^--pair has been performed in a first part. The expected amount of e^+e^--pairs produced by Standard Model processes has been estimated using Monte Carlo simulations and methods based on data. The signal processes have afterwards been unfolded to obtain the cross sections. Two different cross sections are provided, as a function of invariant mass and absolute rapidity, and as a function of invariant mass and absolute pseudorapidity separation. An uncertainty reduction of the measurements was achieved by a combination with a measurement of the $\mu^+\mu^-$ cross sections.[1] The combined cross section reaches a precision below 1% in the regions of low invariant mass. At high invariant masses the measurement is limited by the statistical uncertainty. A comparison to theory calculations showed that this level of accuracy is higher than the accuracy on the theory predictions. The measurement will hence be an important input for the extraction of parton density functions. A first study using a χ^2 minimization showed an uncertainty reduction especially for the part describing the strange-quarks and the photons in the proton. The sensitivity to the photon part has been further studied using a Bayesian reweighting technique and it was found that the measurement can strongly reduce the uncertainty on this part. The data collected at $\sqrt{s} = 13$ TeV is expected to have a significant higher integrated luminosity and will thus be a valuable data set to perform similar measurements. Since a measurement using this data would take place at a higher center of mass energy, the x values covered by the range 116–1500GeVwould be smaller by a factor of approximately two. Higher values of x could be reached by extending the measurement to higher invariant masses.

[1]The measurement of the $\mu^+\mu^-$ cross section is not part of this thesis.

Appendix A
W' Search: Signal Reweighting

In the following appendix, the reweighting methodology for the flat W' signal sample is briefly discussed. For the signal sample, the Breit-Wigner term has been removed from the event generation. This leads to the production of a flat falling spectrum, similar to the off-shell tail of the W process. In addition, the square of the matrix element has been divided by a function of $m_{\ell\nu}$ [1]

$$f(m_{l\nu}) = \exp\left(\frac{-p_1 m_{l\nu}}{\sqrt{s}}\right)\left(\frac{m_{l\nu}}{\sqrt{s}}\right)^{p_2} \quad (\sqrt{s} = 13000\ \text{GeV}), \tag{A.1}$$

where p_1 and p_2 are determined from a fit. This is done to avoid a fast drop in cross section as a function of $m_{\ell\nu}$. The resulting samples which are approximately flat in $\log(m_{\ell\nu})$ can be reweighted to any pole mass $m_{W'}$ using the following formula [1]:

$$w = \begin{cases} 10^{12} \times 102.77 \exp\left(-11.5 m_{l\nu}/\sqrt{s}\right) \times W_{BW} & m_{l\nu} < 299\ \text{GeV}, \\ 10^{12} \times \exp\left(-16.1 m_{l\nu}/\sqrt{s}\right) \times (m_{l\nu}/\sqrt{s})^{1.2} \times W_{BW} & m_{l\nu} \geq 299\ \text{GeV}, m_{l\nu} < 3003\ \text{GeV}, \\ 10^{16} \times 1.8675 \exp\left(-31.7 m_{l\nu}/\sqrt{s}\right) \times (m_{l\nu}/\sqrt{s})^{4.6} \times W_{BW} & m_{l\nu} \geq 3003\ \text{GeV}, \end{cases}$$

where $m_{l\nu}$ is the lepton-neutrino invariant mass in GeV. Three arbitrary fit regions were chosen which describe the spectra. The quantities W_{BW} and Γ are determined as:

$$W_{BW} = \frac{1}{(m_{l\nu}^2 - m_{W'}^2)^2 + m_{W'}^2 \Gamma^2}$$

$$m_{W'} < m_t + m_b : \Gamma = 10.14861 \times 10^{-2} m_{W'}$$

$$m_{W'} > m_t + m_b : \Gamma = 3.382870 \times 10^{-2} m_{W'} \left(3 + \left(1 + \frac{1}{2}\left(\frac{m_t}{m_{W'}}\right)^2\right)\frac{1}{4}\left(1 - \left(\frac{m_t}{m_{W'}}\right)^2\right)^2\right),$$

where $m_{W'}$ is the required pole mass in GeV and $m_t = 172.5$ GeV is the mass of the top quark. The mass of the top quark is needed as the decay $W' \rightarrow tb$ is allowed for masses $m_{W'} > m_t + m_b$.

M. Zinser, *Search for New Heavy Charged Bosons and Measurement of High-Mass Drell-Yan Production in Proton-Proton Collisions*, Springer Theses,
https://doi.org/10.1007/978-3-030-00650-1

Reference

1. Chizhov M private communication

Appendix B
W' Search: Detailed Information About MC Samples

The following appendix provides detailed tables with information about the Monte Carlo samples used in the W' search. The Monte Carlo samples are described in detail in Sect. 9.1. Table B.1 list information about the W' signal samples. Tables B.2, B.3, and B.4 list informations about the samples of the leading W background. The background samples from the Z/γ^* process are listed in Tables B.5, B.6, and B.7. Finally, information about the samples for backgrounds arising from top-quark and diboson processes is listed in Table B.8.

Table B.1 Monte Carlo W' signal samples used for this analysis. For each dataset, the following is listed: the ATLAS Monte Carlo run number, the physics process (including the pole mass in TeV when appropriate), the number of generated events, the cross section times branching ratio, K-factor ($K(m)$ denotes a mass dependent K-factor is used)

Dataset ID	Process	N_{evt} [k]	Generator σB [nb]	K-factor
301533	$W' \to e\nu$ (Flat)	1000	0.024960	$K(m)$
301534	$W' \to \mu\nu$ (Flat)	1000	0.024944	$K(m)$
301242	$W' \to e\nu$ (2 TeV)	20	0.00011010	$K(m)$
301243	$W' \to e\nu$ (3 TeV)	20	0.000011358	$K(m)$
301244	$W' \to e\nu$ (4 TeV)	20	0.0000017915	$K(m)$
301245	$W' \to e\nu$ (5 TeV)	20	0.00000040860	$K(m)$
301246	$W' \to \mu\nu$ (2 TeV)	20	0.00010993	$K(m)$
301247	$W' \to \mu\nu$ (3 TeV)	20	0.000011380	$K(m)$
301248	$W' \to \mu\nu$ (4 TeV)	20	0.000001775	$K(m)$
301249	$W' \to \mu\nu$ (5 TeV)	20	0.00000040933	$K(m)$

M. Zinser, *Search for New Heavy Charged Bosons and Measurement of High-Mass Drell-Yan Production in Proton-Proton Collisions*, Springer Theses,
https://doi.org/10.1007/978-3-030-00650-1

Table B.2 Monte Carlo samples for backgrounds that contribute to the electron channel. For each dataset, the following is listed: the ATLAS Monte Carlo run number, the physics process (including the mass range in GeV when appropriate), the number of generated events, the cross section times branching ratio times $\varepsilon_{\text{filt}}$ (the filter efficiency reported by the generator), K-factor ($K(m)$ denotes a mass dependent K-factor is used), and the equivalent integrated luminosity $L_{\text{int}} = N_{\text{evt}}/(\sigma B)$

Dataset ID	Process	N_{evt} [k]	Generator $\sigma B \varepsilon_{\text{filt}}$ [pb]	K-factor	L_{int} [fb^{-1}]
Inclusive and mass binned $W \to e\nu$					
361100	$W^+ \to e\nu$	29979	11306.0	$K(m)$	2.65e+00
361103	$W^- \to e\nu$	39974	8282.6	$K(m)$	4.83e+00
301060	$W^+(120, 180) \to e\nu$	500	32.053	$K(m)$	1.56e+01
301061	$W^+(180, 250) \to e\nu$	250	5.0029	$K(m)$	5.00e+01
301062	$W^+(250, 400) \to e\nu$	140	1.7543	$K(m)$	7.98e+01
301063	$W^+(400, 600) \to e\nu$	100	0.31235	$K(m)$	3.20e+02
301064	$W^+(600, 800) \to e\nu$	50	0.060793	$K(m)$	8.22e+02
301065	$W^+(800, 1000) \to e\nu$	50	0.017668	$K(m)$	2.83e+03
301066	$W^+(1000, 1250) \to e\nu$	50	0.0072895	$K(m)$	6.86e+03
301067	$W^+(1250, 1500) \to e\nu$	50	0.0025071	$K(m)$	1.99e+04
301068	$W^+(1500, 1750) \to e\nu$	50	0.00098628	$K(m)$	5.07e+04
301069	$W^+(1750, 2000) \to e\nu$	40	0.0004245	$K(m)$	9.42e+04
301070	$W^+(2000, 2250) \to e\nu$	50	0.00019463	$K(m)$	2.57e+05
301071	$W^+(2250, 2500) \to e\nu$	50	9.3349e-05	$K(m)$	5.36e+05
301072	$W^+(2500, 2750) \to e\nu$	50	4.6259e-05	$K(m)$	1.08e+06
301073	$W^+(2750, 3000) \to e\nu$	50	2.3476e-05	$K(m)$	2.13e+06
301074	$W^+(3000, 3500) \to e\nu$	50	1.845e-05	$K(m)$	2.71e+06
301075	$W^+(3500, 4000) \to e\nu$	50	5.0968e-06	$K(m)$	9.81e+06
301076	$W^+(4000, 4500) \to e\nu$	50	1.4307e-06	$K(m)$	3.49e+07
301077	$W^+(4500, 5000) \to e\nu$	50	4.0127e-07	$K(m)$	1.25e+08
301078	$W^+(> 5000) \to e\nu$	50	1.5346e-07	$K(m)$	3.26e+08
301080	$W^-(120, 180) \to e\nu$	500	22.198	$K(m)$	2.25e+01
301081	$W^-(180, 250) \to e\nu$	250	3.2852	$K(m)$	7.61e+01
301082	$W^-(250, 400) \to e\nu$	150	1.0832	$K(m)$	1.38e+02
301083	$W^-(400, 600) \to e\nu$	100	0.17541	$K(m)$	5.70e+02
301084	$W^-(600, 800) \to e\nu$	50	0.03098	$K(m)$	1.61e+03
301085	$W^-(800, 1000) \to e\nu$	50	0.0082865	$K(m)$	6.03e+03
301086	$W^-(1000, 1250) \to e\nu$	50	0.0031594	$K(m)$	1.58e+04
301087	$W^-(1250, 1500) \to e\nu$	50	0.0010029	$K(m)$	4.99e+04
301088	$W^-(1500, 1750) \to e\nu$	50	0.00036812	$K(m)$	1.36e+05
301089	$W^-(1750, 2000) \to e\nu$	50	0.00014945	$K(m)$	3.35e+05
301090	$W^-(2000, 2250) \to e\nu$	50	6.5311e-05	$K(m)$	7.66e+05
301091	$W^-(2250, 2500) \to e\nu$	50	3.0167e-05	$K(m)$	1.66e+06
301092	$W^-(2500, 2750) \to e\nu$	50	1.4549e-05	$K(m)$	3.44e+06

(continued)

Table B.2 (continued)

Dataset ID	Process	N_{evt} [k]	Generator $\sigma B \varepsilon_{filt}$ [pb]	K-factor	L_{int} [fb^{-1}]
Inclusive and mass binned $W \rightarrow e\nu$					
301093	$W^-(2750, 3000) \rightarrow e\nu$	50	7.2592e-06	$K(m)$	6.89e+06
301094	$W^-(3000, 3500) \rightarrow e\nu$	50	5.6692e-06	$K(m)$	8.82e+06
301095	$W^-(3500, 4000) \rightarrow e\nu$	50	1.5975e-06	$K(m)$	3.13e+07
301096	$W^-(4000, 4500) \rightarrow e\nu$	50	4.721e-07	$K(m)$	1.06e+08
301097	$W^-(4500, 5000) \rightarrow e\nu$	50	1.4279e-07	$K(m)$	3.50e+08
301098	$W^-(> 5000) \rightarrow e\nu$	50	6.1624e-08	$K(m)$	8.11e+08

Table B.3 Monte Carlo samples for backgrounds that contribute to the muon channel. For each dataset, the following is listed: the ATLAS Monte Carlo run number, the physics process (including the mass range in GeV when appropriate), the number of generated events, the cross section times branching ratio, K-factor ($K(m)$ denotes a mass dependent K-factor is used), and the equivalent integrated luminosity $L_{int} = N_{evt}/(\sigma B)$

Dataset ID	Process	N_{evt} [k]	Generator σB [pb]	K-factor	L_{int} [fb^{-1}]
Inclusive and mass binned $W \rightarrow \mu\nu$					
361101	$W^+ \rightarrow \mu\nu$	29972	11306.0	$K(m)$	2.65e+00
361104	$W^- \rightarrow \mu\nu$	19984	8282.6	$K(m)$	2.41e+00
301100	$W^+(120, 180) \rightarrow \mu\nu$	500	32.053	$K(m)$	1.56e+01
301101	$W^+(180, 250) \rightarrow \mu\nu$	250	5.0029	$K(m)$	5.00e+01
301102	$W^+(250, 400) \rightarrow \mu\nu$	150	1.7543	$K(m)$	8.55e+01
301103	$W^+(400, 600) \rightarrow \mu\nu$	100	0.31235	$K(m)$	3.20e+02
301104	$W^+(600, 800) \rightarrow \mu\nu$	50	0.060793	$K(m)$	8.22e+02
301105	$W^+(800, 1000) \rightarrow \mu\nu$	50	0.017668	$K(m)$	2.83e+03
301106	$W^+(1000, 1250) \rightarrow \mu\nu$	50	0.0072895	$K(m)$	6.86e+03
301107	$W^+(1250, 1500) \rightarrow \mu\nu$	50	0.0025071	$K(m)$	1.99e+04
301108	$W^+(1500, 1750) \rightarrow \mu\nu$	50	0.00098628	$K(m)$	5.07e+04
301109	$W^+(1750, 2000) \rightarrow \mu\nu$	50	0.00042457	$K(m)$	1.18e+05
301110	$W^+(2000, 2250) \rightarrow \mu\nu$	50	0.00019463	$K(m)$	2.57e+05
301111	$W^+(2250, 2500) \rightarrow \mu\nu$	50	9.3349e-05	$K(m)$	5.36e+05
301112	$W^+(2500, 2750) \rightarrow \mu\nu$	50	4.6259e-05	$K(m)$	1.08e+06
301113	$W^+(2750, 3000) \rightarrow \mu\nu$	50	2.3476e-05	$K(m)$	2.13e+06
301114	$W^+(3000, 3500) \rightarrow \mu\nu$	50	1.845e-05	$K(m)$	2.71e+06
301115	$W^+(3500, 4000) \rightarrow \mu\nu$	50	5.0968e-06	$K(m)$	9.81e+06
301116	$W^+(4000, 4500) \rightarrow \mu\nu$	50	1.4307e-06	$K(m)$	3.49e+07
301117	$W^+(4500, 5000) \rightarrow \mu\nu$	50	4.0127e-07	$K(m)$	1.25e+08
301118	$W^+(> 5000) \rightarrow \mu\nu$	50	1.5346e-07	$K(m)$	3.26e+08
301120	$W^-(120, 180) \rightarrow \mu\nu$	500	22.198	$K(m)$	2.25e+01

(continued)

Table B.3 (continued)

Dataset ID	Process	N_{evt} [k]	Generator σB [pb]	K-factor	L_{int} [fb^{-1}]
Inclusive and mass binned $W \rightarrow \mu\nu$					
301121	$W^{-}(180, 250) \rightarrow \mu\nu$	250	3.2853	$K(m)$	7.61e+01
301122	$W^{-}(250, 400) \rightarrow \mu\nu$	150	1.0832	$K(m)$	1.38e+02
301123	$W^{-}(400, 600) \rightarrow \mu\nu$	100	0.17541	$K(m)$	5.70e+02
301124	$W^{-}(600, 800) \rightarrow \mu\nu$	50	0.03098	$K(m)$	1.61e+03
301125	$W^{-}(800, 1000) \rightarrow \mu\nu$	50	0.0082865	$K(m)$	6.03e+03
301126	$W^{-}(1000, 1250) \rightarrow \mu\nu$	50	0.0031594	$K(m)$	1.58e+04
301127	$W^{-}(1250, 1500) \rightarrow \mu\nu$	50	0.0010029	$K(m)$	4.99e+04
301128	$W^{-}(1500, 1750) \rightarrow \mu\nu$	50	0.00036812	$K(m)$	1.36e+05
301129	$W^{-}(1750, 2000) \rightarrow \mu\nu$	50	0.00014945	$K(m)$	3.35e+05
301130	$W^{-}(2000, 2250) \rightarrow \mu\nu$	50	6.5311e-05	$K(m)$	7.66e+05
301131	$W^{-}(2250, 2500) \rightarrow \mu\nu$	50	3.0167e-05	$K(m)$	1.66e+06
301132	$W^{-}(2500, 2750) \rightarrow \mu\nu$	50	1.4549e-05	$K(m)$	3.44e+06
301133	$W^{-}(2750, 3000) \rightarrow \mu\nu$	50	7.2592e-06	$K(m)$	6.89e+06
301134	$W^{-}(3000, 3500) \rightarrow \mu\nu$	50	5.6692e-06	$K(m)$	8.82e+06
301135	$W^{-}(3500, 4000) \rightarrow \mu\nu$	50	1.5975e-06	$K(m)$	3.13e+07
301136	$W^{-}(4000, 4500) \rightarrow \mu\nu$	50	4.721e-07	$K(m)$	1.06e+08
301137	$W^{-}(4500, 5000) \rightarrow \mu\nu$	50	1.4279e-07	$K(m)$	3.50e+08
301138	$W^{-}(> 5000) \rightarrow \mu\nu$	50	6.1624e-08	$K(m)$	8.11e+08

Table B.4 Monte Carlo samples for backgrounds that contribute to both, the electron and muon channels. For each dataset, the following is listed: the ATLAS Monte Carlo run number, the physics process (including the mass range in GeV when appropriate), the number of generated events, the cross section times branching ratio, K-factor ($K(m)$ denotes a mass dependent K-factor is used), and the equivalent integrated luminosity $L_{\text{int}} = N_{\text{evt}}/(\sigma B)$

Dataset ID	Process	N_{evt} [k]	Generator σB [pb]	K-factor	L_{int} [fb^{-1}]
Inclusive and mass binned $W \to \tau\nu$					
361102	$W^+ \to \tau\nu$	29980	11306.0	$K(m)$	2.65e+00
361105	$W^- \to \tau\nu$	19961	8282.6	$K(m)$	2.41e+00
301140	$W^+(120, 180) \to \tau\nu$	500	32.053	$K(m)$	1.56e+01
301141	$W^+(180, 250) \to \tau\nu$	250	5.0029	$K(m)$	5.00e+01
301142	$W^+(250, 400) \to \tau\nu$	150	1.7543	$K(m)$	8.55e+01
301143	$W^+(400, 600) \to \tau\nu$	100	0.31235	$K(m)$	3.20e+02
301144	$W^+(600, 800) \to \tau\nu$	50	0.060793	$K(m)$	8.22e+02
301145	$W^+(800, 1000) \to \tau\nu$	50	0.017668	$K(m)$	2.83e+03
301146	$W^+(1000, 1250) \to \tau\nu$	50	0.0072895	$K(m)$	6.86e+03
301147	$W^+(1250, 1500) \to \tau\nu$	50	0.0025071	$K(m)$	1.99e+04
301148	$W^+(1500, 1750) \to \tau\nu$	50	0.00098628	$K(m)$	5.07e+04
301149	$W^+(1750, 2000) \to \tau\nu$	50	0.00042457	$K(m)$	1.18e+05
301150	$W^+(2000, 2250) \to \tau\nu$	50	0.00019463	$K(m)$	2.57e+05
301151	$W^+(2250, 2500) \to \tau\nu$	50	9.3349e-05	$K(m)$	5.36e+05
301152	$W^+(2500, 2750) \to \tau\nu$	50	4.6259e-05	$K(m)$	1.08e+06
301153	$W^+(2750, 3000) \to \tau\nu$	50	2.3476e-05	$K(m)$	2.13e+06
301154	$W^+(3000, 3500) \to \tau\nu$	50	1.845e-05	$K(m)$	2.71e+06
301155	$W^+(3500, 4000) \to \tau\nu$	50	5.0968e-06	$K(m)$	9.81e+06
301156	$W^+(4000, 4500) \to \tau\nu$	50	1.4307e-06	$K(m)$	3.49e+07
301157	$W^+(4500, 5000) \to \tau\nu$	50	4.0127e-07	$K(m)$	1.25e+08

(continued)

Table B.4 (continued)

Dataset ID	Process	N_{evt} [k]	Generator σB [pb]	K-factor	L_{int} [fb^{-1}]
Inclusive and mass binned $W \to \tau\nu$					
301158	$W^+(> 5000) \to \tau\nu$	50	1.5346e-07	$K(m)$	3.26e+08
301160	$W^-(120, 180) \to \tau\nu$	500	22.198	$K(m)$	2.25e+01
301161	$W^-(180, 250) \to \tau\nu$	250	3.2852	$K(m)$	7.61e+01
301162	$W^-(250, 400) \to \tau\nu$	150	1.0832	$K(m)$	1.38e+02
301163	$W^-(400, 600) \to \tau\nu$	100	0.17541	$K(m)$	5.70e+02
301164	$W^-(600, 800) \to \tau\nu$	50	0.03098	$K(m)$	1.61e+03
301165	$W^-(800, 1000) \to \tau\nu$	50	0.0082865	$K(m)$	6.03e+03
301166	$W^-(1000, 1250) \to \tau\nu$	50	0.0031594	$K(m)$	1.58e+04
301167	$W^-(1250, 1500) \to \tau\nu$	50	0.0010029	$K(m)$	4.99e+04
301168	$W^-(1500, 1750) \to \tau\nu$	50	0.00036812	$K(m)$	1.36e+05
301169	$W^-(1750, 2000) \to \tau\nu$	50	0.00014945	$K(m)$	3.35e+05
301170	$W^-(2000, 2250) \to \tau\nu$	50	6.5311e-05	$K(m)$	7.66e+05
301171	$W^-(2250, 2500) \to \tau\nu$	50	3.0167e-05	$K(m)$	1.66e+06
301172	$W^-(2500, 2750) \to \tau\nu$	50	1.4549e-05	$K(m)$	3.44e+06
301173	$W^-(2750, 3000) \to \tau\nu$	50	7.2592e-06	$K(m)$	6.89e+06
301174	$W^-(3000, 3500) \to \tau\nu$	50	5.6692e-06	$K(m)$	8.82e+06
301175	$W^-(3500, 4000) \to \tau\nu$	50	1.5975e-06	$K(m)$	3.13e+07
301176	$W^-(4000, 4500) \to \tau\nu$	50	4.721e-07	$K(m)$	1.06e+08
301177	$W^-(4500, 5000) \to \tau\nu$	50	1.4279e-07	$K(m)$	3.50e+08
301178	$W^-(> 5000) \to \tau\nu$	50	6.1624e-08	$K(m)$	8.11e+08

Table B.5 Monte Carlo samples for backgrounds that contribute to the electron channel. For each dataset, the following is listed: the ATLAS Monte Carlo run number, the physics process (including the mass range in GeV when appropriate), the number of generated events, the cross section times branching ratio, K-factor ($K(m)$ denotes a mass dependent K-factor is used), and the equivalent integrated luminosity $L_{\text{int}} = N_{\text{evt}}/(\sigma B)$

Run	Process	N_{evt} [k]	Generator σB [pb]	K-factor	L_{int} [fb^{-1}]
Inclusive and mass binned $Z \to ee$					
361106	$Z \to ee$	19993	1901.2	$K(m)$	1.05e+01
301000	$Z(120, 180) \to ee$	500	17.478	$K(m)$	2.86e+01
301001	$Z(180, 250) \to ee$	250	2.9212	$K(m)$	8.56e+01
301002	$Z(250, 400) \to ee$	150	1.082	$K(m)$	1.39e+02
301003	$Z(400, 600) \to ee$	100	0.1955	$K(m)$	5.12e+02
301004	$Z(600, 800) \to ee$	145	0.037401	$K(m)$	3.88e+03
301005	$Z(800, 1000) \to ee$	50	0.010607	$K(m)$	4.71e+03
301006	$Z(1000, 1250) \to ee$	50	0.0042582	$K(m)$	1.17e+04
301007	$Z(1250, 1500) \to ee$	50	0.0014219	$K(m)$	3.52e+04
301008	$Z(1500, 1750) \to ee$	50	0.00054521	$K(m)$	9.17e+04
301009	$Z(1750, 2000) \to ee$	50	0.00022991	$K(m)$	2.17e+05
301010	$Z(2000, 2250) \to ee$	50	0.00010387	$K(m)$	4.81e+05
301011	$Z(2250, 2500) \to ee$	50	4.94e-05	$K(m)$	1.01e+06
301012	$Z(2500, 2750) \to ee$	50	2.4452e-05	$K(m)$	2.04e+06
301013	$Z(2750, 3000) \to ee$	50	1.2487e-05	$K(m)$	4.00e+06
301014	$Z(3000, 3500) \to ee$	10	1.0025e-05	$K(m)$	9.98e+05
301014	$Z(3000, 3500) \to ee$	50	1.0029e-05	$K(m)$	4.99e+06
301015	$Z(3500, 4000) \to ee$	50	2.9342e-06	$K(m)$	1.70e+07
301016	$Z(4000, 4500) \to ee$	50	8.9764e-07	$K(m)$	5.57e+07
301017	$Z(4500, 5000) \to ee$	50	2.8071e-07	$K(m)$	1.78e+08
301018	$Z(> 5000) \to ee$	50	1.2649e-07	$K(m)$	3.95e+08

Table B.6 Monte Carlo samples for backgrounds that contribute to the muon channel. For each dataset, the following is listed: the ATLAS Monte Carlo run number, the physics process (including the mass range in GeV when appropriate), the number of generated events, the cross section times branching ratio, K-factor ($K(m)$ denotes a mass dependent K-factor is used), and the equivalent integrated luminosity $L_{\text{int}} = N_{\text{evt}}/(\sigma B)$

Run	Process	N_{evt} [k]	Generator σB [pb]	K-factor	L_{int} [fb^{-1}]
Inclusive and mass binned $Z \to \mu\mu$					
361107	$Z \to \mu\mu$	19981	1901.2	$K(m)$	1.05e+01
301020	$Z(120, 180) \to \mu\mu$	500	17.478	$K(m)$	2.86e+01
301021	$Z(180, 250) \to \mu\mu$	250	2.9212	$K(m)$	8.56e+01
301022	$Z(250, 400) \to \mu\mu$	150	1.082	$K(m)$	1.39e+02
301023	$Z(400, 600) \to \mu\mu$	100	0.1955	$K(m)$	5.12e+02
301024	$Z(600, 800) \to \mu\mu$	50	0.037399	$K(m)$	1.34e+03
301025	$Z(800, 1000) \to \mu\mu$	50	0.010607	$K(m)$	4.71e+03
301026	$Z(1000, 1250) \to \mu\mu$	50	0.0042582	$K(m)$	1.17e+04
301027	$Z(1250, 1500) \to \mu\mu$	50	0.0014219	$K(m)$	3.52e+04
301028	$Z(1500, 1750) \to \mu\mu$	50	0.00054521	$K(m)$	9.17e+04
301029	$Z(1750, 2000) \to \mu\mu$	50	0.00022991	$K(m)$	2.17e+05
301030	$Z(2000, 2250) \to \mu\mu$	50	0.00010387	$K(m)$	4.81e+05
301031	$Z(2250, 2500) \to \mu\mu$	50	4.94e-05	$K(m)$	1.01e+06
301032	$Z(2500, 2750) \to \mu\mu$	50	2.4452e-05	$K(m)$	2.04e+06
301033	$Z(2750, 3000) \to \mu\mu$	50	1.2487e-05	$K(m)$	4.00e+06
301034	$Z(3000, 3500) \to \mu\mu$	50	1.0029e-05	$K(m)$	4.99e+06
301035	$Z(3500, 4000) \to \mu\mu$	50	2.9342e-06	$K(m)$	1.70e+07
301036	$Z(4000, 4500) \to \mu\mu$	50	8.9764e-07	$K(m)$	5.57e+07
301037	$Z(4500, 5000) \to \mu\mu$	50	2.8071e-07	$K(m)$	1.78e+08
301038	$Z(> 5000) \to \mu\mu$	50	1.2649e-07	$K(m)$	3.95e+08

Table B.7 Monte Carlo samples for backgrounds that contribute to both, the electron and muon channels. For each dataset, the following is listed: the ATLAS Monte Carlo run number, the physics process (including the mass range in GeV when appropriate), the number of generated events, the cross section times branching ratio times $\varepsilon_{\text{filt}}$ (the filter efficiency reported by the generator), K-factor ($K(m)$ denotes a mass dependent K-factor is used), and the equivalent integrated luminosity $L_{\text{int}} = N_{\text{evt}}/(\sigma B)$

Run	Process	N_{evt} [k]	Generator $\sigma B\varepsilon_{\text{filt}}$ [pb]	K-factor	L_{int} [fb^{-1}]
Inclusive and mass binned $Z \to \tau\tau$					
361108	$Z \to \tau\tau$	19742	1901.2	$K(m)$	1.04e+01
301040	$Z(120, 180) \to \tau\tau$	150	17.48	$K(m)$	8.58e+00
301041	$Z(180, 250) \to \tau\tau$	150	2.9209	$K(m)$	5.14e+01
301042	$Z(250, 400) \to \tau\tau$	150	1.082	$K(m)$	1.39e+02
301043	$Z(400, 600) \to \tau\tau$	150	0.1955	$K(m)$	7.67e+02
301044	$Z(600, 800) \to \tau\tau$	150	0.037401	$K(m)$	4.01e+03
301045	$Z(800, 1000) \to \tau\tau$	150	0.010607	$K(m)$	1.41e+04
301046	$Z(1000, 1250) \to \tau\tau$	150	0.0042584	$K(m)$	3.52e+04
301047	$Z(1250, 1500) \to \tau\tau$	150	0.001422	$K(m)$	1.05e+05
301048	$Z(1500, 1750) \to \tau\tau$	50	0.00054521	$K(m)$	9.17e+04
301049	$Z(1750, 2000) \to \tau\tau$	50	0.00022991	$K(m)$	2.17e+05
301050	$Z(2000, 2250) \to \tau\tau$	50	0.00010387	$K(m)$	4.81e+05
301051	$Z(2250, 2500) \to \tau\tau$	50	4.94e-05	$K(m)$	1.01e+06
301052	$Z(2500, 2750) \to \tau\tau$	50	2.4452e-05	$K(m)$	2.04e+06
301053	$Z(2750, 3000) \to \tau\tau$	50	1.2487e-05	$K(m)$	4.00e+06
301054	$Z(3000, 3500) \to \tau\tau$	50	1.0029e-05	$K(m)$	4.99e+06
301055	$Z(3500, 4000) \to \tau\tau$	50	2.9342e-06	$K(m)$	1.70e+07
301056	$Z(4000, 4500) \to \tau\tau$	50	8.9764e-07	$K(m)$	5.57e+07
301057	$Z(4500, 5000) \to \tau\tau$	50	2.8071e-07	$K(m)$	1.78e+08
301058	$Z(> 5000) \to \tau\tau$	50	1.2649e-07	$K(m)$	3.95e+08

Table B.8 Monte Carlo samples for backgrounds that contribute to both, the electron and muon channels. For each dataset, the following is listed: the ATLAS Monte Carlo run number, the physics process (including the mass range in GeV when appropriate), the number of generated events, the cross section times branching ratio times $\varepsilon_{\mathrm{filt}}$ (the filter efficiency reported by the generator), K-factor ($K(m)$ denotes a mass dependent K-factor is used), and the equivalent integrated luminosity $L_{\mathrm{int}} = N_{\mathrm{evt}}/(\sigma B)$

Run	Process	N_{evt} [k]	Generator $\sigma B \times \varepsilon_{\mathrm{filt}}$ [pb]	K-factor	L_{int} [fb^{-1}]
Diboson					
361063	$ZZ \to \ell\ell\ell\ell$	17993	12.849×1.0	$K(m)$	1.40e+03
361064	$WZ \to \ell\ell\ell\nu$ (SFMinus)	450	1.8442×1.0	$K(m)$	2.44e+02
361065	$WZ \to \ell\ell\ell\nu$ (OFMinus)	900	3.6254×1.0	$K(m)$	2.48e+02
361066	$WZ \to \ell\ell\ell\nu$ (SFPlus)	600	2.5618×1.0	$K(m)$	2.34e+02
361067	$WZ \to \ell\ell\ell\nu$ (OFPlus)	1200	5.0248×1.0	$K(m)$	2.39e+02
361068	$VV \to \ell\ell\nu\nu$	5942	14.0×1.0	$K(m)$	4.24e+02
361088	$WZ \to \ell\nu\nu\nu$	2000	3.4001×1.0	$K(m)$	5.88e+02
361091	$W^+W^- \to \ell\nu qq$	2000	24.885×1.0	$K(m)$	8.04e+01
361092	$W^+W^- \to qq\ell\nu$	2000	24.857×1.0	$K(m)$	8.05e+01
361093	$WZ \to \ell\nu qq$	2000	11.494×1.0	$K(m)$	1.74e+02
361094	$WZ \to qq\ell\ell$	500	3.4234×1.0	$K(m)$	1.46e+02
361096	$ZZ \to qq\ell\ell$	500	16.445×0.143	$K(m)$	3.04e+01
361097	$ZZ \to qq\nu\nu$	500	16.432×0.282	$K(m)$	3.04e+01
Top					
410000	$t\bar{t} \to \ell X$	49974	696.11×0.543	1.195	7.18e+01
410011	t-channel $t \to \ell X$	5000	43.739×1.0	1.0	1.14e+02
410012	t-channel $\bar{t} \to \ell X$	5000	25.778×1.0	1.0	1.94e+02
410013	s-channel Wt	5000	34.009×1.0	1.0	1.47e+02
410014	s-channel $W\bar{t}$	5000	33.989×1.0	1.0	1.47e+02

Appendix C
W' Search: Cut Efficiencies

The following appendix provides detailed tables with information about the cut efficiencies for signals and backgrounds in the W' search. The cuts are described in detail in Sect. 9.2. Tables C.1 and C.2 show the cut efficiencies for the backgrounds which are determined from Monte Carlo simulation for the electron and muon selection, respectively. Each line shows the efficiency relative to the previous line. The efficiencies for W' bosons with masses of 2, 3, 4, and 5 TeV are shown in Tables C.3 and C.4 for the electron and muon selection, respectively. The numbers are not including efficiency corrections accounting for the differences observed between data and simulation for the lepton trigger, reconstruction, identification, and isolation efficiencies as these are only defined for the final selection and not for the intermediate selection steps.

Figure C.1 shows the trigger, identification, and isolation efficiencies for the electron and muon channel as a function of η, ϕ, and p_{T} of the lepton. The left hand side shows the efficiencies in the electron channel and the right hand side the efficiencies of the muon channel. The efficiencies were determined using the flat W' MC samples. The studied lepton candidate is required to match the generated lepton by requiring $\Delta R < 0.2$. The isolation efficiency is shown with respect to all previous selection steps, including the trigger and identification (*high-*p_{T} selection for muons and *tight* likelihood identification for electrons). It is for both channels around 98% at lower p_{T} values and rising to above 99% for higher p_{T} values. No dependency on η or ϕ is observed. The identification efficiency is defined with respect to the previous selection steps, i.e. the denominator includes all events which pass the selection up to the identification criteria and the numerator all events which pass the identification criteria. The likelihood *tight* identification efficiency in the electron channel is rising in p_{T} from around 93% to 96% at around 200 GeV. It is afterwards slightly dropping again to about 86% at a p_{T} of 3 TeV. The identification efficiency is lowest in the very central region and around the transition region from the barrel electromagnetic calorimeter to the endcap electromagnetic calorimeter. No dependency in ϕ is observed. The efficiency of the muon *high-*p_{T} selection is with about 87% highest at lower p_{T} value and slowly decreasing to about 80% at a p_{T} of 3 TeV. The

M. Zinser, *Search for New Heavy Charged Bosons and Measurement of High-Mass Drell-Yan Production in Proton-Proton Collisions*, Springer Theses,
https://doi.org/10.1007/978-3-030-00650-1

Table C.1 Cut efficiencies for all backgrounds in the electron channel. The efficiencies are with respect to the previous line in the table. All samples were preselected with the requirement of at least one electron or muon. The efficiency numbers shown are without accounting for the differences observed between data and simulation for the lepton trigger, reconstruction, identification, and isolation efficiencies

Selection step	W [%]	Z [%]	Top-quark [%]	Diboson [%]
Total/GRL	-	-	-	-
Event cleaning	100.0	100.0	100.0	100.0
Trigger	41.6	43.8	48.0	48.2
$\|\eta\| < 1.37$ or $1.52 < \|\eta\| < 2.47$	98.6	99.6	99.9	99.5
Electron cleaning	99.9	100.0	100.0	100.0
$p_T > 55$ GeV	65.8	61.2	82.1	77.9
d_0 significance	99.6	99.5	98.9	99.5
Likelihood identification	94.8	93.6	92.3	93.3
Isolation	98.8	98.8	97.6	98.1
Additional electron veto	100.0	40.6	94.3	84.1
Additional muon veto	100.0	99.9	91.4	92.0
$E_T^{miss} > 55$ GeV	16.7	3.7	47.1	34.1
$m_T > 110$ GeV	49.9	37.2	30.7	49.1
Total efficiency	2.0	0.1	4.9	4.2

Table C.2 Cut efficiencies for all backgrounds in the muon channel. The efficiencies are with respect to the previous line in the table. All samples were preselected with the requirement of at least one electron or muon. The efficiency numbers shown are without accounting for the differences observed between data and simulation for the lepton trigger, reconstruction, identification, and isolation efficiencies

Selection step	W [%]	Z [%]	Top-quark [%]	Diboson [%]
Event cleaning	100.0	100.0	100.0	100.0
Trigger	39.1	40.2	43.1	41.5
$p_T > 55$ GeV	68.2	64.7	85.2	81.9
High-p_T Selection	87.2	87.4	86.5	87.2
d_0 significance	99.3	99.4	97.5	99.1
$\|z_0\| \sin(\theta)$	99.8	99.9	100.0	99.9
Isolation	99.1	99.1	96.9	98.1
Additional muon veto	100.0	30.0	93.0	81.6
Additional electron veto	100.0	99.8	92.5	92.7
$E_T^{miss} > 55$ GeV	17.9	12.1	46.9	34.4
$m_T > 110$ GeV	55.8	69.2	34.4	50.0
Total efficiency	2.0	0.5	4.2	3.3

Table C.3 Cut efficiencies for SSM W' signal in the electron channel. The efficiencies are with respect to the previous line in the table. Besides for the last line in the table, the efficiency numbers shown are without accounting for the differences observed between data and simulation for the lepton trigger, reconstruction, identification, and isolation efficiencies

Selection step	SSM W' (2 TeV) [%]	SSM W' (3 TeV) [%]	SSM W' (4 TeV) [%]	SSM W' (5 TeV) [%]
Total/GRL	-	-	-	-
Event cleaning	100.0	100.0	100.0	100.0
Trigger	92.2	91.9	89.8	86.0
$\lvert\eta\rvert < 1.37$ or $1.52 < \lvert\eta\rvert < 2.47$	99.3	99.2	99.1	99.1
Electron cleaning	99.9	99.9	99.9	100.0
$p_T > 55$ GeV	96.6	96.1	95.0	93.6
d_0 significance	99.8	99.9	99.9	99.8
Likelihood identification	95.1	94.2	93.4	93.8
Isolation	99.1	99.3	98.9	99.2
Additional electron veto	99.9	99.9	99.9	99.9
Additional muon veto	100.0	100.0	100.0	100.0
$E_T^{miss} > 55$ GeV	99.9	99.7	99.4	99.0
$m_T > 110$ GeV	99.9	99.9	99.8	99.8
Total efficiency	82.4	79.4	74.8	70.6

Table C.4 Cut efficiencies for SSM W' signal in the muon channel. The efficiencies are with respect to the previous line in the table. Besides for the last line in the table, the efficiency numbers shown are without accounting for the differences observed between data and simulation for the lepton trigger, reconstruction, identification, and isolation efficiencies

Selection step	SSM W' (2 TeV) [%]	SSM W' (3 TeV) [%]	SSM W' (4 TeV) [%]	SSM W' (5 TeV) [%]
Total/GRL	-	-	-	-
Event cleaning	100.0	100.0	100.0	100.0
Trigger	75.3	74.0	72.4	69.0
$p_T > 55$ GeV	99.6	99.4	98.9	98.3
High-p_T Selection	83.8	82.7	83.1	83.3
d_0 significance	99.8	99.8	99.8	99.8
$\lvert z_0\rvert \sin(\theta)$	99.3	99.5	99.3	99.3
Isolation	99.5	99.6	99.5	99.4
Additional muon veto	100.0	100.0	100.0	100.0
Additional electron veto	99.9	99.9	99.9	99.9
$E_T^{miss} > 55$ GeV	99.8	99.6	99.5	98.9
$m_T > 110$ GeV	99.9	99.9	99.9	99.7
Total efficiency	54.6	52.7	50.2	48.8

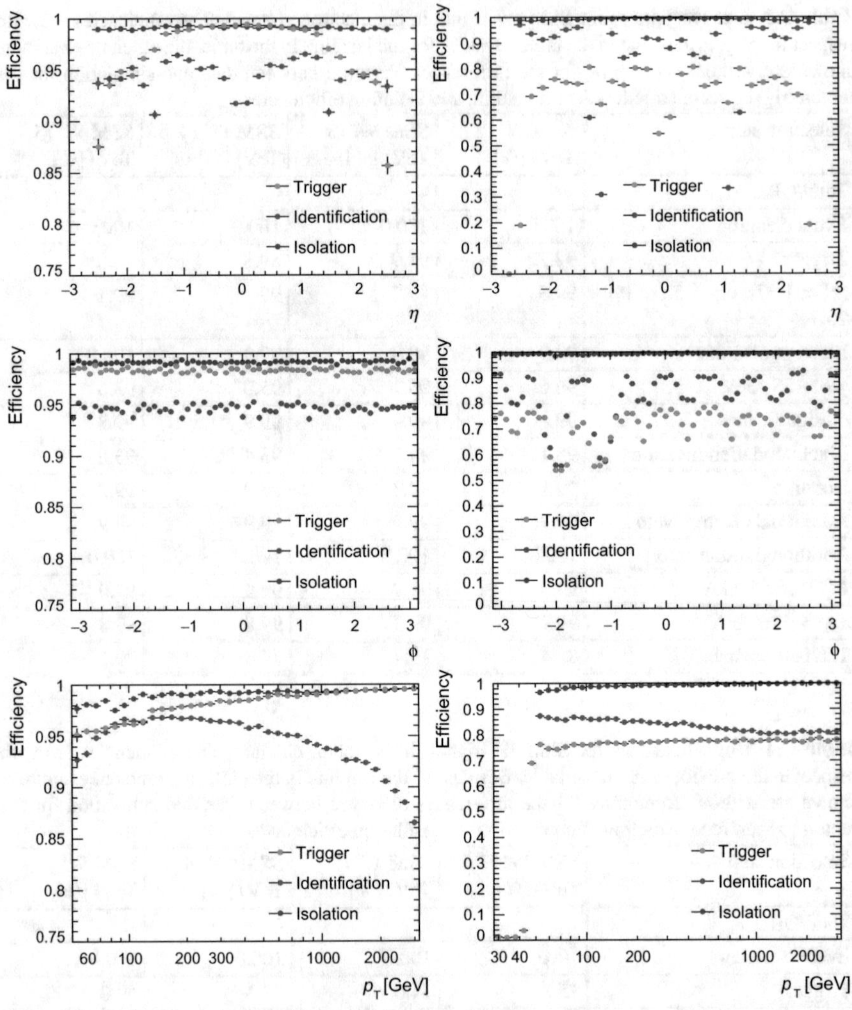

Fig. C.1 Trigger, identification, and isolation efficiencies for the electron and muon channel as a function of η (top), ϕ (middle), and p_T (bottom). The left hand side shows the efficiencies in the electron channel and the right hand side the efficiencies of the muon channel. See text for a definition of the efficiency numerator and denominator

identification efficiency is slightly lower in the central region ($|\eta| < 1.05$) due to the barrel toroid. The lowest efficiency is in the region $1.01 < |\eta| < 1.3$ due to the veto of the transition region and not well aligned muon chambers. In ϕ an efficiency modulation can be observed due to the barrel toroid coils. The two larger dips in the region $-2.0 < |\eta| < -1.0$ correspond to an inefficiency due to the structure on which the ATLAS detector is placed. The trigger efficiency is defined in different ways for the two channels. In the electron channel it is, due to the applied identification criteria, defined with respect to the final selection (number of events passing the full selection divided by number of events passing the full selection without a trigger requirement). In the muon channel the efficiency is defined with respect to all candidates (number of events that pass the trigger divided by all events). The trigger efficiency in the electron channel is around 99% in the central region, dropping to about 94% at larger values of $|\eta|$. No dependence on ϕ is observed. The efficiency is slightly rising with p_T from 95% to above 99%. A small discontinuity can be observed at $p_T \approx 120$ GeV, where an additional trigger starts to contribute. In the muon channel, the efficiency is much lower. It is around 70% in the barrel region and slightly higher in the endcap region. The same structure as for the identification efficiency can be observed in ϕ. The trigger efficiency shows a sharp turn-on at the trigger threshold of $p_T = 50$ GeV, staying afterwards flat at around 76%.

Appendix D
W' Search: Top-Quark Background Control Region

A top-quark background control region has been defined in order to check for modeling problems of the top-quark background. The top-quark and diboson backgrounds are enriched by inverting in the electron channel selection the additional muon veto. Instead of rejecting events which have an additional muon, an additional muon is required to be present in the event. This suppresses to a very high extend the W background. The dominant contribution in this control region is coming from dileptonic $t\bar{t}$ background, followed by dileptonic diboson backgrounds. Dileptonic $t\bar{t}$ is not the dominant source of the top-quark background in the signal region, but mismodeling of any relevant kinematics should be seen for both, semileptonic $t\bar{t}$ and dileptonic $t\bar{t}$. Figure D.1 shows the electron η, ϕ, $p_{\rm T}$ $E_{\rm T}^{\rm miss}$ and $m_{\rm T}$ distribution for this selection without any $E_{\rm T}^{\rm miss}$ or $m_{\rm T}$ cut applied. Figure D.2 shows the same distributions with these cuts applied. No multijet background has been estimated for this control region. The background is assumed to be negligibly small, as the probability to fake an electron and a muon simultaneously is very small. All electron kinematics are well described by the MC and show no indication for any mismodeling. Therefore no additional uncertainty arises from the modeling of the top-quark background.

M. Zinser, *Search for New Heavy Charged Bosons and Measurement of High-Mass Drell-Yan Production in Proton-Proton Collisions*, Springer Theses,
https://doi.org/10.1007/978-3-030-00650-1

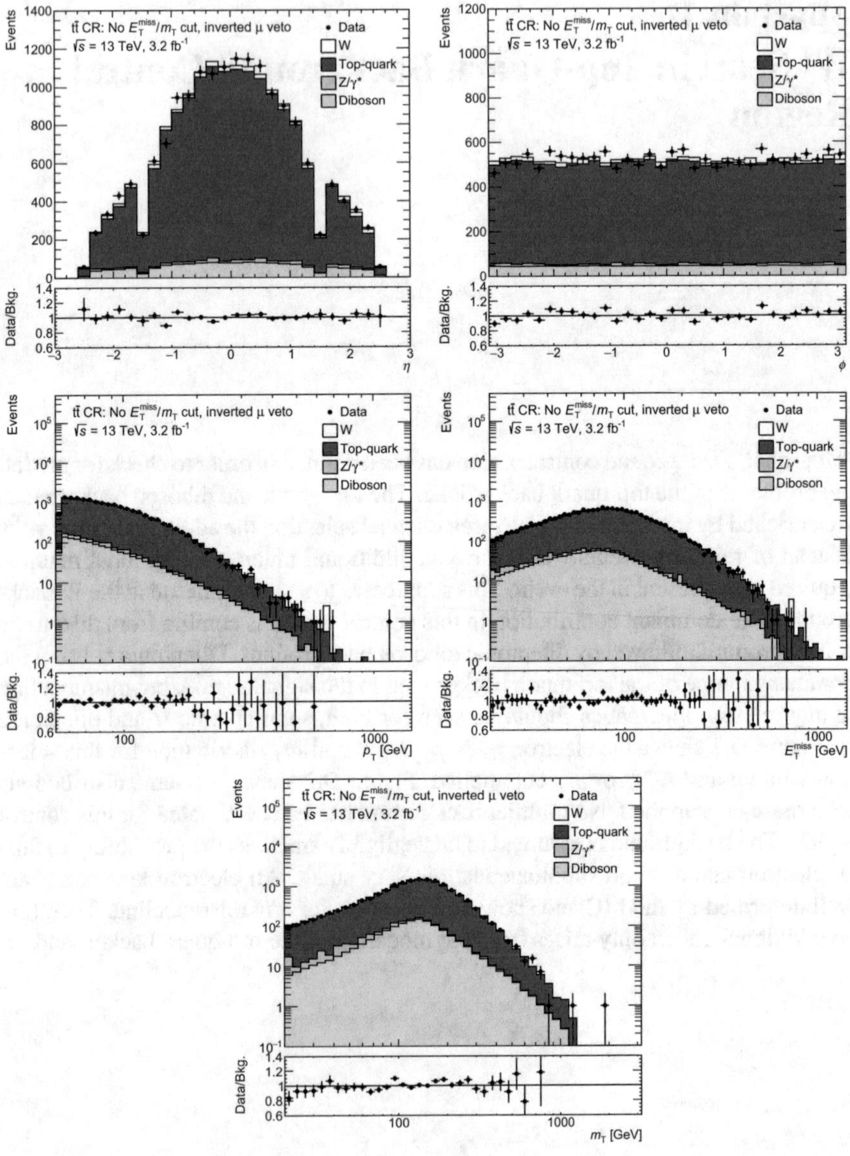

Fig. D.1 Electron η, ϕ, p_T, E_T^{miss} and missing E_T distributions in the $t\bar{t}$ control region. No E_T^{miss} or m_T requirements are applied

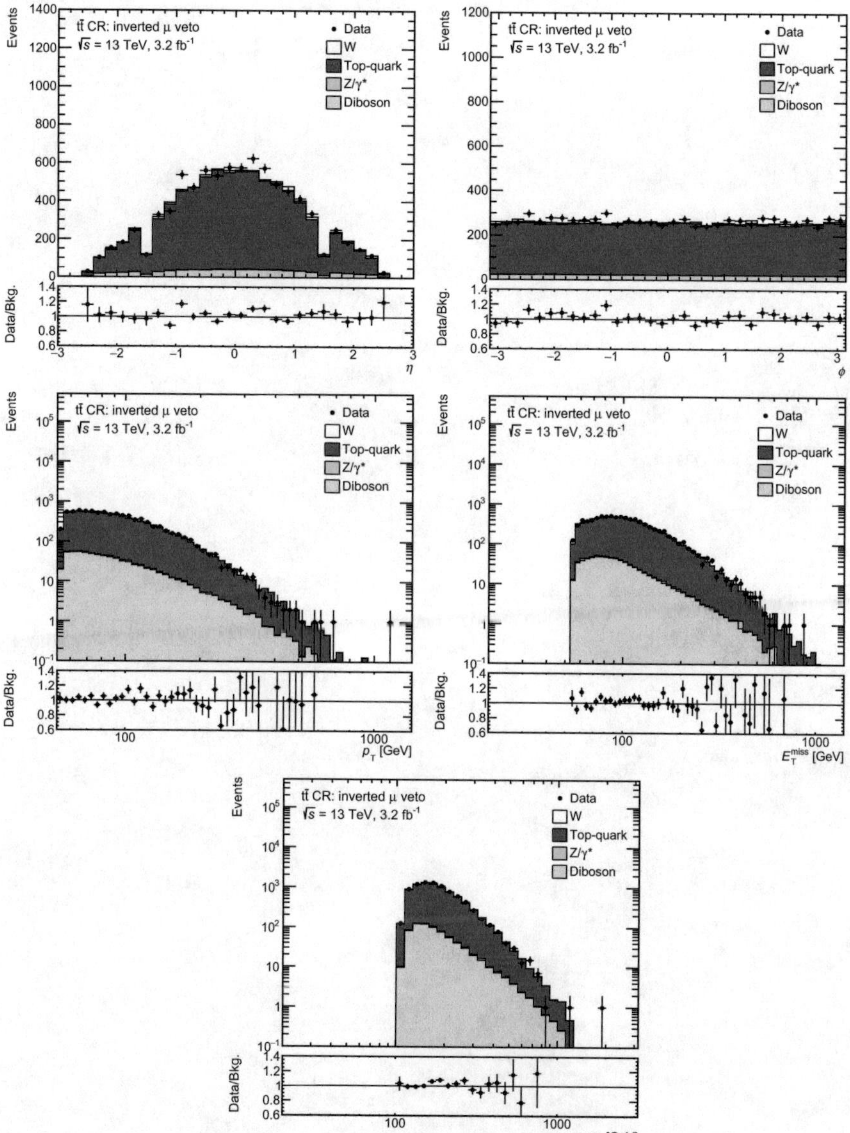

Fig. D.2 Electron η, ϕ, p_{T}, $E_{\mathrm{T}}^{\mathrm{miss}}$ and missing E_{T} distributions in the $t\bar{t}$ control region

Appendix E
W' Search: High-p_{T} Control Region

The electron p_{T} distribution in Fig. 9.33 shows a deficit of data in the region $p_{\mathrm{T}} > 400$ GeV. In this region also the contribution of the multijet background becomes large. The region $p_{\mathrm{T}} > 400$ GeV has therefore been further studied to exclude a mismodeling of the multijet background. Figure E.1 shows the η, ϕ, $E_{\mathrm{T}}^{\mathrm{miss}}$, m_{T}, $|\Delta\phi_{e,E_{\mathrm{T}}^{\mathrm{miss}}}|$ and $p_{\mathrm{T}}/E_{\mathrm{T}}^{\mathrm{miss}}$ distributions for the region $p_{\mathrm{T}} > 400$ GeV. The η and ϕ distributions look as expected. The shape of the data looks similar to the distributions in the full signal selection. The multijet background contributes to about 50% to the total background. In η or ϕ no localized deficits are visible, although some bins are clearly below data. The $E_{\mathrm{T}}^{\mathrm{miss}}$ distribution shows a maximum at low $E_{\mathrm{T}}^{\mathrm{miss}}$ values and around 400 GeV. The two maxima correspond to boosted W bosons (peak at low $E_{\mathrm{T}}^{\mathrm{miss}}$) and W bosons with a low boost (peak at 400 GeV). The multijet background is dominantly distributed at low $E_{\mathrm{T}}^{\mathrm{miss}}$ values. This behavior is expected, as the background originates from high-p_{T} jets which fake $E_{\mathrm{T}}^{\mathrm{miss}}$. The $E_{\mathrm{T}}^{\mathrm{miss}}$ is therefore most likely closer to the cut value in the analysis. The m_{T} distribution shows three distinct peaks. The first peak at m_{T} values below 200 GeV corresponds to W bosons and multijet background events in which the $E_{\mathrm{T}}^{\mathrm{miss}}$ is pointing in the same direction as the electron. The resulting transverse mass is therefore small. The second peak around 400 GeV corresponds to events with a back-to-back topology in which the $E_{\mathrm{T}}^{\mathrm{miss}}$ is still small and therefore the m_{T} is dominated by p_{T}. The third peak around 900 GeV corresponds to the off-shell W bosons which are produced with similar p_{T} and $E_{\mathrm{T}}^{\mathrm{miss}}$ values. The expected background and data are in reasonable agreement. Single bins show a deficit, but these are not in regions in which the multijet background dominates. The distributions show therefore no sign of a background mismodeling. The $|\Delta\phi_{e,E_{\mathrm{T}}^{\mathrm{miss}}}|$ and $p_{\mathrm{T}}/E_{\mathrm{T}}^{\mathrm{miss}}$ distributions support the above observations. The $p_{\mathrm{T}}/E_{\mathrm{T}}^{\mathrm{miss}}$ distribution shows a good separation between the boosted W bosons and the multijet background. Also here no evidence for any mismodeling of the multijet background is observed. One of the largest deficits is observed in the second bin of the distribution in which the W background dominates. The third bin shows again good agreement. The observed deficit is therefore expected to come from a statistical fluctuation.

M. Zinser, *Search for New Heavy Charged Bosons and Measurement of High-Mass Drell-Yan Production in Proton-Proton Collisions*, Springer Theses,
https://doi.org/10.1007/978-3-030-00650-1

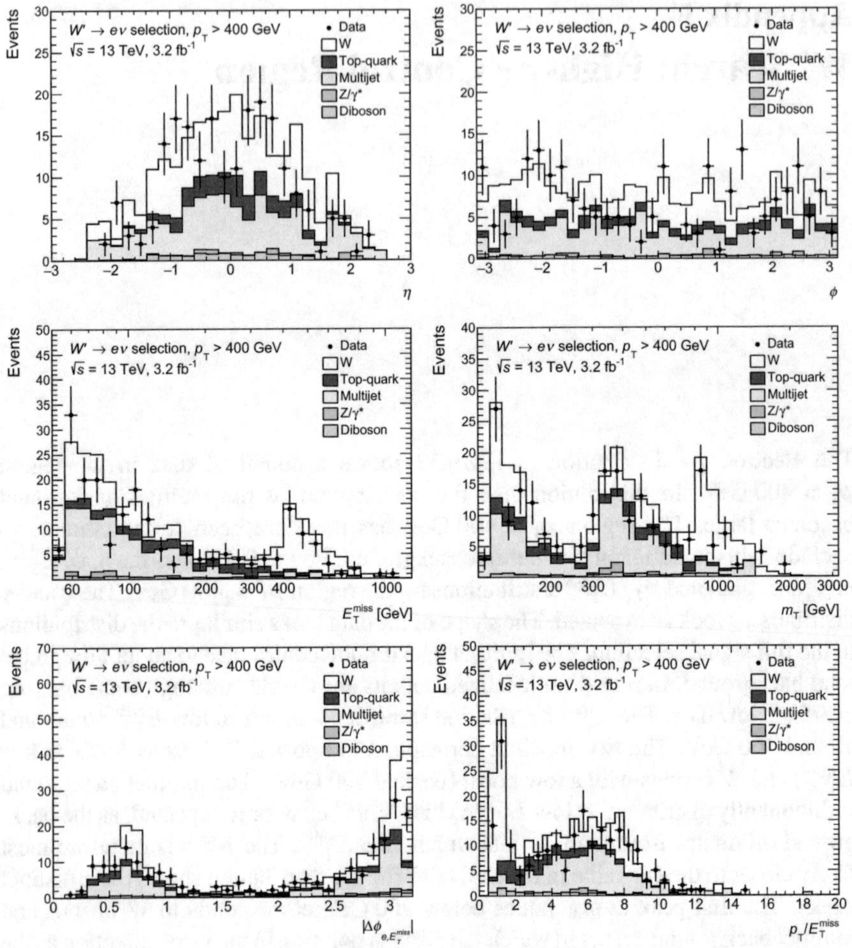

Fig. E.1 Shown are the distributions of η, ϕ, $E_{\mathrm{T}}^{\mathrm{miss}}$, m_{T}, $|\Delta\phi_{e,E_{\mathrm{T}}^{\mathrm{miss}}}|$ and $p_{\mathrm{T}}/E_{\mathrm{T}}^{\mathrm{miss}}$ for the signal selection with the requirement $p_{\mathrm{T}} > 400$ GeV

Appendix F
W' Search: Cross Section Limits

The following appendix contains detailed tables with the 95% CL cross section limits calculated with the procedure described in Sect. 10.4. Tables F.1, F.2, and F.3 contain the cross section limits for the electron, muon, and combined channel, respectively.

M. Zinser, *Search for New Heavy Charged Bosons and Measurement of High-Mass Drell-Yan Production in Proton-Proton Collisions*, Springer Theses,
https://doi.org/10.1007/978-3-030-00650-1

Table F.1 Observed and expected electron channel 95% CL limits on the cross section of a W'

$m_{W'}$ [TeV]	Exp. [pb]	Obs. [pb]	$m_{W'}$ [TeV]	Exp. [pb]	Obs. [pb]	$m_{W'}$ [TeV]	Exp. [pb]	Obs. [pb]
0.15	2.1	1.7	2.15	0.0028	0.0034	4.15	0.0024	0.0026
0.2	0.57	0.53	2.2	0.0027	0.0032	4.2	0.0025	0.0026
0.25	0.31	0.35	2.25	0.0027	0.0032	4.25	0.0025	0.0026
0.3	0.24	0.15	2.3	0.0026	0.0031	4.3	0.0025	0.0027
0.35	0.12	0.084	2.35	0.0026	0.003	4.35	0.0026	0.0028
0.4	0.079	0.086	2.4	0.0025	0.0029	4.4	0.0027	0.0028
0.45	0.059	0.062	2.45	0.0024	0.0028	4.45	0.0028	0.0029
0.5	0.046	0.056	2.5	0.0023	0.0027	4.5	0.0028	0.0029
0.55	0.037	0.054	2.55	0.0023	0.0027	4.55	0.0029	0.003
0.6	0.031	0.05	2.6	0.0023	0.0027	4.6	0.0029	0.0031
0.65	0.025	0.039	2.65	0.0023	0.0026	4.65	0.003	0.0031
0.7	0.022	0.029	2.7	0.0022	0.0025	4.7	0.0031	0.0032
0.75	0.019	0.019	2.75	0.0022	0.0025	4.75	0.0032	0.0033
0.8	0.017	0.014	2.8	0.0022	0.0025	4.8	0.0033	0.0034
0.85	0.016	0.013	2.85	0.0022	0.0024	4.85	0.0034	0.0035
0.9	0.013	0.0089	2.9	0.0021	0.0024	4.9	0.0034	0.0035
0.95	0.012	0.0062	2.95	0.0021	0.0023	4.95	0.0035	0.0037
1.0	0.01	0.0055	3.0	0.0021	0.0023	5.0	0.0036	0.0037
1.05	0.0094	0.0055	3.05	0.0021	0.0023	5.05	0.0037	0.0038
1.1	0.0086	0.0049	3.1	0.0021	0.0023	5.1	0.0039	0.0039
1.15	0.0079	0.0044	3.15	0.0021	0.0022	5.15	0.004	0.004
1.2	0.0074	0.0039	3.2	0.0021	0.0023	5.2	0.0042	0.0041
1.25	0.0066	0.0036	3.25	0.002	0.0023	5.25	0.0042	0.0043
1.3	0.006	0.0033	3.3	0.0021	0.0023	5.3	0.0044	0.0044
1.35	0.0059	0.0031	3.35	0.0021	0.0023	5.35	0.0046	0.0046
1.4	0.0055	0.0035	3.4	0.0021	0.0023	5.4	0.0048	0.0047
1.45	0.0049	0.004	3.45	0.0021	0.0023	5.45	0.0048	0.0048
1.5	0.0047	0.0043	3.5	0.0021	0.0023	5.5	0.0051	0.005
1.55	0.0044	0.0042	3.55	0.0021	0.0023	5.55	0.0051	0.0052
1.6	0.0042	0.0039	3.6	0.0021	0.0022	5.6	0.0053	0.0053
1.65	0.0041	0.0036	3.65	0.0021	0.0023	5.65	0.0054	0.0054
1.7	0.004	0.0034	3.7	0.0021	0.0023	5.7	0.0056	0.0056
1.75	0.0038	0.0034	3.75	0.0022	0.0023	5.75	0.0058	0.0057
1.8	0.0036	0.0036	3.8	0.0022	0.0023	5.8	0.0061	0.0059
1.85	0.0035	0.0041	3.85	0.0022	0.0024	5.85	0.0063	0.006
1.9	0.0034	0.0042	3.9	0.0022	0.0024	5.9	0.0065	0.0062
1.95	0.0033	0.0041	3.95	0.0023	0.0024	5.95	0.0066	0.0064
2.0	0.0032	0.004	4.0	0.0023	0.0025	6.0	0.0066	0.0064
2.05	0.003	0.0038	4.05	0.0023	0.0025			
2.1	0.0029	0.0036	4.1	0.0024	0.0025			

Table F.2 Observed and expected muon channel 95% CL limits on the cross section of a W'

$m_{W'}$ [TeV]	Exp. [pb]	Obs. [pb]	$m_{W'}$ [TeV]	Exp. [pb]	Obs. [pb]	$m_{W'}$ [TeV]	Exp. [pb]	Obs. [pb]
0.15	2.5	2.3	2.15	0.0045	0.0067	4.15	0.0042	0.0057
0.2	0.63	0.32	2.2	0.0046	0.0067	4.2	0.0042	0.0059
0.25	0.39	0.36	2.25	0.0044	0.0066	4.25	0.0043	0.006
0.3	0.23	0.24	2.3	0.0043	0.0063	4.3	0.0044	0.006
0.35	0.15	0.27	2.35	0.0043	0.0062	4.35	0.0044	0.0061
0.4	0.11	0.12	2.4	0.0042	0.0063	4.4	0.0045	0.0062
0.45	0.084	0.079	2.45	0.004	0.0061	4.45	0.0046	0.0065
0.5	0.068	0.057	2.5	0.004	0.0059	4.5	0.0048	0.0066
0.55	0.053	0.051	2.55	0.004	0.0059	4.55	0.005	0.0068
0.6	0.046	0.047	2.6	0.0039	0.0058	4.6	0.0051	0.007
0.65	0.039	0.038	2.65	0.0038	0.0057	4.65	0.0051	0.0072
0.7	0.034	0.034	2.7	0.0038	0.0056	4.7	0.0051	0.0073
0.75	0.03	0.033	2.75	0.0037	0.0055	4.75	0.0055	0.0075
0.8	0.026	0.03	2.8	0.0037	0.0055	4.8	0.0055	0.0076
0.85	0.023	0.024	2.85	0.0037	0.0054	4.85	0.0056	0.0078
0.9	0.02	0.019	2.9	0.0037	0.0054	4.9	0.0059	0.0082
0.95	0.018	0.014	2.95	0.0036	0.0054	4.95	0.0061	0.0083
1.0	0.017	0.012	3.0	0.0037	0.0053	5.0	0.0062	0.0086
1.05	0.015	0.011	3.05	0.0036	0.0053	5.05	0.0065	0.0088
1.1	0.014	0.0098	3.1	0.0036	0.0053	5.1	0.0067	0.0092
1.15	0.013	0.0099	3.15	0.0036	0.0052	5.15	0.0069	0.0093
1.2	0.012	0.0095	3.2	0.0035	0.0051	5.2	0.007	0.0096
1.25	0.011	0.0095	3.25	0.0035	0.005	5.25	0.0072	0.0098
1.3	0.01	0.0097	3.3	0.0035	0.0051	5.3	0.0073	0.01
1.35	0.0097	0.0098	3.35	0.0035	0.0051	5.35	0.0077	0.01
1.4	0.009	0.01	3.4	0.0036	0.0051	5.4	0.0077	0.011
1.45	0.0083	0.0097	3.45	0.0036	0.0051	5.45	0.0079	0.011
1.5	0.0078	0.0094	3.5	0.0035	0.0051	5.5	0.0085	0.011
1.55	0.0073	0.0094	3.55	0.0037	0.005	5.55	0.0088	0.012
1.6	0.0069	0.0095	3.6	0.0036	0.0051	5.6	0.0091	0.012
1.65	0.0067	0.0092	3.65	0.0036	0.0051	5.65	0.009	0.012
1.7	0.0064	0.009	3.7	0.0037	0.0051	5.7	0.0096	0.013
1.75	0.0062	0.0088	3.75	0.0037	0.0052	5.75	0.0096	0.013
1.8	0.0061	0.0085	3.8	0.0037	0.0052	5.8	0.0097	0.013
1.85	0.0058	0.0083	3.85	0.0038	0.0053	5.85	0.01	0.014
1.9	0.0055	0.0081	3.9	0.0038	0.0053	5.9	0.01	0.014
1.95	0.0053	0.0078	3.95	0.0039	0.0054	5.95	0.01	0.014
2.0	0.0052	0.0077	4.0	0.004	0.0055	6.0	0.011	0.015
2.05	0.0051	0.0072	4.05	0.0041	0.0056			
2.1	0.0047	0.0069	4.1	0.004	0.0056			

Table F.3 Observed and expected combined channel 95% CL limits on the cross section of a W'

$m_{W'}$ [TeV]	Exp. [pb]	Obs. [pb]	$m_{W'}$ [TeV]	Exp. [pb]	Obs. [pb]	$m_{W'}$ [TeV]	Exp. [pb]	Obs. [pb]
0.15	1.7	1.4	2.15	0.0021	0.0031	4.15	0.0017	0.002
0.2	0.43	0.26	2.2	0.002	0.003	4.2	0.0017	0.0021
0.25	0.25	0.26	2.25	0.002	0.0029	4.25	0.0017	0.0021
0.3	0.14	0.12	2.3	0.002	0.0028	4.3	0.0018	0.0021
0.35	0.093	0.11	2.35	0.0019	0.0027	4.35	0.0018	0.0022
0.4	0.065	0.07	2.4	0.0019	0.0026	4.4	0.0018	0.0022
0.45	0.049	0.048	2.45	0.0018	0.0025	4.45	0.0019	0.0023
0.5	0.037	0.04	2.5	0.0018	0.0025	4.5	0.0019	0.0023
0.55	0.031	0.04	2.55	0.0017	0.0024	4.55	0.002	0.0024
0.6	0.026	0.038	2.6	0.0017	0.0024	4.6	0.002	0.0025
0.65	0.021	0.029	2.65	0.0017	0.0023	4.65	0.0021	0.0025
0.7	0.018	0.022	2.7	0.0016	0.0022	4.7	0.0021	0.0026
0.75	0.016	0.016	2.75	0.0016	0.0022	4.75	0.0022	0.0026
0.8	0.014	0.013	2.8	0.0016	0.0022	4.8	0.0022	0.0027
0.85	0.012	0.011	2.85	0.0016	0.0021	4.85	0.0023	0.0027
0.9	0.011	0.0073	2.9	0.0015	0.0021	4.9	0.0024	0.0028
0.95	0.0095	0.0049	2.95	0.0015	0.0021	4.95	0.0024	0.0029
1.0	0.0086	0.0042	3.0	0.0015	0.002	5.0	0.0025	0.003
1.05	0.0079	0.0041	3.05	0.0015	0.002	5.05	0.0026	0.0031
1.1	0.007	0.0037	3.1	0.0015	0.002	5.1	0.0027	0.0032
1.15	0.0065	0.0033	3.15	0.0015	0.0019	5.15	0.0028	0.0033
1.2	0.006	0.0031	3.2	0.0014	0.0019	5.2	0.0029	0.0034
1.25	0.0056	0.003	3.25	0.0014	0.0019	5.25	0.0029	0.0035
1.3	0.005	0.0028	3.3	0.0014	0.0019	5.3	0.0031	0.0036
1.35	0.0047	0.0028	3.35	0.0015	0.0019	5.35	0.0032	0.0037
1.4	0.0044	0.0032	3.4	0.0015	0.0019	5.4	0.0033	0.0039
1.45	0.0041	0.0038	3.45	0.0014	0.0019	5.45	0.0035	0.004
1.5	0.0039	0.0039	3.5	0.0015	0.0019	5.5	0.0035	0.0041
1.55	0.0036	0.0039	3.55	0.0014	0.0019	5.55	0.0036	0.0042
1.6	0.0034	0.0038	3.6	0.0014	0.0019	5.6	0.0038	0.0043
1.65	0.0032	0.0035	3.65	0.0015	0.0019	5.65	0.0039	0.0045
1.7	0.0031	0.0034	3.7	0.0015	0.0019	5.7	0.004	0.0047
1.75	0.0029	0.0034	3.75	0.0015	0.0019	5.75	0.0042	0.0048
1.8	0.0029	0.0035	3.8	0.0015	0.0019	5.8	0.0043	0.005
1.85	0.0027	0.0038	3.85	0.0015	0.0019	5.85	0.0044	0.0052
1.9	0.0026	0.0039	3.9	0.0015	0.0019	5.9	0.0046	0.0053
1.95	0.0025	0.0038	3.95	0.0016	0.0019	5.95	0.0047	0.0054
2.0	0.0024	0.0037	4.0	0.0016	0.0019	6.0	0.0047	0.0056
2.05	0.0023	0.0034	4.05	0.0016	0.002			
2.1	0.0021	0.0032	4.1	0.0016	0.002			

Appendix G
High-Mass Drell-Yan: Quark Distributions

This appendix contains the fractional flavor contribution to the high-mass Drell-Yan measurement as a function of absolute rapidity and the Bjorken-x distribution of the quarks in different ranges of invariant mass and absolute rapidity. The distributions were obtained from the Drell-Yan MC sample described in Sect. 14.1. Figures G.1 and G.2 show the fractional flavor contribution as a function of absolute rapidity in different invariant mass ranges for the quarks and anti-quarks, respectively. Figure G.3 shows the Bjorken-x distribution of the quarks and anti-quarks in three different regions of invariant mass. Figures G.4 and G.5 show the same distributions for a low and high absolute rapidity region in the first and last bin of the two dimensional measurement, respectively.

M. Zinser, *Search for New Heavy Charged Bosons and Measurement of High-Mass Drell-Yan Production in Proton-Proton Collisions*, Springer Theses,
https://doi.org/10.1007/978-3-030-00650-1

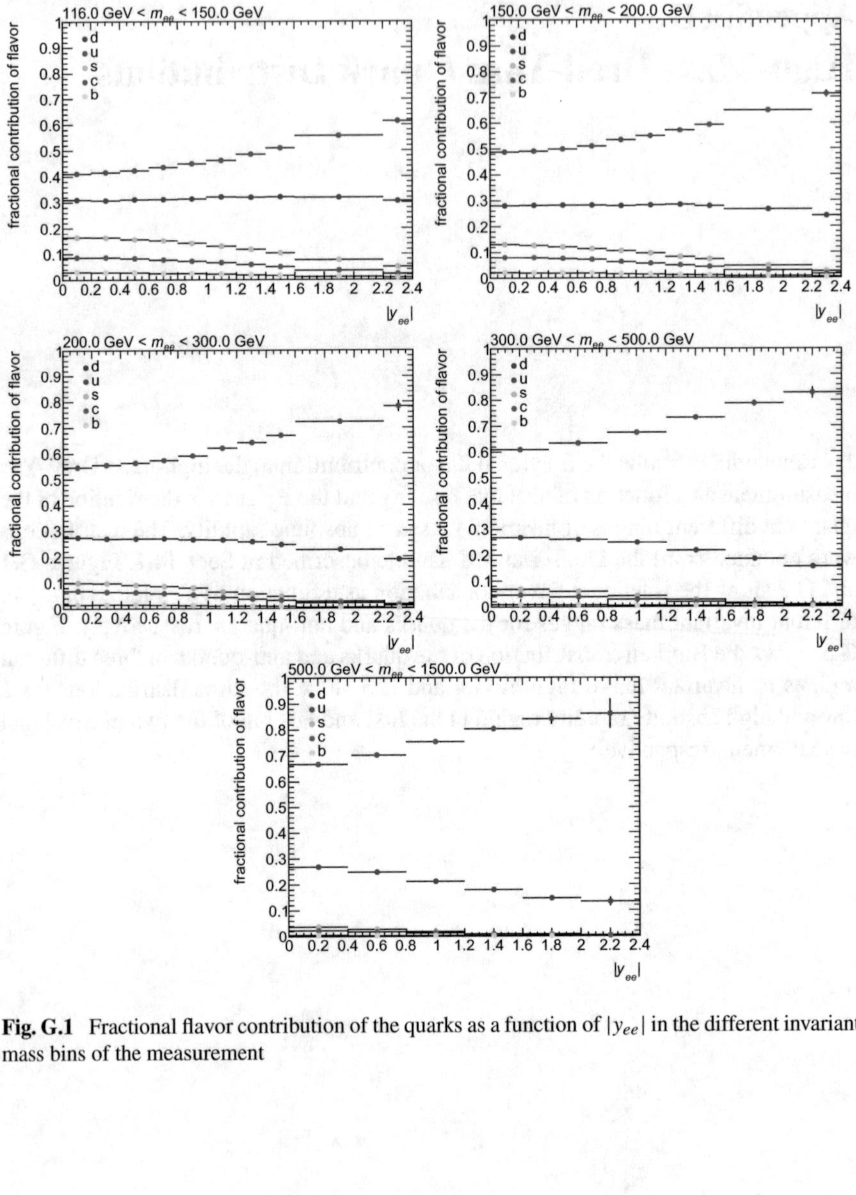

Fig. G.1 Fractional flavor contribution of the quarks as a function of $|y_{ee}|$ in the different invariant mass bins of the measurement

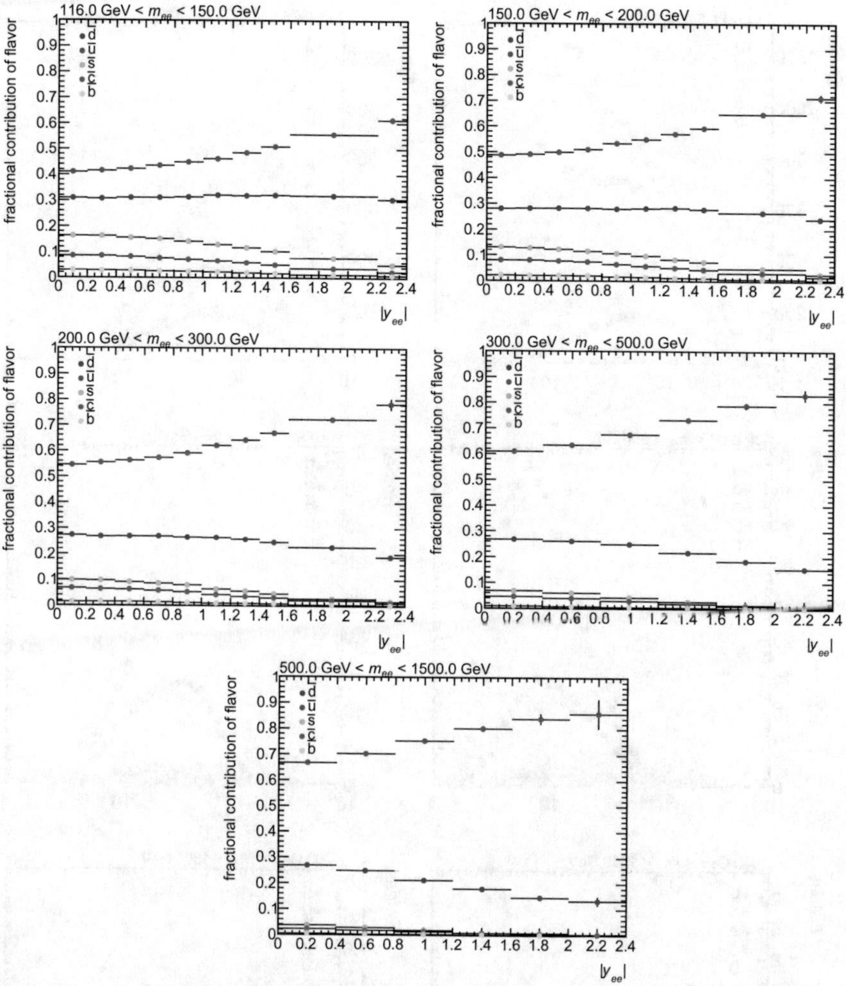

Fig. G.2 Fractional flavor contribution of the anti-quarks as a function of $|y_{ee}|$ in the different invariant mass bins of the measurement

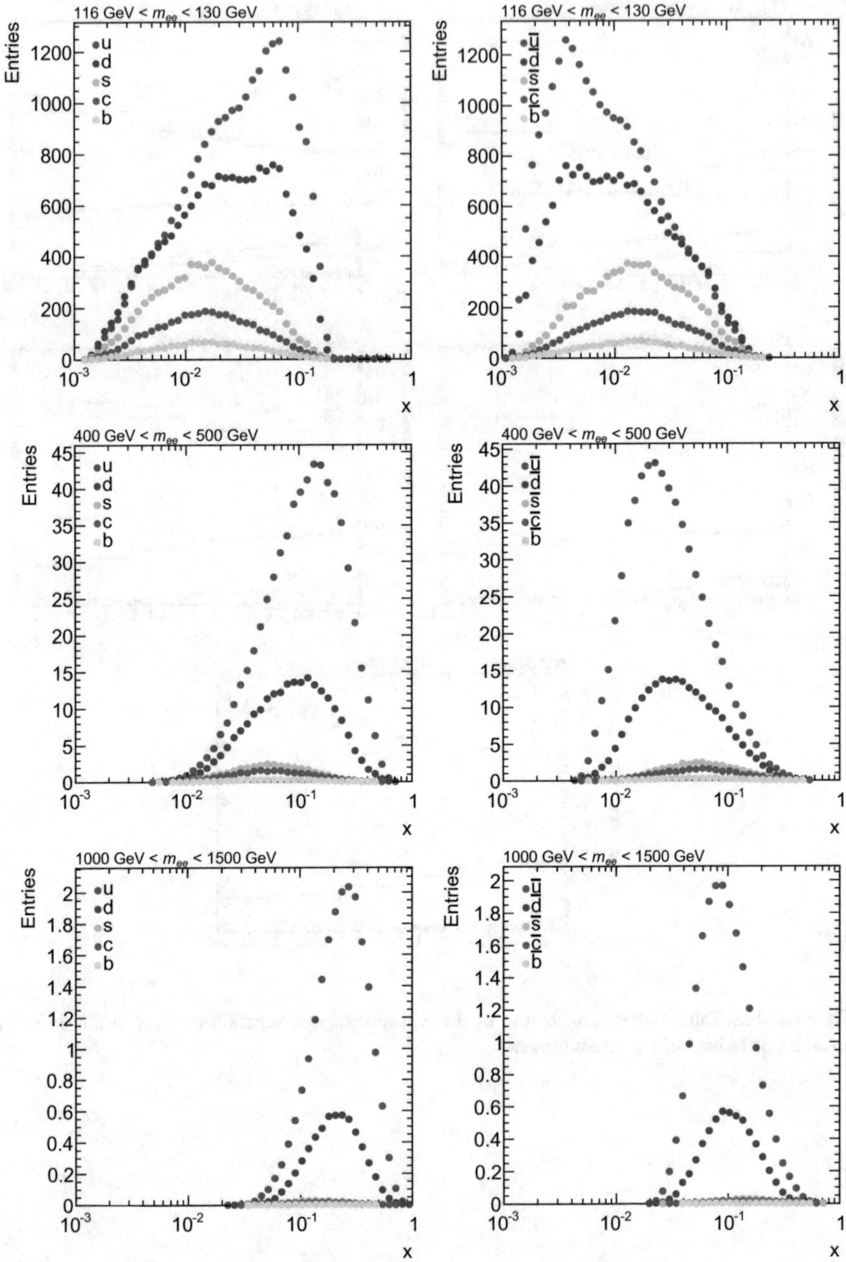

Fig. G.3 Bjorken-x distribution separated for flavors in different ranges of invariant mass m_{ee}

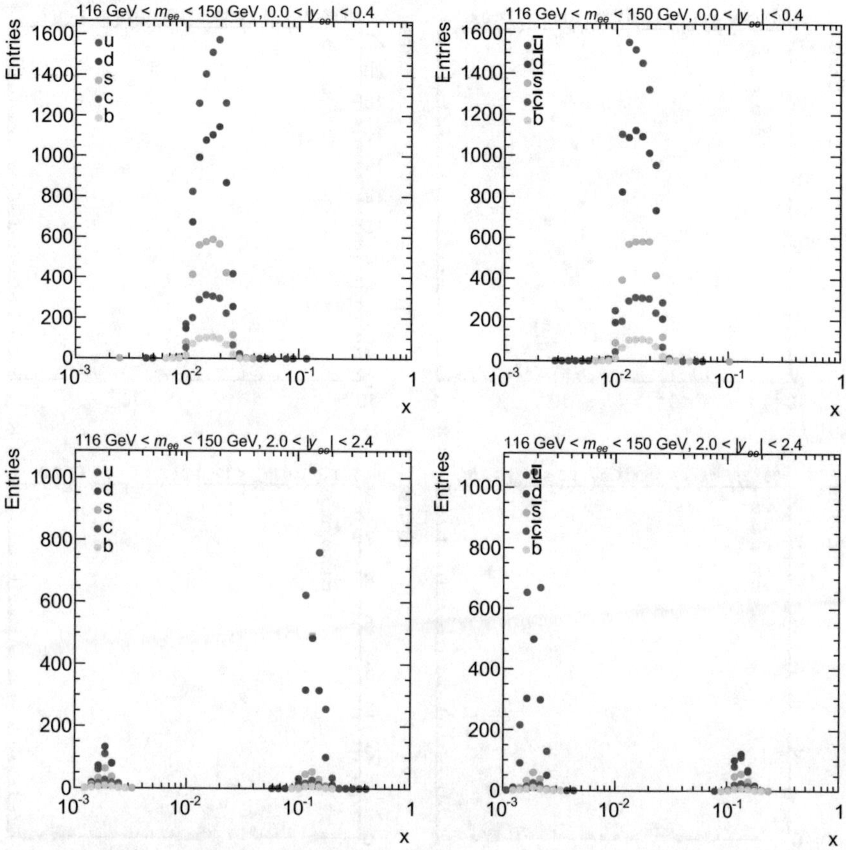

Fig. G.4 Bjorken-x distribution separated for flavors in different ranges of absolute rapidity in the invariant mass range 116 GeV $< m_{ee} <$ 150 GeV

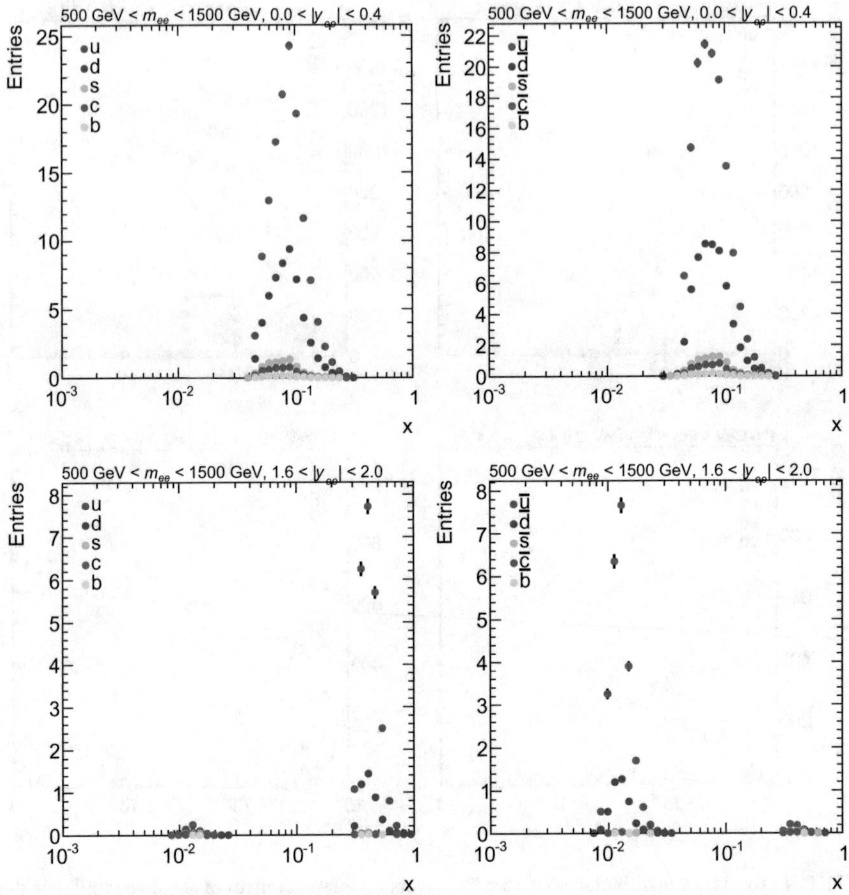

Fig. G.5 Bjorken-x distribution separated for flavors in different ranges of absolute rapidity in the invariant mass range 500 GeV $< m_{ee} <$ 1500 GeV

Appendix H
High-Mass Drell-Yan: Detailed Information About MC Samples

The following appendix provides detailed tables with information about the Monte Carlo samples used in the high-mass Drell-Yan analysis. The Monte Carlo samples are described in detail in Sect. 14.1. Tables H.1 and H.2 list information about the nominal Drell-Yan Monte Carlo signal samples and the samples used for generator uncertainty studies. Table H.3 lists informations about the signal samples simulating the photon induced process. The background samples are listed in Tables H.4 and H.5. Finally, information about the W samples, which are only used for studies in Sect. 14.3, is listed in Table H.6.

M. Zinser, *Search for New Heavy Charged Bosons and Measurement of High-Mass Drell-Yan Production in Proton-Proton Collisions*, Springer Theses,
https://doi.org/10.1007/978-3-030-00650-1

Table H.1 Drell-Yan POWHEG-PYTHIA8 [1, 2] Monte Carlo samples used in the dielectron channel of the analysis. The CT10 PDF set [3] is used and the AU2 tune [4]. The first column gives the mass range in which the Drell-Yan process was simulated, the second the internal ATLAS run number. For each sample the cross section times branching ratio with which the Powheg generator produced, the efficiency ϵ_F with which the sample was filtered and the number of produced events are given. In last column, the integrated luminosity $L_{MC} = N_{evt}/(\epsilon_F \sigma Br)$ of each sample is given

Signature	m_{ee} [GeV]	MC run number	σBr [pb] POWHEG	ϵ_F [%]	N_{evt} [k]	L_{MC} [fb^{-1}]
$Z \to ee$	60-	129680	1.1099E+3	55.65	50000	81
$Z \to ee$	60-	129685	1.1099E+3	31.47	10000	29
$Z \to ee$	60-	129696	1.1099E+3	12.89	3000	21
$Z \to ee$	120-180	129504	9.8460E-00	100.0	5000	508
$Z \to ee$	180-250	129505	1.5710E-00	100.0	1000	637
$Z \to ee$	250-400	129506	5.4920E-01	100.0	600	1093
$Z \to ee$	400-600	129507	8.9660E-02	100.0	400	4461
$Z \to ee$	600-800	129508	1.5100E-02	100.0	100	6623
$Z \to ee$	800-1000	129509	3.7500E-03	100.0	100	26667
$Z \to ee$	1000-1250	129510	1.2930E-03	100.0	100	77340
$Z \to ee$	1250-1500	129511	3.5770E-04	100.0	100	279564
$Z \to ee$	1500-1750	129512	1.1230E-04	100.0	100	890472
$Z \to ee$	1750-2000	129513	3.8380E-05	100.0	100	2605524
$Z \to ee$	2000-2250	129514	1.3890E-05	100.0	100	7199424
$Z \to ee$	2250-2500	129515	5.2260E-06	100.0	100	19135094
$Z \to ee$	2500-2750	129516	2.0170E-06	100.0	100	49578582
$Z \to ee$	2750-3000	129517	7.8910E-07	100.0	100	126726651
$Z \to ee$	3000-	129518	5.0390E-07	100.0	100	198452074

Table H.2 Drell-Yan MC@NLO-HERWIG++ [5, 6] Monte Carlo samples used in the dielectron channel of the analysis. The CT10 PDF set [3] is used and the UE-EE3 tune [6]. The first column gives the mass range in which the Drell-Yan process was simulated, the second the internal ATLAS run number. For each sample the cross section times branching ratio with which the MC@NLO generator produced, the efficiency ϵ_F with which the sample was filtered and the number of produced events are given. In last column, the integrated luminosity $L_{MC} = N_{evt}/(\epsilon_F \sigma Br)$ of each sample is given

Signature	m_{ee} [GeV]	MC run number	σBr [pb] MC@NLO	ϵ_F [%]	N_{evt} [k]	L_{MC} [fb^{-1}]
$Z \to ee$	60-	129766	1.1966E+03	92.39	9999.886	8.35692E+00
$Z \to ee$	120-180	129714	9.9056E-00	100.0	4994.787	5.04239E+02
$Z \to ee$	180-250	129715	1.5959E-00	100.0	999.995	6.26603E+02
$Z \to ee$	250-400	129716	5.6252E-01	100.0	599.998	1.06663E+03
$Z \to ee$	400-600	129717	9.2863E-02	100.0	399.998	4.30740E+03
$Z \to ee$	600-800	129718	1.5774E-02	100.0	99.998	6.33942E+03
$Z \to ee$	800-1000	129719	3.9426E-03	100.0	99.997	2.53632E+04
$Z \to ee$	1000-1250	129720	1.3659E-03	100.0	99.997	7.32096E+04
$Z \to ee$	1250-1500	129721	3.7982E-04	100.0	99.999	2.63280E+05
$Z \to ee$	1500-1750	129722	1.1976E-04	100.0	99.998	8.34987E+05
$Z \to ee$	1750-2000	129723	4.1064E-05	100.0	99.998	2.43517E+06
$Z \to ee$	2000-2250	129724	1.4900E-05	100.0	99.993	6.71094E+06
$Z \to ee$	2250-2500	129725	5.6224E-06	100.0	99.998	1.77856E+07
$Z \to ee$	2500-2750	129726	2.1736E-06	100.0	99.995	4.60043E+07
$Z \to ee$	2750-3000	129727	8.5247E-07	100.0	99.994	1.17299E+08
$Z \to ee$	3000-	129728	5.4570E-07	100.0	99.993	1.83238E+08

Table H.3 Photon induced PYTHIA8 [2] Monte Carlo samples used in the dielectron channel of the analysis. The MRST2004qed PDF set [7] is used and the 4C tune [8]. The first column gives the mass range in which the photon induced process was simulated, the second the internal ATLAS run number. For each sample the cross section times branching ratio with which the PYTHIA generator produced, the efficiency ϵ_F with which the sample was filtered and the number of produced events are given. In last column, the integrated luminosity $L_{MC} = N_{evt}/(\epsilon_F \sigma Br)$ of each sample is given

Signature	m_{ee} [GeV]	MC run number	σBr [pb] POWHEG	ϵ_F [%]	N_{evt} [k]	L_{MC} [fb^{-1}]
$\gamma\gamma \to ee$	60-200	129652	2.6976E-00	100.0	500	185
$\gamma\gamma \to ee$	200-600	129653	1.2184E-01	100.0	200	1642
$\gamma\gamma \to ee$	600-1500	129654	3.4933E-03	100.0	100	28626
$\gamma\gamma \to ee$	1500-2500	129655	5.8593E-05	100.0	100	1706689
$\gamma\gamma \to ee$	2500-	129656	2.2978E-06	100.0	100	43519889

Table H.4 Top MC@NLO- HERWIG++ [5, 6] and POWHEG- PYTHIA8 [1, 2] Monte Carlo samples used in the analysis. The CT10 PDF set [3] is used and the AUET2 tune [9]. The first column gives the internal ATLAS run number. For each sample the cross section times branching ratio with which the generator produced the sample. Also given is σBr at NNLO which was used for the normalization, the efficiency ϵ_F with which the sample was filtered and the number of produced events. In last column, the integrated luminosity $L_{MC} = N_{evt}/(\epsilon_F \sigma Br)$ of each sample is given

Signature	MC run number	σBr [pb]		ϵ_F [%]	N_{evt} [k]	L_{MC} [fb^{-1}]
		MC@NLO/ POWHEG	NNLO			
$t\bar{t} \to \ell X$	105200	208.13	252.89	54.26	28747	256
$t\bar{t} \to \ell X$	110404	210.84	252.89	54.30	50000	441
$Wt \to X$	108346	20.67	22.37	100.00	2000	97

Table H.5 Diboson HERWIG [10] Monte Carlo samples used in the analysis. The CTEQ6L1 PDF set [11] is used and AUET2 tune [9]. The first column gives the mass range in which the diboson processes were simulated, the second the internal ATLAS run number. For each sample the cross section times branching ratio with which the HERWIG generator produced the sample. Also given is σBr at NLO which was used for the normalization, the efficiency ϵ_F with which the sample was filtered and the number of produced events. In last column, the integrated luminosity $L_{MC} = N_{evt}/(\epsilon_F \sigma Br)$ of each sample is given. † Note that the selection on m_{ee} given in this table applies to the two highest p_T leptons in the event at the truth Born level

Signature	$m_{ee}^{\dagger}$ [GeV]	MC run number	σBr [pb]		ϵ_F [%]	N_{evt} [k]	L_{MC} [fb^{-1}]
			HERWIG	NLO			
$WW \to eX$		105985	32.501	70.4	38.21	2500	201
$ZZ \to eX$		105986	4.6914	7.2	21.17	245	252
$WZ \to eX$		105987	12.009	20.3	30.55	1000	273
$WW \to e\nu e\nu$	400–1000	180451	0.37892	0.8207	0.72	10	37701
$WW \to e\nu e\nu$	1000-	180452	0.37895	0.8207	0.01	10	263887
$ZZ \to ee$	400–1000	180455	0.34574	0.5307	0.13	10	22249
$ZZ \to ee$	1000-	180456	0.34574	0.5307	0.003	10	997361
$WZ \to ee$	400-1000	180453	0.46442	0.7853	0.31	10	6975
$WZ \to ee$	1000-	180454	0.46442	0.7853	0.011	10	188879

Table H.6 *W* POWHEG- PYTHIA8 [1, 2] Monte Carlo samples used in the analysis. The CT10 PDF set [3] is used and the AU2 tune [4]. The first column gives the internal ATLAS run number. For each sample the cross section times branching ratio with which the POWHEG generator produced the sample. Also given is σBr at NNLO which was used for the normalization and the number of produced events. In last column, the integrated luminosity $L_{MC} = N_{evt}/(\sigma Br)$ of each sample is given

Signature	MC run number	σBr [pb]		N_{evt} [k]	L_{MC} [fb^{-1}]
		POWHEG	NNLO		
$W^+ \to e\nu$	147800	6891.0	7073.8	23000	3.25
$W^- \to e\nu$	147803	4790.2	5016.2	17000	3.39

References

1. Alioli S, Nason P, Oleari C, Re E (2010) A general framework for implementing NLO calculations in shower Monte Carlo programs: the POWHEG BOX. JHEP 06:043. https://doi.org/10.1007/JHEP06(2010)043, arXiv:1002.2581 [hep-ph]
2. Sjöstrand T, Mrenna S, Skands PZ (2008) A brief introduction to PYTHIA 8.1. Comput Phys Commun 178:852. https://doi.org/10.1016/j.cpc.2008.01.036, arXiv:0710.3820 [hep-ph]
3. Lai H-L et al (2010) New parton distributions for collider physics. Phys Rev D 82:074024. https://doi.org/10.1103/PhysRevD.82.074024, arXiv:1007.2241 [hep-ph]
4. Summary of ATLAS Pythia 8 tunes (2012) Technical report ATL-PHYS-PUB-2012-003. Geneva: CERN. https://cds.cern.ch/record/1474107
5. Frixione S, Webber BR (2002) Matching NLO QCD computations and parton shower simulations. JHEP 06:029. https://doi.org/10.1088/1126-6708/2002/06/029, arXiv:hep-ph/0204244 [hep-ph]
6. Bahr M et al (2008) Herwig++ physics and manual. Eur Phys J C 58:639. https://doi.org/10.1140/epjc/s10052-008-0798-9, arXiv:0803.0883 [hep-ph]
7. Martin AD et al (2005) Parton distributions incorporating QED contributions. Eur Phys J C 39:155. https://doi.org/10.1140/epjc/s2004-02088-7, arXiv:hep-ph/0411040 [hep-ph]
8. Corke R, Sjostrand T (2011) Interleaved parton showers and tuning prospects. JHEP 03:032. https://doi.org/10.1007/JHEP03(2011)032, arXiv:1011.1759 [hep-ph]
9. New ATLAS event generator tunes to 2010 data (2011) Technical report ATL-PHYS-PUB-2011-008. Geneva: CERN. https://cds.cern.ch/record/1345343
10. Corcella G et al (2001) HERWIG 6: an event generator for hadron emission reactions with interfering gluons (including supersymmetric processes). JHEP 01:010. https://doi.org/10.1088/1126-6708/2001/01/010, arXiv:hep-ph/0011363 [hep-ph]
11. Pumplin J et al (2002) New generation of parton distributions with uncertainties from global QCD analysis. JHEP 0207:012. https://doi.org/10.1088/1126-6708/2002/07/012, arXiv:hep-ph/0201195 [hep-ph]

Appendix I
High-Mass Drell-Yan: Event Yield Table

Table I.1 shows the events passing the electron selection of the high-mass Drell-Yan analysis. The selection is described in more detail in Sect. 14.2. Shown are also the statistical uncertainties and systematic uncertainties of the backgrounds. A detailed description of the systematic uncertainties can be found in Sect. 15.3.

M. Zinser, *Search for New Heavy Charged Bosons and Measurement of High-Mass Drell-Yan Production in Proton-Proton Collisions*, Springer Theses,
https://doi.org/10.1007/978-3-030-00650-1

Table I.1 Number of selected events from all estimated processes in bins of invariant mass. Shown are data, expected signal and background. Given are also statistical and systematic uncertainties

m_{ee}^{min}-m_{ee}^{max}	Data	Drell-Yan	Photon induced	Top-quark
66–116	4428035 ± 2104(stat.)	4245053 ± 1052(stat.)±184124(syst.)	1108 ± 9(stat.)±443(syst.)	6543 ± 23(stat.)±392(syst.)
116–130	43428 ± 208(stat.)	38039 ± 63(stat.)±1633(syst.)	261 ± 4(stat.)±104(syst.)	1766 ± 12(stat.)±106(syst.)
130–150	30423 ± 174(stat.)	25561 ± 32(stat.)±1092(syst.)	271 ± 4(stat.)±108(syst.)	2136 ± 13(stat.)±128(syst.)
150–175	19905 ± 141(stat.)	15495 ± 25(stat.)±659(syst.)	226 ± 4(stat.)±90(syst.)	1981 ± 13(stat.)±118(syst.)
175–200	11038 ± 105(stat.)	8190 ± 17(stat.)±347(syst.)	152 ± 3(stat.)±60(syst.)	1370 ± 11(stat.)±82(syst.)
200–230	7492 ± 86(stat.)	5489 ± 13(stat.)±232(syst.)	116 ± 1(stat.)±46(syst.)	1080 ± 9(stat.)±64(syst.)
230–260	4407 ± 66(stat.)	3121 ± 9(stat.)±131(syst.)	71 ± 0(stat.)±28(syst.)	691 ± 7(stat.)±41(syst.)
260–300	3320 ± 57(stat.)	2349 ± 7(stat.)±99(syst.)	58 ± 0(stat.)±23(syst.)	520 ± 6(stat.)±31(syst.)
300–380	2791 ± 52(stat.)	2096 ± 6(stat.)±89(syst.)	57 ± 0(stat.)±22(syst.)	465 ± 6(stat.)±27(syst.)
380–500	1340 ± 36(stat.)	1024 ± 2(stat.)±44(syst.)	30 ± 0(stat.)±12(syst.)	190 ± 4(stat.)±11(syst.)
500–700	533 ± 23(stat.)	423 ± 1(stat.)±19(syst.)	14 ± 0(stat.)±5(syst.)	48 ± 2(stat.)±2(syst.)
700–1000	136 ± 11(stat.)	112 ± 0(stat.)±5(syst.)	4 ± 0(stat.)±1(syst.)	7 ± 0(stat.)±0(syst.)
1000–1500	27 ± 5(stat.)	22 ± 0(stat.)±1(syst.)	1 ± 0(stat.)±0(syst.)	1 ± 0(stat.)±0(syst.)
m_{ee}^{min}-m_{ee}^{max}	Multijet	WW	WZ	ZZ
66–116	18197 ± 85(stat.)±8332(syst.)	1337 ± 20(stat.)±133(syst.)	3537 ± 24(stat.)±141(syst.)	2493 ± 20(stat.)±104(syst.)
116–130	1088 ± 13(stat.)±187(syst.)	346 ± 10(stat.)±34(syst.)	105 ± 4(stat.)±4(syst.)	46 ± 2(stat.)±1(syst.)
130–150	1106 ± 13(stat.)±183(syst.)	426 ± 11(stat.)±42(syst.)	106 ± 4(stat.)±4(syst.)	41 ± 2(stat.)±1(syst.)
150–175	872 ± 11(stat.)±137(syst.)	362 ± 10(stat.)±36(syst.)	97 ± 4(stat.)±3(syst.)	35 ± 2(stat.)±1(syst.)
175–200	562 ± 9(stat.)±83(syst.)	263 ± 9(stat.)±26(syst.)	72 ± 3(stat.)±2(syst.)	23 ± 2(stat.)±0(syst.)

(continued)

Table I.1 (continued)

m_{ee}^{min}-m_{ee}^{max}	Data	Drell-Yan	Photon induced	Top-quark
200–230	433 ± 8(stat.)±69(syst.)	191 ± 7(stat.)±19(syst.)	53 ± 3(stat.)±2(syst.)	17 ± 1(stat.)±0(syst.)
230–260	262 ± 6(stat.)±38(syst.)	121 ± 6(stat.)±12(syst.)	35 ± 2(stat.)±1(syst.)	12 ± 1(stat.)±0(syst.)
260–300	198 ± 5(stat.)±32(syst.)	116 ± 5(stat.)±11(syst.)	34 ± 2(stat.)±1(syst.)	9 ± 1(stat.)±0(syst.)
300–380	149 ± 4(stat.)±24(syst.)	97 ± 5(stat.)±9(syst.)	31 ± 2(stat.)±1(syst.)	8 ± 1(stat.)±0(syst.)
380–500	68 ± 2(stat.)±14(syst.)	46 ± 2(stat.)±4(syst.)	19 ± 1(stat.)±0(syst.)	7 ± 0(stat.)±0(syst.)
500–700	21 ± 1(stat.)±3(syst.)	18 ± 0(stat.)±1(syst.)	9 ± 0(stat.)±0(syst.)	3 ± 0(stat.)±0(syst.)
700–1000	5 ± 0(stat.)±0(syst.)	4 ± 0(stat.)±0(syst.)	3 ± 0(stat.)±0(syst.)	0 ± 0(stat.)±0(syst.)
1000–1500	0 ± 0(stat.)±0(syst.)	0 ± 0(stat.)±0(syst.)	1 ± 0(stat.)±0(syst.)	0 ± 0(stat.)±0(syst.)

Appendix J
High-Mass Drell-Yan: Fake Efficiency Measurement

The fake efficiency $f = N_{\text{tight}}^{\text{fake}} / N_{\text{loose}}^{\text{fake}}$ and fake factor $F_i^{FTM} = \frac{N_{tight}^{fake}}{N_{fail\,track\,match}^{fake}}$ cannot be reliably calculated with simulation and therefore need to be measured from data. Three different methods are performed which aim to obtain a jet enriched control region in which the fake efficiencies and fake factors can be calculated. Two of the methods use single jet triggers and one the same trigger as for the signal selection. The different methods are discussed and the resulting fake efficiencies and fake factors are compared to each other.

J.1 Single Object Method

The default method is based on objects which have been recorded by single jet triggers. Jets appear very often in a hadron collider. Hence, it is not possible to record very event in which a jet occurs. Eleven different triggers[1] with different requirements on the jet p_{T} are used. Each of the triggers collected a different amount of integrated luminosity. The higher the p_{T} requirement, the more luminosity was collected. Starting from $p_{\text{T}} > 360$ GeV, the full integrated luminosity of 20.3 fb^{-1} was collected. Besides the trigger, all events criteria are the same as for the signal section discussed in Sect. 14.2.2.

The jets in the selected events are reconstructed with the anti-k_{t} algorithm [1] with a radius parameter of $R = 0.4$. Basic quality criteria,[2] such as cuts against background from cosmic muons, quality cuts on the hadronic calorimeter and cuts on the fraction of energy in the electromagnetic calorimeter are applied. The reconstructed jet is required to match also a reconstructed electron candidate within a cone of $\Delta R < 0.1$.

---

[1] EF_jX_a4tchad ($X = 25, 35, 45, 55, 80, 110, 145, 180, 220, 280, 360$), X corresponds to the p_{T} cut.

[2] *medium* jet cleaning.

M. Zinser, *Search for New Heavy Charged Bosons and Measurement of High-Mass Drell-Yan Production in Proton-Proton Collisions*, Springer Theses,
https://doi.org/10.1007/978-3-030-00650-1

These electron candidates matched to a jet are used to measure the fake efficiency and fake factor. They have to fulfill the same selection cuts regarding reconstruction algorithm, object quality and phase space (as discussed in Sect. 14.2.3).

Real electrons can still enter the selected events as the jet triggers have only very loose identification requirements. Cuts are applied to get a clean jet-enriched sample by reducing dilution from real electrons. Events in which two electron candidates fulfill the *medium* identification are vetoed in order to suppress dilution from Drell-Yan. To further reduce dilution from the Z-resonance, also all events are vetoed in which two candidates fulfill the *loose* identification requirement and fall into an invariant mass window of 20 GeV around the Z mass. A cut of $E_{\mathrm{T}}^{\mathrm{miss}} < 25$ GeV is applied to further reduce dilution from $W \rightarrow e\nu$ decays.

All candidates which pass the vetoes are divided into the following categories to calculate the fake efficiencies and fake factors: N^{fake}_{loose}, $N^{fake}_{tight,\ leading}$, $N^{fake}_{tight,\ subleading}$ and $N^{fake}_{fail\,track\,match}$. For each of the triggers a fake efficiency is calculated. The final fake efficiency is calculated by building the weighted average of all separate fake efficiencies

$$f = \frac{\sum_{i=1}^{n_{trig}} f_i/\Delta f_i^2}{\sum_{i=1}^{n_{trig}} 1/\Delta f_i^2}, \qquad \Delta f^2 = \frac{1}{\sum_{i=1}^{n_{trig}} 1/\Delta f_i^2}, \tag{J.1}$$

where Δf_i is the statistical uncertainty of each fake efficiency and Δf the statistical uncertainty of the averaged fake efficiency. The same averaging is done for the fake factor. The resulting fake efficiencies and fake factors are discussed together with the results from the other methods in Sect. 14.3.3.

J.2 Reverse Tag and Probe Method

An additional method, the reverse tag and probe method, is used to measure the fake efficiencies and fake factors. The idea is to tag di-jet events by requiring one object to fail a certain identification level and to probe a second object which is assumed to be also a jet. The method is performed in two different ways using two different sets of triggers. First the same single jet triggers as in the default method and in addition using the same trigger as for the signal selection.

Jet trigger The object which is used to tag the event as a di-jet event is required to have $p_{\mathrm{T}} > 25$ GeV and to fail the *loose* identification. If such a tag object is found in an event, all other reconstructed electron candidates are assumed to be a jet and considered as probes. In these selected events are still dilutions from processes including real electrons. Similar cleaning cuts as in the default method are applied to reduce contribution from Drell-Yan. The tag and the probe are required to have an invariant mass $|m_{ee} - 91\ \mathrm{GeV}| > 20$ GeV. Tag and probe object additionally have to have the same charge. This is a powerful requirement to strongly suppress dilution from Drell-Yan. A veto on events with $E_{\mathrm{T}}^{\mathrm{miss}} > 25$ GeV is applied to further suppress dilution from W-boson production. All selected probes are divided into the different

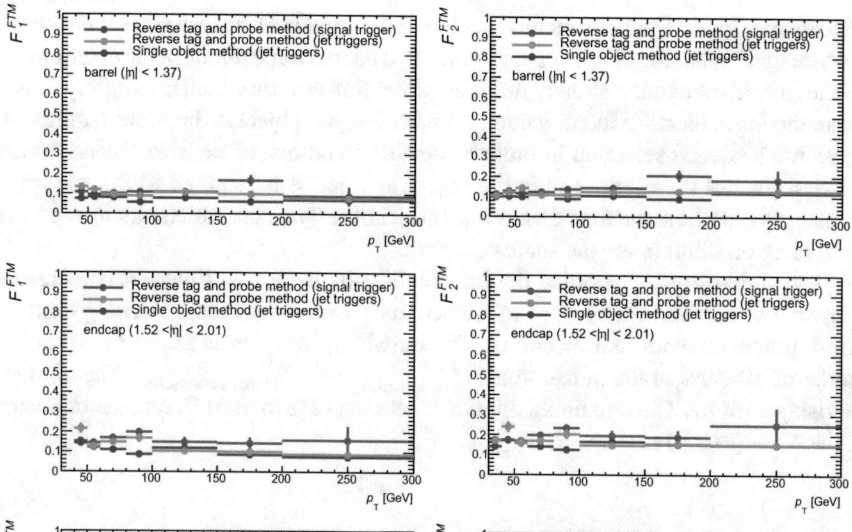

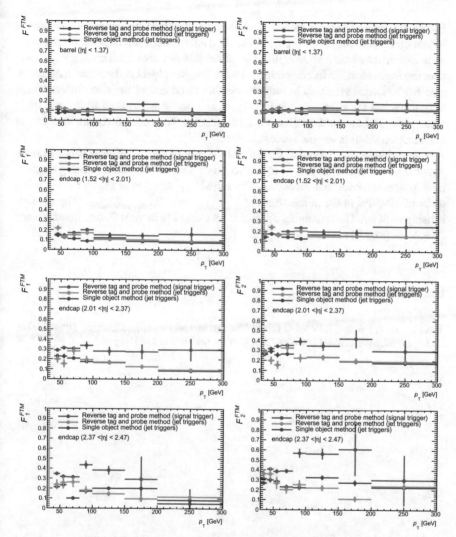

Fig. J.1 Comparison of the fake factors F_i^{FTM} calculated with the three different methods (reverse tag and probe signal trigger, reverse tag and probe single jet triggers and single object method with single jet triggers). The upper row shows the fake factors for the barrel region ($|\eta| < 1.37$). The corresponding fake factors for the endcap regions ($1.52 < |\eta| < 2.01$, $2.01 < |\eta| < 2.37$ and $2.37 < |\eta| < 2.47$) are shown from the second to the fourth row. The fake factors for the leading object are shown on the left side and for the subleading object on the right side

categories to measure the fake efficiencies and fake factors: N^{fake}_{loose}, $N^{fake}_{tight,\ leading}$, $N^{fake}_{tight,\ subleading}$ and $N^{fake}_{fail\,track\,match}$. For each of the jet triggers a fake efficiency and fake factor is calculated. The final fake efficiency and fake factor is calculated by building the weighted average of all separate fake efficiencies (as in Eq. J.1).

Electron trigger The same method has been performed with the trigger which is used in the signal selection. The trigger requires two energy depositions in the electromagnetic calorimeter which already fulfill identification criteria which are slightly looser than the *loose* electron identification criteria. The tag object is therefore required to pass the loosened selection in order to be able to be one of the two objects which have triggered the event. To tag the object as a jet, it is required to fail the track matching cut of the *medium* electron identification. The cuts which aim to suppress real electron dilution are the same as before.

The signal trigger recorded the full luminosity of the 2012 data. It is therefore possible to use MC simulations to further study and correct the remaining dilution from processes with real electrons. The remaining dilution is found to be on the order of 10–30% in the numerators $N^{fake}_{tight,\ leading}$ and $N^{fake}_{tight,\ subleading}$. The dilution is rising with p_{T}. The remaining dilution is less than 1% in the different denominator categories (Fig. J.1).

Reference

1. Cacciari M, Salam GP, Soyez G (2008) The anti-k(t) jet clustering algorithm. JHEP 04:063. https://doi.org/10.1088/1126-6708/2008/04/063, arXiv:0802.1189 [hep-ph]

Appendix K
High-Mass Drell-Yan: Multijet & W+jets Background Systematic Uncertainty

This appendix contains additional plots showing the ratio between the default multijet & W+jets background estimate and all other method variations. The differences between the methods is used to asses the systematic uncertainty on this background and further described in Sect. 14.3.6. Figure K.1 shows the ratio in all five invariant mass bins of the measurement as a function of $|y_{ee}|$ and Fig. K.2 as a function of $|\Delta\eta_{ee}|$.

M. Zinser, *Search for New Heavy Charged Bosons and Measurement of High-Mass Drell-Yan Production in Proton-Proton Collisions*, Springer Theses,
https://doi.org/10.1007/978-3-030-00650-1

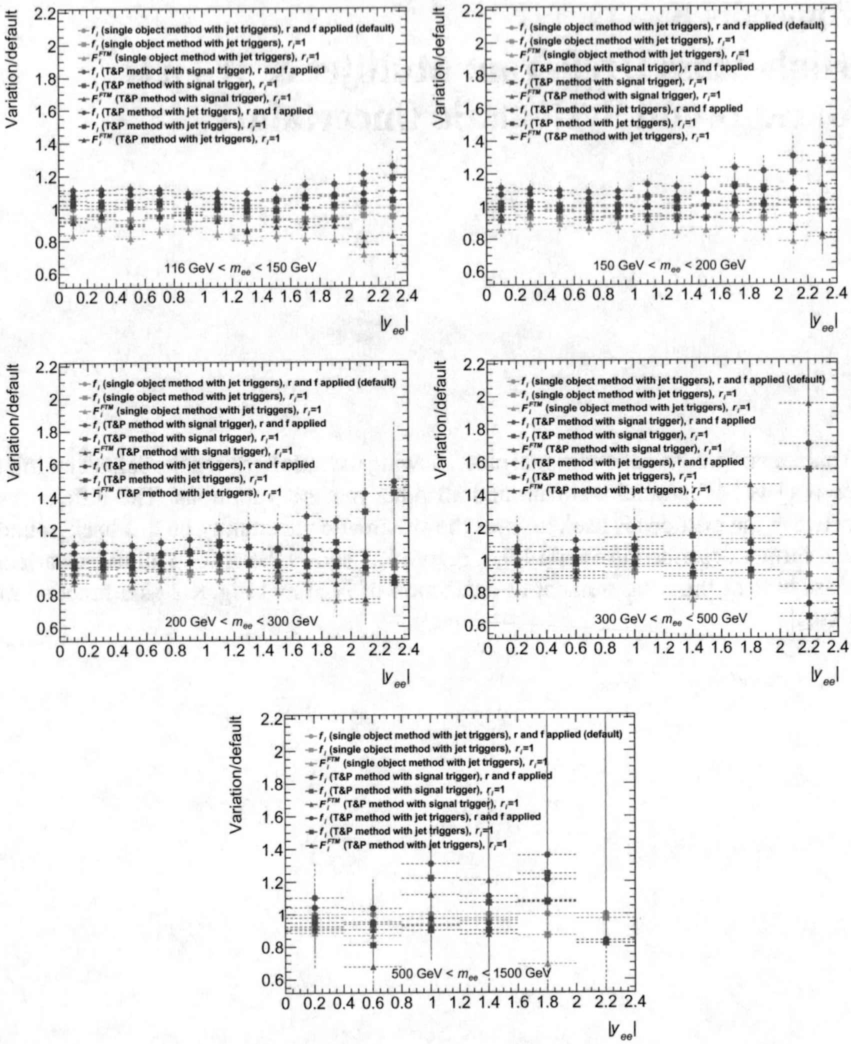

Fig. K.1 Ratio of the final multijet & W+jets background estimate of all method variations to the default method as a function of $|y_{ee}|$ in all invariant mass bins of the measurement

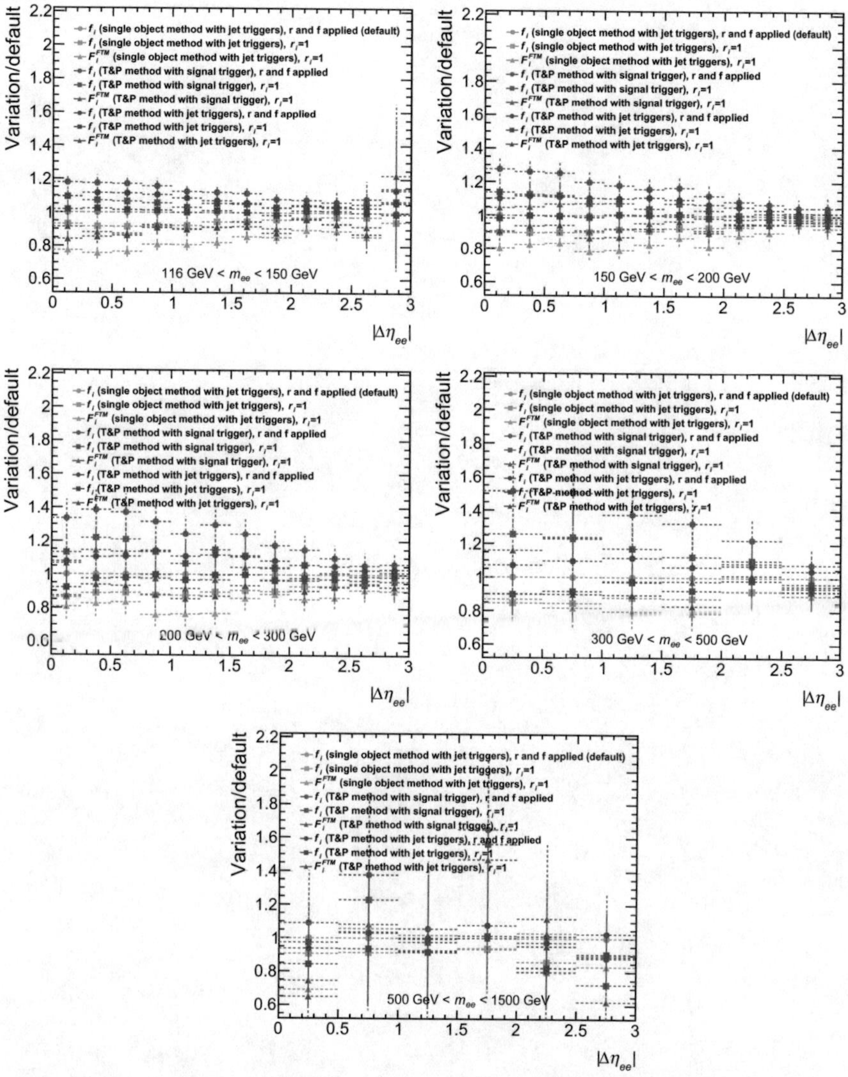

Fig. K.2 Ratio of the final multijet & W+jets background estimate of all method variations to the default method as a function of $|\Delta\eta_{ee}|$ in all invariant mass bins of the measurement

Appendix L
High-Mass Drell-Yan: Top-Quark Background Studies

In this appendix additional studies to validate the background description of the top-quark background are shown. The number of b-jet and the $E_{\mathrm{T}}^{\mathrm{miss}}$ distributions have been studied which are dominated by the top background for $N_{b-jet} > 0$ and large $E_{\mathrm{T}}^{\mathrm{miss}}$. Figure L.1 shows the number of b-jet distribution and Fig. L.2 the $E_{\mathrm{T}}^{\mathrm{miss}}$ distribution in the five invariant mass bins of the double-differential measurement Good agreement of the top background within the normalization uncertainty has been found. The default POWHEG $t\bar{t}$ MC sample has been compared to a MC sample generated with MC@NLO to check for uncertainties on the modeling of the measured observables. The ratio of the two MC samples for the two dimensional observables in the five invariant mass bins of the measurement is shown in Figs. L.3 and L.4. A linear fit to the ratio is also shown. No systematic uncertainty is added since the differences are within the statistical fluctuations which are propagated to the measurement.

M. Zinser, *Search for New Heavy Charged Bosons and Measurement of High-Mass Drell-Yan Production in Proton-Proton Collisions*, Springer Theses,
https://doi.org/10.1007/978-3-030-00650-1

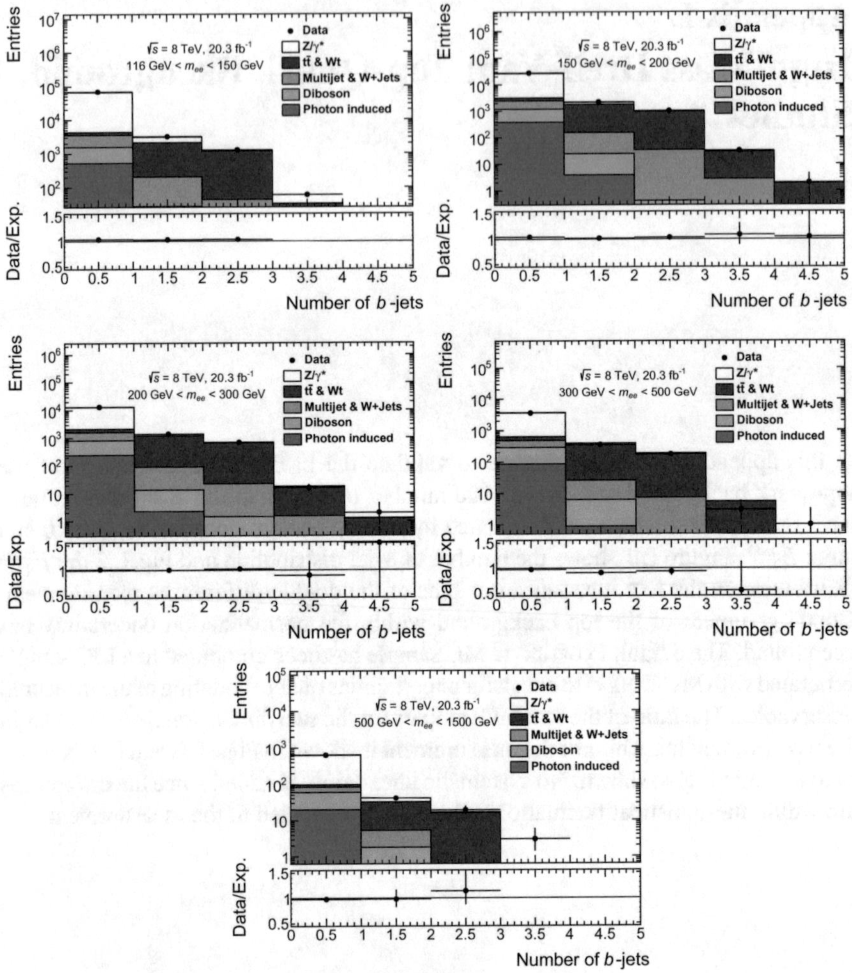

Fig. L.1 Number of b-jets in five bins of invariant mass after the final high-mass Drell-Yan signal selection. Shown for data (solid points) and expectation (stacked histogram). The lower panels show the ratio of data to the expectation

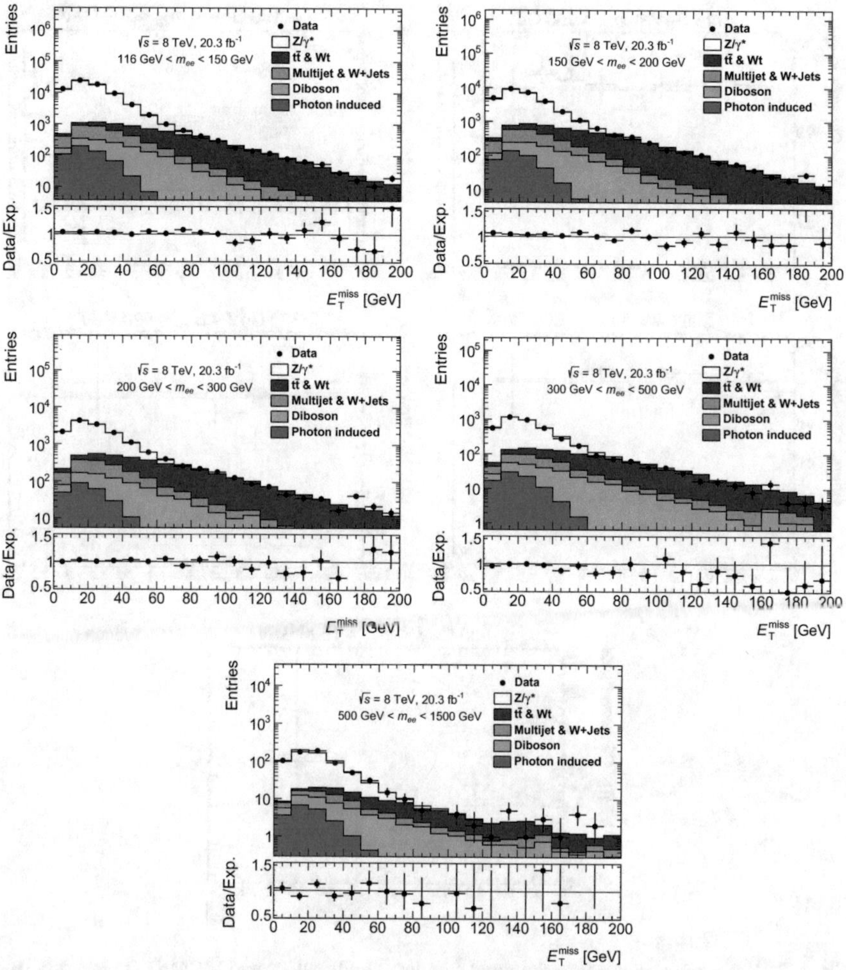

Fig. L.2 $E_{\mathrm{T}}^{\mathrm{miss}}$ distribution in five bins of invariant mass after the final high-mass Drell-Yan signal selection. Shown for data (solid points) and expectation (stacked histogram). The lower panels show the ratio of data to the expectation

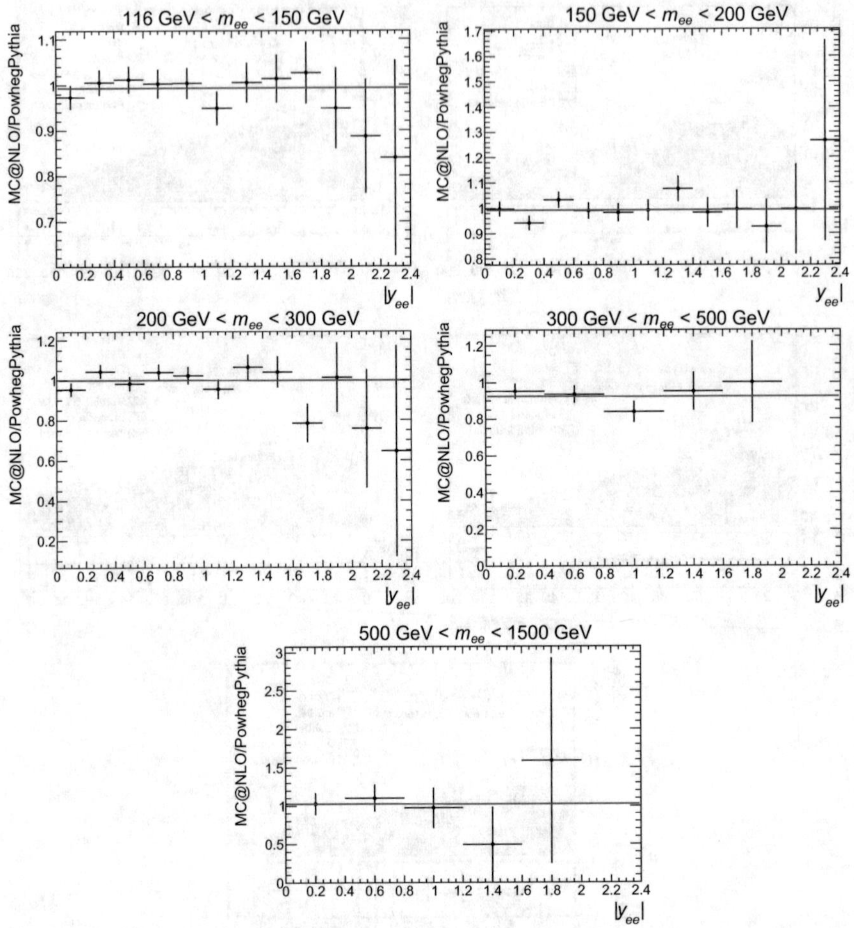

Fig. L.3 Ratio between the $t\bar{t}$ background simulated by POWHEG and MC@NLO binned in $|y_{ee}|$ for the five invariant mass bins of the high-mass Drell-Yan measurement. A linear fit has been performed to the ratio

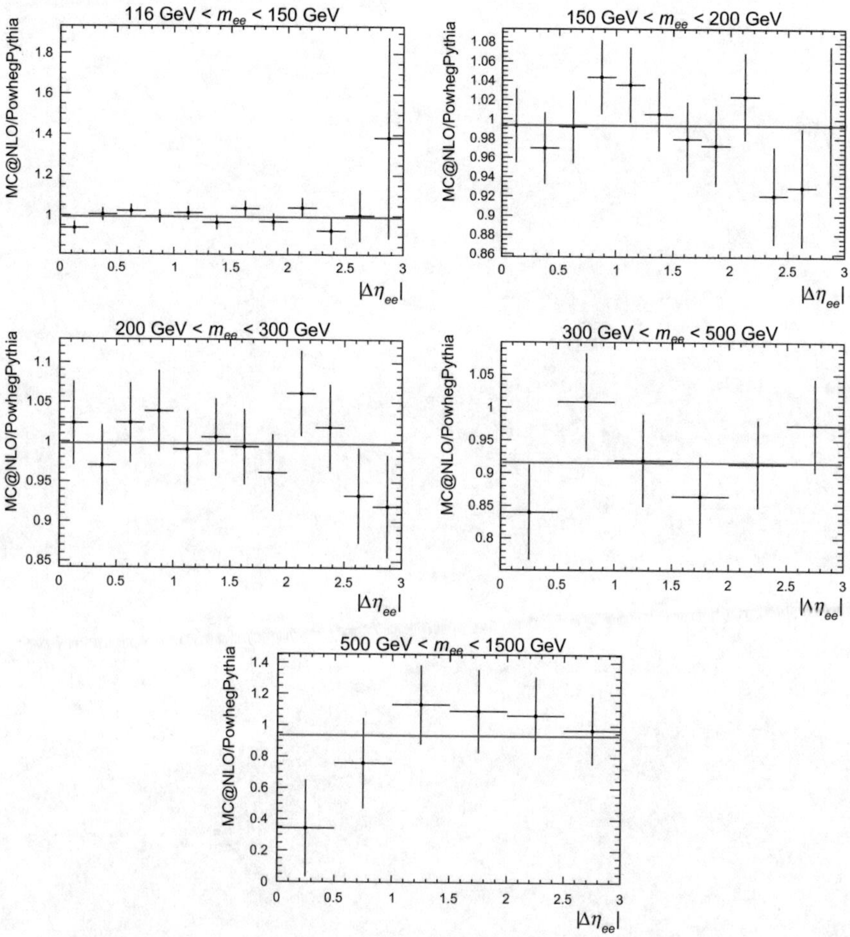

Fig. L.4 Ratio between the $t\bar{t}$ background simulated by POWHEG and MC@NLO binned in $|\Delta\eta_{ee}|$ for the five invariant mass bins of the high-mass Drell-Yan measurement. A linear fit has been performed to the ratio

Appendix M
High-Mass Drell-Yan: Generator Uncertainty Study

To study a potential uncertainty from the choice of a specific Monte Carlo generator, the unfolding factor C_{DY} was calculated with a different generator. Hence, the default generator POWHEG interfaced with PYTHIA is compared to MC@NLO. The uncertainty is described in more detail in Sect. 15.3. Figure M.1 shows the ratio of C_{DY} as a function of invariant mass for both generators together with a linear fit. Both, the Born level and dressed level C_{DY} are compared. Figures M.2 and M.3 show the same comparison for C_{DY} as a function of invariant mass and absolute rapidity. The C_{DY} as a function of invariant mass and absolute pseudorapidity separation is shown in Figs. M.4 and M.5.

M. Zinser, *Search for New Heavy Charged Bosons and Measurement of High-Mass Drell-Yan Production in Proton-Proton Collisions*, Springer Theses,
https://doi.org/10.1007/978-3-030-00650-1

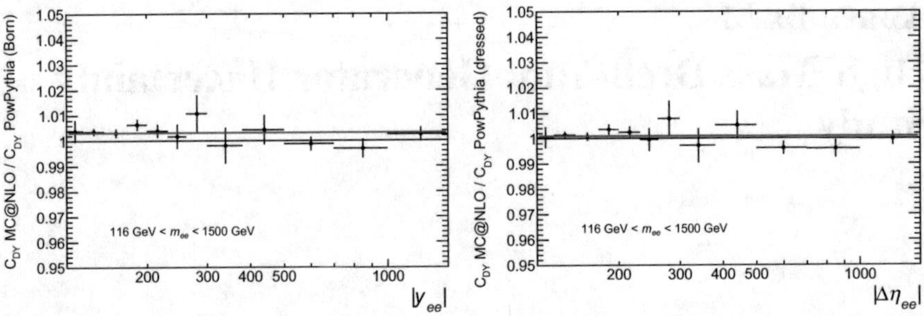

Fig. M.1 Ratio between C_{DY} calculated with MC@NLO to C_{DY} calculated with POWHEG interfaced with PYTHIA at Born and dressed level

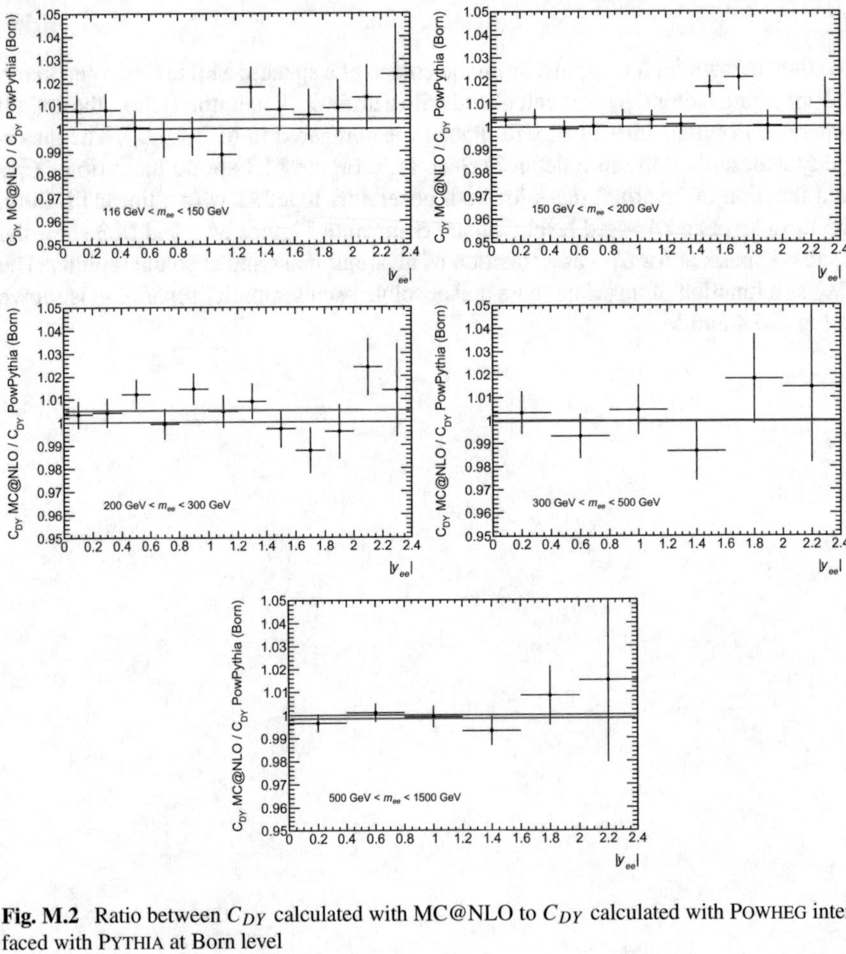

Fig. M.2 Ratio between C_{DY} calculated with MC@NLO to C_{DY} calculated with POWHEG interfaced with PYTHIA at Born level

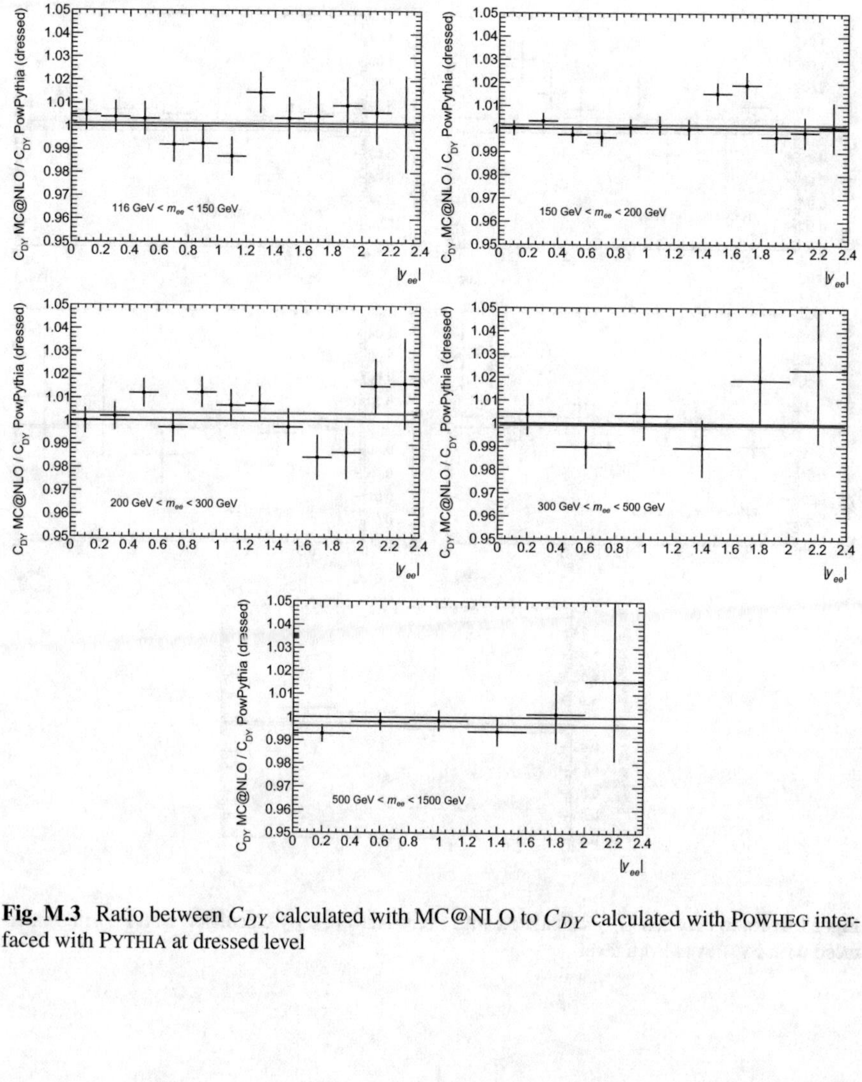

Fig. M.3 Ratio between C_{DY} calculated with MC@NLO to C_{DY} calculated with POWHEG interfaced with PYTHIA at dressed level

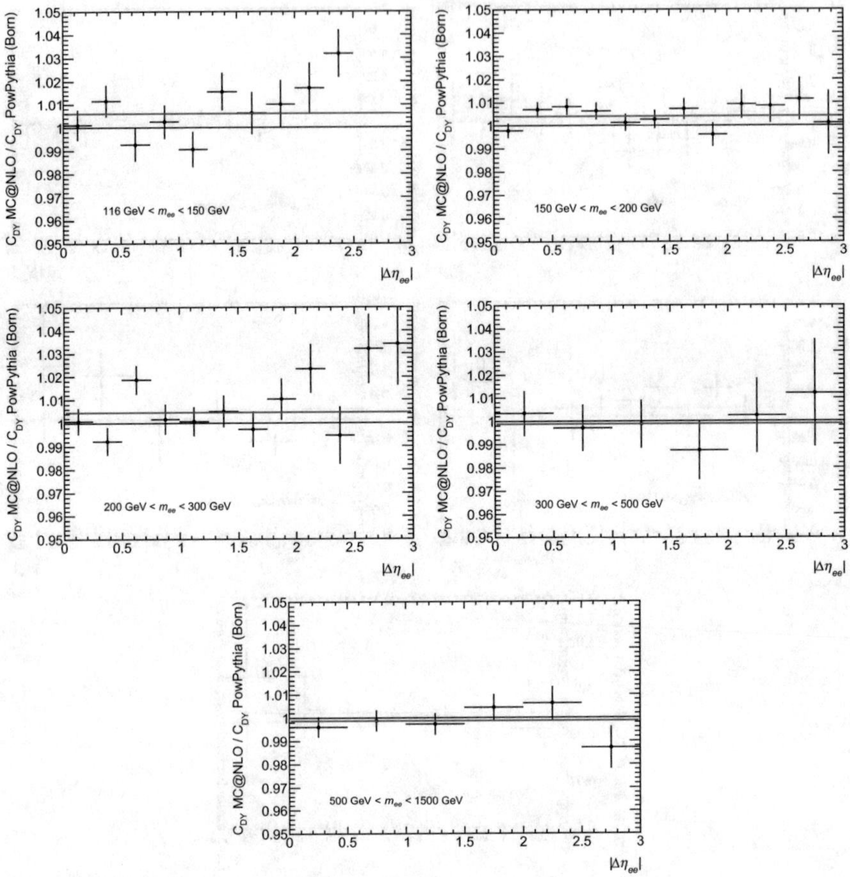

Fig. M.4 Ratio between C_{DY} calculated with MC@NLO to C_{DY} calculated with POWHEG interfaced with PYTHIA at Born level

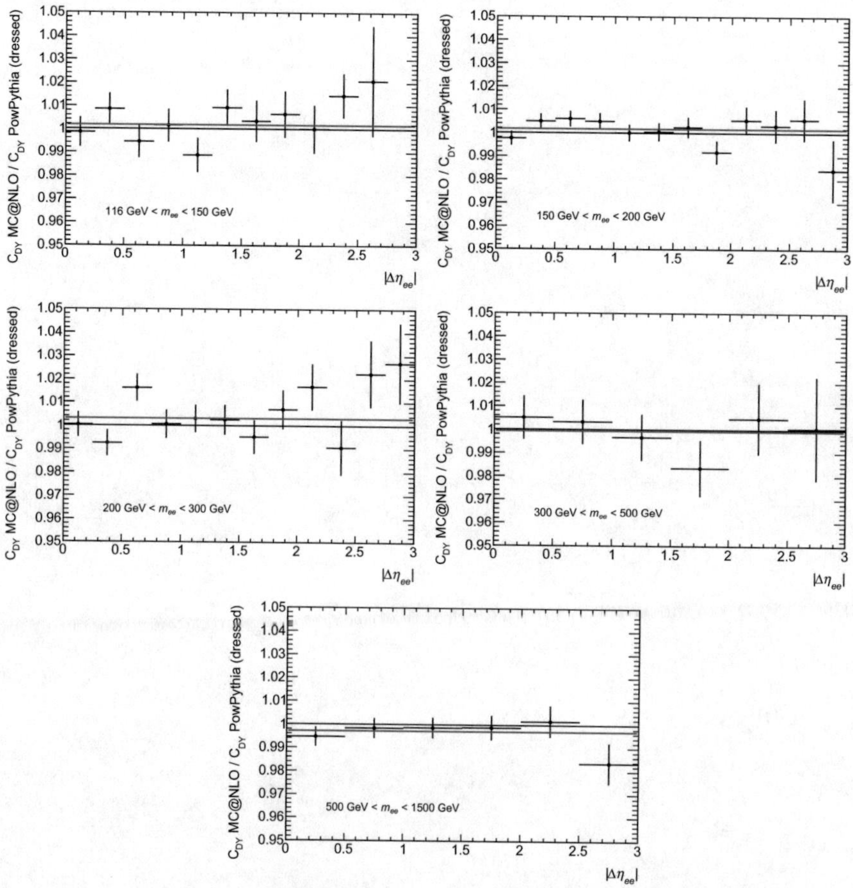

Fig. M.5 Ratio between C_{DY} calculated with with MC@NLO to C_{DY} calculated with POWHEG interfaced with PYTHIA at dressed level

Appendix N
High-Mass Drell-Yan: Bayesian Unfolding Method

In the following appendix, the measured cross section is compared to a cross section unfolded with an alternative iterative Bayesian unfolding method. It is described in more detail in Reference [1]. It has been applied using the RooUnfold package [2]. Figure N.1 shows the ratio of the Born level cross sections unfolded with the default method and with the Bayesian method using 3, 4, and 5 iterations. The differences observed are negligible compared to the statistical precision of the measurement. Figures N.2 and N.3 show the same ratios for the double-differential cross sections. Also here, the observed differences are negligible compared to the statistical uncertainty of the data.

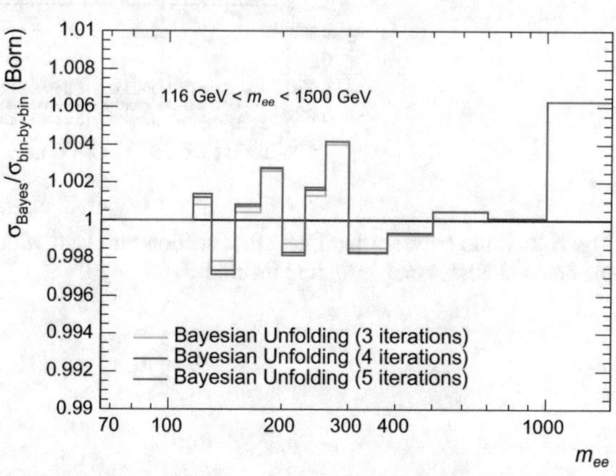

Fig. N.1 Ratio between the final cross section unfolded with the standard bin-by-bin unfolding method and a Bayesian unfolding method

M. Zinser, *Search for New Heavy Charged Bosons and Measurement of High-Mass Drell-Yan Production in Proton-Proton Collisions*, Springer Theses,
https://doi.org/10.1007/978-3-030-00650-1

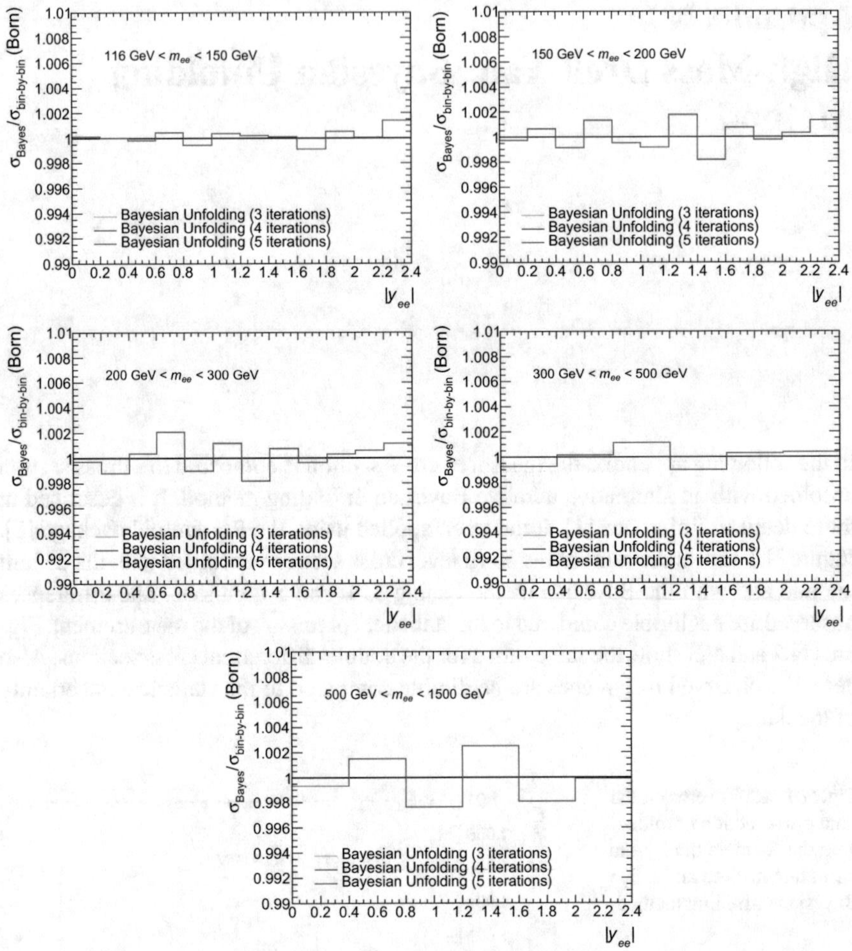

Fig. N.2 Ratio between the final cross section unfolded with the standard bin-by-bin unfolding method and a Bayesian unfolding method

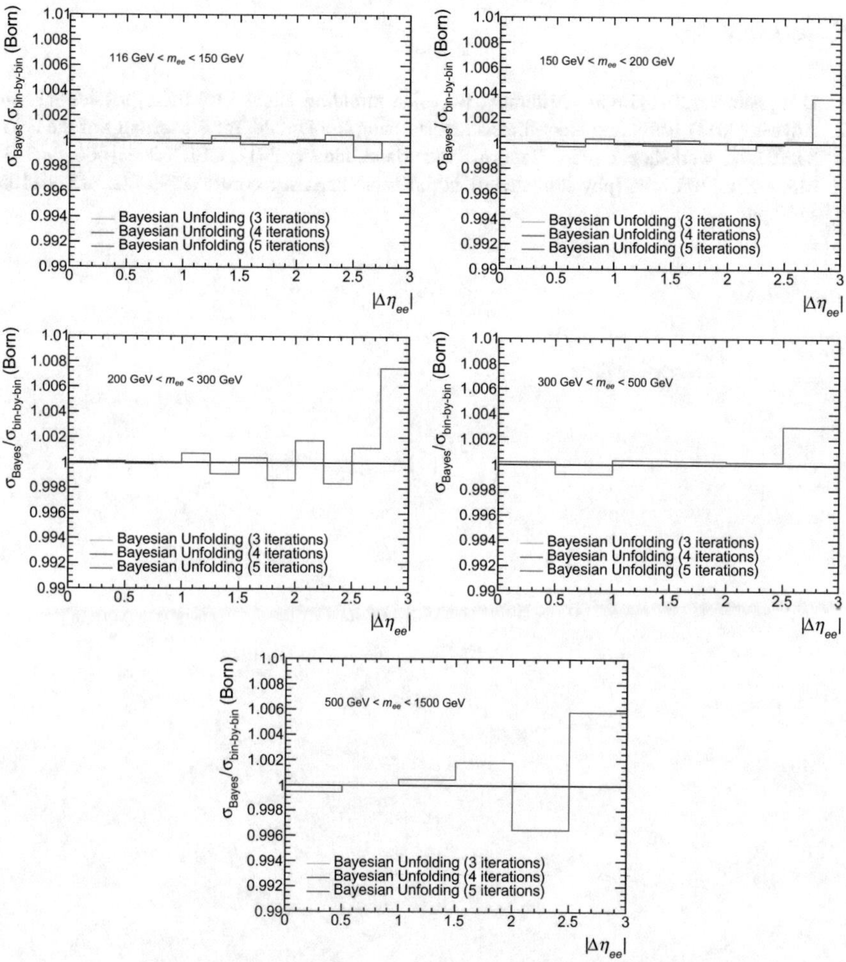

Fig. N.3 Ratio between the final cross section unfolded with the standard bin-by-bin unfolding method and a Bayesian unfolding method

References

1. D'Agostini G (2010) Improved iterative Bayesian unfolding. arXiv:1010.0632 [physics.data-an]
2. Adye T (2011) Unfolding algorithms and tests using RooUnfold. In: Proceedings of the PHY-STAT 2011 workshop, CERN, Geneva, Switzerland, January 2011, CERN-2011-006, pp 313–318. arXiv:1105.1160 [physics.data-an], http://inspirehep.net/record/898599/files/arXiv:1105.1160.pdf

Appendix O
High-Mass Drell-Yan: Electron Energy Scale/Resolution Uncertainties

The following appendix lists in the Tables O.1, O.2, O.3, O.4, O.5, O.6, O.7, O.8, O.9, O.10 and O.11 the breakdown of the energy scale and energy resolution uncertainties for the single-differential and double-differential cross section measurements. A description of the separate sources is given in Sect. 15.3.

M. Zinser, *Search for New Heavy Charged Bosons and Measurement of High-Mass Drell-Yan Production in Proton-Proton Collisions*, Springer Theses,
https://doi.org/10.1007/978-3-030-00650-1

Table O.1 Systematic energy scale and resolution uncertainties, as a function of m_{ee}. The given separate sources are the maximum between up an down variation. For the total uncertainty first the up/down variation of each source is added in quadrature and then the maximum is taken

m_{ee}^{min}-m_{ee}^{max} [GeV] [%]	66–116	116–130	130–150	150–175	175–200	200–230	230–260
E-res ZSmearing	−0.018	0.103	0.070	0.023	0.002	−0.027	0.038
E-res SamplingTerm	0.003	−0.008	−0.006	−0.008	0.006	0.029	−0.036
E-res MaterialID	−0.005	0.045	0.013	0.015	0.007	−0.023	0.024
E-res MaterialCalo	−0.001	0.010	0.003	−0.000	0.002	−0.010	0.023
E-res MaterialGap	−0.004	0.014	−0.001	0.001	0.012	−0.006	0.011
E-res MaterialCryo	−0.001	0.013	−0.002	0.006	0.003	−0.010	0.001
E-res PileUp	0.001	−0.018	−0.002	0.005	−0.007	0.011	−0.016
E-scale stat.	−0.000	0.002	−0.002	−0.009	0.010	−0.019	0.028
E-scale Zee syst.	0.156	0.424	0.310	0.255	0.242	0.219	0.264
E-scale PS	0.018	−0.066	−0.052	−0.063	−0.081	−0.086	−0.164
E-scale S12	0.010	−0.037	−0.032	−0.045	−0.062	−0.043	−0.101
E-scale MatID	−0.015	0.041	0.044	0.046	0.070	0.061	0.121
E-scale MatCryo	−0.033	0.103	0.084	0.105	0.106	0.146	0.220
E-scale MatCalo	−0.009	0.030	0.023	0.036	0.032	0.033	0.088
E-scale LArCalib	0.023	−0.082	−0.074	−0.084	−0.125	−0.106	−0.223
E-scale LArUnconvCalib	−0.008	0.023	0.022	0.034	0.028	0.034	0.063
E-scale LArElecUnconv	0.039	−0.129	−0.104	−0.137	−0.144	−0.178	−0.294
E-scale LArElecCalib	−0.002	0.005	0.006	0.008	0.012	0.003	0.027
E-scale L1Gain	0.001	0.006	0.013	0.006	0.023	0.016	0.075
E-scale L2Gain	0.040	−0.188	−0.151	−0.235	−0.333	−0.502	−0.804
E-scale G4	−0.004	0.011	0.009	0.013	0.015	0.020	0.012
E-scale Pedestal	0.012	−0.018	−0.022	−0.030	−0.023	−0.030	−0.052
E-res syst.	0.025	0.128	0.100	0.052	0.044	0.086	0.140

(continued)

Table O.1 (continued)

m_{ee}^{min}-m_{ee}^{max} [GeV] [%]	66–116	116–130	130–150	150–175	175–200	200–230	230–260
E-scale syst.	0.174	0.517	0.391	0.426	0.521	0.658	1.019
m_{ee}^{min}-m_{ee}^{max} [GeV] [%]	260–300	300–380	380–500	500–700	700–1000	1000–1500	
E-res ZSmearing	−0.041	0.077	−0.006	−0.027	0.031	−0.018	
E-res SamplingTerm	0.025	−0.047	0.030	0.017	−0.023	−0.015	
E-res MaterialID	−0.034	0.031	0.016	−0.019	0.009	0.006	
E-res MaterialCalo	−0.007	−0.004	0.011	−0.007	−0.001	0.008	
E-res MaterialGap	−0.012	0.006	−0.000	−0.008	0.005	0.001	
E-res MaterialCryo	−0.021	0.011	0.011	−0.010	−0.007	−0.004	
E-res PileUp	0.006	−0.011	−0.001	0.013	0.003	−0.000	
E-scale stat.	−0.002	0.007	−0.004	−0.008	0.010	0.001	
E-scale Zee syst.	0.209	0.234	0.238	0.280	0.236	0.249	
E-scale PS	−0.111	−0.180	−0.211	−0.317	−0.357	−0.500	
E-scale S12	−0.088	−0.130	−0.142	−0.206	−0.234	−0.297	
E-scale MatID	0.083	0.124	0.129	0.214	0.210	0.232	
E-scale MatCryo	0.160	0.253	0.268	0.363	0.376	0.444	
E-scale MatCalo	0.048	0.075	0.079	0.142	0.138	0.209	
E-scale LArCalib	−0.177	−0.265	−0.324	−0.447	−0.562	−0.772	
E-scale LArUnconvCalib	0.047	0.062	0.073	0.131	0.122	0.171	
E-scale LArElecUnconv	−0.204	−0.344	−0.371	−0.520	−0.616	−0.773	
E-scale LArElecCalib	0.014	0.019	0.021	0.024	0.022	0.024	
E-scale L1Gain	0.031	0.092	0.125	0.181	0.183	0.166	
E-scale L2Gain	−0.761	−1.103	−1.353	−1.700	−1.972	−2.392	
E-scale G4	0.028	0.034	0.029	0.061	0.054	0.054	
E-scale Pedestal	−0.039	−0.070	−0.054	−0.105	−0.089	−0.095	
E-res syst.	0.084	0.141	0.087	0.060	0.049	0.037	
E-scale syst.	0.878	1.276	1.554	1.986	2.308	2.803	

Table O.2 Systematic energy scale and resolution uncertainties, as a function of $|y_{ee}|$ in the 116-150 GeV mass region. The given separate sources are the maximum between up an down variation. For the total uncertainty first the up/down variation of each source is added in quadrature and then the maximum is taken

$\|y_{ee}^{min}\|$-$\|y_{ee}^{max}\|$ [%]	0.0–0.2	0.2–0.4	0.4–0.6	0.6–0.8	0.8–1.0	1.0–1.2
E-res ZSmearing	0.015	0.043	0.110	0.023	0.148	0.036
E-res SamplingTerm	−0.011	0.014	−0.010	0.041	−0.040	−0.017
E-res MaterialID	0.001	−0.022	0.119	−0.084	0.054	0.055
E-res MaterialCalo	0.025	0.026	0.018	−0.042	0.055	−0.009
E-res MaterialGap	−0.001	−0.011	−0.009	−0.004	−0.007	0.003
E-res MaterialCryo	0.002	0.007	0.035	−0.051	0.014	0.026
E-res PileUp	0.006	−0.026	−0.023	0.011	−0.035	0.002
E-scale stat.	−0.014	−0.035	−0.007	0.032	−0.015	−0.021
E-scale Zee syst.	0.150	0.199	0.233	0.260	0.425	0.417
E-scale PS	−0.010	−0.058	−0.027	−0.084	−0.117	−0.076
E-scale S12	−0.017	−0.046	−0.032	−0.063	−0.042	−0.038
E-scale MatID	−0.030	−0.030	−0.014	0.039	0.035	0.122
E-scale MatCryo	−0.007	0.036	0.026	0.052	0.064	0.136
E-scale MatCalo	0.004	0.029	0.027	0.096	0.069	0.083
E-scale LArCalib	−0.057	−0.111	−0.075	−0.079	−0.062	−0.060
E-scale LArUnconvCalib	−0.006	0.025	0.012	0.073	0.065	0.073
E-scale LArElecUnconv	0.028	0.008	−0.052	−0.108	−0.103	−0.245
E-scale LArElecCalib	0.000	0.000	−0.001	−0.003	−0.013	0.000
E-scale L1Gain	0.000	0.000	0.001	0.002	0.000	−0.005
E-scale L2Gain	−0.039	−0.142	−0.176	−0.090	0.128	−0.100
E-scale G4	0.010	0.014	0.008	0.030	0.035	0.024
E-scale Pedestal	−0.010	−0.030	−0.025	−0.065	−0.042	−0.027
E-res syst.	0.094	0.085	0.258	0.233	0.254	0.136

(continued)

Table O.2 (continued)

$\|y_{ee}^{min}\|$-$\|y_{ee}^{max}\|$ [%]	0.0–0.2	0.2–0.4	0.4–0.6	0.6–0.8	0.8–1.0	1.0–1.2
E-scale syst.	0.196	0.332	0.339	0.395	0.523	0.568
$\|y_{ee}^{min}\|$-$\|y_{ee}^{max}\|$ [%]	1.2–1.4	1.4–1.6	1.6–1.8	1.8–2.0	2.0–2.2	2.2–2.4
E-res ZSmearing	0.230	0.064	0.082	0.153	0.345	−0.044
E-res SamplingTerm	−0.050	0.011	−0.012	0.005	0.054	−0.018
E-res MaterialID	0.091	−0.078	0.079	0.014	0.104	0.124
E-res MaterialCalo	0.002	−0.003	0.031	−0.005	0.011	0.000
E-res MaterialGap	0.060	−0.016	0.083	0.028	0.001	−0.004
E-res MaterialCryo	−0.032	0.030	−0.007	0.043	0.019	−0.041
E-res PileUp	−0.008	−0.018	−0.025	0.033	0.052	−0.003
E-scale stat.	0.012	0.017	−0.015	−0.038	0.020	0.005
E-scale Zee syst.	0.497	0.529	0.713	0.748	0.617	1.065
E-scale PS	−0.097	−0.055	−0.041	−0.046	−0.010	0.000
E-scale S12	0.001	0.002	−0.027	−0.089	−0.145	−0.201
E-scale MatID	0.071	0.112	0.109	0.078	0.076	0.052
E-scale MatCryo	0.228	0.171	0.215	0.311	0.033	0.041
E-scale MatCalo	0.036	0.007	−0.004	0.016	−0.013	0.000
E-scale LArCalib	−0.027	−0.042	−0.074	−0.127	−0.197	−0.164
E-scale LArUnconvCalib	0.041	0.018	−0.002	0.023	−0.013	0.000
E-scale LArElecUnconv	−0.293	−0.255	−0.188	−0.209	0.021	0.000
E-scale LArElecCalib	0.011	−0.003	0.015	0.073	0.076	0.025
E-scale L1Gain	−0.006	−0.018	0.045	0.025	0.201	0.259
E-scale L2Gain	−0.258	−0.537	−0.527	−0.607	−0.244	−0.114
E-scale G4	0.018	−0.010	−0.007	0.027	0.040	−0.012
E-scale Pedestal	−0.015	0.007	0.010	−0.021	−0.010	0.003
E-res syst.	0.336	0.161	0.206	0.325	0.413	0.371
E-scale syst.	0.708	0.836	0.961	1.058	0.761	1.189

Table O.3 Systematic energy scale and resolution uncertainties, as a function of $|y_{ee}|$ in the 150-200 GeV mass region. The given separate sources are the maximum between up an down variation. For the total uncertainty first the up/down variation of each source is added in quadrature and then the maximum is taken

$\|y_{ee}^{min}\|$-$\|y_{ee}^{max}\|$ [%]	0.0–0.2	0.2–0.4	0.4–0.6	0.6–0.8	0.8–1.0	1.0–1.2
E-res ZSmearing	−0.017	0.026	0.014	−0.069	0.021	0.086
E-res SamplingTerm	0.035	0.004	−0.007	−0.003	−0.035	−0.024
E-res MaterialID	−0.032	0.031	−0.024	0.079	−0.037	0.050
E-res MaterialCalo	−0.010	0.005	−0.012	0.051	0.014	−0.017
E-res MaterialGap	0.001	0.002	0.031	0.012	−0.007	0.031
E-res MaterialCryo	−0.013	−0.008	−0.000	0.029	−0.001	0.025
E-res PileUp	0.005	−0.010	−0.002	−0.003	−0.005	−0.003
E-scale stat.	−0.003	0.016	0.005	−0.045	−0.008	−0.040
E-scale Zee syst.	0.103	0.119	0.164	0.236	0.196	0.223
E-scale PS	−0.043	−0.074	−0.052	−0.103	−0.061	−0.112
E-scale S12	−0.059	−0.045	−0.028	−0.077	−0.007	−0.021
E-scale MatID	−0.038	−0.004	0.003	0.055	0.065	0.099
E-scale MatCryo	0.035	0.043	0.021	0.099	0.058	0.150
E-scale MatCalo	0.016	0.051	0.025	0.081	0.026	0.068
E-scale LArCalib	−0.118	−0.118	−0.070	−0.137	−0.014	−0.061
E-scale LArUnconvCalib	0.007	0.027	0.020	0.083	0.029	0.069
E-scale LArElecUnconv	0.003	−0.015	−0.081	−0.132	−0.204	−0.231
E-scale LArElecCalib	0.000	−0.005	−0.004	−0.005	−0.006	−0.018
E-scale L1Gain	0.000	0.002	0.002	0.001	−0.013	−0.027
E-scale L2Gain	−0.128	−0.195	−0.099	−0.084	−0.092	−0.367
E-scale G4	0.028	0.006	−0.003	0.009	0.009	0.018
E-scale Pedestal	−0.044	−0.041	−0.018	−0.039	−0.020	−0.052
E-res syst.	0.107	0.111	0.085	0.245	0.124	0.240

(continued)

Table O.3 (continued)

$\lvert y_{ee}^{min}\rvert$-$\lvert y_{ee}^{max}\rvert$ [%]	0.0–0.2	0.2–0.4	0.4–0.6	0.6–0.8	0.8–1.0	1.0–1.2
E-scale syst.	0.238	0.292	0.246	0.462	0.392	0.571
$\lvert y_{ee}^{min}\rvert$-$\lvert y_{ee}^{max}\rvert$ [%]	1.2–1.4	1.4–1.6	1.6–1.8	1.8–2.0	2.0–2.2	2.2–2.4
E-res ZSmearing	−0.020	−0.010	−0.023	0.198	0.065	0.048
E-res SamplingTerm	0.002	0.041	−0.002	0.014	0.006	0.019
E-res MaterialID	−0.038	0.018	−0.009	0.090	0.152	0.038
E-res MaterialCalo	0.000	−0.020	0.004	−0.006	0.000	0.000
E-res MaterialGap	0.022	−0.005	−0.012	−0.008	−0.000	0.011
E-res MaterialCryo	−0.006	−0.003	0.024	0.021	0.050	0.009
E-res PileUp	−0.003	0.028	−0.029	−0.021	0.027	0.019
E-scale stat.	0.035	0.001	0.014	−0.006	0.007	−0.006
E-scale Zee syst.	0.303	0.471	0.601	0.512	0.384	1.090
E-scale PS	−0.046	−0.095	−0.061	−0.008	0.028	0.000
E-scale S12	−0.028	−0.043	−0.069	−0.072	−0.119	−0.292
E-scale MatID	0.104	0.105	0.155	0.026	0.039	0.109
E-scale MatCryo	0.178	0.283	0.315	0.143	−0.030	0.039
E-scale MatCalo	0.016	0.050	0.018	−0.002	0.000	−0.000
E-scale LArCalib	−0.019	−0.097	−0.125	−0.143	−0.137	−0.222
E-scale LArUnconvCalib	0.014	0.068	0.013	−0.008	−0.004	−0.000
E-scale LArElecUnconv	−0.236	−0.350	−0.261	−0.098	0.075	0.000
E-scale LArElecCalib	0.014	0.038	0.040	0.036	0.058	−0.001
E-scale L1Gain	−0.020	−0.002	0.065	0.046	0.102	0.245
E-scale L2Gain	−0.512	−0.723	−0.899	−0.564	0.010	−0.114
E-scale G4	0.006	0.035	0.029	0.021	0.019	0.015
E-scale Pedestal	0.002	−0.051	−0.009	−0.011	−0.002	−0.002
E-res syst.	0.097	0.119	0.158	0.305	0.228	0.160
E-scale syst.	0.696	1.009	1.186	0.841	0.458	1.313

Table O.4 Systematic energy scale and resolution uncertainties, as a function of $|y_{ee}|$ in the 200-300 GeV mass region. The given separate sources are the maximum between up an down variation. For the total uncertainty first the up/down variation of each source is added in quadrature and then the maximum is taken

$\|y_{ee}^{min}\|$-$\|y_{ee}^{max}\|$ [%]	0.0–0.2	0.2–0.4	0.4–0.6	0.6–0.8	0.8–1.0	1.0–1.2
E-res ZSmearing	0.006	−0.088	0.089	0.035	−0.013	−0.067
E-res SamplingTerm	0.013	0.028	−0.038	0.016	0.028	0.033
E-res MaterialID	−0.002	−0.008	0.046	−0.038	−0.003	−0.084
E-res MaterialCalo	−0.008	−0.003	0.033	−0.030	0.002	−0.003
E-res MaterialGap	−0.005	0.020	−0.006	−0.007	−0.019	−0.037
E-res MaterialCryo	0.005	−0.008	0.019	−0.018	−0.042	−0.015
E-res PileUp	0.004	0.001	0.005	0.023	0.005	0.046
E-scale stat.	−0.007	−0.006	−0.006	−0.005	0.037	0.026
E-scale Zee syst.	0.157	0.122	0.110	0.163	0.240	0.283
E-scale PS	−0.106	−0.095	−0.075	−0.161	−0.193	−0.176
E-scale S12	−0.057	−0.069	−0.046	−0.048	−0.049	−0.058
E-scale MatID	0.027	0.024	0.032	0.062	0.103	0.123
E-scale MatCryo	0.060	0.046	0.054	0.185	0.183	0.225
E-scale MatCalo	0.059	0.056	0.049	0.064	0.096	0.075
E-scale LArCalib	−0.153	−0.129	−0.127	−0.145	−0.193	−0.167
E-scale LArUnconvCalib	0.055	0.034	0.036	0.060	0.073	0.072
E-scale LArElecUnconv	−0.075	−0.067	−0.091	−0.221	−0.279	−0.362
E-scale LArElecCalib	0.000	−0.002	0.001	−0.004	0.002	0.007
E-scale L1Gain	0.000	−0.000	−0.003	−0.008	−0.019	−0.005
E-scale L2Gain	−0.383	−0.289	−0.410	−0.580	−0.751	−1.087
E-scale G4	0.018	0.010	0.040	0.017	0.036	0.017
E-scale Pedestal	−0.051	−0.045	−0.054	−0.038	−0.042	−0.051
E-res syst.	0.107	0.171	0.125	0.095	0.164	0.163

(continued)

Table O.4 (continued)

$\lvert y_{ee}^{min}\rvert$-$\lvert y_{ee}^{max}\rvert$ [%]	0.0–0.2	0.2–0.4	0.4–0.6	0.6–0.8	0.8–1.0	1.0–1.2
E-scale syst.	0.542	0.402	0.485	0.768	0.996	1.341
$\lvert y_{ee}^{min}\rvert$-$\lvert y_{ee}^{max}\rvert$ [%]	1.2–1.4	1.4–1.6	1.6–1.8	1.8–2.0	2.0–2.2	2.2–2.4
E-res ZSmearing	−0.071	−0.037	0.118	−0.147	0.119	0.318
E-res SamplingTerm	−0.046	−0.011	0.034	0.009	−0.033	−0.019
E-res MaterialID	0.059	−0.013	−0.070	−0.004	−0.019	0.058
E-res MaterialCalo	0.014	0.005	−0.002	0.011	0.000	0.000
E-res MaterialGap	−0.028	0.022	−0.001	0.042	−0.006	0.015
E-res MaterialCryo	0.027	−0.019	−0.050	−0.003	−0.075	0.060
E-res PileUp	−0.070	−0.027	0.030	−0.029	−0.043	−0.032
E-scale stat.	−0.007	−0.001	−0.015	0.013	0.001	−0.012
E-scale Zee syst.	0.252	0.424	0.358	0.405	0.266	0.939
E-scale PS	−0.112	−0.118	−0.101	−0.008	0.007	0.000
E-scale S12	−0.003	−0.080	−0.103	−0.185	−0.170	−0.366
E-scale MatID	0.103	0.204	0.124	0.178	0.071	0.122
E-scale MatCryo	0.272	0.280	0.281	0.350	0.074	0.101
E-scale MatCalo	0.053	0.028	0.011	0.008	0.006	0.000
E-scale LArCalib	−0.127	−0.155	−0.162	−0.235	−0.279	−0.230
E-scale LArUnconvCalib	0.054	0.026	0.025	0.021	0.016	0.000
E-scale LArElecUnconv	−0.403	−0.390	−0.298	−0.165	0.052	0.015
E-scale LArElecCalib	0.012	0.003	0.059	0.098	0.106	0.086
E-scale L1Gain	−0.055	0.106	0.148	0.225	0.440	0.704
E-scale L2Gain	−1.205	−0.900	−1.148	−1.113	−0.091	−0.117
E-scale G4	0.005	−0.010	−0.002	0.039	0.017	0.055
E-scale Pedestal	0.000	−0.015	−0.009	−0.040	−0.012	−0.002
E-res syst.	0.149	0.135	0.293	0.341	0.329	0.685
E-scale syst.	1.374	1.171	1.506	1.508	0.683	1.401

Table O.5 Systematic energy scale and resolution uncertainties, as a function of $|y_{ee}|$ in the 300-500 GeV mass region. The given separate sources are the maximum between up an down variation. For the total uncertainty first the up/down variation of each source is added in quadrature and then the maximum is taken

$\|y_{ee}^{min}\|$-$\|y_{ee}^{max}\|$ [%]	0.0–0.4	0.4–0.8	0.8–1.2
E-res ZSmearing	0.053	0.017	0.080
E-res SamplingTerm	−0.086	−0.013	−0.009
E-res MaterialID	0.055	−0.004	0.022
E-res MaterialCalo	0.019	0.000	−0.007
E-res MaterialGap	0.020	−0.001	−0.002
E-res MaterialCryo	0.015	0.004	0.002
E-res PileUp	−0.023	−0.014	0.003
E-scale stat.	−0.003	−0.005	0.020
E-scale Zee syst.	0.095	0.143	0.227
E-scale PS	−0.123	−0.193	−0.200
E-scale S12	−0.115	−0.090	−0.100
E-scale MatID	0.013	0.059	0.154
E-scale MatCryo	0.083	0.188	0.258
E-scale MatCalo	0.072	0.087	0.069
E-scale LArCalib	−0.251	−0.264	−0.213
E-scale LArUnconvCalib	0.046	0.070	0.060
E-scale LArElecUnconv	−0.120	−0.351	−0.433
E-scale LArElecCalib	0.002	−0.007	0.013
E-scale L1Gain	−0.000	−0.016	−0.036
E-scale L2Gain	−0.730	−1.217	−1.650
E-scale G4	0.023	0.008	0.023
E-scale Pedestal	−0.083	−0.062	−0.044
E-res syst.	0.150	0.053	0.229
E-scale syst.	0.867	1.367	2.014
$\|y_{ee}^{min}\|$-$\|y_{ee}^{max}\|$ [%]	1.2–1.6	1.6–2.0	2.0–2.4
E-res ZSmearing	−0.068	0.065	−0.141
E-res SamplingTerm	0.019	−0.061	−0.037
E-res MaterialID	0.010	0.046	−0.044
E-res MaterialCalo	−0.004	−0.011	0.000
E-res MaterialGap	0.005	0.041	−0.061
E-res MaterialCryo	−0.007	−0.021	0.069
E-res PileUp	0.022	0.004	0.027
E-scale stat.	0.003	0.021	−0.011
E-scale Zee syst.	0.409	0.594	0.848
E-scale PS	−0.263	−0.156	−0.092
E-scale S12	−0.137	−0.238	−0.586
E-scale MatID	0.235	0.266	0.360

(continued)

Table O.5 (continued)

$\|y_{ee}^{min}\|$-$\|y_{ee}^{max}\|$ [%]	1.2–1.6	1.6–2.0	2.0–2.4
E-scale MatCryo	0.437	0.609	0.174
E-scale MatCalo	0.060	0.003	0.020
E-scale LArCalib	−0.324	−0.449	−0.814
E-scale LArUnconvCalib	0.069	0.008	0.020
E-scale LArElecUnconv	−0.583	−0.607	−0.181
E-scale LArElecCalib	0.042	0.089	0.156
E-scale L1Gain	0.151	0.441	1.814
E-scale L2Gain	−1.726	−2.432	−0.168
E-scale G4	0.050	0.025	0.089
E-scale Pedestal	−0.070	−0.022	−0.049
E-res syst.	0.090	0.190	0.431
E-scale syst.	2.010	2.839	2.501

Table O.6 Systematic energy scale and resolution uncertainties, as a function of $|y_{ee}|$ in the 500-1500 GeV mass region. The given separate sources are the maximum between up an down variation. For the total uncertainty first the up/down variation of each source is added in quadrature and then the maximum is taken

$\|y_{ee}^{min}\|$-$\|y_{ee}^{max}\|$ [%]	0.0–0.4	0.4–0.8	0.8–1.2
E-res ZSmearing	−0.003	−0.035	−0.090
E-res SamplingTerm	0.005	−0.024	0.042
E-res MaterialID	−0.034	0.004	−0.018
E-res MaterialCalo	−0.032	0.029	−0.019
E-res MaterialGap	0.001	−0.026	−0.024
E-res MaterialCryo	−0.028	0.005	−0.024
E-res PileUp	0.041	−0.028	0.036
E-scale stat.	0.003	−0.020	0.021
E-scale Zee syst.	0.132	0.239	0.285
E-scale PS	−0.265	−0.363	−0.320
E-scale S12	−0.206	−0.177	−0.146
E-scale MatID	0.105	0.203	0.227
E-scale MatCryo	0.184	0.311	0.379
E-scale MatCalo	0.155	0.190	0.124
E-scale LArCalib	−0.474	−0.414	−0.412
E-scale LArUnconvCalib	0.102	0.188	0.123
E-scale LArElecUnconv	−0.259	−0.441	−0.638
E-scale LArElecCalib	0.003	−0.004	0.022
E-scale L1Gain	−0.005	−0.015	0.041
E-scale L2Gain	−1.267	−1.848	−2.322

(continued)

Table O.6 (continued)

$\|y_{ee}^{min}\|$-$\|y_{ee}^{max}\|$ [%]	0.0–0.4	0.4–0.8	0.8–1.2
E-scale G4	0.053	0.069	0.048
E-scale Pedestal	−0.130	−0.122	−0.064
E-res syst.	0.100	0.137	0.128
E-scale syst.	1.516	2.073	2.697
$\|y_{ee}^{min}\|$-$\|y_{ee}^{max}\|$ [%]	1.2–1.6	1.6–2.0	2.0–2.4
E-res ZSmearing	−0.005	0.112	0.199
E-res SamplingTerm	−0.066	0.045	−0.394
E-res MaterialID	−0.039	0.085	0.451
E-res MaterialCalo	−0.010	−0.017	0.000
E-res MaterialGap	0.005	0.019	0.522
E-res MaterialCryo	−0.013	0.026	−0.069
E-res PileUp	0.010	−0.010	−0.065
E-scale stat.	−0.010	−0.029	−0.546
E-scale Zee syst.	0.480	0.782	2.317
E-scale PS	−0.386	−0.416	−0.265
E-scale S12	−0.259	−0.488	−1.563
E-scale MatID	0.355	0.452	0.712
E-scale MatCryo	0.668	0.970	0.448
E-scale MatCalo	0.114	0.064	0.000
E-scale LArCalib	−0.549	−0.809	−1.878
E-scale LArUnconvCalib	0.132	0.064	0.000
E-scale LArElecUnconv	−0.956	−0.968	−0.201
E-scale LArElecCalib	0.068	0.097	0.498
E-scale L1Gain	0.442	1.855	3.803
E-scale L2Gain	−2.681	−2.703	−1.720
E-scale G4	0.043	0.102	0.217
E-scale Pedestal	−0.049	−0.097	−0.000
E-res syst.	0.123	0.317	1.098
E-scale syst.	3.238	3.863	5.690

Table O.7 Systematic energy scale and resolution uncertainties, as a function of $|\Delta\eta_{ee}|$ in the 116-150 GeV mass region. The given separate sources are the maximum between up an down variation. For the total uncertainty first the up/down variation of each source is added in quadrature and then the maximum is taken

$\|\Delta\eta^{min}\|$-$\|\Delta\eta^{max}\|$ [%]	0.0–0.2	0.2–0.5	0.5–0.8	0.8–1.0	1.0–1.2	1.2–1.5
E-res ZSmearing	0.137	0.116	0.077	0.074	0.053	0.067
E-res SamplingTerm	−0.013	−0.005	−0.011	−0.011	0.002	−0.011
E-res MaterialID	0.036	0.050	0.011	0.040	−0.004	0.025
E-res MaterialCalo	0.017	0.015	−0.000	−0.001	0.002	0.014
E-res MaterialGap	0.008	0.029	0.017	0.023	0.009	0.002
E-res MaterialCryo	0.012	0.024	−0.009	−0.001	0.008	0.020
E-res PileUp	−0.010	−0.019	−0.022	−0.018	−0.005	−0.005
E-scale stat.	0.015	−0.013	0.001	0.012	−0.023	0.003
E-scale Zee syst.	0.339	0.444	0.422	0.339	0.358	0.368
E-scale PS	−0.103	−0.108	−0.099	−0.061	−0.080	−0.009
E-scale S12	−0.060	−0.077	−0.062	−0.026	−0.038	−0.005
E-scale MatID	0.064	0.083	0.075	0.042	0.060	0.002
E-scale MatCryo	0.149	0.198	0.165	0.100	0.100	0.040
E-scale MatCalo	0.038	0.056	0.048	0.026	0.041	−0.004
E-scale LArCalib	−0.143	−0.165	−0.125	−0.065	−0.103	−0.009
E-scale LArUnconvCalib	0.040	0.049	0.040	0.014	0.040	−0.013
E-scale LArElecUnconv	−0.186	−0.232	−0.204	−0.127	−0.119	−0.044
E-scale LArElecCalib	0.008	0.021	0.010	−0.002	0.006	0.004
E-scale L1Gain	0.014	0.026	0.011	0.003	0.011	−0.002
E-scale L2Gain	−0.329	−0.394	−0.349	−0.181	−0.106	0.009
E-scale G4	0.017	0.020	0.012	0.004	0.023	−0.002
E-scale Pedestal	−0.036	−0.049	−0.033	−0.011	−0.026	0.000
E-res syst.	0.158	0.188	0.118	0.102	0.179	0.120

(continued)

Table O.7 (continued)

$\lvert\Delta\eta^{min}\rvert$-$\lvert\Delta\eta^{max}\rvert$ [%]	0.0–0.2	0.2–0.5	0.5–0.8	0.8–1.0	1.0–1.2	1.2–1.5
E-scale syst.	0.580	0.720	0.654	0.455	0.448	0.383
$\lvert\Delta\eta^{min}\rvert$-$\lvert\Delta\eta^{max}\rvert$ [%]	1.5–1.8	1.8–2.0	2.0–2.2	2.2–2.5	2.5–2.8	2.8–3.0
E-res ZSmearing	0.133	0.034	0.065	0.120	0.002	−1.855
E-res SamplingTerm	−0.015	0.008	0.023	−0.038	0.081	0.792
E-res MaterialID	0.100	−0.010	0.048	−0.015	0.115	3.147
E-res MaterialCalo	0.009	−0.001	−0.011	0.015	0.141	0.754
E-res MaterialGap	0.000	−0.034	−0.059	0.024	−0.190	0.000
E-res MaterialCryo	0.016	−0.020	−0.012	−0.011	0.129	0.533
E-res PileUp	−0.002	−0.004	−0.004	0.016	0.017	0.000
E-scale stat.	−0.011	0.019	0.019	−0.021	0.073	0.532
E-scale Zee syst.	0.382	0.388	0.343	0.350	0.083	−0.531
E-scale PS	−0.009	0.037	0.067	0.086	0.045	0.000
E-scale S12	−0.002	0.019	0.055	0.061	−0.035	−0.532
E-scale MatID	0.015	−0.025	−0.080	−0.065	−0.079	0.532
E-scale MatCryo	0.004	−0.063	−0.141	−0.187	0.013	−0.000
E-scale MatCalo	0.003	−0.010	−0.041	−0.037	0.036	−0.000
E-scale LArCalib	−0.000	0.058	0.113	0.113	0.135	−0.532
E-scale LArUnconvCalib	0.004	−0.012	−0.033	−0.031	0.036	−0.000
E-scale LArElecUnconv	−0.014	0.064	0.141	0.207	0.092	1.103
E-scale LArElecCalib	0.002	−0.008	−0.020	−0.027	0.016	0.532
E-scale L1Gain	−0.001	−0.004	−0.006	0.011	−0.000	0.532
E-scale L2Gain	0.069	0.140	0.178	0.259	0.143	0.475
E-scale G4	0.007	−0.003	−0.010	−0.024	−0.002	−0.000
E-scale Pedestal	−0.003	0.019	0.050	0.042	0.014	0.000
E-res syst.	0.243	0.150	0.171	0.195	0.420	6.745
E-scale syst.	0.432	0.537	0.481	0.598	0.454	2.382

Table O.8 Systematic energy scale and resolution uncertainties, as a function of $|\Delta\eta_{ee}|$ in the 150-200 GeV mass region. The given separate sources are the maximum between up an down variation. For the total uncertainty first the up/down variation of each source is added in quadrature and then the maximum is taken

$\lvert\Delta\eta^{min}\rvert$-$\lvert\Delta\eta^{max}\rvert$ [%]	0.0–0.2	0.2–0.5	0.5–0.8	0.8–1.0	1.0–1.2	1.2–1.5
E-res ZSmearing	0.016	0.033	−0.040	0.056	0.068	0.006
E-res SamplingTerm	0.013	−0.004	0.016	−0.017	−0.038	0.016
E-res MaterialID	0.018	0.003	−0.002	0.034	0.018	0.024
E-res MaterialCalo	−0.016	0.005	0.001	0.005	0.000	0.015
E-res MaterialGap	−0.003	−0.003	−0.004	0.010	0.014	0.011
E-res MaterialCryo	−0.002	0.005	0.030	0.007	0.007	−0.011
E-res PileUp	−0.001	−0.005	0.000	0.004	−0.008	0.018
E-scale stat.	0.008	−0.001	−0.005	−0.004	0.014	−0.016
E-scale Zee syst.	0.243	0.259	0.238	0.260	0.280	0.260
E-scale PS	−0.116	−0.107	−0.068	−0.091	−0.103	−0.064
E-scale S12	−0.091	−0.074	−0.045	−0.061	−0.079	−0.054
E-scale MatID	0.080	0.089	0.045	0.079	0.086	0.061
E-scale MatCryo	0.177	0.145	0.144	0.142	0.142	0.092
E-scale MatCalo	0.054	0.055	0.026	0.044	0.057	0.040
E-scale LArCalib	−0.157	−0.154	−0.095	−0.135	−0.133	−0.105
E-scale LArUnconvCalib	0.048	0.050	0.022	0.039	0.052	0.040
E-scale LArElecUnconv	−0.187	−0.222	−0.179	−0.197	−0.184	−0.125
E-scale LArElecCalib	0.016	0.011	0.005	0.015	0.021	0.013
E-scale L1Gain	0.041	0.005	0.007	0.008	0.014	0.017
E-scale L2Gain	−0.498	−0.510	−0.426	−0.332	−0.292	−0.151
E-scale G4	0.025	0.020	0.008	0.014	0.023	0.013
E-scale Pedestal	−0.048	−0.046	−0.022	−0.038	−0.050	−0.011
E-res syst.	0.052	0.055	0.098	0.076	0.113	0.067

(continued)

Table O.8 (continued)

$\lvert\Delta\eta^{min}\rvert$-$\lvert\Delta\eta^{max}\rvert$ [%]	0.0–0.2	0.2–0.5	0.5–0.8	0.8–1.0	1.0–1.2	1.2–1.5
E-scale syst.	0.661	0.679	0.566	0.562	0.522	0.391
$\lvert\Delta\eta^{min}\rvert$-$\lvert\Delta\eta^{max}\rvert$ [%]	1.5–1.8	1.8–2.0	2.0–2.2	2.2–2.5	2.5–2.8	2.8–3.0
E-res ZSmearing	0.013	0.013	−0.094	−0.013	−0.030	0.185
E-res SamplingTerm	−0.017	−0.050	−0.007	0.038	0.018	0.142
E-res MaterialID	0.013	0.022	−0.030	−0.050	0.008	0.096
E-res MaterialCalo	0.004	0.000	0.027	−0.021	−0.035	0.043
E-res MaterialGap	−0.002	0.023	0.033	−0.005	0.040	−0.113
E-res MaterialCryo	0.025	0.022	−0.015	−0.023	−0.036	−0.098
E-res PileUp	−0.007	−0.014	0.006	−0.000	−0.001	0.126
E-scale stat.	−0.012	−0.015	−0.028	−0.006	0.015	0.076
E-scale Zee syst.	0.222	0.247	0.255	0.216	0.176	0.252
E-scale PS	−0.039	−0.023	0.005	0.035	0.041	0.158
E-scale S12	−0.027	−0.014	−0.006	0.015	0.061	0.070
E-scale MatID	0.023	0.017	−0.007	−0.028	−0.084	−0.017
E-scale MatCryo	0.050	0.032	−0.001	−0.049	−0.126	−0.214
E-scale MatCalo	0.021	0.010	−0.005	−0.019	−0.025	−0.046
E-scale LArCalib	−0.061	−0.016	0.005	0.047	0.081	0.131
E-scale LArUnconvCalib	0.021	0.011	−0.000	−0.015	−0.025	−0.020
E-scale LArElecUnconv	−0.080	−0.035	−0.011	0.070	0.091	0.216
E-scale LArElecCalib	0.006	0.008	−0.004	−0.002	−0.028	−0.029
E-scale L1Gain	0.013	0.007	−0.016	−0.008	−0.024	−0.011
E-scale L2Gain	−0.044	0.023	0.098	0.160	0.212	0.302
E-scale G4	0.015	0.005	−0.001	−0.004	−0.017	−0.010
E-scale Pedestal	−0.014	−0.012	−0.001	0.018	0.026	0.046
E-res syst.	0.115	0.083	0.159	0.106	0.131	0.556
E-scale syst.	0.283	0.276	0.303	0.305	0.386	0.734

Table O.9 Systematic energy scale and resolution uncertainties, as a function of $|\Delta\eta_{ee}|$ in the 200-300 GeV mass region. The given separate sources are the maximum between up an down variation. For the total uncertainty first the up/down variation of each source is added in quadrature and then the maximum is taken

$\|\Delta\eta^{min}\|$-$\|\Delta\eta^{max}\|$ [%]	0.0–0.2	0.2–0.5	0.5–0.8	0.8–1.0	1.0–1.2	1.2–1.5
E-res ZSmearing	0.012	0.004	0.005	−0.063	−0.040	−0.015
E-res SamplingTerm	−0.015	0.010	0.006	0.052	0.027	−0.009
E-res MaterialID	0.018	−0.011	−0.026	−0.061	−0.023	0.023
E-res MaterialCalo	0.011	−0.016	0.003	−0.014	0.006	0.011
E-res MaterialGap	0.009	0.006	−0.004	−0.016	−0.015	−0.016
E-res MaterialCryo	0.005	−0.029	−0.006	−0.029	−0.011	0.018
E-res PileUp	−0.015	0.019	0.014	0.010	0.018	−0.021
E-scale stat.	0.004	−0.001	−0.009	0.008	−0.010	−0.017
E-scale Zee syst.	0.219	0.231	0.250	0.214	0.174	0.207
E-scale PS	−0.129	−0.148	−0.160	−0.132	−0.101	−0.134
E-scale S12	−0.098	−0.087	−0.088	−0.070	−0.055	−0.081
E-scale MatID	0.104	0.102	0.116	0.114	0.076	0.074
E-scale MatCryo	0.224	0.243	0.223	0.183	0.161	0.163
E-scale MatCalo	0.065	0.046	0.075	0.070	0.046	0.056
E-scale LArCalib	−0.209	−0.184	−0.207	−0.158	−0.146	−0.187
E-scale LArUnconvCalib	0.058	0.042	0.068	0.060	0.040	0.039
E-scale LArElecUnconv	−0.284	−0.272	−0.324	−0.274	−0.189	−0.215
E-scale LArElecCalib	0.028	0.014	0.015	0.005	0.005	0.009
E-scale L1Gain	0.077	0.049	0.037	0.038	0.037	0.024
E-scale L2Gain	−0.907	−1.012	−0.995	−0.755	−0.605	−0.571
E-scale G4	0.027	0.011	0.032	0.034	0.015	0.013
E-scale Pedestal	−0.046	−0.034	−0.059	−0.050	−0.034	−0.042
E-res syst.	0.061	0.080	0.054	0.148	0.105	0.095

(continued)

Table O.9 (continued)

$\lvert\Delta\eta^{min}\rvert$-$\lvert\Delta\eta^{max}\rvert$ [%]	0.0–0.2	0.2–0.5	0.5–0.8	0.8–1.0	1.0–1.2	1.2–1.5
E-scale syst.	1.050	1.179	1.154	0.959	0.777	0.740
$\lvert\Delta\eta^{min}\rvert$-$\lvert\Delta\eta^{max}\rvert$ [%]	1.5–1.8	1.8–2.0	2.0–2.2	2.2–2.5	2.5–2.8	2.8–3.0
E-res ZSmearing	−0.004	0.029	0.009	−0.002	0.016	−0.151
E-res SamplingTerm	0.018	0.021	0.026	−0.022	0.007	−0.087
E-res MaterialID	−0.018	0.015	−0.026	0.008	0.037	−0.052
E-res MaterialCalo	0.004	0.001	−0.021	0.017	0.030	−0.028
E-res MaterialGap	−0.010	−0.000	−0.004	0.006	−0.002	0.092
E-res MaterialCryo	−0.033	−0.018	−0.014	−0.009	0.042	0.072
E-res PileUp	0.006	0.003	−0.001	−0.006	0.030	−0.089
E-scale stat.	0.018	0.014	0.021	−0.005	−0.054	−0.062
E-scale Zee syst.	0.206	0.264	0.292	0.315	0.302	0.287
E-scale PS	−0.084	−0.092	−0.068	−0.067	0.017	−0.072
E-scale S12	−0.070	−0.053	−0.041	−0.031	−0.001	0.004
E-scale MatID	0.075	0.056	0.075	0.041	0.030	−0.093
E-scale MatCryo	0.132	0.141	0.103	0.079	0.023	0.021
E-scale MatCalo	0.049	0.046	0.035	0.032	0.002	−0.002
E-scale LArCalib	−0.129	−0.115	−0.088	−0.083	−0.029	−0.001
E-scale LArUnconvCalib	0.040	0.042	0.032	0.028	0.002	−0.007
E-scale LArElecUnconv	−0.151	−0.163	−0.108	−0.090	−0.022	−0.019
E-scale LArElecCalib	0.014	0.006	0.019	0.004	0.005	0.001
E-scale L1Gain	0.018	0.010	0.032	0.013	0.013	−0.008
E-scale L2Gain	−0.422	−0.334	−0.198	−0.105	0.136	0.089
E-scale G4	0.014	0.016	0.014	0.013	0.014	−0.004
E-scale Pedestal	−0.042	−0.039	−0.026	−0.021	0.007	0.016
E-res syst.	0.153	0.118	0.128	0.051	0.210	0.402
E-scale syst.	0.616	0.531	0.419	0.408	0.437	0.453

Table O.10 Systematic energy scale and resolution uncertainties, as a function of $|\Delta\eta_{ee}|$ in the 300-500 GeV mass region. The given separate sources are the maximum between up an down variation. For the total uncertainty first the up/down variation of each source is added in quadrature and then the maximum is taken

$\|\Delta\eta^{min}\|$-$\|\Delta\eta^{max}\|$ [%]	0.0–0.5	0.5–1.0	1.0–1.5
E-res ZSmearing	0.085	0.020	0.029
E-res SamplingTerm	−0.037	−0.007	−0.019
E-res MaterialID	0.050	−0.004	0.004
E-res MaterialCalo	0.010	−0.009	−0.005
E-res MaterialGap	−0.011	−0.001	0.007
E-res MaterialCryo	0.040	−0.018	−0.016
E-res PileUp	−0.022	0.006	0.005
E-scale stat.	−0.011	0.006	0.015
E-scale Zee syst.	0.278	0.223	0.211
E-scale PS	−0.283	−0.229	−0.181
E-scale S12	−0.180	−0.143	−0.138
E-scale MatID	0.185	0.142	0.110
E-scale MatCryo	0.341	0.286	0.237
E-scale MatCalo	0.126	0.091	0.072
E-scale LArCalib	−0.431	−0.293	−0.273
E-scale LArUnconvCalib	0.107	0.082	0.068
E-scale LArElecUnconv	−0.448	−0.439	−0.332
E-scale LArElecCalib	0.030	0.014	0.020
E-scale L1Gain	0.181	0.078	0.102
E-scale L2Gain	−1.510	−1.415	−1.137
E-scale G4	0.052	0.037	0.034
E-scale Pedestal	−0.100	−0.077	−0.062
E-res syst.	0.140	0.059	0.093
E-scale syst.	1.842	1.642	1.391
$\|\Delta\eta^{min}\|$-$\|\Delta\eta^{max}\|$ [%]	1.5–2.0	2.0–2.5	2.5–3.0
E-res ZSmearing	0.028	0.095	0.094
E-res SamplingTerm	−0.008	−0.038	−0.025
E-res MaterialID	−0.008	0.099	0.073
E-res MaterialCalo	−0.004	0.009	0.019
E-res MaterialGap	0.017	0.023	0.008
E-res MaterialCryo	−0.001	0.054	0.032
E-res PileUp	−0.002	−0.027	−0.031
E-scale stat.	−0.003	0.008	−0.002
E-scale Zee syst.	0.204	0.195	0.296
E-scale PS	−0.132	−0.041	−0.113
E-scale S12	−0.111	−0.066	−0.090
E-scale MatID	0.087	0.049	0.108

(continued)

Table O.10 (continued)

$\|\Delta\eta^{min}\|$-$\|\Delta\eta^{max}\|$ [%]	1.5–2.0	2.0–2.5	2.5–3.0
E-scale MatCryo	0.232	0.138	0.191
E-scale MatCalo	0.037	0.020	0.041
E-scale LArCalib	−0.242	−0.117	−0.122
E-scale LArUnconvCalib	0.018	0.017	0.036
E-scale LArElecUnconv	−0.279	−0.204	−0.193
E-scale LArElecCalib	0.018	0.008	0.033
E-scale L1Gain	0.079	0.030	0.067
E-scale L2Gain	−1.022	−0.661	−0.643
E-scale G4	0.016	0.010	0.004
E-scale Pedestal	−0.032	−0.018	−0.049
E-res syst.	0.078	0.181	0.165
E-scale syst.	1.197	0.777	1.010

Table O.11 Systematic energy scale and resolution uncertainties, as a function of $|\Delta\eta_{ee}|$ in the 500-1500 GeV mass region. The given separate sources are the maximum between up an down variation. For the total uncertainty first the up/down variation of each source is added in quadrature and then the maximum is taken

$\|\Delta\eta^{min}\|$-$\|\Delta\eta^{max}\|$ [%]	0.0–0.5	0.5–1.0	1.0–1.5
E-res ZSmearing	−0.015	0.022	0.005
E-res SamplingTerm	0.035	−0.014	−0.019
E-res MaterialID	−0.010	0.007	0.004
E-res MaterialCalo	−0.010	−0.006	−0.002
E-res MaterialGap	0.000	0.017	−0.009
E-res MaterialCryo	−0.018	0.011	0.007
E-res PileUp	0.007	0.012	0.014
E-scale stat.	−0.005	−0.031	−0.002
E-scale Zee syst.	0.245	0.281	0.250
E-scale PS	−0.369	−0.394	−0.346
E-scale S12	−0.239	−0.235	−0.212
E-scale MatID	0.218	0.240	0.219
E-scale MatCryo	0.399	0.421	0.372
E-scale MatCalo	0.178	0.180	0.147
E-scale LArCalib	−0.528	−0.545	−0.520
E-scale LArUnconvCalib	0.156	0.169	0.125
E-scale LArElecUnconv	−0.671	−0.625	−0.567
E-scale LArElecCalib	0.015	0.024	0.022
E-scale L1Gain	0.124	0.119	0.235
E-scale L2Gain	−1.965	−1.968	−1.907

(continued)

Table O.11 (continued)

$\lvert\Delta\eta^{min}\rvert$-$\lvert\Delta\eta^{max}\rvert$ [%]	0.0–0.5	0.5–1.0	1.0–1.5
E-scale G4	0.078	0.092	0.047
E-scale Pedestal	−0.132	−0.135	−0.092
E-res syst.	0.073	0.095	0.043
E-scale syst.	2.345	2.303	2.201
$\lvert\Delta\eta^{min}\rvert$-$\lvert\Delta\eta^{max}\rvert$ [%]	1.5–2.0	2.0–2.5	2.5–3.0
E-res ZSmearing	−0.010	−0.128	−0.006
E-res SamplingTerm	0.032	0.050	−0.061
E-res MaterialID	−0.029	−0.072	0.036
E-res MaterialCalo	−0.010	−0.006	0.022
E-res MaterialGap	−0.029	−0.022	−0.010
E-res MaterialCryo	−0.011	−0.026	−0.037
E-res PileUp	−0.001	0.024	−0.010
E-scale stat.	0.012	0.026	0.017
E-scale Zee syst.	0.294	0.293	0.246
E-scale PS	−0.326	−0.248	−0.158
E-scale S12	−0.230	−0.168	−0.141
E-scale MatID	0.225	0.180	0.141
E-scale MatCryo	0.317	0.351	0.228
E-scale MatCalo	0.129	0.094	0.047
E-scale LArCalib	−0.449	−0.406	−0.272
E-scale LArUnconvCalib	0.119	0.090	0.054
E-scale LArElecUnconv	−0.464	−0.402	−0.287
E-scale LArElecCalib	0.030	0.035	0.017
E-scale L1Gain	0.250	0.253	0.157
E-scale L2Gain	−1.618	−1.459	−1.321
E-scale G4	0.045	0.020	0.035
E-scale Pedestal	−0.097	−0.055	−0.035
E-res syst.	0.104	0.183	0.122
E-scale syst.	1.904	1.724	1.555

Appendix P
High-Mass Drell-Yan: Electron Cross Section Tables

This appendix contains detailed tables with the measured electron channel cross sections and their uncertainties. Tables P.1, P.2, P.3 contain the cross sections for the single-differential cross section, the double-differential cross section as a function of absolute rapidity, and the double-differential cross section as a function of absolute pseudorapidity separation, respectively. For the electron energy scale and energy resolution uncertainties a single nuisance parameter is given. A detailed breakdown of these uncertainties is provided in Appendix O.

M. Zinser, *Search for New Heavy Charged Bosons and Measurement of High-Mass Drell-Yan Production in Proton-Proton Collisions*, Springer Theses,
https://doi.org/10.1007/978-3-030-00650-1

Table P.1 The electron channel Born level single-differential cross section $\frac{d\sigma}{dm_{ee}}$. The measurements are listed together with the statistical (δ^{stat}), systematic (δ^{sys}) and total (δ^{tot}) uncertainties. In addition the contributions from the individual correlated (cor) and uncorrelated (unc) systematic error sources are also provided consisting of the trigger efficiency (δ^{trig}), electron reconstruction efficiency (δ^{reco}), electron identification efficiency (δ^{id}), the isolation efficiency (δ^{iso}), the electron energy resolution (δ^{Eres}), the electron energy scale (δ^{Escale}), the multijet and W+jets background ($\delta^{\text{mult.}}$), the top and diboson background normalisation (δ^{top}, δ^{diboson}), the top and diboson background MC statistical uncertainty (δ^{bgMC}), and the signal MC statistical uncertainty (δ^{MC}). The ratio of the dressed level to Born level predictions (k_{dressed}) is also provided. The luminosity uncertainty of 1.9% is not shown and not included in the overall systematic and total uncertainties

m_{ee} [GeV]	$\frac{d\sigma}{dm_{ee}}$ [pb/GeV]	δ^{stat} [%]	δ^{sys} [%]	δ^{tot} [%]	$\delta^{\text{trig}}_{\text{cor}}$ [%]	$\delta^{\text{trig}}_{\text{unc}}$ [%]	$\delta^{\text{reco}}_{\text{cor}}$ [%]	$\delta^{\text{id}}_{\text{cor}}$ [%]	$\delta^{\text{iso}}_{\text{cor}}$ [%]	$\delta^{\text{iso}}_{\text{unc}}$ [%]	$\delta^{\text{Eres}}_{\text{cor}}$ [%]	$\delta^{\text{Escale}}_{\text{cor}}$ [%]	$\delta^{\text{mult.}}_{\text{cor}}$ [%]	$\delta^{\text{mult.}}_{\text{unc}}$ [%]	$\delta^{\text{top}}_{\text{cor}}$ [%]	$\delta^{\text{diboson}}_{\text{cor}}$ [%]	$\delta^{\text{bgMC}}_{\text{unc}}$ [%]	$\delta^{\text{MC}}_{\text{unc}}$ [%]	k_{dressed} [%]
116–130	2.31×10^{-1}	0.5	0.8	1.0	−0.1	0.0	0.0	−0.3	0.0	0.0	0.1	0.5	−0.5	0.1	−0.3	−0.1	0.0	0.1	1.047
130–150	1.05×10^{-1}	0.7	1.0	1.2	−0.1	0.0	−0.1	−0.4	0.0	0.1	0.1	0.4	−0.7	0.2	−0.5	−0.2	0.1	0.1	1.046
150–175	5.06×10^{-2}	0.8	1.3	1.6	0.0	0.1	−0.1	−0.5	0.0	0.1	0.1	0.4	−0.8	0.3	−0.7	−0.2	0.1	0.1	1.047
175–200	2.60×10^{-2}	1.2	1.6	2.0	−0.1	0.1	−0.1	−0.6	0.0	0.1	0.0	0.5	−0.9	0.3	−0.9	−0.3	0.2	0.1	1.052
200–230	1.39×10^{-2}	1.5	2.0	2.5	−0.1	0.1	−0.1	−0.7	0.0	0.2	0.1	0.7	−1.2	0.4	−1.1	−0.4	0.2	0.2	1.053
230–260	7.95×10^{-3}	2.0	2.2	3.0	−0.1	0.1	−0.2	−0.7	−0.1	0.2	0.1	1.0	−1.1	0.4	−1.3	−0.4	0.3	0.2	1.056
260–300	4.43×10^{-3}	2.4	2.3	3.3	−0.1	0.1	−0.2	−0.7	−0.1	0.2	0.1	0.9	−1.3	0.5	−1.3	−0.6	0.4	0.2	1.058
300–380	1.84×10^{-3}	2.6	2.5	3.6	−0.1	0.2	−0.2	−0.8	−0.1	0.3	0.1	1.3	−1.1	0.4	−1.4	−0.6	0.4	0.2	1.063
380–500	5.99×10^{-4}	3.6	2.7	4.5	−0.1	0.2	−0.2	−0.8	−0.2	0.5	0.1	1.6	−1.4	0.5	−1.1	−0.6	0.5	0.2	1.067
500–700	1.52×10^{-4}	5.3	2.6	6.0	−0.1	0.2	−0.2	−0.8	−0.2	0.7	0.1	2.0	−0.7	0.5	−0.7	−0.6	0.5	0.3	1.075
700–1000	2.64×10^{-5}	10.2	3.3	10.7	−0.2	0.4	−0.2	−0.8	−0.3	1.4	0.1	2.3	−0.6	0.8	−0.4	−0.6	0.7	0.4	1.085
1000–1500	3.23×10^{-6}	22.5	5.8	23.2	−0.7	0.9	−0.2	−0.8	−0.3	3.5	0.0	2.8	−1.9	1.6	−0.3	−0.6	2.1	0.2	1.100

Table P.2 The electron channel Born level double-differential cross section $\frac{d^2\sigma}{dm_{ee}d|y_{ee}|}$. The measurements are listed together with the statistical (δ^{stat}), systematic (δ^{sys}) and total (δ^{tot}) uncertainties. In addition the contributions from the individual correlated (cor) and uncorrelated (unc) systematic error sources are also provided consisting of the trigger efficiency (δ^{trig}), electron reconstruction efficiency (δ^{reco}), electron identification efficiency (δ^{id}), the isolation efficiency (δ^{iso}), the electron energy resolution (δ^{Eres}), the electron energy scale (δ^{Escale}), the multijet and W+jets background ($\delta^{mult.}$), the top and diboson background normalisation (δ^{top}, $\delta^{diboson}$), the top and diboson background MC statistical uncertainty (δ^{bgMC}), and the signal MC statistical uncertainty (δ^{MC}). The ratio of the dressed level to Born level predictions ($k_{dressed}$) is also provided. The luminosity uncertainty of 1.9% is not shown and not included in the overall systematic and total uncertainties

m_{ee} [GeV]	$\|y_{ee}\|$	$\frac{d^2\sigma}{dm_{ee}d\|y_{ee}\|}$ [pb/GeV]	δ^{stat} [%]	δ^{sys} [%]	δ^{tot} [%]	δ^{trig}_{cor} [%]	δ^{trig}_{unc} [%]	δ^{reco}_{cor} [%]	δ^{id}_{cor} [%]	δ^{iso}_{cor} [%]	δ^{iso}_{unc} [%]	δ^{Eres}_{cor} [%]	δ^{Escale}_{cor} [%]	$\delta^{mult.}_{cor}$ [%]	$\delta^{mult.}_{unc}$ [%]	δ^{top}_{cor} [%]	$\delta^{diboson}_{cor}$ [%]	δ^{bgMC}_{unc} [%]	δ^{MC}_{unc} [%]	$k_{dressed}$
116–150	0.0–0.2	4.15×10^{-2}	1.1	0.8	1.4	−0.1	0.0	−0.1	−0.3	0.0	0.0	0.1	0.2	−0.3	0.1	−0.5	−0.1	0.1	0.2	1.048
116–150	0.2–0.4	4.11×10^{-2}	1.2	0.8	1.4	−0.1	0.0	−0.1	−0.3	0.0	0.0	0.1	0.3	−0.3	0.1	−0.5	−0.1	0.1	0.2	1.048
116–150	0.4–0.6	4.09×10^{-2}	1.2	0.9	1.5	−0.1	0.0	−0.1	−0.3	0.0	0.0	0.3	0.3	−0.4	0.1	−0.5	−0.1	0.1	0.3	1.047
116–150	0.6–0.8	4.09×10^{-2}	1.2	0.9	1.5	−0.1	0.0	−0.1	−0.3	0.0	0.0	0.2	0.4	−0.3	0.1	−0.4	−0.1	0.1	0.3	1.048
116–150	0.8–1.0	3.97×10^{-2}	1.3	0.9	1.6	−0.1	0.0	−0.1	−0.3	0.0	0.0	0.2	0.5	−0.3	0.2	−0.4	−0.1	0.1	0.3	1.047
116–150	1.0–1.2	3.97×10^{-2}	1.3	1.0	1.6	−0.1	0.0	−0.1	−0.3	0.0	0.0	0.1	0.6	−0.5	0.2	−0.3	−0.1	0.1	0.3	1.047
116–150	1.2–1.4	3.86×10^{-2}	1.3	1.2	1.8	−0.1	0.0	−0.1	−0.3	0.0	0.0	0.3	0.7	−0.6	0.2	−0.3	−0.1	0.1	0.3	1.046
116–150	1.4–1.6	3.44×10^{-2}	1.4	1.3	1.9	−0.1	0.0	−0.1	−0.4	0.0	0.0	0.2	0.8	−0.7	0.2	−0.2	−0.1	0.1	0.3	1.046
116–150	1.6–1.8	2.86×10^{-2}	1.6	1.5	2.2	-0.1	0.0	−0.1	−0.5	0.0	0.0	0.2	1.0	−0.9	0.3	−0.2	−0.1	0.1	0.4	1.044
116–150	1.8–2.0	2.29×10^{-2}	1.8	1.6	2.4	−0.1	0.0	−0.1	−0.6	0.0	0.1	0.3	1.1	−0.9	0.3	−0.1	−0.1	0.1	0.4	1.043
116–150	2.0–2.2	1.49×10^{-2}	2.1	2.0	2.9	−0.1	0.0	−0.1	−0.6	0.0	0.1	0.4	0.8	−1.5	0.4	−0.1	−0.1	0.1	0.5	1.044
116–150	2.2–2.4	7.05×10^{-3}	3.3	3.1	4.5	0.0	0.0	−0.2	−0.7	0.0	0.1	0.4	1.2	−2.5	0.6	−0.1	−0.1	0.2	0.8	1.045
150–200	0.0–0.2	1.06×10^{-2}	2.0	1.5	2.5	−0.1	0.1	−0.1	−0.5	0.0	0.1	0.1	0.2	−0.6	0.2	−1.2	−0.3	0.3	0.2	1.052
150–200	0.2–0.4	1.06×10^{-2}	2.0	1.5	2.6	−0.1	0.1	−0.1	−0.5	0.0	0.1	0.1	0.3	−0.6	0.3	−1.1	−0.3	0.3	0.2	1.050
150–200	0.4–0.6	1.05×10^{-2}	2.1	1.5	2.6	−0.1	0.1	−0.1	−0.5	0.0	0.1	0.1	0.2	−0.6	0.3	−1.1	−0.4	0.3	0.3	1.052
150–200	0.6–0.8	1.06×10^{-2}	2.1	1.5	2.6	−0.1	0.1	−0.1	−0.5	0.0	0.1	0.2	0.5	−0.7	0.3	−1.0	−0.3	0.3	0.3	1.053
150–200	0.8–1.0	1.02×10^{-2}	2.1	1.5	2.6	−0.1	0.1	−0.1	−0.5	0.0	0.1	0.1	0.4	−0.8	0.4	−0.8	−0.3	0.3	0.3	1.050
150–200	1.0–1.2	9.71×10^{-3}	2.2	1.7	2.8	−0.1	0.1	−0.1	−0.5	0.0	0.1	0.2	0.6	−1.1	0.4	−0.7	−0.3	0.3	0.3	1.050
150–200	1.2–1.4	9.25×10^{-3}	2.3	1.5	2.7	−0.1	0.1	−0.1	−0.5	0.0	0.1	0.1	0.7	−0.8	0.3	−0.6	−0.3	0.3	0.3	1.048
150–200	1.4–1.6	7.60×10^{-3}	2.5	1.8	3.1	0.0	0.1	−0.1	−0.6	0.0	0.1	0.1	1.0	−1.0	0.4	−0.5	−0.3	0.3	0.3	1.046
150–200	1.6–1.8	6.66×10^{-3}	2.8	1.9	3.3	0.0	0.1	−0.2	−0.7	0.0	0.1	0.2	1.2	−1.1	0.4	−0.3	−0.2	0.3	0.4	1.043

(continued)

Table P.2 (continued)

m_{ee} [GeV]	$\|y_{ee}\|$	$\frac{d^2\sigma}{dm_{ee}d\|y_{ee}\|}$ [pb/GeV]	δ^{stat} [%]	δ^{sys} [%]	δ^{tot} [%]	δ^{trig}_{cor} [%]	δ^{trig}_{unc} [%]	δ^{reco}_{cor} [%]	δ^{id}_{cor} [%]	δ^{iso}_{cor} [%]	δ^{iso}_{unc} [%]	δ^{Eres}_{cor} [%]	δ^{Escale}_{cor} [%]	$\delta^{mult.}_{cor}$ [%]	$\delta^{mult.}_{unc}$ [%]	δ^{top}_{cor} [%]	$\delta^{diboson}_{cor}$ [%]	δ^{bgMC}_{unc} [%]	δ^{MC}_{unc} [%]	$k_{dressed}$
150–200	1.8–2.0	4.94×10^{-3}	3.1	1.7	3.6	0.0	0.1	−0.2	−0.8	0.0	0.1	0.3	0.8	−1.0	0.4	−0.2	−0.2	0.3	0.5	1.043
150–200	2.0–2.2	3.30×10^{-3}	3.5	1.9	4.0	0.0	0.1	−0.4	−0.8	0.0	0.1	0.2	0.5	−1.4	0.4	−0.1	−0.1	0.3	0.5	1.038
150–200	2.2–2.4	1.52×10^{-3}	5.5	3.2	6.3	−0.1	0.1	−0.6	−0.9	0.0	0.1	0.2	1.3	−2.4	0.8	−0.1	−0.1	0.3	0.8	1.038
200–300	0.0–0.2	2.33×10^{-3}	3.2	2.5	4.1	−0.1	0.1	−0.2	−0.7	−0.1	0.2	0.1	0.5	−0.9	0.5	−1.9	−0.5	0.6	0.3	1.063
200–300	0.2–0.4	2.34×10^{-3}	3.2	2.4	4.0	−0.1	0.1	−0.2	−0.7	−0.1	0.2	0.2	0.4	−0.9	0.5	−1.8	−0.5	0.6	0.3	1.063
200–300	0.4–0.6	2.49×10^{-3}	3.2	2.4	4.0	−0.1	0.1	−0.2	−0.7	−0.1	0.2	0.1	0.5	−1.3	0.6	−1.6	−0.6	0.6	0.3	1.063
200–300	0.6–0.8	2.54×10^{-3}	3.1	2.3	3.9	−0.1	0.1	−0.1	−0.7	−0.1	0.2	0.1	0.8	−1.2	0.6	−1.4	−0.5	0.5	0.3	1.060
200–300	0.8–1.0	2.29×10^{-3}	3.3	2.3	4.0	−0.1	0.1	−0.1	−0.7	−0.1	0.2	0.2	1.0	−1.1	0.6	−1.3	−0.5	0.5	0.3	1.056
200–300	1.0–1.2	2.14×10^{-3}	3.4	2.4	4.1	−0.1	0.1	−0.2	−0.7	−0.1	0.2	0.2	1.3	−1.3	0.5	−1.0	−0.5	0.5	0.4	1.053
200–300	1.2–1.4	1.83×10^{-3}	3.6	2.4	4.4	−0.1	0.1	−0.2	−0.7	−0.1	0.2	0.1	1.4	−1.4	0.5	−0.8	−0.4	0.5	0.4	1.049
200–300	1.4–1.6	1.63×10^{-3}	3.7	2.1	4.3	−0.1	0.1	−0.2	−0.8	−0.1	0.2	0.1	1.2	−1.2	0.4	−0.6	−0.3	0.5	0.4	1.044
200–300	1.6–1.8	1.32×10^{-3}	4.2	2.3	4.8	-0.1	0.1	−0.3	−0.8	−0.1	0.2	0.3	1.5	−1.2	0.4	−0.4	−0.3	0.5	0.5	1.041
200–300	1.8–2.0	9.87×10^{-4}	4.8	2.4	5.4	−0.1	0.1	−0.4	−0.9	−0.1	0.2	0.3	1.5	−1.2	0.5	−0.2	−0.2	0.5	0.6	1.044
200–300	2.0–2.2	6.13×10^{-4}	5.6	2.3	6.1	−0.1	0.1	−0.6	−1.0	−0.1	0.2	0.3	0.7	−1.6	0.5	−0.1	−0.1	0.4	0.6	1.044
200–300	2.2–2.4	2.51×10^{-4}	9.1	3.2	9.6	−0.1	0.2	−0.9	−1.1	−0.1	0.2	0.7	1.4	−1.8	1.1	−0.1	−0.1	0.5	1.1	1.042
300–500	0.0–0.4	3.23×10^{-4}	4.6	3.3	5.7	−0.1	0.2	−0.2	−0.8	−0.1	0.4	0.1	0.9	−1.8	0.6	−2.2	−0.8	0.8	0.3	1.080
300–500	0.4–0.8	3.34×10^{-4}	4.3	2.8	5.1	−0.1	0.2	−0.2	−0.8	−0.1	0.4	0.1	1.4	−1.1	0.6	−1.6	−0.7	0.7	0.3	1.072
300–500	0.8–1.2	3.16×10^{-4}	4.3	2.8	5.2	−0.1	0.2	−0.2	−0.8	−0.1	0.4	0.2	2.0	−0.9	0.5	−1.1	−0.6	0.7	0.3	1.058
300–500	1.2–1.6	2.30×10^{-4}	4.9	2.9	5.7	−0.1	0.2	−0.2	−0.8	−0.1	0.4	0.1	2.0	−1.6	0.5	−0.6	−0.4	0.6	0.4	1.053
300–500	1.6–2.0	1.31×10^{-4}	6.5	3.2	7.3	−0.1	0.2	−0.4	−0.9	−0.2	0.4	0.2	2.8	−0.3	0.4	−0.2	−0.2	0.5	0.6	1.047
300–500	2.0–2.4	3.62×10^{-5}	11.5	3.5	12.0	−0.1	0.2	−0.6	−1.0	−0.2	0.4	0.4	2.5	−1.3	1.0	0.0	−0.1	0.8	0.9	1.046
500–1500	0.0–0.4	1.45×10^{-5}	8.9	2.8	9.4	−0.2	0.3	−0.2	−0.8	−0.2	1.0	0.1	1.5	−0.7	0.8	−1.0	−0.7	1.0	0.3	1.096
500–1500	0.4–0.8	1.45×10^{-5}	8.5	2.9	9.0	−0.2	0.3	−0.2	−0.8	−0.2	1.0	0.1	2.1	−0.3	0.6	−0.6	−0.6	0.7	0.5	1.083
500–1500	0.8–1.2	1.05×10^{-5}	10.0	3.5	10.6	−0.1	0.3	−0.2	−0.8	−0.2	0.9	0.1	2.7	−1.1	0.8	−0.5	−0.5	0.9	0.5	1.067
500–1500	1.2–1.6	7.86×10^{-6}	11.1	3.6	11.7	−0.1	0.2	−0.2	−0.8	−0.2	0.9	0.1	3.2	−0.3	0.7	−0.1	−0.2	0.4	0.4	1.055
500–1500	1.6–2.0	2.29×10^{-6}	21.4	4.3	21.8	−0.1	0.2	−0.4	−0.9	−0.3	0.8	0.3	3.9	−0.4	0.7	−0.1	−0.2	0.7	0.9	1.056
500–1500	2.0–2.4	2.51×10^{-7}	60.4	7.8	60.9	−0.1	0.2	−0.6	−1.0	−0.3	0.8	1.1	5.7	−2.7	2.7	−0.1	−0.1	2.3	2.4	1.067

Table P.3 The electron channel Born level double-differential cross section $\frac{d^2\sigma}{dm_{ee}d|\Delta\eta_{ee}|}$. The measurements are listed together with the statistical (δ^{stat}), systematic (δ^{sys}) and total (δ^{tot}) uncertainties. In addition the contributions from the individual correlated (cor) and uncorrelated (unc) systematic error sources are also provided consisting of the trigger efficiency (δ^{trig}), electron reconstruction efficiency (δ^{reco}), electron identification efficiency (δ^{id}), the isolation efficiency (δ^{iso}), the electron energy resolution (δ^{Eres}), the electron energy scale (δ^{Escale}), the multijet and W+jets background ($\delta^{mult.}$), the top and diboson background normalisation (δ^{top}, $\delta^{diboson}$), the top and diboson background MC statistical uncertainty (δ^{bgMC}), and the signal MC statistical uncertainty (δ^{MC}). The ratio of the dressed level to Born level predictions ($k_{dressed}$) is also provided. The luminosity uncertainty of 1.9% is not shown and not included in the overall systematic and total uncertainties

m_{ee} [GeV]	$\|\Delta\eta_{ee}\|$	$\frac{d^2\sigma}{dm_{ee}d\|\Delta\eta_{ee}\|}$ [pb/GeV]	δ^{stat} [%]	δ^{sys} [%]	δ^{tot} [%]	δ^{trig}_{cor} [%]	δ^{trig}_{unc} [%]	δ^{reco}_{cor} [%]	δ^{id}_{cor} [%]	δ^{iso}_{cor} [%]	δ^{iso}_{unc} [%]	δ^{Eres}_{cor} [%]	δ^{Escale}_{cor} [%]	$\delta^{mult.}_{cor}$ [%]	$\delta^{mult.}_{unc}$ [%]	δ^{top}_{cor} [%]	$\delta^{diboson}_{cor}$ [%]	δ^{bgMC}_{unc} [%]	δ^{MC}_{unc} [%]	$k_{dressed}$
116–150	0.00–0.25	4.99×10^{-2}	1.0	1.0	1.4	0.0	0.1	-0.1	-0.4	0.0	0.1	0.2	0.6	-0.5	0.1	−0.3	−0.1	0.1	0.2	1.043
116–150	0.25–0.50	4.72×10^{-2}	1.0	1.1	1.5	−0.1	0.0	−0.1	−0.4	0.0	0.1	0.2	0.7	−0.5	0.1	−0.3	−0.1	0.1	0.2	1.044
116–150	0.50–0.75	4.40×10^{-2}	1.1	1.0	1.5	−0.1	0.0	−0.1	−0.4	0.0	0.1	0.1	0.7	−0.5	0.1	−0.3	−0.1	0.1	0.2	1.044
116–150	0.75–1.00	4.05×10^{-2}	1.1	0.9	1.4	−0.1	0.0	−0.1	−0.3	0.0	0.0	0.1	0.5	−0.4	0.1	−0.3	−0.1	0.1	0.3	1.045
116–150	1.00–1.25	3.59×10^{-2}	1.2	0.9	1.5	−0.1	0.0	−0.1	−0.3	0.0	0.0	0.2	0.5	−0.5	0.2	−0.4	−0.1	0.1	0.3	1.048
116–150	1.25–1.50	3.25×10^{-2}	1.3	0.9	1.5	−0.1	0.0	0.0	−0.3	0.0	0.0	0.1	0.4	−0.5	0.2	−0.4	−0.1	0.1	0.3	1.050
116–150	1.50–1.75	2.60×10^{-2}	1.4	1.0	1.8	−0.1	0.0	0.0	−0.2	0.0	0.0	0.2	0.4	−0.6	0.2	−0.4	−0.2	0.1	0.3	1.050
116–150	1.75–2.00	2.03×10^{-2}	1.7	1.1	2.0	−0.1	0.0	0.0	−0.3	0.0	0.0	0.1	0.5	−0.7	0.3	−0.4	−0.2	0.2	0.4	1.055
116–150	2.00–2.25	1.20×10^{-2}	2.2	1.4	2.6	−0.1	0.0	0.0	−0.3	0.0	0.0	0.2	0.5	−0.9	0.5	−0.5	−0.3	0.3	0.4	1.055
116–150	2.25–2.50	4.25×10^{-3}	4.0	2.2	4.5	−0.1	0.0	0.0	−0.3	0.0	0.0	0.2	0.6	−1.4	0.9	−0.8	−0.3	0.5	0.6	1.047
116–150	2.50–2.75	6.70×10^{-4}	11.4	5.3	12.5	−0.1	0.0	−0.1	−0.4	0.0	0.0	0.4	0.5	−3.6	2.4	−1.4	−0.8	1.8	1.7	1.044
150–200	0.00–0.25	1.08×10^{-2}	1.7	1.3	2.2	−0.1	0.1	−0.1	−0.7	0.0	0.1	0.1	0.7	−0.7	0.2	−0.6	−0.1	0.2	0.2	1.042
150–200	0.25–0.50	1.04×10^{-2}	1.8	1.3	2.2	−0.1	0.1	−0.1	−0.6	0.0	0.1	0.1	0.7	−0.6	0.2	−0.6	−0.2	0.2	0.2	1.042
150–200	0.50–0.75	9.63×10^{-3}	1.9	1.3	2.3	0.0	0.1	−0.1	−0.6	0.0	0.1	0.1	0.6	−0.7	0.2	−0.7	−0.2	0.2	0.2	1.043
150–200	0.75–1.00	9.38×10^{-3}	2.0	1.3	2.4	0.0	0.1	−0.1	−0.6	0.0	0.1	0.1	0.6	−0.6	0.2	−0.7	−0.2	0.2	0.3	1.044
150–200	1.00–1.25	8.24×10^{-3}	2.0	1.4	2.5	0.0	0.1	−0.1	−0.5	0.0	0.1	0.1	0.5	−0.7	0.2	−0.8	−0.2	0.3	0.3	1.046
150–200	1.25–1.50	7.14×10^{-3}	2.2	1.4	2.7	0.0	0.1	−0.1	−0.5	0.0	0.1	0.1	0.4	−0.8	0.3	−0.9	−0.3	0.3	0.3	1.049
150–200	1.50–1.75	6.21×10^{-3}	2.5	1.5	2.9	−0.1	0.0	−0.1	−0.4	0.0	0.1	0.1	0.3	−0.9	0.3	−1.0	−0.3	0.3	0.3	1.054
150–200	1.75–2.00	4.95×10^{-3}	2.9	1.9	3.4	−0.1	0.0	−0.1	−0.3	0.0	0.0	0.1	0.3	−1.2	0.5	−1.1	−0.5	0.5	0.4	1.058
150–200	2.00–2.25	3.74×10^{-3}	3.5	2.1	4.1	−0.1	0.0	0.0	−0.3	0.0	0.0	0.2	0.3	−1.3	0.7	−1.3	−0.6	0.6	0.4	1.064
150–200	2.25–2.50	2.94×10^{-3}	4.0	2.5	4.8	−0.1	0.0	0.0	−0.3	0.0	0.0	0.1	0.3	−1.7	1.0	−1.2	−0.6	0.6	0.5	1.071

(continued)

Table P.3 (continued)

m_{ee} [GeV]	$\lvert\Delta\eta_{ee}\rvert$	$\frac{d^2\sigma}{dm_{ee}d\lvert\Delta\eta_{ee}\rvert}$ [pb/GeV]	δ^{stat} [%]	δ^{sys} [%]	δ^{tot} [%]	δ^{trig}_{cor} [%]	δ^{trig}_{unc} [%]	δ^{reco}_{cor} [%]	δ^{id}_{cor} [%]	δ^{iso}_{cor} [%]	δ^{iso}_{unc} [%]	δ^{Eres}_{cor} [%]	δ^{Escale}_{cor} [%]	$\delta^{mult.}_{cor}$ [%]	$\delta^{mult.}_{unc}$ [%]	δ^{top}_{cor} [%]	$\delta^{diboson}_{cor}$ [%]	δ^{bgMC}_{unc} [%]	δ^{MC}_{unc} [%]	$k_{dressed}$
150–200	2.50–2.75	2.01×10^{-3}	5.1	2.9	5.9	−0.1	0.0	0.0	−0.3	0.0	0.0	0.1	0.4	−1.8	1.6	−1.2	−0.7	0.8	0.7	1.073
150–200	2.75–3.00	9.24×10^{-4}	8.0	4.9	9.4	−0.1	0.0	0.0	−0.4	0.0	0.0	0.6	0.7	−3.4	2.7	−1.4	−0.8	1.2	1.0	1.070
200–300	0.00–0.25	2.11×10^{-3}	2.8	1.8	3.3	−0.1	0.2	−0.2	−0.8	−0.1	0.2	0.1	1.1	−0.7	0.2	−0.7	−0.2	0.3	0.3	1.043
200–300	0.25–0.50	2.08×10^{-3}	2.9	1.9	3.4	−0.1	0.2	−0.2	−0.8	−0.1	0.2	0.1	1.2	−0.7	0.2	−0.8	−0.2	0.4	0.3	1.044
200–300	0.50–0.75	1.98×10^{-3}	3.0	1.9	3.6	−0.1	0.1	−0.2	−0.8	−0.1	0.2	0.1	1.1	−0.8	0.2	−0.9	−0.2	0.4	0.3	1.044
200–300	0.75–1.00	1.89×10^{-3}	3.1	1.9	3.6	−0.1	0.1	−0.2	−0.8	−0.1	0.2	0.1	1.0	−0.9	0.2	−0.9	−0.3	0.4	0.3	1.047
200–300	1.00–1.25	1.74×10^{-3}	3.1	1.8	3.6	−0.1	0.1	−0.2	−0.8	−0.1	0.2	0.1	0.8	−0.8	0.2	−1.0	−0.3	0.4	0.3	1.048
200–300	1.25–1.50	1.40×10^{-3}	3.6	2.2	4.2	−0.1	0.1	−0.2	−0.7	−0.1	0.2	0.1	0.7	−1.2	0.3	−1.3	−0.4	0.6	0.3	1.049
200–300	1.50–1.75	1.25×10^{-3}	3.9	2.2	4.5	−0.1	0.1	−0.1	−0.7	0.0	0.1	0.1	0.6	−1.1	0.4	−1.4	−0.4	0.6	0.4	1.057
200–300	1.75–2.00	1.02×10^{-3}	4.6	2.5	5.2	0.0	0.1	−0.1	−0.6	0.0	0.1	0.1	0.5	−1.2	0.5	−1.8	−0.6	0.8	0.4	1.060
200–300	2.00–2.25	9.44×10^{-4}	4.9	2.8	5.6	0.0	0.1	−0.1	−0.5	0.0	0.1	0.1	0.4	−1.5	0.7	−1.8	−0.7	0.9	0.5	1.068
200–300	2.25–2.50	6.59×10^{-4}	6.3	3.8	7.4	0.0	0.1	−0.1	−0.5	0.0	0.1	0.1	0.4	−2.2	1.2	−2.3	−1.1	1.3	0.6	1.078
200–300	2.50–2.75	5.75×10^{-4}	7.0	3.6	7.8	−0.1	0.0	−0.1	−0.4	0.0	0.1	0.2	0.4	−1.9	1.4	−2.1	−1.2	1.4	0.7	1.087
200–300	2.75–3.00	4.31×10^{-4}	8.5	5.2	10.0	−0.1	0.0	−0.1	−0.3	0.0	0.0	0.4	0.5	−3.5	2.4	−2.2	−1.5	1.7	1.0	1.110
300–500	0.00–0.50	2.60×10^{-4}	3.9	2.4	4.6	−0.1	0.2	−0.2	−0.8	−0.2	0.5	0.1	1.8	−0.7	0.2	−0.6	−0.2	0.5	0.3	1.048
300–500	0.50–1.00	2.28×10^{-4}	4.3	2.3	4.9	−0.1	0.2	−0.2	−0.8	−0.2	0.4	0.1	1.6	−0.9	0.2	−0.8	−0.3	0.6	0.3	1.048
300–500	1.00–1.50	2.18×10^{-4}	4.4	2.3	5.0	−0.1	0.2	−0.2	−0.8	−0.1	0.4	0.1	1.4	−0.9	0.3	−1.0	−0.4	0.7	0.3	1.057
300–500	1.50–2.00	1.64×10^{-4}	5.4	2.7	6.1	−0.1	0.2	−0.2	−0.8	−0.1	0.3	0.1	1.2	−1.4	0.4	−1.5	−0.5	0.9	0.4	1.064
300–500	2.00–2.50	1.04×10^{-4}	7.4	3.5	8.2	−0.1	0.1	−0.2	−0.8	−0.1	0.2	0.2	0.8	−1.7	0.8	−2.3	−0.8	1.4	0.6	1.082
300–500	2.50–3.00	5.21×10^{-5}	12.7	6.5	14.3	−0.1	0.1	−0.2	−0.7	0.0	0.1	0.2	1.0	−3.2	2.3	−4.1	−2.2	2.9	0.8	1.107
500–1500	0.00–0.50	7.69×10^{-6}	9.8	3.1	10.3	−0.2	0.3	−0.2	−0.8	−0.3	1.3	0.1	2.4	−0.6	0.6	−0.2	−0.3	0.7	0.3	1.054
500–1500	0.50–1.00	8.74×10^{-6}	9.3	2.9	9.7	−0.2	0.3	−0.2	−0.8	−0.3	1.2	0.1	2.3	−0.3	0.4	−0.3	−0.3	0.8	0.3	1.058
500–1500	1.00–1.50	8.68×10^{-6}	9.3	2.7	9.7	−0.1	0.3	−0.2	−0.8	−0.2	1.0	0.0	2.2	−0.1	0.4	−0.4	−0.3	0.6	0.4	1.063
500–1500	1.50–2.00	6.99×10^{-6}	10.8	2.7	11.1	−0.1	0.2	−0.2	−0.8	−0.2	0.7	0.1	1.9	−1.1	0.6	−0.5	−0.5	0.7	0.4	1.078
500–1500	2.00–2.50	2.92×10^{-6}	19.2	4.1	19.6	−0.1	0.2	−0.2	−0.8	−0.2	0.5	0.2	1.7	−0.5	1.7	−1.6	−1.2	2.6	0.5	1.095
500–1500	2.50–3.00	1.90×10^{-6}	26.3	6.0	27.0	−0.1	0.2	−0.3	−0.8	−0.1	0.4	0.1	1.6	−0.9	3.0	−2.6	−1.9	3.9	1.2	1.120

Appendix Q
High-Mass Drell-Yan: Correlation Matrices

The following appendix contains the statistical correlations between the three measured cross sections in the electron channel. The correlations were obtained using the bootstrap method [1] on the electron data with 10000 replica. The bootstrap method is based on pseudo-experiments. Aside the measured data spectrum, 10000 toy spectra are created. For each event, a set of 10000 weights is generated according to the Poisson distribution with mean equal to 1. If an event contributes to the nominal spectrum, it is filled to each toy spectrum weighted by the corresponding weight. The correlations between the bins are extracted by building the correlation matrix from all toys. Figure Q.1 shows the statistical correlation between the measurement bins in mass, absolute rapidity $|y_{ee}|$ and absolute pseudorapidity separation $|\Delta\eta_{ee}|$ in the electron channel. Each plot shows the correlation in a single two dimensional mass bin. Within each histogram or two dimensional mass bin, the first 2-3 bins show the correlation with the single-differential cross section which fall within the same mass range, then correlation with $|y_{ee}|$ bins is shown and finally the correlation with the $|\Delta\eta_{ee}|$ bins. The binning is defined in Table Q.1. The Tables Q.2, Q.3, Q.4, Q.5 and Q.6 list the correlations in detail. The correlation coefficients are sometimes negative in bins where the correlation should be zero. This is due to the limited statistical precision.

M. Zinser, *Search for New Heavy Charged Bosons and Measurement of High-Mass Drell-Yan Production in Proton-Proton Collisions*, Springer Theses,
https://doi.org/10.1007/978-3-030-00650-1

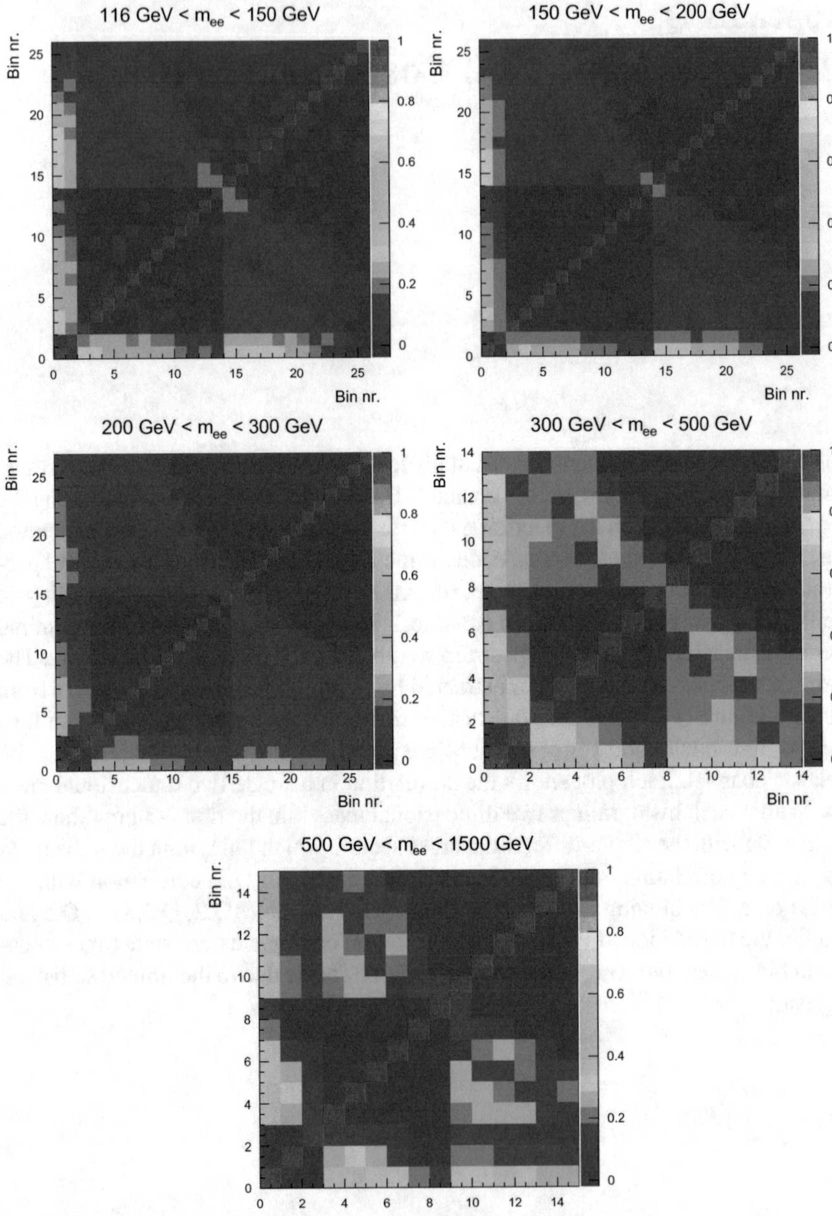

Fig. Q.1 Statistical correlations in the electron channel between the single-differential and the two double-differential cross section bins

Table Q.1 Binning used to show the statistical correlation in the electron channel between the three cross section measurements

Bin nr.	Variable	Range
m_{ee}: 116–150 GeV		
1–2	m_{ee}	(116–130, 130–150)
3–14	y_{ee}	(0.0–2.4 in 0.2 steps)
15–26	$\|\Delta\eta_{ee}\|$	(0.0–3.0 in 0.25 steps)
m_{ee}: 150–200 GeV		
1–2	m_{ee}	(150–175, 175–200)
3–14	y_{ee}	(0.0–2.4 in 0.2 steps)
15–26	$\|\Delta\eta_{ee}\|$	(0.0–3.0 in 0.25 steps)
m_{ee}: 200–300 GeV		
1–3	m_{ee}	(200–230, 230–260, 260–300)
4–15	y_{ee}	(0.0–2.4 in 0.2 steps)
16–27	$\|\Delta\eta_{ee}\|$	(0.0–3.0 in 0.25 steps)
m_{ee}: 300–500 GeV		
1–2	m_{ee}	(300–380, 380–500)
3–8	y_{ee}	(0.0–2.4 in 0.4 steps)
9–14	$\|\Delta\eta_{ee}\|$	(0.0–3.0 in 0.5 steps)
m_{ee}: 500–1500 GeV		
1–3	m_{ee}	(500–700, 700–1000, 1000–1500)
4–9	y_{ee}	(0.0–2.4 in 0.4 steps)
10–15	$\|\Delta\eta_{ee}\|$	(0.0–3.0 in 0.5 steps)

Table Q.2 Statistical correlation between the measurement bins in mass, absolute rapidity and absolute pseudorapidity separation in the electron channel for 116 GeV $< m_{ee} <$ 150 GeV

Bin	116 GeV $< m_{ee} <$ 130 GeV	130 GeV $< m_{ee} <$ 150 GeV	0.0 $< \lvert y_{ee}\rvert <$ 0.2	0.2 $< \lvert y_{ee}\rvert <$ 0.4	0.4 $< \lvert y_{ee}\rvert <$ 0.6	0.6 $< \lvert y_{ee}\rvert <$ 0.8	0.8 $< \lvert y_{ee}\rvert <$ 1.0	1.0 $< \lvert y_{ee}\rvert <$ 1.2	1.2 $< \lvert y_{ee}\rvert <$ 1.4	1.4 $< \lvert y_{ee}\rvert <$ 1.6	1.6 $< \lvert y_{ee}\rvert <$ 1.8	1.8 $< \lvert y_{ee}\rvert <$ 2.0	2.0 $< \lvert y_{ee}\rvert <$ 2.2	2.2 $< \lvert y_{ee}\rvert <$ 2.4
116 GeV $< m_{ee} <$ 130 GeV	1.00	0.02	0.29	0.27	0.27	0.26	0.25	0.24	0.23	0.23	0.19	0.16	0.17	0.09
130 GeV $< m_{ee} <$ 150 GeV	0.02	1.00	0.22	0.24	0.24	0.22	0.22	0.22	0.19	0.19	0.17	0.13	0.11	0.06
0.0 $< \lvert y_{ee}\rvert <$ 0.2	0.29	0.22	1.00	0.00	0.01	0.00	-0.01	0.01	-0.01	0.01	-0.00	-0.02	0.02	-0.00
0.2 $< \lvert y_{ee}\rvert <$ 0.4	0.27	0.24	0.00	1.00	0.02	0.01	0.00	0.03	-0.00	0.02	-0.00	-0.01	0.00	-0.04
0.4 $< \lvert y_{ee}\rvert <$ 0.6	0.27	0.24	0.01	0.02	1.00	0.03	0.02	0.00	-0.00	0.00	-0.00	0.01	0.01	0.00
0.6 $< \lvert y_{ee}\rvert <$ 0.8	0.26	0.22	0.00	0.01	0.03	1.00	-0.00	-0.01	0.01	0.01	0.01	-0.02	0.01	-0.00
0.8 $< \lvert y_{ee}\rvert <$ 1.0	0.25	0.22	-0.01	0.00	0.02	-0.00	1.00	-0.00	0.00	0.01	0.02	-0.00	0.00	0.01
1.0 $< \lvert y_{ee}\rvert <$ 1.2	0.24	0.22	0.01	0.03	0.00	-0.01	-0.00	1.00	-0.00	-0.00	0.00	0.00	0.01	-0.01
1.2 $< \lvert y_{ee}\rvert <$ 1.4	0.23	0.19	-0.01	-0.00	-0.00	0.01	0.00	-0.00	1.00	0.01	-0.01	0.00	0.01	0.01
1.4 $< \lvert y_{ee}\rvert <$ 1.6	0.23	0.19	0.01	0.02	0.00	0.01	0.01	-0.00	0.01	1.00	-0.01	-0.01	0.00	-0.01
1.6 $< \lvert y_{ee}\rvert <$ 1.8	0.19	0.17	-0.00	-0.00	-0.00	0.01	0.02	0.00	-0.01	-0.01	1.00	0.02	0.01	0.01
1.8 $< \lvert y_{ee}\rvert <$ 2.0	0.16	0.13	-0.02	-0.01	0.01	-0.02	-0.00	0.00	0.00	-0.01	0.02	1.00	-0.02	0.00
2.0 $< \lvert y_{ee}\rvert <$ 2.2	0.17	0.11	0.02	0.00	0.01	0.01	0.00	0.01	0.01	0.00	0.01	-0.02	1.00	0.00
2.2 $< \lvert y_{ee}\rvert <$ 2.4	0.09	0.06	-0.00	-0.04	0.00	-0.00	0.01	-0.01	0.01	-0.01	0.01	0.00	0.00	1.00
0.0 $< \lvert\Delta\eta_{ee}\rvert <$ 0.25	0.32	0.25	0.11	0.12	0.14	0.12	0.13	0.13	0.07	0.03	0.13	0.14	0.19	0.21
0.25 $< \lvert\Delta\eta_{ee}\rvert <$ 0.5	0.31	0.25	0.12	0.13	0.14	0.13	0.12	0.12	0.07	0.10	0.07	0.14	0.18	0.10
0.5 $< \lvert\Delta\eta_{ee}\rvert <$ 0.75	0.30	0.23	0.13	0.14	0.14	0.13	0.14	0.05	0.09	0.13	0.08	0.13	0.12	-0.00
0.75 $< \lvert\Delta\eta_{ee}\rvert <$ 1.0	0.28	0.24	0.14	0.12	0.14	0.16	0.09	0.08	0.13	0.12	0.14	0.06	0.03	0.02
1.0 $< \lvert\Delta\eta_{ee}\rvert <$ 1.25	0.28	0.20	0.12	0.13	0.11	0.11	0.07	0.12	0.12	0.15	0.15	0.06	0.01	-0.02
1.25 $< \lvert\Delta\eta_{ee}\rvert <$ 1.5	0.26	0.23	0.13	0.15	0.13	0.07	0.11	0.14	0.13	0.14	0.09	0.02	-0.01	-0.00
1.5 $< \lvert\Delta\eta_{ee}\rvert <$ 1.75	0.22	0.20	0.12	0.13	0.11	0.08	0.10	0.13	0.12	0.13	0.04	-0.01	0.00	-0.01
1.75 $< \lvert\Delta\eta_{ee}\rvert <$ 2.0	0.22	0.17	0.12	0.10	0.06	0.10	0.12	0.13	0.11	0.07	0.02	0.01	0.01	-0.00
2.0 $< \lvert\Delta\eta_{ee}\rvert <$ 2.25	0.11	0.19	0.08	0.05	0.07	0.10	0.11	0.09	0.10	0.03	0.00	0.02	0.01	-0.01
2.25 $< \lvert\Delta\eta_{ee}\rvert <$ 2.5	0.03	0.14	0.04	0.04	0.08	0.06	0.06	0.06	0.04	-0.01	0.00	-0.02	-0.00	-0.00
2.5 $< \lvert\Delta\eta_{ee}\rvert <$ 2.75	0.00	0.05	0.01	0.03	0.02	0.03	0.02	0.01	-0.02	-0.00	0.01	-0.01	0.00	-0.00
2.75 $< \lvert\Delta\eta_{ee}\rvert <$ 3.0	-0.00	0.03	0.01	0.01	-0.00	0.02	0.03	-0.01	0.01	0.00	0.01	-0.01	-0.01	-0.02

Bin	0.0 $< \lvert\Delta\eta_{ee}\rvert <$ 0.25	0.25 $< \lvert\Delta\eta_{ee}\rvert <$ 0.5	0.5 $< \lvert\Delta\eta_{ee}\rvert <$ 0.75	0.75 $< \lvert\Delta\eta_{ee}\rvert <$ 1.0	1.0 $< \lvert\Delta\eta_{ee}\rvert <$ 1.25	1.25 $< \lvert\Delta\eta_{ee}\rvert <$ 1.5	1.5 $< \lvert\Delta\eta_{ee}\rvert <$ 1.75	1.75 $< \lvert\Delta\eta_{ee}\rvert <$ 2.0	2.0 $< \lvert\Delta\eta_{ee}\rvert <$ 2.25	2.25 $< \lvert\Delta\eta_{ee}\rvert <$ 2.5	2.5 $< \lvert\Delta\eta_{ee}\rvert <$ 2.75	2.75 $< \lvert\Delta\eta_{ee}\rvert <$ 3.0
116 GeV $< m_{ee} <$ 130 GeV	0.32	0.31	0.30	0.28	0.28	0.26	0.22	0.22	0.11	0.03	0.00	-0.00
130 GeV $< m_{ee} <$ 150 GeV	0.25	0.25	0.23	0.24	0.20	0.23	0.20	0.17	0.19	0.14	0.05	0.03
0.0 $< \lvert y_{ee}\rvert <$ 0.2	0.11	0.12	0.13	0.14	0.12	0.13	0.12	0.12	0.08	0.04	0.01	0.01
0.2 $< \lvert y_{ee}\rvert <$ 0.4	0.12	0.13	0.14	0.12	0.13	0.15	0.13	0.10	0.05	0.04	0.03	0.01
0.4 $< \lvert y_{ee}\rvert <$ 0.6	0.14	0.14	0.14	0.14	0.11	0.13	0.11	0.06	0.07	0.08	0.02	-0.00
0.6 $< \lvert y_{ee}\rvert <$ 0.8	0.12	0.13	0.13	0.16	0.11	0.07	0.08	0.10	0.10	0.06	0.03	0.02
0.8 $< \lvert y_{ee}\rvert <$ 1.0	0.13	0.12	0.14	0.09	0.07	0.11	0.10	0.12	0.11	0.06	0.02	0.03
1.0 $< \lvert y_{ee}\rvert <$ 1.2	0.13	0.12	0.05	0.08	0.12	0.14	0.13	0.13	0.09	0.06	0.01	-0.01
1.2 $< \lvert y_{ee}\rvert <$ 1.4	0.07	0.07	0.09	0.13	0.12	0.13	0.12	0.11	0.10	0.04	-0.02	0.01
1.4 $< \lvert y_{ee}\rvert <$ 1.6	0.03	0.10	0.13	0.12	0.15	0.14	0.13	0.07	0.03	-0.01	-0.00	0.00
1.6 $< \lvert y_{ee}\rvert <$ 1.8	0.13	0.07	0.08	0.14	0.15	0.09	0.04	0.02	0.00	0.00	0.01	0.01
1.8 $< \lvert y_{ee}\rvert <$ 2.0	0.14	0.14	0.13	0.06	0.06	0.02	-0.01	0.01	0.02	-0.02	-0.01	-0.01
2.0 $< \lvert y_{ee}\rvert <$ 2.2	0.19	0.18	0.12	0.03	0.01	-0.01	0.00	0.01	0.01	-0.00	0.00	-0.01
2.2 $< \lvert y_{ee}\rvert <$ 2.4	0.21	0.10	-0.00	0.02	-0.02	-0.00	-0.01	-0.00	-0.01	-0.00	-0.00	-0.02
0.0 $< \lvert\Delta\eta_{ee}\rvert <$ 0.25	1.00	0.00	0.00	0.00	-0.02	0.02	0.01	0.02	0.01	-0.00	0.00	-0.01
0.25 $< \lvert\Delta\eta_{ee}\rvert <$ 0.5	0.00	1.00	0.01	0.01	-0.01	-0.00	0.02	0.01	0.01	-0.01	-0.02	0.01
0.5 $< \lvert\Delta\eta_{ee}\rvert <$ 0.75	0.00	0.01	1.00	0.01	0.01	-0.02	-0.01	-0.00	0.01	-0.01	-0.00	0.01
0.75 $< \lvert\Delta\eta_{ee}\rvert <$ 1.0	0.00	0.01	0.01	1.00	0.01	0.01	-0.02	0.01	0.01	0.01	-0.01	0.02
1.0 $< \lvert\Delta\eta_{ee}\rvert <$ 1.25	-0.02	-0.01	0.01	0.01	1.00	0.01	0.02	-0.00	0.00	0.01	0.00	-0.01
1.25 $< \lvert\Delta\eta_{ee}\rvert <$ 1.5	0.02	-0.00	-0.02	0.01	0.01	1.00	0.01	0.02	0.01	-0.00	0.00	-0.00
1.5 $< \lvert\Delta\eta_{ee}\rvert <$ 1.75	0.01	0.02	-0.01	-0.02	0.02	0.01	1.00	0.00	-0.02	-0.01	-0.02	0.00
1.75 $< \lvert\Delta\eta_{ee}\rvert <$ 2.0	0.02	0.01	-0.00	0.01	-0.00	0.02	0.00	1.00	-0.00	-0.01	-0.02	0.00
2.0 $< \lvert\Delta\eta_{ee}\rvert <$ 2.25	0.01	0.01	0.01	0.01	0.00	0.01	-0.02	-0.00	1.00	0.01	-0.00	0.00
2.25 $< \lvert\Delta\eta_{ee}\rvert <$ 2.5	-0.00	-0.01	-0.01	0.01	0.01	-0.00	-0.01	-0.01	0.01	1.00	-0.01	-0.03
2.5 $< \lvert\Delta\eta_{ee}\rvert <$ 2.75	0.00	-0.02	-0.00	-0.01	0.00	0.00	-0.02	-0.02	-0.00	-0.01	1.00	-0.01
2.75 $< \lvert\Delta\eta_{ee}\rvert <$ 3.0	-0.01	0.01	0.01	0.02	-0.01	-0.00	0.00	0.00	0.00	-0.03	-0.01	1.00

Table Q.3 Statistical correlation between the measurement bins in mass, absolute rapidity and absolute pseudorapidity separation in the electron channel for 150 GeV $< m_{ee} <$ 200 GeV

Bin	150 GeV $< m_{ee} <$ 175 GeV	175 GeV $< m_{ee} <$ 200 GeV	$0.0 < \lvert y_{ee}\rvert < 0.2$	$0.2 < \lvert y_{ee}\rvert < 0.4$	$0.4 < \lvert y_{ee}\rvert < 0.6$	$0.6 < \lvert y_{ee}\rvert < 0.8$	$0.8 < \lvert y_{ee}\rvert < 1.0$	$1.0 < \lvert y_{ee}\rvert < 1.2$	$1.2 < \lvert y_{ee}\rvert < 1.4$	$1.4 < \lvert y_{ee}\rvert < 1.6$	$1.6 < \lvert y_{ee}\rvert < 1.8$	$1.8 < \lvert y_{ee}\rvert < 2.0$	$2.0 < \lvert y_{ee}\rvert < 2.2$	$2.2 < \lvert y_{ee}\rvert < 2.4$
150 GeV $< m_{ee} <$ 175 GeV	1.00	0.01	0.28	0.30	0.27	0.27	0.27	0.25	0.23	0.22	0.19	0.18	0.16	[illegible]
175 GeV $< m_{ee} <$ 200 GeV	0.01	1.00	0.22	0.22	0.22	0.21	0.20	0.18	0.18	0.15	0.15	0.11	0.10	[illegible]
$0.0 < \lvert y_{ee}\rvert < 0.2$	0.28	0.22	1.00	0.01	0.01	-0.01	0.02	0.00	-0.01	0.01	-0.01	-0.00	-0.00	[illegible]
$0.2 < \lvert y_{ee}\rvert < 0.4$	0.30	0.22	0.01	1.00	0.00	0.01	-0.00	-0.00	0.02	0.01	-0.00	0.01	0.01	[illegible]
$0.4 < \lvert y_{ee}\rvert < 0.6$	0.27	0.22	0.01	0.00	1.00	-0.01	-0.00	-0.00	-0.01	0.00	0.01	0.00	0.00	[illegible]
$0.6 < \lvert y_{ee}\rvert < 0.8$	0.27	0.21	-0.01	0.01	-0.01	1.00	-0.01	0.00	0.01	-0.00	-0.01	0.00	0.01	[illegible]
$0.8 < \lvert y_{ee}\rvert < 1.0$	0.27	0.20	0.02	-0.00	-0.00	-0.01	1.00	-0.01	-0.02	-0.01	0.02	0.00	0.01	0.00
$1.0 < \lvert y_{ee}\rvert < 1.2$	0.25	0.18	0.00	-0.00	-0.00	0.00	-0.01	1.00	-0.01	0.01	0.00	0.00	-0.01	0.01
$1.2 < \lvert y_{ee}\rvert < 1.4$	0.23	0.18	-0.01	0.02	-0.01	0.01	-0.02	-0.01	1.00	0.00	-0.01	-0.00	-0.00	-0.02
$1.4 < \lvert y_{ee}\rvert < 1.6$	0.22	0.15	0.01	0.01	0.00	-0.00	-0.01	0.01	0.00	1.00	0.00	-0.02	0.02	-0.01
$1.6 < \lvert y_{ee}\rvert < 1.8$	0.19	0.15	-0.01	-0.00	0.01	-0.01	0.02	0.00	-0.01	0.00	1.00	0.01	0.02	0.01
$1.8 < \lvert y_{ee}\rvert < 2.0$	0.18	0.11	-0.00	0.01	0.00	0.00	0.00	0.00	-0.00	-0.02	0.01	1.00	-0.00	0.01
$2.0 < \lvert y_{ee}\rvert < 2.2$	0.16	0.10	-0.00	0.01	0.00	0.01	0.01	-0.01	-0.00	0.02	0.02	-0.00	1.00	-0.01
$2.2 < \lvert y_{ee}\rvert < 2.4$	0.09	0.07	0.01	-0.00	-0.01	-0.00	0.00	0.01	-0.02	-0.01	0.01	0.01	-0.01	1.00
$0.0 < \lvert\Delta\eta_{ee}\rvert < 0.25$	0.30	0.21	0.09	0.13	0.11	0.10	0.12	0.11	0.04	0.01	0.14	0.15	0.16	0.20
$0.25 < \lvert\Delta\eta_{ee}\rvert < 0.5$	0.30	0.21	0.14	0.11	0.11	0.12	0.13	0.09	0.06	0.10	0.07	0.15	0.15	0.08
$0.5 < \lvert\Delta\eta_{ee}\rvert < 0.75$	0.29	0.20	0.10	0.13	0.13	0.12	0.09	0.05	0.11	0.12	0.09	0.12	0.12	0.01
$0.75 < \lvert\Delta\eta_{ee}\rvert < 1.0$	0.27	0.18	0.11	0.10	0.12	0.12	0.10	0.08	0.13	0.12	0.12	0.03	0.03	-0.01
$1.0 < \lvert\Delta\eta_{ee}\rvert < 1.25$	0.26	0.20	0.13	0.13	0.11	0.09	0.05	0.11	0.12	0.12	0.13	0.08	-0.00	0.01
$1.25 < \lvert\Delta\eta_{ee}\rvert < 1.5$	0.24	0.18	0.13	0.12	0.12	0.07	0.08	0.10	0.10	0.11	0.09	0.01	0.01	0.00
$1.5 < \lvert\Delta\eta_{ee}\rvert < 1.75$	0.24	0.19	0.12	0.12	0.10	0.08	0.09	0.11	0.12	0.14	0.04	0.02	0.03	0.01
$1.75 < \lvert\Delta\eta_{ee}\rvert < 2.0$	0.21	0.16	0.12	0.11	0.08	0.09	0.11	0.12	0.09	0.07	0.03	-0.00	-0.01	-0.01
$2.0 < \lvert\Delta\eta_{ee}\rvert < 2.25$	0.19	0.12	0.09	0.08	0.07	0.10	0.10	0.11	0.09	0.04	0.01	-0.02	0.02	0.01
$2.25 < \lvert\Delta\eta_{ee}\rvert < 2.5$	0.16	0.13	0.07	0.08	0.09	0.10	0.10	0.10	0.07	0.01	0.00	0.01	-0.00	0.01
$2.5 < \lvert\Delta\eta_{ee}\rvert < 2.75$	0.14	0.13	0.06	0.08	0.08	0.11	0.11	0.06	0.04	-0.02	0.00	0.02	0.03	-0.00
$2.75 < \lvert\Delta\eta_{ee}\rvert < 3.0$	0.05	0.12	0.05	0.07	0.06	0.06	0.05	0.03	0.01	0.02	-0.01	0.00	-0.00	-0.00

Bin	$0.0 < \lvert\Delta\eta_{ee}\rvert < 0.25$	$0.25 < \lvert\Delta\eta_{ee}\rvert < 0.5$	$0.5 < \lvert\Delta\eta_{ee}\rvert < 0.75$	$0.75 < \lvert\Delta\eta_{ee}\rvert < 1.0$	$1.0 < \lvert\Delta\eta_{ee}\rvert < 1.25$	$1.25 < \lvert\Delta\eta_{ee}\rvert < 1.5$	$1.5 < \lvert\Delta\eta_{ee}\rvert < 1.75$	$1.75 < \lvert\Delta\eta_{ee}\rvert < 2.0$	$2.0 < \lvert\Delta\eta_{ee}\rvert < 2.25$	$2.25 < \lvert\Delta\eta_{ee}\rvert < 2.5$	$2.5 < \lvert\Delta\eta_{ee}\rvert < 2.75$	$2.75 < \lvert\Delta\eta_{ee}\rvert < 3.0$
150 GeV $< m_{ee} <$ 175 GeV	0.30	0.30	0.29	0.27	0.26	0.24	0.24	0.21	0.19	0.16	0.14	0.05
175 GeV $< m_{ee} <$ 200 GeV	0.21	0.21	0.20	0.18	0.20	0.18	0.19	0.16	0.12	0.13	0.13	0.12
$0.0 < \lvert y_{ee}\rvert < 0.2$	0.09	0.14	0.10	0.11	0.13	0.13	0.12	0.12	0.09	0.07	0.06	0.05
$0.2 < \lvert y_{ee}\rvert < 0.4$	0.13	0.11	0.13	0.10	0.13	0.12	0.12	0.11	0.08	0.08	0.08	0.07
$0.4 < \lvert y_{ee}\rvert < 0.6$	0.11	0.11	0.13	0.12	0.11	0.12	0.10	0.08	0.07	0.09	0.08	0.06
$0.6 < \lvert y_{ee}\rvert < 0.8$	0.10	0.12	0.12	0.12	0.09	0.07	0.08	0.09	0.10	0.10	0.11	0.06
$0.8 < \lvert y_{ee}\rvert < 1.0$	0.12	0.13	0.09	0.10	0.05	0.08	0.09	0.11	0.10	0.10	0.11	0.05
$1.0 < \lvert y_{ee}\rvert < 1.2$	0.11	0.09	0.05	0.08	0.11	0.10	0.11	0.12	0.11	0.10	0.06	0.03
$1.2 < \lvert y_{ee}\rvert < 1.4$	0.04	0.06	0.11	0.13	0.12	0.10	0.12	0.09	0.09	0.07	0.04	0.01
$1.4 < \lvert y_{ee}\rvert < 1.6$	0.01	0.10	0.12	0.12	0.12	0.11	0.14	0.07	0.04	0.01	-0.02	0.02
$1.6 < \lvert y_{ee}\rvert < 1.8$	0.14	0.07	0.09	0.12	0.13	0.09	0.04	0.03	0.01	0.00	0.00	-0.01
$1.8 < \lvert y_{ee}\rvert < 2.0$	0.15	0.15	0.12	0.03	0.08	0.01	0.02	-0.00	-0.02	0.01	0.02	0.00
$2.0 < \lvert y_{ee}\rvert < 2.2$	0.16	0.15	0.12	0.03	-0.00	0.01	0.03	-0.01	0.02	-0.00	0.03	-0.00
$2.2 < \lvert y_{ee}\rvert < 2.4$	0.20	0.08	0.01	-0.01	0.01	0.00	0.01	-0.01	0.01	0.01	-0.00	-0.00
$0.0 < \lvert\Delta\eta_{ee}\rvert < 0.25$	1.00	-0.01	0.00	-0.01	-0.01	-0.02	0.01	-0.01	-0.01	0.00	-0.00	-0.00
$0.25 < \lvert\Delta\eta_{ee}\rvert < 0.5$	-0.01	1.00	-0.01	0.00	0.00	-0.01	0.01	0.02	0.00	-0.00	0.02	0.01
$0.5 < \lvert\Delta\eta_{ee}\rvert < 0.75$	0.00	-0.01	1.00	-0.02	0.01	-0.01	0.01	-0.00	0.01	0.02	0.00	-0.00
$0.75 < \lvert\Delta\eta_{ee}\rvert < 1.0$	-0.01	0.00	-0.02	1.00	-0.01	-0.00	0.01	0.00	-0.02	0.00	0.00	-0.02
$1.0 < \lvert\Delta\eta_{ee}\rvert < 1.25$	-0.01	0.00	0.01	-0.01	1.00	0.01	0.01	-0.01	0.01	-0.00	-0.00	-0.00
$1.25 < \lvert\Delta\eta_{ee}\rvert < 1.5$	-0.02	-0.01	-0.01	-0.00	0.01	1.00	-0.01	0.01	-0.00	0.00	-0.00	0.02
$1.5 < \lvert\Delta\eta_{ee}\rvert < 1.75$	0.01	0.01	0.01	0.01	0.01	-0.01	1.00	0.00	-0.01	-0.00	0.01	0.01
$1.75 < \lvert\Delta\eta_{ee}\rvert < 2.0$	-0.01	0.02	-0.00	0.00	-0.01	0.01	0.00	1.00	0.01	-0.01	0.02	-0.01
$2.0 < \lvert\Delta\eta_{ee}\rvert < 2.25$	-0.01	0.00	0.01	-0.02	0.01	-0.00	-0.01	0.01	1.00	0.01	-0.00	-0.02
$2.25 < \lvert\Delta\eta_{ee}\rvert < 2.5$	0.00	-0.00	0.02	0.00	-0.00	0.00	-0.00	-0.01	0.01	1.00	0.00	-0.01
$2.5 < \lvert\Delta\eta_{ee}\rvert < 2.75$	-0.00	0.02	0.00	0.00	-0.00	-0.00	0.01	0.02	-0.00	0.00	1.00	0.01
$2.75 < \lvert\Delta\eta_{ee}\rvert < 3.0$	-0.00	0.01	-0.00	-0.02	-0.00	0.02	0.01	-0.01	-0.02	-0.01	0.01	1.00

Table Q.4 Statistical correlation between the measurement bins in mass, absolute rapidity and absolute pseudorapidity separation in the electron channel for 200 GeV $< m_{ee} <$ 300 GeV

Bin	200 GeV $< m_{ee} <$ 230 GeV	230 GeV $< m_{ee} <$ 260 GeV	260 GeV $< m_{ee} <$ 300 GeV	$0.0 < \lvert y_{ee}\rvert < 0.2$	$0.2 < \lvert y_{ee}\rvert < 0.4$	$0.4 < \lvert y_{ee}\rvert < 0.6$	$0.6 < \lvert y_{ee}\rvert < 0.8$	$0.8 < \lvert y_{ee}\rvert < 1.0$	$1.0 < \lvert y_{ee}\rvert < 1.2$	$1.2 < \lvert y_{ee}\rvert < 1.4$	$1.4 < \lvert y_{ee}\rvert < 1.6$	$1.6 < \lvert y_{ee}\rvert < 1.8$	$1.8 < \lvert y_{ee}\rvert < 2.0$	$2.0 < \lvert y_{ee}\rvert < 2.2$	$2.2 < \lvert y_{ee}\rvert < 2.4$	$0.0 < \lvert\Delta\eta_{ee}\rvert < 0.25$	$0.25 < \lvert\Delta\eta_{ee}\rvert < 0.5$	$0.5 < \lvert\Delta\eta_{ee}\rvert < 0.75$	$0.75 < \lvert\Delta\eta_{ee}\rvert < 1.0$	$1.0 < \lvert\Delta\eta_{ee}\rvert < 1.25$	$1.25 < \lvert\Delta\eta_{ee}\rvert < 1.5$	$1.5 < \lvert\Delta\eta_{ee}\rvert < 1.75$	$1.75 < \lvert\Delta\eta_{ee}\rvert < 2.0$	$2.0 < \lvert\Delta\eta_{ee}\rvert < 2.25$	$2.25 < \lvert\Delta\eta_{ee}\rvert < 2.5$	$2.5 < \lvert\Delta\eta_{ee}\rvert < 2.75$	$2.75 < \lvert\Delta\eta_{ee}\rvert < 3.0$
200 GeV $< m_{ee} <$ 230 GeV	1.00	0.01	0.00	0.26	0.25	0.26	0.27	0.24	0.22	0.20	0.19	0.16	0.14	0.11	0.07	0.26	0.25	0.24	0.24	0.23	0.22	0.20	0.18	0.16	0.14	0.13	0.12
230 GeV $< m_{ee} <$ 260 GeV	0.01	1.00	0.02	0.18	0.21	0.21	0.18	0.18	0.18	0.17	0.13	0.12	0.10	0.09	0.06	0.18	0.19	0.17	0.19	0.16	0.15	0.15	0.13	0.14	0.12	0.11	0.11
260 GeV $< m_{ee} <$ 300 GeV	0.00	0.02	1.00	0.17	0.16	0.19	0.19	0.17	0.13	0.14	0.13	0.11	0.07	0.08	0.06	0.16	0.16	0.15	0.15	0.17	0.14	0.13	0.11	0.13	0.11	0.10	0.10
$0.0 < \lvert y_{ee}\rvert < 0.2$	0.26	0.18	0.17	1.00	0.01	0.00	-0.01	-0.01	-0.01	0.01	0.01	-0.00	0.01	-0.01	0.02	0.12	0.10	0.11	0.13	0.12	0.10	0.12	0.09	0.09	0.08	0.07	0.08
$0.2 < \lvert y_{ee}\rvert < 0.4$	0.25	0.21	0.16	0.01	1.00	0.01	-0.01	0.00	0.01	0.01	0.00	-0.01	-0.01	0.01	0.00	0.10	0.11	0.09	0.12	0.13	0.12	0.12	0.07	0.12	0.06	0.10	0.07
$0.4 < \lvert y_{ee}\rvert < 0.6$	0.26	0.21	0.19	0.00	0.01	1.00	0.02	0.01	-0.00	0.00	0.01	0.00	0.00	0.00	0.00	0.13	0.13	0.12	0.13	0.11	0.12	0.09	0.05	0.07	0.10	0.08	0.11
$0.6 < \lvert y_{ee}\rvert < 0.8$	0.27	0.18	0.19	-0.01	-0.01	0.02	1.00	0.00	0.00	0.01	0.01	0.02	0.01	0.02	0.02	0.12	0.14	0.12	0.13	0.11	0.06	0.06	0.09	0.12	0.09	0.09	0.12
$0.8 < \lvert y_{ee}\rvert < 1.0$	0.24	0.18	0.17	-0.01	0.00	0.01	0.00	1.00	-0.00	0.01	-0.01	0.01	0.00	0.00	-0.01	0.10	0.11	0.10	0.09	0.06	0.08	0.10	0.13	0.11	0.11	0.10	0.09
$1.0 < \lvert y_{ee}\rvert < 1.2$	0.22	0.18	0.13	-0.01	0.01	-0.00	0.00	-0.00	1.00	0.00	-0.01	0.00	-0.01	0.00	-0.01	0.10	0.10	0.05	0.04	0.11	0.12	0.12	0.10	0.13	0.09	0.09	0.03
$1.2 < \lvert y_{ee}\rvert < 1.4$	0.20	0.17	0.14	0.01	0.01	0.00	0.01	0.01	0.00	1.00	0.01	-0.00	-0.00	0.00	0.01	0.07	0.06	0.11	0.10	0.13	0.13	0.10	0.12	0.08	0.06	0.03	0.03
$1.4 < \lvert y_{ee}\rvert < 1.6$	0.19	0.13	0.13	0.01	0.00	0.01	0.01	-0.01	-0.01	0.01	1.00	0.01	0.00	0.00	0.00	0.03	0.09	0.11	0.13	0.13	0.11	0.13	0.08	0.02	0.03	0.01	0.01
$1.6 < \lvert y_{ee}\rvert < 1.8$	0.16	0.12	0.11	-0.00	-0.01	0.00	0.02	0.01	0.00	-0.00	0.01	1.00	-0.00	-0.01	0.00	0.12	0.06	0.08	0.15	0.13	0.10	0.06	0.02	0.01	-0.01	-0.01	0.00
$1.8 < \lvert y_{ee}\rvert < 2.0$	0.14	0.10	0.07	0.01	-0.01	0.00	0.01	0.00	-0.01	-0.00	0.00	-0.00	1.00	-0.00	0.02	0.16	0.16	0.12	0.06	0.06	0.01	-0.00	0.00	0.01	-0.01	0.02	-0.02
$2.0 < \lvert y_{ee}\rvert < 2.2$	0.11	0.09	0.08	-0.01	0.01	0.00	0.02	0.00	0.00	0.00	0.00	-0.01	-0.00	1.00	-0.00	0.15	0.17	0.11	0.04	-0.00	0.00	-0.01	0.01	-0.01	0.01	0.01	-0.00
$2.2 < \lvert y_{ee}\rvert < 2.4$	0.07	0.06	0.06	0.02	0.00	0.00	0.02	-0.01	-0.01	0.01	0.00	0.00	0.02	-0.00	1.00	0.18	0.10	0.03	0.01	0.00	-0.00	0.01	-0.00	-0.01	0.01	-0.00	0.01
$0.0 < \lvert\Delta\eta_{ee}\rvert < 0.25$	0.26	0.18	0.16	0.12	0.10	0.13	0.12	0.10	0.10	0.07	0.03	0.12	0.16	0.15	0.18	1.00	0.01	0.01	0.01	0.01	0.01	-0.00	0.01	0.01	-0.01	-0.01	-0.01
$0.25 < \lvert\Delta\eta_{ee}\rvert < 0.5$	0.25	0.19	0.16	0.10	0.11	0.13	0.14	0.11	0.10	0.06	0.09	0.06	0.16	0.17	0.10	0.01	1.00	0.01	0.02	-0.00	0.02	0.00	-0.00	-0.00	0.00	-0.02	0.01
$0.5 < \lvert\Delta\eta_{ee}\rvert < 0.75$	0.24	0.17	0.15	0.11	0.09	0.12	0.12	0.10	0.05	0.11	0.11	0.08	0.12	0.11	0.03	0.01	0.01	1.00	0.00	-0.01	-0.00	-0.00	-0.00	0.01	0.00	-0.00	-0.01
$0.75 < \lvert\Delta\eta_{ee}\rvert < 1.0$	0.24	0.19	0.15	0.13	0.12	0.13	0.13	0.09	0.04	0.10	0.13	0.15	0.06	0.04	0.01	0.01	0.02	0.00	1.00	-0.00	-0.01	0.01	0.00	0.01	-0.02	0.02	0.01
$1.0 < \lvert\Delta\eta_{ee}\rvert < 1.25$	0.23	0.16	0.17	0.12	0.13	0.11	0.11	0.06	0.11	0.13	0.13	0.13	0.06	-0.00	0.00	0.01	-0.00	-0.01	-0.00	1.00	0.00	0.01	0.01	0.01	0.01	0.00	-0.00
$1.25 < \lvert\Delta\eta_{ee}\rvert < 1.5$	0.22	0.15	0.14	0.10	0.12	0.12	0.06	0.08	0.12	0.13	0.11	0.10	0.01	0.00	-0.00	0.01	0.02	-0.00	-0.01	0.00	1.00	-0.00	-0.03	-0.00	-0.00	0.01	0.00
$1.5 < \lvert\Delta\eta_{ee}\rvert < 1.75$	0.20	0.15	0.13	0.12	0.12	0.09	0.06	0.10	0.12	0.10	0.13	0.06	-0.00	-0.01	0.01	-0.00	0.00	-0.00	0.01	0.01	-0.00	1.00	-0.01	-0.01	-0.01	0.02	0.01
$1.75 < \lvert\Delta\eta_{ee}\rvert < 2.0$	0.18	0.13	0.11	0.09	0.07	0.05	0.09	0.13	0.10	0.12	0.08	0.02	0.00	0.01	-0.00	0.01	-0.00	-0.00	0.00	0.01	-0.03	-0.01	1.00	-0.00	-0.00	0.00	0.01
$2.0 < \lvert\Delta\eta_{ee}\rvert < 2.25$	0.16	0.14	0.13	0.09	0.12	0.07	0.12	0.11	0.13	0.08	0.02	0.01	0.01	-0.01	-0.01	0.01	-0.00	0.01	0.01	0.01	-0.00	-0.01	-0.00	1.00	0.01	-0.01	-0.02
$2.25 < \lvert\Delta\eta_{ee}\rvert < 2.5$	0.14	0.12	0.11	0.08	0.06	0.10	0.09	0.11	0.09	0.06	0.03	-0.01	-0.01	0.01	0.01	-0.01	0.00	0.00	-0.02	0.01	-0.00	-0.01	-0.00	0.01	1.00	-0.01	0.00
$2.5 < \lvert\Delta\eta_{ee}\rvert < 2.75$	0.13	0.11	0.10	0.07	0.10	0.08	0.09	0.10	0.09	0.03	0.01	-0.01	0.02	0.01	-0.00	-0.01	-0.02	-0.00	0.02	0.00	0.01	0.02	0.00	-0.01	-0.01	1.00	-0.01
$2.75 < \lvert\Delta\eta_{ee}\rvert < 3.0$	0.12	0.11	0.10	0.08	0.07	0.11	0.12	0.09	0.03	0.03	0.01	0.00	-0.02	-0.00	0.01	-0.01	0.01	-0.01	0.01	-0.00	0.00	0.01	0.01	-0.02	0.00	-0.01	1.00

Table Q.5 Statistical correlation between the measurement bins in mass, absolute rapidity and absolute pseudorapidity separation in the electron channel for 300 GeV $< m_{ee} <$ 500 GeV

Bin	300 GeV $< m_{ee} <$ 380 GeV	380 GeV $< m_{ee} <$ 500 GeV	0.0 $< \|y_{ee}\| <$ 0.4	0.4 $< \|y_{ee}\| <$ 0.8	0.8 $< \|y_{ee}\| <$ 1.2	1.2 $< \|y_{ee}\| <$ 1.6	1.6 $< \|y_{ee}\| <$ 2.0	2.0 $< \|y_{ee}\| <$ 2.4	0.0 $< \|\Delta\eta_{ee}\| <$ 0.5	0.5 $< \|\Delta\eta_{ee}\| <$ 1.0	1.0 $< \|\Delta\eta_{ee}\| <$ 1.5	1.5 $< \|\Delta\eta_{ee}\| <$ 2.0	2.0 $< \|\Delta\eta_{ee}\| <$ 2.5	2.5 $< \|\Delta\eta_{ee}\| <$ 3.0
300 GeV $< m_{ee} <$ 380 GeV	1.00	0.01	0.43	0.41	0.40	0.31	0.24	0.14	0.36	0.36	0.36	0.34	0.29	0.23
380 GeV $< m_{ee} <$ 500 GeV	0.01	1.00	0.32	0.32	0.27	0.19	0.12	0.06	0.28	0.24	0.27	0.21	0.17	0.19
0.0 $< \|y_{ee}\| <$ 0.4	0.43	0.32	1.00	0.00	0.01	−0.00	0.00	−0.00	0.22	0.19	0.24	0.22	0.20	0.17
0.4 $< \|y_{ee}\| <$ 0.8	0.41	0.32	0.00	1.00	0.01	−0.01	0.01	0.01	0.23	0.21	0.20	0.17	0.18	0.19
0.8 $< \|y_{ee}\| <$ 1.2	0.40	0.27	0.01	0.01	1.00	0.00	0.01	−0.01	0.20	0.14	0.18	0.24	0.22	0.20
1.2 $< \|y_{ee}\| <$ 1.6	0.31	0.19	−0.00	−0.01	0.00	1.00	−0.01	0.00	0.10	0.24	0.24	0.20	0.11	0.02
1.6 $< \|y_{ee}\| <$ 2.0	0.24	0.12	0.00	0.01	0.01	−0.01	1.00	0.02	0.22	0.20	0.16	0.03	−0.02	0.02
2.0 $< \|y_{ee}\| <$ 2.4	0.14	0.06	−0.00	0.01	−0.01	0.00	0.02	1.00	0.23	0.11	0.00	−0.00	−0.00	0.01
0.0 $< \|\Delta\eta_{ee}\| <$ 0.5	0.36	0.28	0.22	0.23	0.20	0.10	0.22	0.23	1.00	0.01	−0.00	−0.01	0.01	0.01

(continued)

Table Q.5 (continued)

Bin	300 GeV $< m_{ee} <$ 380 GeV	380 GeV $< m_{ee} <$ 500 GeV	0.0 $< \lvert y_{ee} \rvert < 0.4$	0.4 $< \lvert y_{ee} \rvert < 0.8$	0.8 $< \lvert y_{ee} \rvert < 1.2$	1.2 $< \lvert y_{ee} \rvert < 1.6$	1.6 $< \lvert y_{ee} \rvert < 2.0$	2.0 $< \lvert y_{ee} \rvert < 2.4$	0.0 $< \lvert \Delta\eta_{ee} \rvert < 0.5$	0.5 $< \lvert \Delta\eta_{ee} \rvert < 1.0$	1.0 $< \lvert \Delta\eta_{ee} \rvert < 1.5$	1.5 $< \lvert \Delta\eta_{ee} \rvert < 2.0$	2.0 $< \lvert \Delta\eta_{ee} \rvert < 2.5$	2.5 $< \lvert \Delta\eta_{ee} \rvert < 3.0$
$0.5 < \lvert \Delta\eta_{ee} \rvert < 1.0$	0.36	0.24	0.19	0.21	0.14	0.24	0.20	0.11	0.01	1.00	0.01	−0.00	0.00	0.01
$1.0 < \lvert \Delta\eta_{ee} \rvert < 1.5$	0.36	0.27	0.24	0.20	0.18	0.24	0.16	0.00	−0.00	0.01	1.00	−0.00	−0.02	0.02
$1.5 < \lvert \Delta\eta_{ee} \rvert < 2.0$	0.34	0.21	0.22	0.17	0.24	0.20	0.03	−0.00	−0.01	−0.00	−0.00	1.00	0.02	0.01
$2.0 < \lvert \Delta\eta_{ee} \rvert < 2.5$	0.29	0.17	0.20	0.18	0.22	0.11	−0.02	−0.00	0.01	0.00	−0.02	0.02	1.00	−0.00
$2.5 < \lvert \Delta\eta_{ee} \rvert < 3.0$	0.23	0.19	0.17	0.19	0.20	0.02	0.02	0.01	0.01	0.01	0.02	0.01	−0.00	1.00

Table Q.6 Statistical correlation between the measurement bins in mass, absolute rapidity and absolute pseudorapidity separation in the electron channel for 500 GeV $< m_{ee} <$ 1500 GeV

Bin	500 GeV $< m_{ee}$ $<$ 700 GeV	700 GeV $< m_{ee}$ $<$ 1000 GeV	1000 GeV $< \lvert y_{ee}\rvert$ $<$ 1500 GeV	0.0 $< \lvert y_{ee}\rvert$ $<$ 0.4	0.4 $< \lvert y_{ee}\rvert$ $<$ 0.8	0.8 $< \lvert y_{ee}\rvert$ $<$ 1.2	1.2 $<$ $\lvert y_{ee}\rvert$ $<$ 1.6	1.6 $< \lvert y_{ee}\rvert$ $<$ 2.0	2.0 $<$ $\lvert\Delta\eta_{ee}\rvert$ $<$ 2.4	0.0 $<$ $\lvert\Delta\eta_{ee}\rvert$ $<$ 0.5	0.5 $<$ $\lvert\Delta\eta_{ee}\rvert$ $<$ 1.0	1.0 $<$ $\lvert\Delta\eta_{ee}\rvert$ $<$ 1.5	1.5 $<$ $\lvert\Delta\eta_{ee}\rvert$ $<$ 2.0	2.0 $<$ $\lvert\Delta\eta_{ee}\rvert$ $<$ 2.5	2.5 $<$ $\lvert\Delta\eta_{ee}\rvert$ $<$ 3.0
500 GeV $< m_{ee} <$ 700 GeV	1.00	0.00	−0.01	0.49	0.45	0.38	0.37	0.19	0.08	0.34	0.38	0.44	0.39	0.26	0.24
700 GeV $< m_{ee} <$ 1000 GeV	0.00	1.00	0.00	0.24	0.30	0.24	0.10	0.01	0.00	0.22	0.22	0.14	0.17	0.15	0.12
1000 GeV $< m_{ee} <$ 1500 GeV	−0.01	0.00	1.00	0.14	0.13	0.06	0.04	−0.01	−0.00	0.15	0.11	0.06	0.04	0.04	0.03
0.0 $< \lvert y_{ee}\rvert <$ 0.4	0.49	0.24	0.14	1.00	0.01	−0.01	0.01	−0.01	0.01	0.20	0.28	0.26	0.25	0.18	0.12
0.4 $< \lvert y_{ee}\rvert <$ 0.8	0.45	0.30	0.13	0.01	1.00	−0.01	−0.01	−0.01	−0.00	0.26	0.24	0.21	0.18	0.18	0.22
0.8 $< \lvert y_{ee}\rvert <$ 1.2	0.38	0.24	0.06	−0.01	−0.01	1.00	0.00	−0.00	−0.00	0.24	0.15	0.17	0.22	0.15	0.16
1.2 $< \lvert y_{ee}\rvert <$ 1.6	0.37	0.10	0.04	0.01	−0.01	0.00	1.00	−0.01	0.00	0.07	0.20	0.26	0.23	0.10	0.01
1.6 $< \lvert y_{ee}\rvert <$ 2.0	0.19	0.01	−0.01	−0.01	−0.01	−0.00	−0.01	1.00	0.01	0.16	0.12	0.12	−0.01	−0.00	0.01
2.0 $< \lvert y_{ee}\rvert <$ 2.4	0.08	0.00	−0.00	0.01	−0.00	−0.00	0.00	0.01	1.00	0.16	0.02	−0.00	−0.00	−0.01	0.00
0.0 $< \lvert\Delta\eta_{ee}\rvert <$ 0.5	0.34	0.22	0.15	0.20	0.26	0.24	0.07	0.16	0.16	1.00	0.00	0.00	0.00	0.00	−0.00
0.5 $< \lvert\Delta\eta_{ee}\rvert <$ 1.0	0.38	0.22	0.11	0.28	0.24	0.15	0.20	0.12	0.02	0.00	1.00	0.00	−0.00	−0.00	0.01
1.0 $< \lvert\Delta\eta_{ee}\rvert <$ 1.5	0.44	0.14	0.06	0.26	0.21	0.17	0.26	0.12	−0.00	0.00	0.00	1.00	0.00	0.00	−0.02
1.5 $< \lvert\Delta\eta_{ee}\rvert <$ 2.0	0.39	0.17	0.04	0.25	0.18	0.22	0.23	−0.01	−0.00	0.00	−0.00	0.00	1.00	0.01	0.00
2.0 $< \lvert\Delta\eta_{ee}\rvert <$ 2.5	0.26	0.15	0.04	0.18	0.18	0.15	0.10	−0.00	−0.01	0.00	−0.00	0.00	0.01	1.00	−0.00
2.5 $< \lvert\Delta\eta_{ee}\rvert <$ 3.0	0.24	0.12	0.03	0.12	0.22	0.16	0.01	0.01	0.00	−0.00	0.01	−0.02	0.00	−0.00	1.00

Reference

1. A Armbruster et al (2014) Practical considerations for unfolding. Technical report ATL-COM-PHYS-2014-277. Geneva: CERN. https://cds.cern.ch/record/1694351

Appendix R
High-Mass Drell-Yan: Combined Cross Section Tables

This appendix contains detailed tables with the combined electron and muon channel cross sections and their uncertainties. Tables R.1, R.2, and R.3 contain the cross sections for the single-differential cross section, the double-differential cross section as a function of absolute rapidity, and the double-differential cross section as a function of absolute pseudorapidity separation, respectively.

M. Zinser, *Search for New Heavy Charged Bosons and Measurement of High-Mass Drell-Yan Production in Proton-Proton Collisions*, Springer Theses,
https://doi.org/10.1007/978-3-030-00650-1

Table R.1 The combined Born level single-differential cross section $\frac{d\sigma}{dm_{\ell\ell}}$. The measurements are listed together with the statistical (δ^{stat}), systematic (δ^{sys}) and total (δ^{tot}) uncertainties. In addition the contributions from the individual correlated (δ^{1}_{cor}-δ^{35}_{cor}) and uncorrelated (δ^{unc}) systematic error sources are also provided. The luminosity uncertainty of 1.9% is not shown and not included in the overall systematic and total uncertainties

$m_{\ell\ell}$ [GeV]	$\frac{d\sigma}{dm_{\ell\ell}}$ [pb/GeV]	δ^{stat} [%]	δ^{sys} [%]	δ^{tot} [%]	δ^{unc} [%]	δ^{1}_{cor} [%]	δ^{2}_{cor} [%]	δ^{3}_{cor} [%]	δ^{4}_{cor} [%]	δ^{5}_{cor} [%]	δ^{6}_{cor} [%]	δ^{7}_{cor} [%]	δ^{8}_{cor} [%]	δ^{9}_{cor} [%]	δ^{10}_{cor} [%]	δ^{11}_{cor} [%]	δ^{12}_{cor} [%]
116–130	2.28×10^{-1}	0.34	0.53	0.63	0.12	0.24	-0.01	0.08	-0.03	-0.03	0.00	0.01	0.01	0.00	0.01	0.00	-0.00
130–150	1.04×10^{-1}	0.44	0.67	0.80	0.13	0.38	-0.00	0.03	-0.05	-0.02	0.03	-0.01	0.00	0.00	0.01	-0.00	-0.00
150–175	4.98×10^{-2}	0.57	0.91	1.08	0.18	0.56	0.01	-0.02	-0.05	-0.01	0.05	-0.03	0.00	0.00	0.01	-0.00	0.00
175–200	2.54×10^{-2}	0.81	1.18	1.43	0.25	0.74	-0.00	-0.06	-0.07	-0.01	0.06	-0.05	-0.01	-0.00	0.01	-0.00	0.00
200–230	1.37×10^{-2}	1.02	1.42	1.75	0.32	0.89	0.02	-0.09	-0.06	-0.01	0.08	-0.05	-0.00	0.00	0.00	-0.00	0.00
230–260	7.89×10^{-3}	1.36	1.59	2.09	0.43	0.99	-0.01	-0.12	-0.08	0.00	0.07	-0.06	-0.01	0.00	0.01	-0.00	0.00
260–300	4.43×10^{-3}	1.58	1.67	2.30	0.46	1.06	0.02	-0.11	-0.05	0.01	0.12	-0.06	-0.00	0.01	0.00	-0.00	-0.00
300–380	1.87×10^{-3}	1.73	1.80	2.50	0.56	1.12	-0.02	-0.11	-0.09	0.01	0.09	-0.07	-0.01	0.00	0.01	-0.01	0.00
380–500	6.20×10^{-4}	2.42	1.71	2.96	0.63	1.03	0.00	-0.08	-0.10	0.04	0.14	-0.07	-0.00	0.01	0.01	-0.00	-0.00
500–700	1.53×10^{-4}	3.65	1.68	4.02	0.57	0.87	-0.08	-0.06	-0.14	0.04	0.15	0.01	0.02	0.01	-0.00	-0.01	0.01
700–1000	2.66×10^{-5}	6.98	1.85	7.22	1.02	0.73	-0.09	0.04	-0.13	0.07	0.17	0.03	0.03	0.03	0.00	-0.01	-0.01
1000–1500	2.66×10^{-6}	17.05	2.95	17.31	2.26	0.71	0.04	0.16	-0.01	0.06	0.33	0.10	0.08	0.05	-0.04	-0.00	-0.00

$m_{\ell\ell}$ [GeV]	δ^{13}_{cor} [%]	δ^{14}_{cor} [%]	δ^{15}_{cor} [%]	δ^{16}_{cor} [%]	δ^{17}_{cor} [%]	δ^{18}_{cor} [%]	δ^{19}_{cor} [%]	δ^{20}_{cor} [%]	δ^{21}_{cor} [%]	δ^{22}_{cor} [%]	δ^{23}_{cor} [%]	δ^{24}_{cor} [%]
116–130	0.00	0.00	-0.01	0.01	-0.00	0.01	-0.02	0.01	-0.02	-0.05	-0.31	0.15
130–150	0.00	0.00	-0.01	-0.01	-0.01	0.02	-0.01	0.01	0.08	-0.08	-0.38	0.10
150–175	0.00	0.00	-0.02	-0.02	-0.01	0.03	0.00	0.00	0.15	-0.10	-0.47	0.06
175–200	0.01	0.00	-0.02	-0.03	-0.01	0.04	0.01	0.00	0.23	-0.11	-0.58	0.02
200–230	0.00	-0.00	-0.02	-0.04	-0.00	0.04	0.02	-0.00	0.29	-0.12	-0.67	-0.01
230–260	0.01	-0.00	-0.02	-0.04	0.00	0.05	0.03	-0.01	0.28	-0.11	-0.74	0.04
260–300	-0.00	-0.00	-0.04	-0.06	0.01	0.07	0.04	0.00	0.35	-0.19	-0.73	0.00
300–380	0.01	-0.00	-0.03	-0.06	-0.00	0.06	0.01	-0.01	0.29	-0.18	-0.79	0.03
380–500	-0.00	0.00	-0.06	-0.08	0.01	0.08	0.01	0.02	0.30	-0.26	-0.69	0.09
500–700	0.03	0.00	-0.05	-0.05	0.03	0.17	0.04	0.12	0.02	-0.21	-0.56	0.10
700–1000	0.02	0.01	-0.07	-0.07	0.05	0.17	-0.01	0.14	-0.15	-0.26	-0.44	0.23
1000–1500	-0.01	-0.01	-0.15	-0.14	0.02	0.22	-0.08	0.16	-0.10	-0.49	-0.32	0.21

$m_{\ell\ell}$ [GeV]	δ^{25}_{cor} [%]	δ^{26}_{cor} [%]	δ^{27}_{cor} [%]	δ^{28}_{cor} [%]	δ^{29}_{cor} [%]	δ^{30}_{cor} [%]	δ^{31}_{cor} [%]	δ^{32}_{cor} [%]	δ^{33}_{cor} [%]	δ^{34}_{cor} [%]	δ^{35}_{cor} [%]
116–130	0.18	0.03	-0.04	-0.01	-0.02	0.11	0.04	0.07	0.07	0.09	-0.05
130–150	0.15	0.04	-0.08	0.01	0.04	0.22	0.03	0.08	0.14	0.08	-0.07
150–175	0.15	0.04	-0.12	-0.00	0.09	0.35	0.04	0.10	0.16	0.10	-0.09
175–200	0.14	0.07	-0.12	-0.00	0.13	0.47	0.03	0.12	0.17	0.10	-0.12
200–230	0.17	0.05	-0.16	0.02	0.16	0.58	0.05	0.16	0.21	0.16	-0.15
230–260	0.19	0.09	-0.14	-0.00	0.23	0.65	0.06	0.23	0.10	0.22	-0.18
260–300	0.17	0.05	-0.15	0.04	0.17	0.68	0.08	0.22	0.18	0.22	-0.19
300–380	0.15	0.08	-0.13	-0.00	0.20	0.76	0.06	0.30	0.03	0.29	-0.20
380–500	0.20	0.05	-0.13	0.05	0.16	0.59	0.06	0.36	0.03	0.39	-0.25
500–700	0.03	0.01	0.06	-0.15	0.06	0.38	0.13	0.96	-0.09	0.35	-0.18
700–1000	0.08	0.06	0.13	-0.09	0.02	0.17	0.19	1.00	-0.17	0.50	-0.17
1000–1500	0.23	0.08	-0.17	0.01	-0.34	0.28	0.32	1.21	-0.03	0.69	-0.35

Table R.2 The combined Born level double-differential cross section $\frac{d^2\sigma}{dm_{\ell\ell}d|y_{\ell\ell}|}$. The measurements are listed together with the statistical (δ^{stat}), systematic (δ^{sys}) and total (δ^{tot}) uncertainties. In addition the contributions from the individual correlated (δ^{1}_{cor}-δ^{35}_{cor}) and uncorrelated (δ^{unc}) systematic error sources are also provided. The luminosity uncertainty of 1.9% is not shown and not included in the overall systematic and total uncertainties

$m_{\ell\ell}$ [GeV]	$\|y_{\ell\ell}\|$	$\frac{d^2\sigma}{dm_{\ell\ell}d\|y_{\ell\ell}\|}$ [pb/GeV]	δ^{stat} [%]	δ^{sys} [%]	δ^{tot} [%]	δ^{unc} [%]	δ^{1}_{cor} [%]	δ^{2}_{cor} [%]	δ^{3}_{cor} [%]	δ^{4}_{cor} [%]	δ^{5}_{cor} [%]	δ^{6}_{cor} [%]	δ^{7}_{cor} [%]	δ^{8}_{cor} [%]	δ^{9}_{cor} [%]	δ^{10}_{cor} [%]	δ^{11}_{cor} [%]	δ^{12}_{cor} [%]	δ^{13}_{cor} [%]	δ^{14}_{cor} [%]	δ^{15}_{cor} [%]	δ^{16}_{cor} [%]	δ^{17}_{cor} [%]	δ^{18}_{cor} [%]	δ^{19}_{cor} [%]	δ^{20}_{cor} [%]	δ^{21}_{cor} [%]	δ^{22}_{cor} [%]	δ^{23}_{cor} [%]	δ^{24}_{cor} [%]	δ^{25}_{cor} [%]	δ^{26}_{cor} [%]	δ^{27}_{cor} [%]	δ^{28}_{cor} [%]	δ^{29}_{cor} [%]	δ^{30}_{cor} [%]	δ^{31}_{cor} [%]	δ^{32}_{cor} [%]	δ^{33}_{cor} [%]	δ^{34}_{cor} [%]	δ^{35}_{cor} [%]
116–150	0.0–0.2	4.10×10^{-2}	0.81	0.62	1.02	0.20	0.00	0.02	0.00	-0.03	0.01	0.06	0.07	-0.05	-0.07	-0.01	0.02	0.02	0.01	0.15	-0.15	-0.02	-0.02	-0.08	0.09	-0.19	-0.08	0.05	-0.14	0.09	-0.21	0.13	-0.19	0.09	-0.15	0.04	0.01	0.25	-0.03	-0.02	0.03
116–150	0.2–0.4	4.04×10^{-2}	0.82	0.63	1.03	0.21	0.00	0.01	0.01	-0.02	0.01	0.06	0.08	-0.05	-0.05	0.00	0.02	0.04	0.02	0.14	-0.16	-0.01	-0.01	-0.10	0.07	-0.20	-0.08	0.06	-0.19	0.05	-0.20	0.10	-0.19	0.10	-0.16	0.01	-0.02	0.24	-0.02	-0.06	0.03
116–150	0.4–0.6	4.01×10^{-2}	0.83	0.62	1.04	0.21	0.01	0.01	0.02	-0.02	-0.01	0.06	0.07	-0.05	-0.05	-0.00	0.02	0.03	0.03	0.13	-0.17	-0.03	-0.03	-0.08	0.09	-0.17	-0.07	0.06	-0.19	0.03	-0.18	0.13	-0.19	0.14	-0.12	-0.01	-0.01	0.26	-0.01	-0.06	0.04
116–150	0.6–0.8	4.04×10^{-2}	0.83	0.59	1.02	0.21	0.01	0.01	0.03	-0.02	-0.02	0.06	0.07	-0.04	-0.06	-0.01	0.02	0.01	0.01	0.13	-0.16	-0.03	-0.03	-0.04	0.09	-0.16	-0.09	0.05	-0.18	0.06	-0.18	0.05	-0.15	0.08	-0.19	0.01	0.05	0.23	0.01	-0.05	0.04
116–150	0.8–1.0	3.97×10^{-2}	0.84	0.59	1.03	0.22	0.02	0.01	0.03	-0.02	-0.02	0.06	0.07	-0.04	-0.05	-0.01	0.01	0.01	0.03	0.11	-0.14	-0.04	-0.04	-0.06	0.04	-0.17	-0.06	0.09	-0.19	-0.01	-0.16	0.06	-0.12	0.15	-0.18	-0.00	0.06	0.25	0.03	0.02	0.03
116–150	1.0–1.2	3.95×10^{-2}	0.84	0.56	1.01	0.22	0.02	0.00	0.04	-0.01	-0.03	0.06	0.07	-0.04	-0.03	0.01	-0.01	0.01	0.03	0.09	-0.12	-0.04	-0.04	-0.05	0.05	-0.12	-0.04	0.06	-0.19	-0.05	-0.10	0.05	-0.13	0.12	-0.20	-0.04	0.07	0.27	0.00	-0.06	0.04
116–150	1.2–1.4	3.84×10^{-2}	0.84	0.58	1.02	0.23	0.02	-0.00	0.03	-0.01	-0.03	0.05	0.07	-0.04	-0.02	0.01	-0.02	0.01	0.02	0.07	-0.11	-0.05	-0.05	-0.03	0.01	-0.12	-0.05	0.05	-0.19	-0.09	-0.06	-0.00	-0.13	0.18	-0.15	-0.02	0.08	0.29	0.01	-0.09	0.06
116–150	1.4–1.6	3.48×10^{-2}	0.89	0.58	1.06	0.24	0.02	-0.01	0.02	-0.01	-0.01	0.06	0.07	-0.04	-0.01	0.01	-0.04	0.02	0.03	0.04	-0.11	-0.05	-0.05	-0.04	-0.02	-0.10	0.01	0.08	-0.17	-0.14	-0.08	-0.03	-0.14	0.07	-0.13	-0.08	0.00	0.31	-0.03	-0.15	0.08
116–150	1.6–1.8	2.92×10^{-2}	1.00	0.64	1.18	0.28	0.02	-0.00	0.01	-0.01	-0.01	0.06	0.08	-0.04	-0.01	0.00	-0.04	0.01	0.03	0.04	-0.11	-0.05	-0.05	-0.01	-0.04	-0.10	-0.00	0.07	-0.19	-0.22	-0.06	-0.02	-0.09	0.11	-0.13	-0.04	-0.00	0.35	-0.04	-0.12	0.10
116–150	1.8–2.0	2.29×10^{-2}	1.14	0.70	1.33	0.31	0.02	-0.01	0.01	-0.01	-0.01	0.07	0.09	-0.05	-0.01	0.01	-0.06	0.01	0.03	0.02	-0.12	-0.07	-0.07	-0.05	-0.04	-0.06	-0.01	0.07	-0.21	-0.28	-0.07	-0.09	-0.09	0.12	-0.13	-0.07	-0.04	0.34	-0.05	-0.14	0.11
116–150	2.0–2.2	1.54×10^{-2}	1.40	0.88	1.65	0.40	0.02	0.00	0.00	-0.00	-0.02	0.06	0.09	-0.07	-0.02	-0.01	-0.05	0.00	0.01	0.07	-0.12	-0.07	-0.07	0.03	-0.03	-0.04	-0.01	0.07	-0.31	-0.15	0.01	-0.15	-0.02	0.25	-0.15	0.18	0.12	0.45	-0.24	-0.01	0.17
116–150	2.2–2.4	7.31×10^{-3}	2.33	1.46	2.75	0.72	0.03	-0.00	-0.03	0.03	-0.01	0.05	0.13	-0.08	0.01	0.02	-0.08	0.02	0.01	0.04	-0.15	-0.07	-0.07	0.03	-0.07	-0.12	-0.00	0.10	-0.48	-0.27	0.13	-0.14	-0.00	0.18	-0.34	0.26	0.26	0.74	-0.46	0.13	0.33
150–200	0.0–0.2	1.04×10^{-2}	1.41	1.31	1.93	0.32	-0.00	0.04	0.00	-0.06	0.01	0.07	0.10	-0.07	-0.12	-0.00	0.12	0.06	0.04	0.31	-0.30	-0.02	-0.02	-0.22	0.29	-0.43	-0.22	0.03	-0.26	0.32	-0.42	0.37	-0.50	0.11	-0.27	-0.06	-0.08	0.42	-0.00	-0.03	0.07
150–200	0.2–0.4	1.04×10^{-2}	1.41	1.28	1.91	0.33	-0.00	0.04	0.01	-0.05	0.00	0.07	0.10	-0.06	-0.13	-0.00	0.11	0.05	0.03	0.30	-0.30	-0.01	-0.01	-0.19	0.27	-0.42	-0.21	0.05	-0.26	0.32	-0.42	0.36	-0.46	0.15	-0.26	-0.03	-0.05	0.42	0.01	-0.05	0.07
150–200	0.4–0.6	1.03×10^{-2}	1.42	1.22	1.87	0.32	-0.00	0.04	0.03	-0.05	0.00	0.07	0.10	-0.06	-0.12	-0.01	0.10	0.05	0.02	0.29	-0.29	-0.03	-0.03	-0.17	0.26	-0.40	-0.19	0.05	-0.23	0.29	-0.38	0.33	-0.45	0.13	-0.28	-0.03	-0.01	0.42	0.01	-0.03	0.07
150–200	0.6–0.8	1.05×10^{-2}	1.39	1.11	1.78	0.32	0.01	0.03	0.02	-0.04	-0.02	0.06	0.09	-0.06	-0.09	-0.01	0.08	0.05	0.04	0.25	-0.25	0.04	-0.04	-0.16	0.23	-0.33	-0.19	0.04	-0.30	0.23	-0.29	0.33	-0.38	0.13	-0.29	-0.05	0.01	0.40	0.00	-0.03	0.07
150–200	0.8–1.0	1.02×10^{-2}	1.42	1.02	1.75	0.33	0.01	0.02	0.03	-0.04	-0.02	0.07	0.08	-0.06	-0.10	-0.01	0.07	0.03	0.02	0.22	-0.24	-0.05	-0.05	-0.15	0.19	-0.29	-0.16	0.06	-0.20	0.20	-0.26	0.25	-0.39	0.10	-0.23	-0.11	0.01	0.41	-0.02	-0.01	0.08
150–200	1.0–1.2	9.68×10^{-3}	1.44	0.94	1.72	0.32	0.02	0.01	0.02	-0.03	-0.03	0.07	0.08	-0.05	-0.07	0.01	0.05	0.03	0.03	0.16	-0.20	-0.05	-0.05	-0.10	0.15	-0.23	-0.16	0.04	-0.26	0.14	-0.19	0.19	-0.34	0.14	-0.19	-0.10	0.04	0.44	-0.08	-0.09	0.12
150–200	1.2–1.4	9.12×10^{-3}	1.47	0.80	1.67	0.31	0.03	0.01	0.02	-0.03	-0.03	0.06	0.07	-0.06	-0.07	-0.02	0.02	0.01	0.02	0.13	-0.18	-0.08	-0.08	-0.05	0.10	-0.20	-0.12	0.05	-0.19	0.07	-0.19	0.14	-0.27	0.09	-0.20	-0.03	0.05	0.38	-0.04	-0.15	0.10
150–200	1.4–1.6	7.79×10^{-3}	1.59	0.78	1.77	0.32	0.03	0.01	0.01	-0.02	-0.02	0.07	0.06	-0.06	-0.05	-0.02	-0.01	0.01	0.03	0.10	-0.18	-0.09	-0.09	-0.06	0.06	-0.13	-0.10	0.04	-0.21	-0.03	-0.17	0.09	-0.21	0.09	-0.17	-0.11	0.00	0.41	-0.04	-0.18	0.14
150–200	1.6–1.8	6.64×10^{-3}	1.73	0.75	1.88	0.33	0.03	-0.00	0.01	-0.01	-0.03	0.07	0.07	-0.07	-0.04	-0.00	-0.02	0.02	0.03	0.07	-0.17	-0.05	-0.05	-0.07	0.01	-0.06	-0.10	0.06	-0.24	-0.12	-0.05	0.03	-0.20	0.07	-0.16	-0.04	-0.04	0.39	-0.05	-0.21	0.15
150–200	1.8–2.0	4.83×10^{-3}	2.03	0.75	2.16	0.38	0.03	-0.00	0.01	-0.02	-0.03	0.08	0.07	-0.07	-0.04	-0.01	-0.04	0.01	0.01	0.08	-0.17	-0.09	-0.09	0.00	-0.02	-0.03	-0.08	0.10	-0.21	-0.15	-0.10	-0.02	-0.14	0.17	-0.12	0.05	0.01	0.37	-0.12	-0.12	0.14
150–200	2.0–2.2	3.35×10^{-3}	2.42	0.88	2.57	0.46	0.03	-0.00	0.00	-0.02	-0.02	0.08	0.07	-0.07	-0.05	-0.03	-0.04	0.00	-0.01	0.07	-0.15	-0.10	-0.10	0.03	-0.03	0.05	-0.09	0.10	-0.22	-0.18	-0.07	-0.02	-0.09	0.16	-0.18	0.10	0.06	0.41	-0.32	0.11	0.17
150–200	2.2–2.4	1.63×10^{-3}	3.91	1.45	4.17	0.81	0.02	-0.01	-0.05	-0.00	-0.02	0.08	0.11	-0.11	-0.01	0.00	-0.07	0.06	0.02	0.06	-0.14	-0.10	-0.10	0.02	-0.02	0.03	-0.12	0.23	-0.42	-0.31	0.09	-0.15	-0.08	0.15	-0.35	0.20	0.06	0.68	-0.39	0.21	0.38
200–300	0.0–0.2	2.42×10^{-3}	2.20	2.14	3.07	0.55	-0.00	0.07	0.02	-0.08	0.02	0.08	0.14	-0.08	-0.20	0.00	0.19	0.09	0.07	0.48	-0.45	0.00	0.00	-0.31	0.49	-0.73	-0.31	0.08	-0.39	0.61	-0.69	0.62	-0.76	0.23	-0.45	-0.02	-0.03	0.67	0.03	-0.12	0.12
200–300	0.2–0.4	2.39×10^{-3}	2.21	2.06	3.03	0.56	-0.00	0.06	0.02	-0.08	0.01	0.08	0.14	-0.09	-0.19	-0.01	0.18	0.08	0.07	0.47	-0.44	-0.01	-0.01	-0.30	0.47	-0.71	-0.31	0.06	-0.37	0.57	-0.65	0.62	-0.75	0.18	-0.45	-0.06	-0.05	0.65	0.00	-0.09	0.11
200–300	0.4–0.6	2.43×10^{-3}	2.18	1.93	2.91	0.56	0.01	0.05	0.01	-0.08	-0.02	0.09	0.13	-0.09	-0.17	-0.01	0.14	0.05	0.06	0.37	-0.40	-0.06	-0.06	-0.26	0.42	-0.63	-0.30	0.06	-0.38	0.54	-0.55	0.57	-0.74	0.21	-0.33	-0.13	-0.07	0.68	-0.05	-0.08	0.15
200–300	0.6–0.8	2.48×10^{-3}	2.12	1.67	2.70	0.52	0.02	0.04	0.01	-0.06	-0.01	0.08	0.12	-0.09	-0.15	0.01	0.12	0.06	0.06	0.34	-0.35	-0.07	-0.07	-0.23	0.34	-0.51	-0.29	0.04	-0.35	0.42	-0.46	0.44	-0.64	0.15	-0.31	-0.16	-0.07	0.61	-0.02	-0.15	0.15
200–300	0.8–1.0	2.29×10^{-3}	2.19	1.51	2.65	0.49	0.02	0.04	0.02	-0.05	-0.02	0.07	0.10	-0.09	-0.14	-0.00	0.08	0.04	0.07	0.31	-0.33	-0.07	-0.07	-0.18	0.28	-0.47	-0.23	0.07	-0.34	0.32	-0.44	0.38	-0.52	0.16	-0.32	-0.12	-0.02	0.56	-0.03	-0.23	0.15
200–300	1.0–1.2	2.16×10^{-3}	2.20	1.29	2.55	0.46	0.02	0.02	0.02	-0.05	-0.02	0.08	0.09	-0.08	-0.13	-0.02	0.04	0.04	0.03	0.23	-0.28	-0.10	-0.10	-0.12	0.19	-0.32	-0.21	0.06	-0.29	0.26	-0.34	0.29	-0.44	0.08	-0.24	-0.13	-0.04	0.56	-0.06	-0.30	0.20
200–300	1.2–1.4	1.88×10^{-3}	2.32	1.14	2.59	0.45	0.03	0.01	0.02	-0.04	-0.02	0.08	0.08	-0.07	-0.08	-0.00	0.01	0.04	0.04	0.17	-0.26	-0.08	-0.08	-0.12	0.13	-0.23	-0.18	0.07	-0.24	0.11	-0.25	0.27	-0.39	0.09	-0.15	-0.18	-0.05	0.53	-0.11	-0.30	0.20
200–300	1.4–1.6	1.66×10^{-3}	2.41	0.95	2.59	0.44	0.04	0.01	0.01	-0.02	-0.03	0.08	0.08	-0.08	-0.08	-0.02	-0.00	0.04	0.02	0.16	-0.22	-0.08	-0.08	-0.04	0.07	-0.15	-0.15	0.08	-0.22	0.03	-0.16	0.12	-0.24	0.12	-0.25	-0.04	0.02	0.46	-0.08	-0.26	0.18
200–300	1.6–1.8	1.34×10^{-3}	2.71	0.90	2.85	0.47	0.04	0.01	0.00	-0.02	-0.02	0.08	0.07	-0.07	-0.08	-0.05	-0.04	0.03	0.02	0.11	-0.20	-0.08	-0.10	-0.02	0.02	-0.07	-0.15	0.09	-0.19	-0.05	-0.14	0.01	-0.21	0.13	-0.15	-0.04	-0.06	0.42	-0.12	-0.30	0.18
200–300	1.8–2.0	1.00×10^{-3}	3.13	0.94	3.27	0.52	0.04	-0.00	0.01	-0.01	-0.01	0.10	0.08	-0.09	-0.05	-0.05	-0.07	0.01	0.01	0.05	-0.17	[illegible]	-0.08	0.00	-0.03	0.01	-0.14	0.08	-0.17	-0.20	-0.13	0.02	-0.11	0.08	-0.21	-0.03	-0.12	0.42	-0.20	-0.29	0.20
200–300	2.0–2.2	6.04×10^{-4}	3.97	1.09	4.12	0.62	0.03	-0.01	-0.00	-0.02	-0.00	0.10	0.09	-0.08	-0.04	-0.04	-0.10	-0.00	0.01	0.09	-0.20	-0.08	-0.08	0.02	-0.06	0.03	-0.15	0.13	-0.17	-0.18	-0.06	-0.12	-0.06	0.24	-0.30	0.06	-0.07	0.43	-0.43	0.07	0.21
200–300	2.2–2.4	2.43×10^{-4}	6.94	1.70	7.15	1.18	0.02	0.01	-0.01	-0.03	0.02	0.09	0.12	-0.12	-0.07	-0.09	-0.07	0.04	0.05	0.04	-0.23	[illegible]	-0.12	0.06	-0.04	0.07	-0.14	0.32	-0.20	-0.30	-0.06	-0.21	0.03	0.38	-0.46	0.26	-0.15	0.59	-0.33	0.17	0.35
300–500	0.0–0.4	3.55×10^{-4}	2.97	2.54	3.90	0.92	0.02	0.03	-0.01	-0.08	-0.01	0.12	0.15	-0.12	-0.19	0.03	0.16	0.09	0.01	0.42	-0.47	[illegible]	-0.09	-0.29	0.49	-0.78	-0.41	0.04	-0.52	0.74	-0.67	0.75	-0.98	0.23	-0.41	-0.18	-0.15	0.89	-0.09	-0.14	0.23
300–500	0.4–0.8	3.36×10^{-4}	2.96	2.10	3.62	0.76	0.04	0.03	0.01	-0.07	-0.02	0.11	0.12	-0.11	-0.18	0.00	0.11	0.07	0.02	0.36	-0.41	[illegible]	-0.14	-0.16	0.35	-0.63	-0.31	0.07	-0.41	0.57	-0.64	0.57	-0.74	0.23	-0.40	-0.08	-0.05	0.71	0.03	-0.43	0.18
300–500	0.8–1.2	3.16×10^{-4}	2.90	1.49	3.26	0.58	0.06	0.02	0.02	-0.04	-0.03	0.09	0.08	-0.10	-0.14	-0.03	0.03	0.05	0.03	0.22	-0.32	[illegible]	-0.13	-0.08	0.22	-0.36	-0.23	0.10	-0.32	0.24	-0.40	0.34	-0.51	0.16	-0.19	-0.16	-0.13	0.54	0.02	-0.52	0.19
300–500	1.2–1.6	2.45×10^{-4}	3.12	1.15	3.33	0.51	0.06	-0.01	-0.00	-0.01	-0.03	0.10	0.05	-0.11	-0.08	-0.02	-0.03	0.05	0.02	0.13	-0.20	[illegible]	-0.11	-0.04	0.04	-0.08	-0.20	0.05	-0.32	0.04	-0.10	0.11	-0.24	0.11	-0.20	-0.12	-0.06	0.55	-0.12	-0.49	0.27
300–500	1.6–2.0	1.31×10^{-4}	4.29	1.24	4.46	0.59	0.07	0.00	-0.01	-0.01	-0.03	0.10	0.05	-0.11	-0.10	-0.06	-0.10	0.04	-0.01	0.09	-0.23	[illegible]	-0.13	0.05	-0.03	0.03	-0.14	0.11	-0.17	-0.17	-0.10	-0.02	-0.02	0.19	-0.18	0.04	-0.23	0.35	0.11	-0.81	0.20
300–500	2.0–2.4	3.81×10^{-5}	7.99	1.78	8.18	1.10	0.05	-0.00	-0.02	-0.03	0.01	0.12	0.09	-0.14	-0.11	-0.08	-0.11	0.03	0.01	0.13	-0.32	[illegible]	-0.14	0.06	-0.10	0.07	-0.03	0.09	-0.24	-0.12	0.17	-0.20	0.35	0.33	-0.82	0.23	-0.45	0.52	-0.25	0.00	0.24
500–1500	0.0–0.4	1.47×10^{-5}	6.15	1.89	6.43	0.99	0.12	-0.04	-0.02	-0.02	-0.05	0.17	0.08	-0.16	-0.16	0.02	-0.02	0.06	-0.07	0.20	-0.28	[illegible]	-0.23	0.04	0.14	-0.34	-0.33	0.01	-0.47	0.44	-0.51	0.41	-0.65	0.18	-0.37	-0.06	-0.16	0.55	0.04	-0.45	0.16
500–1500	0.4–0.8	1.38×10^{-5}	6.08	1.61	6.29	0.90	0.14	-0.02	0.02	-0.01	-0.04	0.14	0.03	-0.12	-0.16	-0.04	-0.07	0.04	-0.04	0.12	-0.26	[illegible]	-0.20	0.10	0.03	-0.21	-0.24	0.09	-0.34	0.22	-0.47	0.29	-0.36	0.15	-0.27	-0.09	-0.14	0.41	0.13	-0.70	0.16
500–1500	0.8–1.2	1.14×10^{-5}	6.46	1.47	6.62	0.85	0.14	-0.02	0.02	-0.00	-0.02	0.11	0.01	-0.13	-0.13	-0.05	-0.11	0.07	-0.01	0.12	-0.23	-0.18	-0.18	0.09	0.00	-0.08	-0.18	0.08	-0.28	0.07	-0.26	0.11	-0.16	0.11	-0.15	-0.09	-0.05	0.52	-0.07	-0.77	0.25
500–1500	1.2–1.6	8.05×10^{-6}	7.39	1.44	7.53	0.71	0.13	-0.02	0.02	0.01	-0.02	0.12	-0.00	-0.13	-0.12	-0.05	-0.15	0.11	-0.02	0.09	-0.23	-0.15	-0.15	0.04	-0.10	0.11	-0.14	0.09	-0.17	-0.16	-0.09	-0.05	0.02	0.15	-0.18	-0.20	-0.30	0.33	0.16	-0.94	0.21
500–1500	1.6–2.0	2.21×10^{-6}	14.30	1.95	14.43	1.04	0.13	-0.01	-0.01	-0.02	-0.00	0.12	0.02	-0.16	-0.20	-0.16	-0.09	0.08	0.06	0.13	-0.38	-0.16	-0.16	0.14	-0.18	0.11	-0.11	0.02	-0.08	-0.04	-0.00	-0.21	0.51	0.43	-0.54	0.22	-0.31	0.56	0.16	-1.00	0.22
500–1500	2.0–2.4	2.87×10^{-7}	35.70	7.63	36.51	6.94	0.04	0.17	0.06	-0.11	0.12	0.14	0.14	-0.26	-0.28	-0.26	-0.28	0.05	-0.15	0.23	-0.54	-0.16	-0.16	0.53	-0.28	-0.24	0.14	0.19	0.07	-0.15	-0.54	-0.40	1.73	0.83	-1.00	0.61	-0.41	1.63	-0.26	-0.47	0.36

Table R.3 The combined Born level double-differential cross section $\frac{d^2\sigma}{dm_{\ell\ell}d|\Delta\eta_{\ell\ell}|}$. The measurements are listed together with the statistical (δ^{stat}), systematic (δ^{sys}) and total (δ^{tot}) uncertainties. In addition the contributions from the individual correlated (δ^{1}_{cor}-δ^{35}_{cor}) and uncorrelated (δ^{unc}) systematic error sources are also provided. The luminosity uncertainty of 1.9% is not shown and not included in the overall systematic and total uncertainties

$m_{\ell\ell}$ [GeV]	$\|\Delta\eta_{\ell\ell}\|$	$\frac{d^2\sigma}{dm_{\ell\ell}d\|\Delta\eta_{\ell\ell}\|}$ [pb/GeV]	δ^{stat} [%]	δ^{sys} [%]	δ^{tot} [%]	δ^{unc} [%]	δ^{1}_{cor} [%]	δ^{2}_{cor} [%]	δ^{3}_{cor} [%]	δ^{4}_{cor} [%]	δ^{5}_{cor} [%]	δ^{6}_{cor} [%]	δ^{7}_{cor} [%]	δ^{8}_{cor} [%]	δ^{9}_{cor} [%]	δ^{10}_{cor} [%]	δ^{11}_{cor} [%]	δ^{12}_{cor} [%]	δ^{13}_{cor} [%]	δ^{14}_{cor} [%]	δ^{15}_{cor} [%]	δ^{16}_{cor} [%]	δ^{17}_{cor} [%]	δ^{18}_{cor} [%]	δ^{19}_{cor} [%]	δ^{20}_{cor} [%]	δ^{21}_{cor} [%]	δ^{22}_{cor} [%]	δ^{23}_{cor} [%]	δ^{24}_{cor} [%]	δ^{25}_{cor} [%]	δ^{26}_{cor} [%]	δ^{27}_{cor} [%]	δ^{28}_{cor} [%]	δ^{29}_{cor} [%]	δ^{30}_{cor} [%]	δ^{31}_{cor} [%]	δ^{32}_{cor} [%]	δ^{33}_{cor} [%]	δ^{34}_{cor} [%]	δ^{35}_{cor} [%]
116–150	0.00–0.25	4.94×10^{-2}	0.66	0.56	0.86	0.16	-0.00	-0.00	0.00	-0.00	-0.03	-0.04	0.03	-0.02	-0.02	-0.06	0.01	-0.01	0.02	0.05	0.02	-0.05	-0.05	-0.06	0.03	0.01	0.04	0.00	-0.11	-0.12	-0.18	-0.03	0.06	0.22	-0.21	0.25	0.12	-0.02	-0.14	0.08	0.07
116–150	0.25–0.50	4.68×10^{-2}	0.68	0.58	0.89	0.18	-0.00	-0.01	-0.00	0.00	-0.03	-0.04	0.03	-0.02	-0.03	-0.06	0.02	-0.01	0.01	0.06	0.03	-0.06	-0.06	-0.06	0.03	0.01	0.06	-0.02	-0.11	-0.12	-0.20	-0.01	0.06	0.20	-0.22	0.26	0.12	-0.06	-0.11	0.11	0.07
116–150	0.50–0.75	4.43×10^{-2}	0.70	0.58	0.91	0.19	-0.00	-0.00	0.00	-0.01	-0.04	-0.04	0.03	-0.03	-0.03	-0.06	0.01	-0.01	0.01	0.06	0.03	-0.05	-0.05	-0.06	0.03	0.01	0.04	-0.02	-0.10	-0.12	-0.20	-0.02	0.02	0.21	-0.22	0.25	0.15	-0.04	-0.12	0.09	0.07
116–150	0.75–1.00	4.03×10^{-2}	0.75	0.58	0.94	0.20	-0.00	-0.01	0.00	-0.01	-0.04	-0.04	0.02	-0.03	-0.03	-0.06	0.01	-0.01	0.01	0.07	0.04	-0.05	-0.05	-0.06	0.03	0.02	0.03	-0.03	-0.09	-0.12	-0.21	-0.03	0.05	0.17	-0.24	0.24	0.17	-0.01	-0.12	0.05	0.04
116–150	1.00–1.25	3.63×10^{-2}	0.78	0.58	0.97	0.21	-0.00	-0.00	0.00	-0.00	-0.04	-0.03	0.02	-0.04	-0.03	-0.06	0.02	-0.03	0.01	0.06	0.03	-0.05	-0.05	-0.06	0.03	0.01	0.04	-0.03	-0.06	-0.11	-0.20	-0.03	0.02	0.17	-0.25	0.27	0.18	-0.02	-0.09	0.03	0.05
116–150	1.25–1.50	3.16×10^{-2}	0.85	0.61	1.04	0.23	-0.00	-0.01	0.00	-0.00	-0.04	-0.04	0.03	-0.03	-0.03	-0.06	0.02	-0.03	0.00	0.07	0.03	-0.05	-0.05	-0.06	0.03	0.02	0.03	-0.05	-0.08	-0.11	-0.21	-0.02	0.04	0.16	-0.29	0.24	0.23	-0.01	-0.07	-0.03	0.04
116–150	1.50–1.75	2.54×10^{-2}	0.95	0.68	1.17	0.26	-0.00	-0.01	0.01	-0.00	-0.04	-0.04	0.02	-0.03	-0.03	-0.06	0.01	-0.03	-0.00	0.07	0.04	-0.05	-0.05	-0.06	0.03	0.03	0.03	-0.04	-0.09	-0.09	-0.24	-0.08	0.06	0.18	-0.29	0.29	0.26	0.01	-0.09	-0.06	0.04
116–150	1.75–2.00	2.03×10^{-2}	1.08	0.72	1.30	0.30	-0.00	-0.00	0.01	-0.00	-0.04	-0.04	0.02	-0.04	-0.03	-0.05	0.02	-0.02	-0.00	0.08	0.02	-0.05	-0.05	-0.04	0.04	0.03	0.03	-0.03	-0.09	-0.10	-0.26	-0.03	0.05	0.19	-0.27	0.31	0.28	0.06	-0.12	-0.10	0.03
116–150	2.00–2.25	1.16×10^{-2}	1.47	0.89	1.72	0.40	-0.00	-0.00	0.01	-0.00	-0.04	-0.04	0.01	-0.04	-0.03	-0.05	0.02	-0.02	-0.01	0.08	0.04	-0.05	-0.05	-0.04	0.04	0.04	0.03	-0.04	-0.09	-0.07	-0.29	-0.02	0.10	0.17	-0.37	0.34	0.40	0.08	-0.15	-0.14	0.04
116–150	2.25–2.50	3.94×10^{-3}	2.61	1.29	2.91	0.67	-0.01	0.00	0.01	0.01	-0.06	-0.03	0.01	-0.05	-0.03	-0.04	0.02	-0.06	-0.02	0.08	-0.01	-0.08	-0.08	-0.02	0.05	-0.01	-0.03	-0.07	0.01	-0.04	-0.29	-0.03	0.10	0.25	-0.54	0.49	0.64	0.11	-0.14	-0.21	0.09
116–150	2.50–2.75	6.18×10^{-4}	7.05	2.70	7.54	1.78	-0.01	0.00	0.01	0.02	-0.04	-0.03	-0.01	-0.04	-0.04	-0.06	0.02	-0.06	-0.01	0.12	-0.04	-0.06	-0.06	-0.00	0.04	0.07	0.04	-0.16	0.09	0.04	-0.32	-0.06	0.24	0.28	-0.88	1.07	1.20	0.10	-0.47	-0.37	0.24
150–200	0.00–0.25	1.09×10^{-2}	1.16	0.76	1.39	0.21	-0.01	-0.00	-0.01	0.00	-0.05	-0.03	0.04	-0.03	-0.03	-0.08	0.03	-0.02	0.02	0.05	0.02	-0.07	-0.07	-0.05	0.05	-0.02	0.07	-0.07	-0.01	-0.17	-0.17	0.02	0.06	0.21	-0.40	0.27	0.30	0.00	-0.16	0.14	0.12
150–200	0.25–0.50	1.04×10^{-2}	1.21	0.80	1.45	0.23	-0.01	0.00	-0.01	-0.01	-0.05	-0.03	0.04	-0.03	-0.03	-0.07	0.02	-0.02	0.02	0.05	0.01	-0.07	-0.07	-0.07	0.05	-0.00	0.04	-0.06	-0.01	-0.17	-0.18	-0.00	0.06	0.23	-0.39	0.31	0.34	0.03	-0.15	0.15	0.10
150–200	0.50–0.75	9.54×10^{-3}	1.28	0.85	1.54	0.25	-0.01	-0.00	-0.01	-0.01	-0.06	-0.03	0.04	-0.04	-0.04	-0.07	0.02	-0.03	0.02	0.05	0.01	-0.07	-0.07	-0.06	0.04	0.02	0.04	-0.11	0.01	-0.18	-0.17	-0.01	0.03	0.22	-0.43	0.32	0.42	0.03	-0.14	0.11	0.10
150–200	0.75–1.00	9.15×10^{-3}	1.32	0.89	1.59	0.26	-0.01	0.00	-0.01	-0.01	-0.06	-0.03	0.04	-0.04	-0.03	-0.08	0.02	-0.02	0.02	0.06	0.00	-0.07	-0.07	-0.06	0.05	0.01	0.02	-0.08	0.03	-0.17	-0.18	-0.05	0.06	0.24	-0.45	0.38	0.43	0.04	-0.11	0.11	0.10
150–200	1.00–1.25	8.03×10^{-3}	1.41	0.99	1.72	0.28	-0.01	-0.00	0.00	-0.00	-0.06	-0.03	0.03	-0.04	-0.03	-0.08	0.02	-0.03	0.02	0.07	-0.00	-0.06	-0.06	-0.06	0.05	0.01	0.01	-0.10	0.06	-0.17	-0.23	-0.05	0.08	0.24	-0.50	0.44	0.48	0.04	-0.12	0.09	0.10
150–200	1.25–1.50	7.02×10^{-3}	1.53	1.09	1.88	0.32	-0.01	-0.00	0.00	-0.00	-0.06	-0.04	0.03	-0.04	-0.04	-0.08	0.02	-0.04	0.01	0.07	0.01	-0.08	-0.08	-0.05	0.04	0.03	0.03	-0.12	0.07	-0.18	-0.24	-0.05	0.08	0.18	-0.56	0.47	0.58	0.05	-0.11	0.04	0.09
150–200	1.50–1.75	6.06×10^{-3}	1.68	1.22	2.07	0.37	-0.01	-0.01	0.01	0.00	-0.06	-0.04	0.03	-0.05	-0.03	-0.08	0.02	-0.04	0.00	0.08	0.00	-0.07	-0.07	-0.05	0.05	0.03	-0.01	-0.14	0.10	-0.17	-0.24	-0.08	0.08	0.21	-0.59	0.54	0.67	0.07	-0.12	-0.02	0.09
150–200	1.75–2.00	4.94×10^{-3}	1.91	1.42	2.38	0.45	-0.01	-0.00	0.01	-0.00	-0.08	-0.03	0.02	-0.06	-0.03	-0.08	0.01	-0.06	-0.01	0.09	0.00	-0.07	-0.07	-0.06	0.05	0.03	-0.04	-0.17	0.11	-0.16	-0.26	-0.12	0.05	0.23	-0.69	0.61	0.82	0.05	-0.14	-0.08	0.11
150–200	2.00–2.25	3.77×10^{-3}	2.26	1.63	2.78	0.58	-0.01	0.00	0.02	-0.00	-0.08	-0.03	0.02	-0.06	-0.04	-0.07	0.03	-0.06	-0.01	0.11	0.01	-0.09	-0.09	-0.04	0.04	0.06	-0.04	-0.19	0.13	-0.16	-0.32	-0.08	0.08	0.19	-0.75	0.72	0.94	0.09	-0.15	-0.10	0.10
150–200	2.25–2.50	2.92×10^{-3}	2.60	1.74	3.13	0.73	-0.02	-0.01	-0.01	-0.00	-0.08	-0.05	0.02	-0.06	-0.02	-0.06	0.02	-0.06	-0.04	0.10	0.03	-0.07	-0.07	-0.03	0.03	0.04	-0.03	-0.13	0.06	-0.09	-0.28	-0.13	0.02	0.17	-0.81	0.72	1.02	0.07	-0.18	-0.18	0.12
150–200	2.50–2.75	1.98×10^{-3}	3.16	1.81	3.64	0.92	-0.02	-0.01	-0.01	-0.00	-0.07	-0.04	0.02	-0.06	-0.02	-0.06	0.01	-0.05	-0.05	0.09	0.04	-0.09	-0.09	-0.03	0.05	0.04	-0.04	-0.12	0.01	-0.04	-0.27	-0.12	0.01	0.14	-0.83	0.68	1.00	0.08	-0.18	-0.18	0.11
150–200	3.00–3.00	8.36×10^{-4}	5.13	2.68	5.79	1.59	-0.02	-0.03	-0.02	0.01	-0.07	-0.05	0.02	-0.07	-0.01	-0.05	0.01	-0.05	-0.06	0.11	0.03	-0.09	-0.09	-0.02	0.02	0.08	0.02	-0.02	-0.06	0.03	-0.33	-0.23	0.09	0.20	-1.02	1.02	1.38	0.08	-0.38	-0.41	0.23
200–300	0.00–0.25	2.19×10^{-3}	1.84	0.97	2.08	0.32	-0.00	-0.00	-0.00	0.01	-0.05	-0.04	0.05	-0.03	-0.03	-0.09	0.02	-0.01	0.03	0.06	-0.00	-0.08	-0.08	-0.07	0.05	-0.03	0.06	-0.09	-0.01	-0.18	-0.19	0.02	0.09	0.28	-0.41	0.35	0.42	0.01	-0.24	0.26	0.15
200–300	0.25–0.50	2.11×10^{-3}	1.90	1.02	2.16	0.34	-0.01	-0.00	-0.01	-0.01	-0.06	-0.04	0.04	-0.04	-0.04	-0.09	0.01	-0.02	0.03	0.05	0.02	-0.07	-0.07	-0.07	0.04	-0.01	0.04	-0.10	0.01	-0.19	-0.21	0.04	0.07	0.29	-0.47	0.32	0.48	-0.01	-0.24	0.28	0.15
200–300	0.50–0.75	1.99×10^{-3}	1.98	1.11	2.27	0.35	-0.01	-0.00	-0.00	-0.00	-0.06	-0.04	0.04	-0.04	-0.04	-0.09	0.01	-0.02	0.03	0.06	0.01	-0.08	-0.08	-0.09	0.04	-0.00	0.03	-0.11	0.04	-0.19	-0.23	0.03	0.07	0.29	-0.52	0.39	0.54	-0.01	-0.23	0.27	0.16
200–300	0.75–1.00	1.91×10^{-3}	2.05	1.16	2.36	0.38	-0.01	-0.00	-0.01	-0.01	-0.07	-0.04	0.04	-0.04	-0.05	-0.09	0.02	-0.02	0.01	0.07	0.02	-0.08	-0.08	-0.08	0.04	0.01	0.05	-0.12	0.06	-0.21	-0.22	0.03	0.04	0.23	-0.60	0.40	0.60	0.02	-0.21	0.21	0.15
200–300	1.00–1.25	1.69×10^{-3}	2.18	1.31	2.55	0.41	-0.01	-0.00	-0.00	0.00	-0.07	-0.03	0.04	-0.04	-0.04	-0.10	0.01	-0.04	0.02	0.07	0.00	-0.09	-0.09	-0.07	0.04	0.01	0.03	-0.16	0.13	-0.23	-0.23	-0.03	0.10	0.28	-0.63	0.51	0.71	0.11	-0.19	0.20	0.14
200–300	1.25–1.50	1.45×10^{-3}	2.42	1.58	2.89	0.48	-0.01	-0.00	0.01	-0.00	-0.08	-0.03	0.04	-0.06	-0.03	-0.09	-0.01	-0.04	0.03	0.09	-0.00	-0.09	-0.09	-0.07	0.07	0.01	-0.01	-0.21	0.18	-0.20	-0.26	-0.08	0.08	0.35	-0.76	0.65	0.88	0.09	-0.22	0.13	0.19
200–300	1.50–1.75	1.27×10^{-3}	2.65	1.77	3.18	0.54	-0.01	-0.00	0.00	0.00	-0.08	-0.03	0.04	-0.05	-0.04	-0.09	0.01	-0.05	0.01	0.10	-0.01	-0.09	-0.09	-0.05	0.05	0.04	-0.02	-0.22	0.24	-0.23	-0.31	-0.05	0.13	0.29	-0.87	0.72	1.02	0.12	-0.15	0.12	0.16
200–300	1.75–2.00	1.06×10^{-3}	3.02	2.11	3.69	0.67	-0.01	-0.01	0.01	0.01	-0.09	-0.04	0.02	-0.06	-0.05	-0.09	0.01	-0.05	0.00	0.11	0.00	-0.09	-0.09	-0.04	0.05	0.08	-0.03	-0.22	0.24	-0.26	-0.36	-0.13	0.20	0.28	-0.94	0.98	1.26	0.17	-0.16	0.10	0.14
200–300	2.00–2.25	9.16×10^{-4}	3.34	2.34	4.08	0.78	-0.02	-0.00	0.01	-0.00	-0.10	-0.04	0.02	-0.07	-0.04	-0.09	0.01	-0.07	-0.02	0.11	-0.00	-0.09	-0.09	-0.04	0.06	0.08	-0.06	-0.27	0.30	-0.24	-0.42	-0.14	0.17	0.30	-1.07	1.04	1.39	0.14	-0.17	0.01	0.17
200–300	2.25–2.50	6.50×10^{-4}	4.17	3.07	5.18	1.09	-0.02	-0.01	0.03	-0.00	-0.12	-0.04	0.01	-0.09	-0.04	-0.07	-0.02	-0.08	-0.03	0.14	-0.01	-0.10	-0.10	-0.03	0.07	0.10	-0.14	-0.34	0.34	-0.15	-0.49	-0.25	0.13	0.36	-1.39	1.38	1.86	0.19	-0.20	-0.06	0.20
200–300	2.50–2.75	5.34×10^{-4}	4.72	3.32	5.77	1.37	-0.03	-0.03	-0.01	0.01	-0.10	-0.06	0.01	-0.08	-0.03	-0.08	0.01	-0.09	-0.05	0.13	0.01	-0.12	-0.12	-0.02	0.03	0.13	-0.05	-0.24	0.18	-0.16	-0.46	-0.30	0.20	0.13	-1.43	1.50	2.04	0.25	-0.12	-0.08	0.15
200–300	3.00–3.00	3.98×10^{-4}	5.65	3.76	6.79	1.81	-0.03	-0.03	0.00	0.01	-0.11	-0.06	0.02	-0.10	-0.01	-0.07	-0.02	-0.10	-0.06	0.14	0.03	-0.12	-0.12	-0.02	0.07	0.07	-0.14	-0.29	0.12	-0.08	-0.43	-0.34	0.04	0.22	-1.54	1.65	2.21	0.15	-0.36	-0.30	0.28
300–500	0.00–0.50	2.67×10^{-4}	2.64	1.11	2.86	0.44	-0.01	-0.01	-0.03	0.01	-0.06	-0.04	0.08	-0.02	-0.02	-0.10	-0.01	0.01	0.05	0.01	-0.02	-0.10	-0.10	-0.07	0.08	-0.09	0.10	-0.03	-0.10	-0.10	-0.15	-0.00	0.03	0.35	-0.39	0.36	0.44	-0.12	-0.28	0.45	0.20
300–500	0.50–1.00	2.52×10^{-4}	2.76	1.16	2.99	0.45	-0.01	-0.00	-0.03	0.00	-0.06	-0.04	0.07	-0.03	-0.04	-0.10	0.01	0.01	0.06	0.03	0.00	-0.09	-0.09	-0.09	0.06	-0.03	0.07	-0.05	-0.04	-0.17	-0.17	0.03	0.07	0.32	-0.43	0.39	0.53	-0.09	-0.33	0.38	0.20
300–500	1.00–1.50	2.19×10^{-4}	3.05	1.41	3.36	0.55	-0.01	-0.01	-0.02	0.01	-0.06	-0.04	0.07	-0.04	-0.04	-0.10	0.01	-0.01	0.04	0.05	-0.00	-0.10	-0.10	-0.07	0.06	-0.03	0.04	-0.10	0.03	-0.19	-0.21	-0.02	0.08	0.31	-0.61	0.52	0.75	-0.02	-0.27	0.33	0.19
300–500	1.50–2.00	1.70×10^{-4}	3.63	1.94	4.12	0.71	-0.02	-0.03	-0.01	-0.01	-0.09	-0.05	0.07	-0.06	-0.03	-0.10	-0.01	-0.03	0.00	0.08	0.02	-0.11	-0.11	-0.06	0.03	0.01	-0.04	-0.24	0.12	-0.18	-0.34	-0.07	0.05	0.33	-0.99	0.68	1.11	-0.01	-0.28	0.26	0.22
300–500	2.00–2.50	1.02×10^{-4}	5.17	3.12	6.04	1.23	-0.03	-0.02	-0.00	0.01	-0.10	-0.06	0.04	-0.08	-0.03	-0.09	-0.02	-0.05	-0.03	0.12	-0.01	-0.12	-0.12	-0.06	0.05	0.07	-0.11	-0.34	0.26	-0.16	-0.46	-0.22	0.21	0.41	-1.42	1.30	1.87	0.24	-0.24	0.18	0.22
300–500	2.50–3.00	6.29×10^{-5}	7.23	4.70	8.62	2.13	-0.03	-0.03	0.01	-0.01	-0.12	-0.07	0.02	-0.09	-0.04	-0.09	-0.04	-0.08	-0.07	0.17	0.03	-0.11	-0.11	-0.04	0.03	0.17	-0.15	-0.46	0.38	-0.23	-0.71	-0.36	0.26	0.28	-2.07	1.95	2.81	0.18	-0.33	0.06	0.28
500–1500	0.00–0.50	8.08×10^{-6}	6.58	1.38	6.72	0.84	-0.03	-0.01	-0.08	-0.01	-0.04	-0.06	0.12	0.01	-0.04	-0.14	0.00	0.10	0.12	-0.03	0.01	-0.13	-0.13	-0.14	0.07	-0.12	0.20	0.11	-0.24	-0.13	-0.06	0.10	0.02	0.33	-0.17	0.32	0.27	-0.22	-0.41	0.59	0.20
500–1500	0.50–1.00	8.79×10^{-6}	6.34	1.33	6.48	0.78	-0.03	-0.01	-0.08	-0.01	-0.05	-0.05	0.10	0.01	-0.05	-0.12	0.01	0.08	0.12	-0.03	0.03	-0.12	-0.12	-0.14	0.06	-0.12	0.21	0.07	-0.22	-0.18	-0.10	0.13	0.06	0.22	-0.25	0.24	0.28	-0.27	-0.31	0.65	0.18
500–1500	1.00–1.50	7.98×10^{-6}	6.76	1.39	6.90	0.81	-0.02	-0.01	-0.06	-0.02	-0.05	-0.05	0.11	0.00	-0.04	-0.14	-0.02	0.07	0.09	-0.03	-0.01	-0.13	-0.13	-0.09	0.06	-0.12	0.20	0.06	-0.20	-0.18	-0.10	0.05	0.04	0.25	-0.25	0.31	0.42	-0.17	-0.28	0.70	0.16
500–1500	1.50–2.00	7.00×10^{-6}	7.44	1.46	7.58	0.83	-0.02	-0.01	-0.05	-0.01	-0.04	-0.05	0.12	-0.01	-0.04	-0.12	-0.02	0.04	0.06	-0.00	-0.01	-0.12	-0.12	-0.07	0.08	-0.12	0.18	0.05	-0.17	-0.13	-0.16	0.01	0.01	0.34	-0.37	0.50	0.54	-0.16	-0.44	0.41	0.25
500–1500	2.00–2.50	4.16×10^{-6}	10.47	2.37	10.74	1.64	-0.02	-0.02	-0.05	0.01	-0.06	-0.06	0.09	-0.03	-0.03	-0.12	0.00	0.00	0.02	0.04	0.01	-0.14	-0.14	-0.03	0.04	-0.04	0.18	-0.01	-0.15	-0.20	-0.23	-0.04	0.08	0.14	-0.70	0.82	1.09	0.03	-0.23	0.51	0.12
500–1500	2.50–3.00	2.50×10^{-6}	14.59	3.74	15.06	2.55	-0.03	-0.05	-0.04	0.02	-0.07	-0.07	0.09	-0.06	0.00	-0.09	-0.04	-0.02	-0.01	0.08	0.00	-0.13	-0.13	-0.04	0.05	-0.01	0.04	-0.06	-0.16	-0.09	-0.30	-0.30	0.04	0.23	-1.21	1.32	1.89	0.24	-0.19	0.46	0.13

Appendix S
High-Mass Drell-Yan: Nuisance Parameter Constraints

This appendix contains the nuisance parameter pulls for the χ^2 minimization comparing the double-differential cross section measurements to theory calculations using different PDFs. Figures S.1, S.2, S.3, S.4 and S.5 show the pulls for the MMHT2014 [1], CT14 [2], HERAPDF2.0 [3], NNPDF3.0 [4], and ABM12 [5] PDF sets using the double-differential measurement as a function of absolute rapidity. Figures S.6, S.7, S.8, S.9 and S.10 show the same pulls for the double-differential measurement as a function of absolute pseudorapidity separation. The nuisance parameter pulls are discussed in Sect. 17.3.1.

M. Zinser, *Search for New Heavy Charged Bosons and Measurement of High-Mass Drell-Yan Production in Proton-Proton Collisions*, Springer Theses,
https://doi.org/10.1007/978-3-030-00650-1

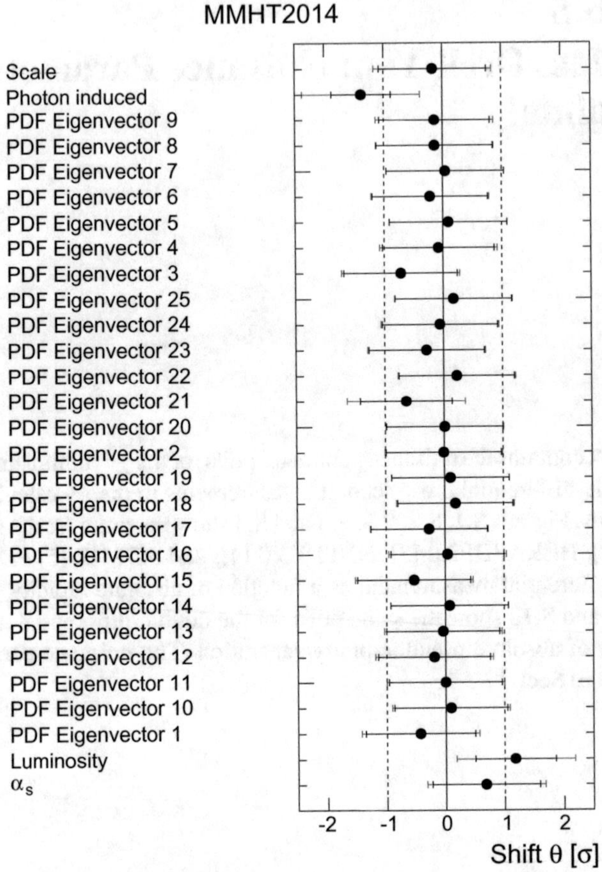

Fig. S.1 Shifts and uncertainty reduction of the theoretical nuisance parameters of the MMHT2014 PDF set and the luminosity from the χ^2 minimization of the double-differential cross section measurement as a function of invariant mass $m_{\ell\ell}$ and absolute dilepton rapidity $|y_{\ell\ell}|$. The red error bars show the original uncertainty while the black error bars show the uncertainty after the combination

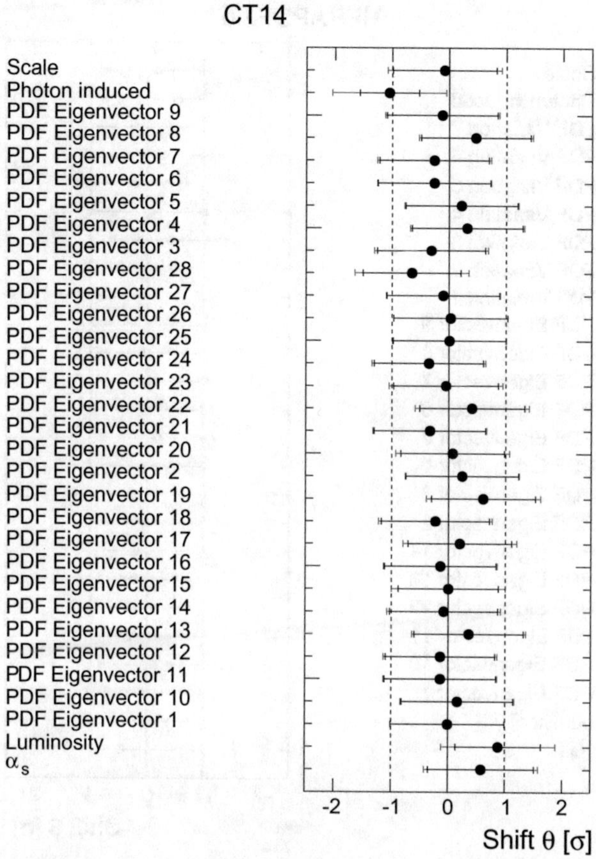

Fig. S.2 Shifts and uncertainty reduction of the theoretical nuisance parameters of the CT14 PDF set and the luminosity from the χ^2 minimization of the double-differential cross section measurement as a function of invariant mass $m_{\ell\ell}$ and absolute dilepton rapidity $|y_{\ell\ell}|$. The red error bars show the original uncertainty while the black error bars show the uncertainty after the combination

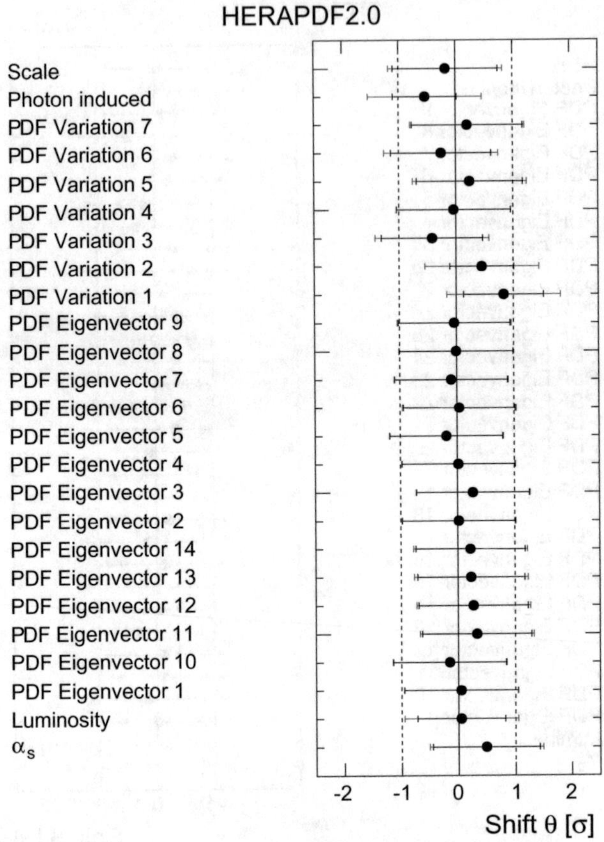

Fig. S.3 Shifts and uncertainty reduction of the theoretical nuisance parameters of the HERAPDF2.0 PDF set and the luminosity from the χ^2 minimization of the double-differential cross section measurement as a function of invariant mass $m_{\ell\ell}$ and absolute dilepton rapidity $|y_{\ell\ell}|$. The red error bars show the original uncertainty while the black error bars show the uncertainty after the combination

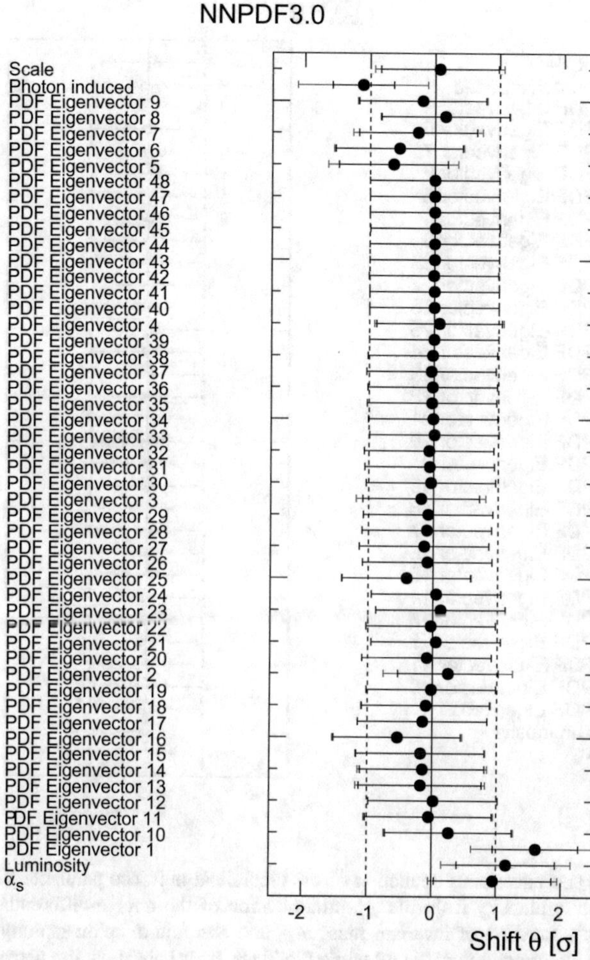

Fig. S.4 Shifts and uncertainty reduction of the theoretical nuisance parameters of the NNPDF3.0 PDF set and the luminosity from the χ^2 minimization of the double-differential cross section measurement as a function of invariant mass $m_{\ell\ell}$ and absolute dilepton rapidity $|y_{\ell\ell}|$. The red error bars show the original uncertainty while the black error bars show the uncertainty after the combination

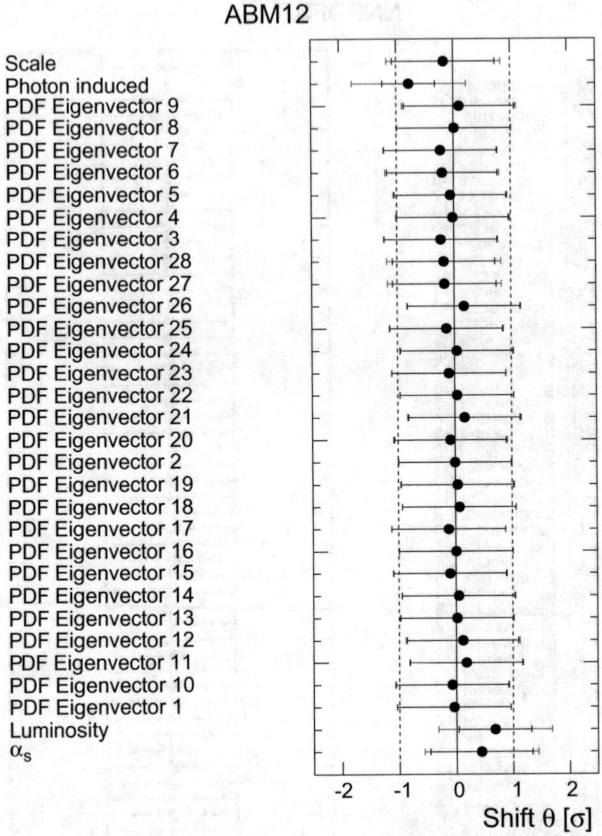

Fig. S.5 Shifts and uncertainty reduction of the theoretical nuisance parameters of the ABM12 PDF set and the luminosity from the χ^2 minimization of the double-differential cross section measurement as a function of invariant mass $m_{\ell\ell}$ and absolute dilepton rapidity $|y_{\ell\ell}|$. The red error bars show the original uncertainty while the black error bars show the uncertainty after the combination

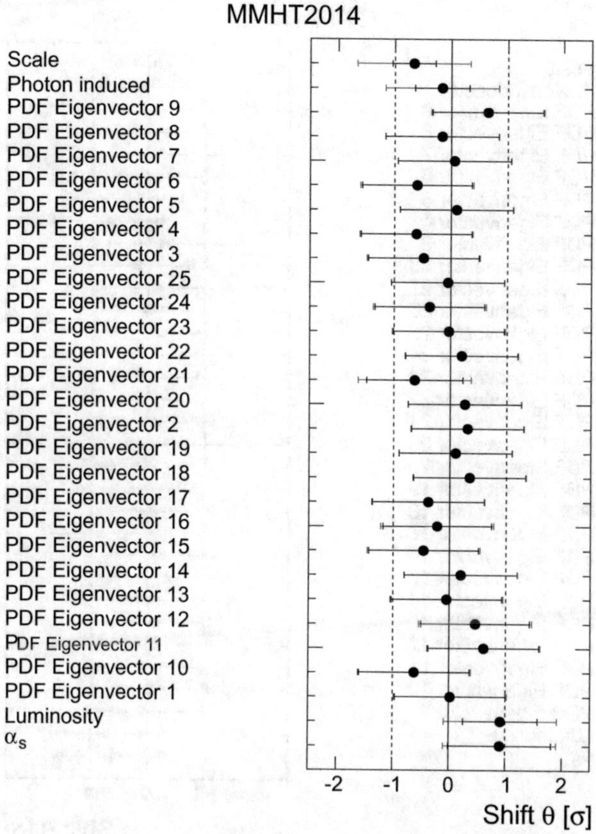

Fig. S.6 Shifts and uncertainty reduction of the theoretical nuisance parameters of the MMHT2014 PDF set and the luminosity from the χ^2 minimization of the double-differential cross section measurement as a function of invariant mass $m_{\ell\ell}$ and absolute lepton pseudorapidity separation $|\Delta\eta_{\ell\ell}|$. The red error bars show the original uncertainty while the black error bars show the uncertainty after the combination

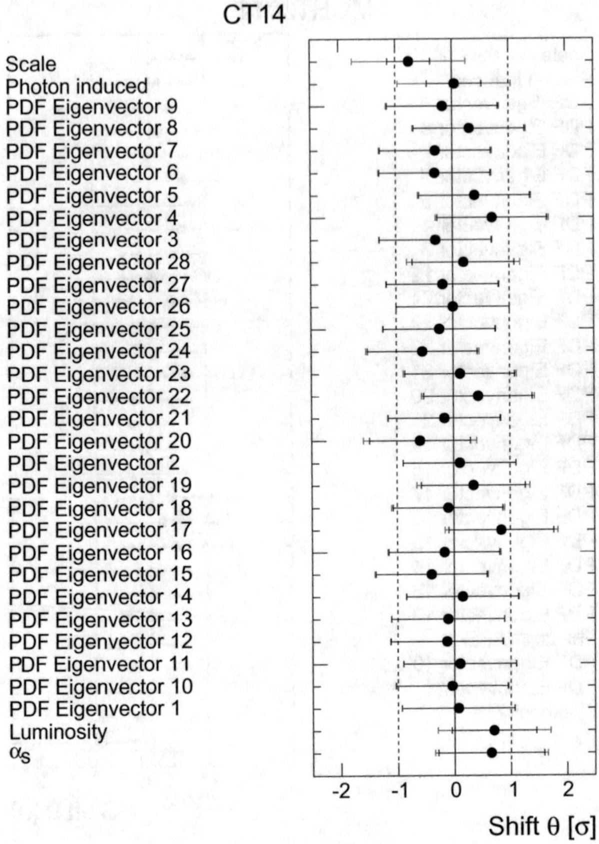

Fig. S.7 Shifts and uncertainty reduction of the theoretical nuisance parameters of the CT14 PDF set and the luminosity from the χ^2 minimization of the double-differential cross section measurement as a function of invariant mass $m_{\ell\ell}$ and absolute lepton pseudorapidity separation $|\Delta\eta_{\ell\ell}|$. The red error bars show the original uncertainty while the black error bars show the uncertainty after the combination

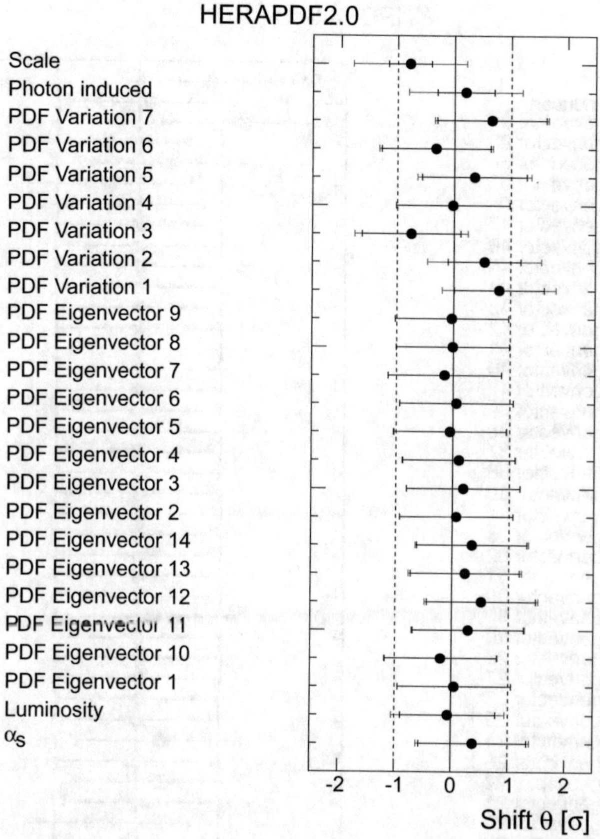

Fig. S.8 Shifts and uncertainty reduction of the theoretical nuisance parameters of the HERA-PDF2.0 PDF set and the luminosity from the χ^2 minimization of the double-differential cross section measurement as a function of invariant mass $m_{\ell\ell}$ and absolute lepton pseudorapidity separation $|\Delta\eta_{\ell\ell}|$. The red error bars show the original uncertainty while the black error bars show the uncertainty after the combination

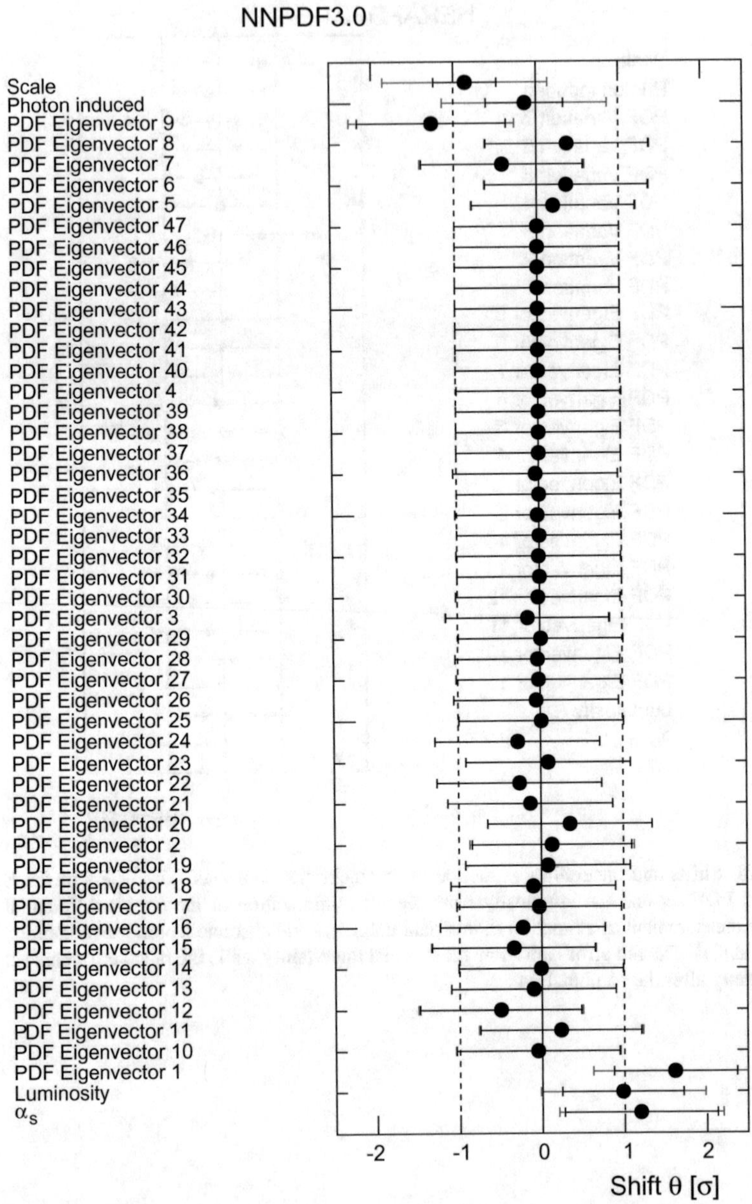

Fig. S.9 Shifts and uncertainty reduction of the theoretical nuisance parameters of the NNPDF3.0 PDF set and the luminosity from the χ^2 minimization of the double-differential cross section measurement as a function of invariant mass $m_{\ell\ell}$ and absolute lepton pseudorapidity separation $|\Delta\eta_{\ell\ell}|$. The red error bars show the original uncertainty while the black error bars show the uncertainty after the combination

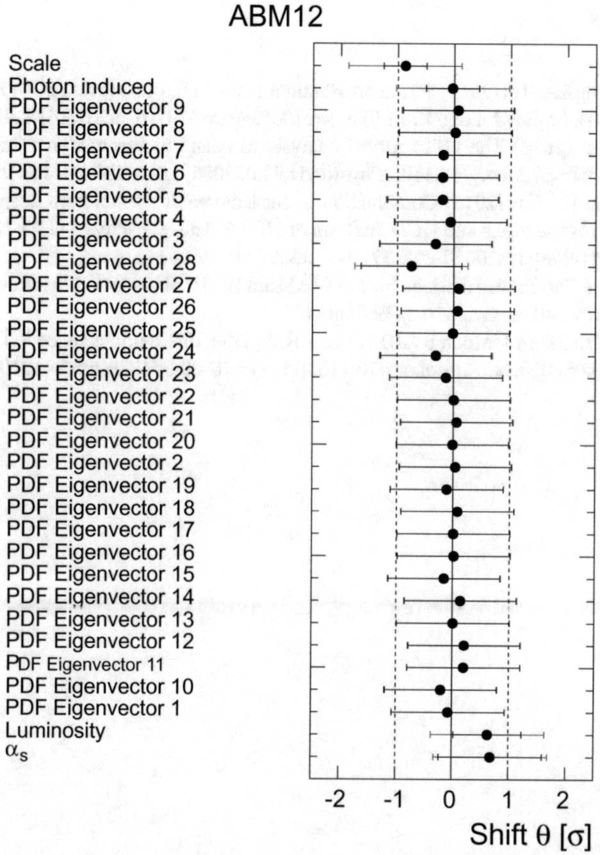

Fig. S.10 Shifts and uncertainty reduction of the theoretical nuisance parameters of the ABM12 PDF set and the luminosity from the χ^2 minimization of the double-differential cross section measurement as a function of invariant mass $m_{\ell\ell}$ and absolute lepton pseudorapidity separation $|\Delta\eta_{\ell\ell}|$. The red error bars show the original uncertainty while the black error bars show the uncertainty after the combination

References

1. Harland-Lang LA et al (2015) Parton distributions in the LHC era: MMHT 2014 PDFs. Eur Phys J C 75(5):204. https://doi.org/10.1140/epjc/s10052-015-3397-6, arXiv:1412.3989 [hep-ph]
2. Dulat S et al (2016) The CT14 global analysis of quantum chromodynamics. Phys Rev D 93:033006. https://doi.org/10.1103/PhysRevD.93.033006, arXiv:1506.07443 [hep-ph]
3. Abramowicz H et al (2015) Combination of measurements of inclusive deep inelastic $e^{\pm}p$ scattering cross sections and QCD analysis of HERA data. Eur Phys J C 75(12):580. https://doi.org/10.1140/epjc/s10052-015-3710-4, arXiv:1506.06042 [hep-ex]
4. Ball RD et al Parton distributions for the LHC Run II'. JHEP 04:040. https://doi.org/10.1007/JHEP04(2015)040, arXiv:1410.8849 [hep-ph]
5. Alekhin S, Blumlein J, Moch S (2014) The ABM parton distributions tuned to LHC data. Phys Rev D 89(5):054028. https://doi.org/10.1103/PhysRevD.89.054028, arXiv:1310.3059 [hep-ph]

Appendix T
High-Mass Drell-Yan: Photon PDF Reweighting

Table T.1 shows the χ^2 values for the 100 replica of the NNPDF2.3qed PDF set [1] which yield as an input for the PDF reweighting described in Sect. 17.3.2.

Table T.1 The χ^2 values for the compatibility of data and theory calculated for each of the 100 replicas of the NNPDF2.3QED NNLO PDF set for the prediction of the PI component and using the central value of the MMHT14 NNLO PDF set for the quark and gluon parton distributions

PDF replica	$\|y_{ee}\|$	$\|\Delta\eta_{ee}\|$	PDF replica	$\|y_{ee}\|$	$\|\Delta\eta_{ee}\|$
0	222.8	243.3	50	59	67
1	65.72	66.37	51	98.27	99.93
2	64.27	65.88	52	205.97	133.18
3	58.72	66.18	53	59.05	66.85
4	151.23	100.5	54	102.09	86.51
5	109.85	80.17	55	59.02	66.9
6	75.78	72.35	56	59.34	66.84
7	59.05	66.86	57	66.38	68.62
8	59.37	66.77	58	60.36	65.04
9	61.76	66.72	59	62.43	65.58
10	61.24	67.65	60	59.4	67.02
11	66.18	66.96	61	65.72	66.97
12	58.99	67.04	62	59.11	66.9
13	119.19	127.2	63	59.2	67.3
14	59.01	67.09	64	59.47	67.1
15	76.78	68.78	65	58.84	66.71
16	59.01	66.92	66	101.84	85.97
17	62.5	67.9	67	59.47	67.3
18	126.02	154.03	68	61.11	66.77

(continued)

M. Zinser, *Search for New Heavy Charged Bosons and Measurement of High-Mass Drell-Yan Production in Proton-Proton Collisions*, Springer Theses,
https://doi.org/10.1007/978-3-030-00650-1

Table T.1 (continued)

PDF replica	$\lvert y_{\ell\ell}\rvert$	$\lvert\Delta\eta_{\ell\ell}\rvert$	PDF replica	$\lvert y_{\ell\ell}\rvert$	$\lvert\Delta\eta_{\ell\ell}\rvert$
19	59.08	66.83	69	65.64	66.9
20	58.7	66.97	70	111.14	101.22
21	59.27	66.88	71	59.32	66.72
22	59.83	66.24	72	173.19	106.88
23	67.09	71.41	73	62.08	65.74
24	95.54	90.67	74	59.43	66.82
25	58.9	67.08	75	60.07	66.93
26	59.02	66.97	76	58.94	66.95
27	59.03	66.44	77	150.34	97.81
28	59.07	66.85	78	59.01	66.96
29	105.19	82.71	79	87.75	73.16
30	59.23	67	80	63.02	67.46
31	58.9	67.18	81	89.29	93.69
32	83.41	72.82	82	65.41	68.5
33	77.3	68.04	83	59	67.1
34	59.12	66.84	84	78.46	79.35
35	189.9	104.33	85	67.18	65.27
36	149.81	127.69	86	81.45	78.24
37	58.91	67.04	87	61.42	68.22
38	67.71	67.06	88	58.94	67.16
39	58.96	66.3	89	59.02	66.98
40	59.4	66.91	90	59.3	66.6
41	156.6	102.34	91	140.1	94.92
42	59.14	66.85	92	72.37	67.36
43	67.6	70.73	93	59.07	67.09
44	58.99	67.03	94	60.37	68.31
45	113.56	103.57	95	59	66.81
46	58.94	66.84	96	59.07	66.85
47	59.32	66.95	97	59.03	66.92
48	59	67.1	98	59.41	67.2
49	58.95	66.85	99	59.49	66.74

Reference

1. Ball RD et al (2013) Parton distributions with QED corrections. Nucl Phys B 877:290–320. https://doi.org/10.1016/j.nuclphysb.2013.10.010. arXiv:1308.0598 [hep-ph]

Curriculum Vitae

Personal Details

Name	Markus Zinser
Date of Birth	09.06.1988
Place of Birth	Worms, Germany
Nationality	German

Education

10/2013–12/2016	PhD in Physics, Johannes Gutenberg-Universität Mainz, Thesis submitted 08/2016, defense 12/2016, Grade: 0.7 (summa cum laude)
10/2011–08/2013	Master of Science in Physics, Johannes Gutenberg-Universität Mainz, Grade: 1.0 (with distinction)
09/2008–02/2012	Bachelor of Science in Physics, Johannes Gutenberg-Universität Mainz, Grade: 1.9
03/2008	Graduation: Abitur, Staatliches Aufbaugymnasium Alzey

M. Zinser, *Search for New Heavy Charged Bosons and Measurement of High-Mass Drell-Yan Production in Proton-Proton Collisions*, Springer Theses,
https://doi.org/10.1007/978-3-030-00650-1

Studies Abroad

08/2012–08/2016	Participation in workshops/conferences, Various stays in Geneva, Grenoble, Lisbon, Tel-Aviv, Vienna, Stanford (up to two weeks)
11/2014–08/2015	Research stay, CERN, Geneva, Switzerland

Scholarships and Awards

05/2017	Award for outstanding PhD thesis of the graduate school "Symmetry Breaking in Fundamental Interactions"
01/2017	Award for outstanding PhD thesis of the Johannes Gutenberg-Universität Mainz
04/2015	Junior member of the Gutenberg Academy
03/2014	Price for outstanding thesis of the physics, mathematics and informatics faculty
03/2014–03/2016	Fellow of the graduate school "Symmetry Breaking in Fundamental Interactions"

Publications with Significant Contribution

ATLAS Collaboration	Search for new high-mass phenomena in the dilepton final state using 36.1 fb^{-1} of proton-proton collision data at $\sqrt{s}$ = 13 TeV with the ATLAS detector, *arXiv:1707.02424, submitted to JHEP*
ATLAS Collaboration	Search for high-mass new phenomena in the dilepton final state using proton-proton collisions at $\sqrt{s}$=13 TeV with the ATLAS detector, *Phys.Lett. B761 (2016) 372-392*
ATLAS Collaboration	Search for new resonances in events with one lepton and missing transverse momentum in *pp* collisions at $\sqrt{s}$=13 TeV with the ATLAS detector, *Phys.Lett. B762 (2016) 334-352*
ATLAS Collaboration	Measurement of the double-differential high-mass Drell-Yan cross section in *pp* collisions at $\sqrt{s}$ = 8 TeV with the ATLAS detector, *JHEP 08 (2016) 009*
Markus Zinser on behalf of the ATLAS Collaboration	Drell-Yan and vector boson plus jets measurements with the ATLAS detector, *PoS EPS-HEP2015 (2015) 481*
E. Simioni et al.	Upgrade of the ATLAS Level-1 Trigger with event topology information, *J.Phys.Conf.Ser. 664 (2015) no.8, 082052*
S. Artz et al.	Upgrade of the ATLAS Central Trigger for LHC Run-2, *JINST 10 (2015) no.02, C02030*
S. Artz et al.	The ATLAS Level-1 Muon Topological Trigger Information for Run 2 of the LHC, *JINST 10 (2015) no.02, C02027*
ATLAS Collaboration	Search for high-mass dilepton resonances in pp collisions at $\sqrt{s}$ = 8 TeV with the ATLAS detector, *Phys. Rev. D 90, 052005 (2014)*

E. Simioni et al. The Topological Processor for the future ATLAS Level-1 Trigger: From design to commissioning, *Real Time Conference (RT), 2014 19th IEEE-NPSS*

Printed in the United States
By Bookmasters